Power Systems

For further volumes:
http://www.springer.com/series/4622

Günter Kessler

Sustainable and Safe Nuclear Fission Energy

Technology and Safety of Fast and Thermal Nuclear Reactors

Springer

Günter Kessler
Jägersteig 4
76297 Stutensee
Germany

ISSN 1612-1287 ISSN 1860-4676 (electronic)
ISBN 978-3-642-11989-7 ISBN 978-3-642-11990-3 (eBook)
DOI 10.1007/978-3-642-11990-3
Springer Heidelberg New York Dordrecht London

Library of Congress Control Number: 2012935791

Printed on acid-free paper

Springer is part of Springer Science+Business Media (www.springer.com)

To Lotte, Birgit, Anne and Jürgen
for their encouragement and patience

Preface

This book was written in an effort to present a complete description and evaluation of the current power generation technology by means of nuclear fission reactions. Similar to an earlier book by the author[1] it covers the entire nuclear fuel cycle, from mining of natural uranium, uranium conversion and enrichment, to the fabrication of fuel elements for the cores of various types of commercial nuclear power plants.

Numerically, light water reactors (LWRs) outweigh all other types of reactors, generating electricity. In most countries of the world applying nuclear energy, electricity is generated in nuclear power plants at a lower cost than in fossil-fueled power plants. Most likely, LWRs will continue to hold the largest share of the market in the coming decades, when the contribution of nuclear power for the generation of electricity in most industrialized countries will rise from its present level. The enrichment of natural uranium as fuel for these nuclear power plants is still achieved by about 50% in gaseous diffusion plants, but gas ultracentrifuge enrichment is becoming more dominating. Only in the future decades will laser enrichment be able to secure a certain share of the uranium enrichment market.

Assessments of the world uranium resources by international organizations such as OECD and IAEA as well as analyses on the natural uranium consumption by nuclear power reactors in the world point to a growing scarcity of natural uranium in the second half of this century. This threat can be counteracted by the replacement, in due time, of today's light water reactors (highest uranium consumption), by advanced reactors, and, above all, by breeder reactors. This will drastically reduce the consumption of natural uranium. Especially, fast breeder reactors operated in symbiosis with light water reactors can curb the uranium requirement enough to assure the world's energy generation. It is technically feasible to introduce fast breeder reactors commercially during the second half of this century. In such a case even today's assured world uranium reserves would be sufficient to meet the requirements over thousands of years. However, the

[1] Nuclear Fission Reactors, Springer Verlag Wien New York, 1983.

commercialization of advanced reactor lines will imply further development efforts and costs, in particular for the development of advanced technologies and processes for fuel fabrication and reprocessing.

Advanced converter and breeder reactors require a closed nuclear fuel cycle in order to get started with plutonium or U-233. These man-made fissile materials must be produced by chemical reprocessing of the spent fuel elements of the present line of commercial nuclear power plants. This directly links with the decision to be taken on the construction of internationally operated reprocessing and refabrication centers and installations for subsequent waste conditioning and final storage of the radioactive waste. The technical availability of reprocessing and refabrication facilities, and the development of chemical processes for the separation of the different minor actinides opens possibilities for transmutation and incineration of plutonium and of the minor actinides. Analyses show that plutonium and the minor actinides can be incinerated except for the chemical losses during reprocessing and refabrication of less than 1% going to the high active nuclear waste. After multi-recycling this eventually results, depending on the reprocessing losses, in an overall utilization of 60–80% of the resources of natural uranium.

Recent results of research programs show that the incineration of the actinides by transmutation processes is technically feasible. However, this requires the development of new chemical separation processes for the spent fuel and advanced fuel fabrication technologies. This is accompanied by advanced material research for high burnup fuels.

The environmental impacts and risks associated with the different types of nuclear power plants and nuclear fuel cycle facilities must remain below the limits set by the International Commission for Radiation Protection (ICRP) and by national authorities. The environmental impacts are due to the release of radioactive substances from various stages of the nuclear fuel cycle, such as uranium ore mining, uranium conversion, enrichment plants, fuel fabrication plants, nuclear power plants, reprocessing plants, and waste conditioning installations.

The objective of reactor safety concepts is to protect the operational personnel, the environment, and the population against radioactivity releases during normal operation and accidents. The safety concept is based on multiple containment structures as well as engineered safeguards components. In addition, other safety measures combined in a staggered-in-depth concept of four safety levels must be incorporated. Design basis accidents must be accommodated by design features of the protection and safety systems, as well as by the emergency cooling systems of the nuclear plant.

Probabilistic safety analysis is a supplement to this deterministic approach. Reactor risk studies which were performed during the 1970s (USA) and 1980s (Europe) had shown that the risk arising from light water reactors as a result of core melt down is well below the risk of other power generating systems or traffic systems. However, the Chernobyl accident in 1986 (water cooled, graphite moderated 1,000 MW(e) reactor of Soviet RBMK design) resulted, in addition to severe radiation exposures to the rescue personal and to the population, also in large-scale land contamination by radioactive cesium isotopes.

As a consequence, new research programs were initiated on severe accident consequences. Their results led to a revision of the results of the early risk studies of the 1980s and to the application of a new safety concept for modern light water reactors, e.g., the European Pressurized Water Reactor (EPR) and the European Boiling Water Reactor (SWR-1000).

The safety design concept of future liquid metal cooled fast breeder reactors (LMFBRs) will also have to follow the basic safety principles (multiple barrier concept and staggered in-depth four level safety concept) as developed for light water reactors. This holds despite the fact that LMFBRs have different design characteristics (fast neutron spectrum, liquid metal as coolant, plutonium–uranium fuel). It has been shown that LMFBRs have a strong negative power coefficient and good power control stability. The main design characteristics of control and shut-off systems do not differ much from those of light water reactors. The excellent cooling and natural convection properties of liquid metals as well as the low system pressure of about 1 bar allow the safe decay heat removal in a number of ways. The consequences of sodium fires or sodium water reactions can be prevented or limited by special design provisions. On the other hand, lead or lead-bismuth-eutecticum (LBE) as coolant do not chemically react either with oxygen in the atmosphere or with water in the failing tubes of a steam generator.

The safety concept of fuel cycle plants, e.g., spent fuel storage facilities, reprocessing facilities, and waste treatment facilities is based on similar multiple barrier and engineered safeguards measures as they are applied to nuclear reactors. However, the risk of these fuel cycle facilities is smaller than for nuclear power plants as the fuel is at much lower temperatures and atmospheric pressure in reprocessing and refabrication plants.

In covering the many interdisciplinary aspects discussed in this book the author was able to make use of the excellent library facilities of the Karlsruhe Institute of Technology. A number of former colleagues of the former Institute of Neutron Physics and Reactor Technology of the former Forschungszentrum Karlsruhe (FZK) now part of KIT, the Karlsruhe Institute of Technology, assisted considerably in completing this work. Their help and support is much appreciated. The author would like to thank explicitly the following scientists and former staff members of this former Institute:

- Dr. E. Kiefhaber for his excellent scientific advice and critical review of all parts of the manuscript
- Mrs. Ch. Kastner for her great efforts in typing the manuscript and preparing it for publication
- Dr. C. H. M. Broeders for providing scientific material, Dr. X. -N. Chen for giving scientific advice and Dr. Shoji Kotake (formerly JAEA, Japan) for reviewing the chapter on fast breeders
- Dr. W. Koelzer for reviewing the chapter on radioactive releases
- Dipl. -Ing. (TU) A. Veser, Dipl. -Ing. (FH) F. Zimmermann, Dipl. -Ing. (FH) F. Lang and W. Goetzmann for their continued support in preparing the numerous figures

- Mrs. M. Wettstein for her assistance in the literature search
- Dr. G. Mueller and Dr. G. Schumacher for their continued interest
- R. Friese for the translation of some chapters of this book written originally in German

The author hopes that this publication will make a helpful contribution to the understanding and advancement of the further nuclear fission reactor deployment.

Stutensee, Germany G. Kessler

Contents

Acronyms

ACR	Advanced CANDU reactor
ADS	Accelerator Driven System
AECL	Atomic Energy of Canada Ltd
AGR	Advanced Gas cooled, graphite moderated Reactor
ALARA	As Low as Reasonably Achievable
ANL	Argonne National Laboratory (USA)
ANS	American Nuclear Society
APS	American Physical Society
AREVA	French company (reactor manufacturer)
ASME	American Society of Mechanical Engineers
ATALANTE	ATelier Alpha et Laboratoires pour ANalyses Transuraniens et Etudes de retraitement (Marcoule, France)
ATRIUM	Boiling water reactor fuel element (AREVA)
ATWS	Anticipated Transient Without Scram
AUC	Ammonium Uranyl Carbonate
AUPuC	Ammonium Uranyl Plutonyl Carbonate
AVLIS	Atomic Vapour Laser Isotope Separation
AVR	Arbeitsgemeinschaft VersuchsReaktor, experimental HTR plant (Jülich, Germany)
BFS	Fast critical zero power test facility (Russia)
BMU	Bundesministerium für Umwelt, Naturschutz und Reaktorsicherheit (Germany)
BN 350	Prototype fast breeder reactor (USSR)
BN 600	Prototype fast breeder reactor (Russia)
BOC	Begin Of Cycle
BOR 60	Soviet experimental fast reactors
BR	Breeding Ratio
BREST	Russian fast breeder design (lead cooling)
BR-1, BR-2, BR-5	Soviet experimental fast reactors
BSK	Container for HLW disposal

BWR	Boiling Water Reactor, cooled and moderated by light water
CANDU	CANadian Deuterium Uranium reactor
CANDU-PHWR	CANadian Deuterium Uranium Pressurized Heavy Water Reactor
CAPRA	Consommation Améliorée du Plutonium dans les Réacteurs Avancés
CASTOR	German fuel transport cask
CC	Combined Cycle
CDA	Core Disruptive Accident
CEA	Commissariat à l'Energie Atomique (France)
CHRS	Containment Heat Removal System
CLEMENTINE	The first Pu-fueled fast research reactor cooled by liquid mercury (USA)
CR	Conversion Ratio
CRBR	Clinch River Breeder Reactor
CRIEPI	Japanese Research Institute
CRDM	Control Rod Drive Mechanism
CW	Cold Worked (steel)
DBA	Design Basis Accident
DFR	Dounreay Fast Reactor, experimental fast test reactor (UK)
DIAMEX	DIAMide EXtraction process (France)
DIDPA	Di-IsoDecylPhosphoric Acid process
DMDBTDMA	DiMethyl-DiButyl-TetraDecyl-MAlonamide process
DNB	Departure from Nucleate Boiling
DNBR	Depature from Nucleate Boiling Ratio
DOE	Department Of Energy (USA)
DOVITA	Russian pyrochemical separation process
DTPA	Diethylene Triamine Pentaacetic Actinide back-extraction process
DWK	Deutsche Gesellschaft für Wiederaufarbeitung von Kernbrennstoffen (Germany)
EBR-I, EBR-II	Experimental Breeder Reactors (USA)
ECCS	Emergency Core Cooling System
EFFBR	Enrico Fermi Fast Breeder Reactor (USA)
EFWS	Emergency Feed Water System
ENDF	Evaluated Nuclear Data File
EOC	End of Cycle
EPR	European Pressurized Water Reactor
EPRI	Electric Power Research Institute (USA)
EPSS	Emergency Power Supply System
EURATOM	European Organisation
EURODIF	Derived from EUROpean DIFfusion (international company for uranium isotope enrichment)
FAIDUS	Fuel Assembly with Inner DUct Structure

FCA	Fast Critical Assembly (Japan)
FCI	Fuel Coolant Interaction
FE	Fuel Element
FEFPL	Fuel Element Failure Propagation Loop
FFTF	Fast Flux Test Facility (USA)
FP	Fission Product
FBR	Fast Breeder Reactor
FR	Fast Reactor
GANEX	Group ActiNide EXtraction process(France)
GCR	Gas Cooled Reactor
GGR	Gas cooled, Graphite moderated Reactor
GNEP	Global Nuclear Energy Partnership
GNS	Gesellschaft für Nuklear-Service (Germany)
GRS	Gesellschaft für Reaktor- und AnlagenSicherheit (Germany)
HAW	High Active Waste
HCDA	Hypothetical Core Disruptive Accident
HEPA	High Efficiency Particulate Air Filter
HEU	Highly Enriched Uranium
HELIKON	Vortex tube enrichment process (South Africa)
HLW	High Level Waste
HLWC	High Level Waste Concentrate
HF	Hydrogen Fluoride
HM	Heavy Metal
HP	High Pressure
HSST	Heavy-Section Steel Technology Program (ORNL)
HTGR	High Temperature Gas cooled, graphite moderated Reactor with prismatic fuel elements
HTR	High temperature gas cooled, graphite moderated reactor with spherical fuel elements (pebble bed reactor)
HTR-Th	HTR with thorium fuel
HWR	Heavy Water Reactor
IAEA	International Atomic Energy Agency (Vienna)
ICRP	International Commission on Radiological Protection
IRIS	Name of the process for the treatment of radioactive wastes stems from the name of the prototype unit: Incineration Research Installation for Solid waste (France)
IRSN	Institut de Radioprotection et de Sûreté Nucléaire
IHX	Intermediate heat exchanger
IFR	Integral Fast Reactor (USA)
INFCE	International Nuclear Fuel Cycle Evaluation
IRWST	In-containment Refueling Water Storage Tank
IUREP	International Uranium Resource Evaluation Project
JAEA	Japanese Atomic Energy agency
JOYO	Japanese experimental fast breeder reactor
JSFR	Japanese Sodium cooled Fast Reactor

KFA	Kernforschungs-Anlage Jülich, Nuclear Research Center, Jülich (Germany)
KfK	Kernforschungszentrum Karlsruhe, Nuclear Research Center, Karlsruhe (Germany), now part of KIT
KHE	Kessler—Hennies—Eibl (persons initiating the new PWR safety concept)
KIT	Karlsruhe Institute of Technology
KNK	Kompakte Natriumgekühlte Kernreaktoranlage (Germany)
KWU	Kraftwerk Union (Germany)
LAB	Lower Axial Blanket
LBE	Lead-Bismuth Eutectic alloy
LEU	Low Enriched Uranium
LGR	Light water cooled Graphite moderated Reactor
LLW	Low Level Waste
LMFBR	Liquid Metal cooled Fast Breeder Reactor
LMFR	Liquid Metal Fast Reactor
LOCA	Loss-Of-Coolant Accident
LOF	Loss-Of-Flow (accident)
LP	Low Pressure
LUCA	Chemical separation process (KFA)
LWBR	Light Water Breeder Reactor
LWGR	Light Water cooled Graphite moderated Reactor
LWR	Light Water cooled and light water moderated Reactor
LWR-Pu	LWR with plutonium fuel
MA	Minor Actinide
MAGNOX	MAGnesium Non-OXidizing alloy gas cooled graphite moderated reactor using a Mg alloy as cladding material and natural uranium metal as fuel
MASURCA	MAquette SURgénératrice de CAdarache
MDC	Moderator Density Coefficient
MELOX	MOX fabrication plant (Marcoule, France)
MEU	Medium Enriched Uranium
MEXT	Japanese Ministry for Industry, Science and Technology
MF	Moderator to Fuel (ratio)
MFCI	Molten Fuel Coolant Interaction
MLIS	Molecular Laser Isotope Separation
MLW	Medium Level Waste
MONJU	Fast breeder prototype reactor (Japan)
MOX	PuO_2/UO_2 Mixed OXide fuel
MSBR	Molten Salt Breeder Reactor
MTC	Moderator Temperature Coefficient
NAS	National Academy of Sciences
NDT	Nil Ductility Temperature
NEA	Nuclear Energy Agency of OECD
NNSA	National Nuclear Security Agency (USA)

NRC	see USNRC
ODS	Oxide Dispersed Steel
OECD	Organization for Economic Cooperation and Development
ORNL	Oak Ridge National Laboratory (USA)
OT	Once-Through (fuel cycle)
OTTO	Once-Through Then Out
PCRV	Prestressed Concrete Reactor Vessel
PFBR	Prototype Fast Breeder Reactor (India)
PFR	Prototype Fast Reactor (Dounreay, UK)
PHENIX	Fast breeder prototype reactor (Marcoule, France)
PHWR	Pressurized Heavy Water moderated and cooled Reactor
POLLUX	Container for HLW storage
PPS	Plant protection system
PWR	Pressurized light Water cooled and light water moderated Reactor
PWR-Pu	PWR with plutonium fuel
PUREX	Plutonium and Uranium Recovery by EXtraction
rad	radiation absorbed dose
RAPSODIE	Réacteur RAPide à SODIum Experimentale (Cadarache, France)
RAR	Reasonably Assured Resources
rem	roentgen equivalent man
Rep-U	Reprocessed uranium
RPV	Reactor Pressure Vessel
RRU	Reenriched Reprocessed Uranium
SANEX	Selective ActiNide (III) EXtraction process (France)
SAP	Service Atelier Pilote, reprocessing plant for FR MOX fuel (Marcoule, France)
SASS	Self-Actuated Shutdown System
SCRAM	Fast Reactor shut down
SCWR	Super Critical pressure Water cooled Reactor
SEFOR	Southwest Experimental Fast Oxide Reactor (USA)
SESAME	Selective Extracting Separation of Americium by Means of Electrolysis
SFBR	Sodium cooled Fast Breeder Reactor
SFI	Sodium Fuel Interaction
SFR	Sodium cooled Fast Reactor
SG	Steam Generator
SGHWR	Steam Generating Heavy Water Reactor
SGR	Self-Generated Recycling
SILEX	Separation of Isotopes by Laser EXcitation enrichment process
SNEAK	Fast critical zero power test facility (Germany), Schnelle NullEnergie-Anlage, Karlsruhe
SNF	Spent Nuclear Fuel

SNR 300	Schneller Natriumgekühlter Reaktor, protoype fast breeder reactor (Federal Republic of Germany/Belgium/ Netherlands)
SS	Stainless Steel
SUPERPHENIX	Fast breeder reactor power plant following after PHENIX in the French breeder reactor program (Creys-Malville, France)
SVBR	Russian LBE-cooled small fast reactor
SWR 1000	Modern boiling water reactor (AREVA), (SWR: Siede-Wasser Reaktor, Germany)
SWU	Separative Work Unit characterizing specific energy demand (for uranium enrichment process)
TBP	Tri-n-Butyl Phosphate (for fuel reprocessing)
THOREX	THOrium Recovery by EXtraction
THTR	Thorium fueled High Temperature gas cooled Reactor
TMI	Three Mile Island on March 28, 1979 a core melt down accident happened at unit 2 of the TMI Nuclear Generation Station near Harrisburg, Pennsylvania, USA
TODGA	Tetra-n-Octyl DiGlycolAmide separation process (Germany)
TOP	Transient OverPower (accident)
TOR	Traitement d'Oxides Rapides (France)
TRPO	TRialkyl Phosphine Oxide process (China)
TRU	TRansUranium actinides
TRUEX	TRans Uranium EXtraction process
UAB	Upper Axial Blanket
UKAEA	United Kingdom Atomic Energy Authority
U/Pu	Uranium/Plutonium
ULOF	Unprotected Loss of Flow (accident)
ULOHS	Unprotected Loss of Heat Sink (accident)
UOX	Uranium diOXide fuel
URENCO	URanium ENrichment COmpany
USDOE	United States Department of Energy
USEPA	United States Environmental Protection Agency
USNRC	United States Nuclear Regulatory Commission
UTOP	Unprotected Transient OverPower (accident)
VHTGR	Very High Temperature Gas cooled Reactor
VVER	Russian pressurized water reactor
WLM	Working level month
ZEBRA	Zero Energy BReeder reactor Assembly (Winfrith, UK)
ZPPR	Zero Power Plutonium Reactor (ANL-West, USA)
ZPR	Zero Power Reactor (ANL-East, USA)

Chapter 1
The Development of Nuclear Energy in the World

Abstract In 2011 there were about 436 commercial nuclear power reactors operating in the world with a total capacity of 370 GW(e). Nuclear energy supplied about 16% of the world electricity. The countries with the largest nuclear energy generating capacities were the USA, France, Japan, Russia, South Korea, UK, Canada, Ukraine, China, Spain. About 81% of the operating nuclear reactors were light water cooled and moderated reactors. About 11% were pressurized heavy water moderated reactors and about 3.4% graphite moderated and gas cooled reactors. Another about 4% light water cooled and graphite moderated reactors of the Chernobyl type were still operating in Russia. The remaining 0.6% were new prototype power reactors. Besides the use of nuclear power for electricity generation, wider application directly using the nuclear heat as process heat with temperatures of about 800°C (gas cooled reactors) is possible in future. In the past BN 350 situated on the shore of the Caspian Sea was already used as a dual purpose plant, supplying in addition to electricity (150 MW(e)) also fresh water (120,000 m^3/d) by desalination. The economic advantages of nuclear power is based on the relatively low fuel cycle costs. However, nuclear power plants have capital costs higher than those of e.g. fossil fired power plants, but a much more pronounced cost degression for larger units. Nuclear power avoids to a large extent the emission of CO_2, SO_2, NO_x and also particulate emissions.

1.1 History of Development and Projections

The first nuclear power station for the generations of electricity with an output of 5 MW(e) for commercial use started to operate at Obninsk, Russia in 1954. It was a graphite moderated, light water cooled reactor. Only two years later, four 50 MW(e) graphite moderated and gas (carbon dioxide) cooled reactors started to operate at Calder Hall, UK. One year later the first electricity producing pressurized light water reactor (PWR) with 60 MW(e) began operation at Shippingport, USA. This initial phase was followed by a rapid growth of nuclear energy production in many countries.

G. Kessler, *Sustainable and Safe Nuclear Fission Energy*, Power Systems,
DOI: 10.1007/978-3-642-11990-3_1, © Springer-Verlag Berlin Heidelberg 2012

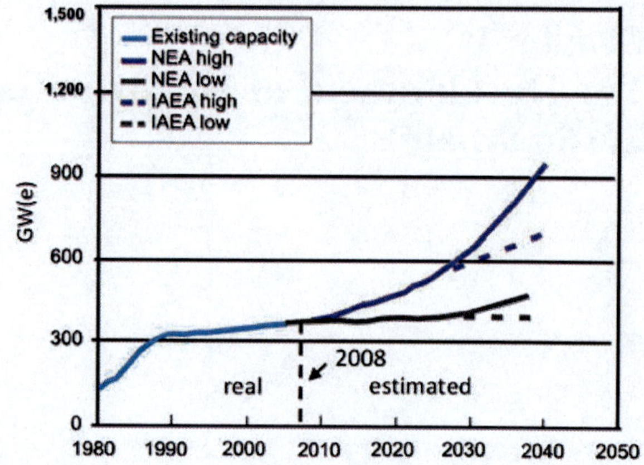

Fig. 1.1 Global nuclear capacity in the OECD NEA and IAEA high and low scenarios [2, 3]

In 2011 there were 436 nuclear reactors operating in the world with a total capacity of 370 GW(e) [1]. Projections for the future development of nuclear energy on a global basis are made by the Nuclear Energy Agency (NEA) as shown in Fig. 1.1 up to the year 2030. A high growth and a low growth scenario are shown [2, 3]. These projections are in broad agreement with those from other international organizations, e.g. IAEA etc..

The future requirement of nuclear generating capacity strongly depends on the level of the future cumulative world energy requirement which, in turn, is determined very much by the growth of the world population and by economic developments in industrialized and developing countries.

The market share that nuclear power will be able to gain will depend on the economically available reserves of nuclear fuels, the reserves of the established sources of primary energy, i.e., coal, oil, natural gas and hydropower as well as on new technologies for the exploitation of renewable energies, especially wind, solar power and biomass. Finally, the rate at which nuclear power will be introduced will be determined also by the solution of the acceptance problems in the public, by the international non-proliferation policy and by associated political decisions.

This multitude of partly conflicting factors, some of which include major uncertainties and/or regional and national differences, makes it extremely difficult to forecast the expected future worldwide nuclear generating capacity.

1.2 Status of Nuclear Energy Generation in 2008

Nuclear energy supplied about 16% of the world electricity in 2011 [1]. The countries that have built up the largest nuclear energy generating capacities are shown in Fig. 1.2. The countries with the highest number of nuclear power reactors were the USA, France, Japan, Russia, South Korea, UK, Canada, Ukraine, China, Sweden and Spain. In the rest of different countries 72 nuclear reactors were operating.

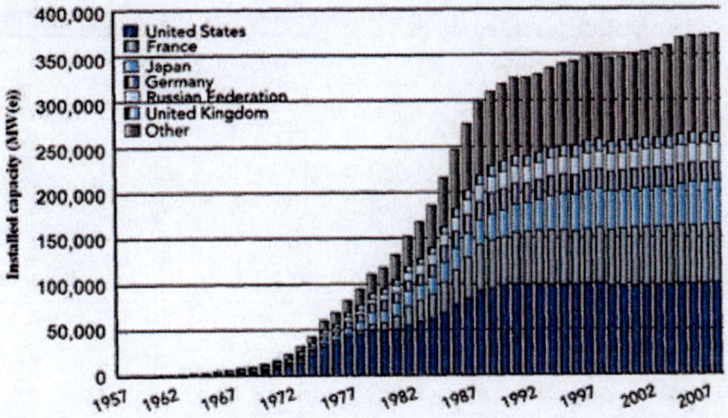

Fig. 1.2 Contributions to global nuclear generating capacity by country, 1957–2007 [2, 3]. Sources based on data taken from NEA (2006a) and NEA (2008a)

Table 1.1 Number of nuclear reactors and their share for electricity production [2, 4]

Country	Number of nuclear reactors	Share (%) for electricity production
USA	104	20
France	58	76
Japan	50	25
Russia	32	17
South Korea	20	36
Germany	9	16
United Kingdom	19	13
Canada	18	15
Ukraine	15	47
China	13	2
Sweden	10	42
Spain	8	18

Table 1.1 shows the number of nuclear reactors and their share (%) for electrical energy production.

The possibility of selecting different materials for moderating the energy of the fission neutrons in the reactor core as well as the different coolants used, lead to different types of nuclear power reactors. This is shown in Table 1.2. Light water cooled and moderated reactors (pressurized water reactors (PWRs) and boiling water reactors (BWRs) are dominating with 82%. Pressurized heavy water moderated and cooled reactors (PHWRs) follow with 10%. Graphite moderated and gas cooled reactors (GCRs) represent about 4%. Light water cooled graphite moderated reactors (LWGRs) (4%) were still operating in Russia around 2009.

Liquid metal cooled fast neutron reactors are developed for future application. They are the only type of reactors which can, by applying a breeding process, indirectly fission also the U-238 and Th-232 isotopes abundant in uranium and thorium ores.

In 2009 about 40 additional nuclear power reactors with a total capacity of 35 GW(e) were under construction. In addition many countries including China,

Table 1.2 Number and fraction of different nuclear reactors in the world [2, 3]

Type	Number of units	Fraction (%)
PWR	262	60.4
BWR	94	21.4
PHWR	44	10.0
GCR	18	4.1
LWGR	16	3.6
FBR	2	0.5
Total	436	100.0

India, Japan, South Korea, the Ukraine and Russia announced ambitious programs to expand nuclear energy capacities in the coming decades [2].

Nuclear power reactors are currently built mainly with an electricity generating capacity of 1,200–1,600 MW(e). They are erected in a construction period of about 60 months and are operating in the base load regime with an availability factor of 85–90%. Their operating life time has been increased from 35 years to about 60 years.

1.3 Technical Applications of Nuclear Fission Energy

1.3.1 Nuclear Power for Electricity and Process Heat Generation

The energy released in the nuclear fuel of the reactor core by the fission of uranium or plutonium nuclei mostly consists of kinetic energy of the fission products with the result that the nuclear fuel is heated. Consequently, the primary energy in nuclear fission is thermal energy (heat), which can be extracted from the reactor core by means of a coolant and used either directly as process heat or be converted into electricity by a thermodynamic water/steam process.

Nuclear power plants equipped with LWRs attain saturated steam conditions slightly below 300°C and approximately 70–78 bar (thermal efficiency: 33–36%). Power plants incorporating advanced types of reactors (advanced gas cooled reactors or liquid metal cooled fast neutron reactors) use superheated steam at slightly more than 500°C and 160 bar (thermal efficiency, approximately 40%).

The use of electricity for lighting, power and direct ohmic heating in industry, transport and private households has become widespread. The percentage fraction of electricity in the total consumption of final energy is likely to increase also in the future. In addition to direct ohmic heating, heating by means of electric heat pumps may well achieve growing importance in the future.

Besides the use of nuclear power for electricity generation, wider applications directly utilizing the heat are quite possible in the future, especially under the

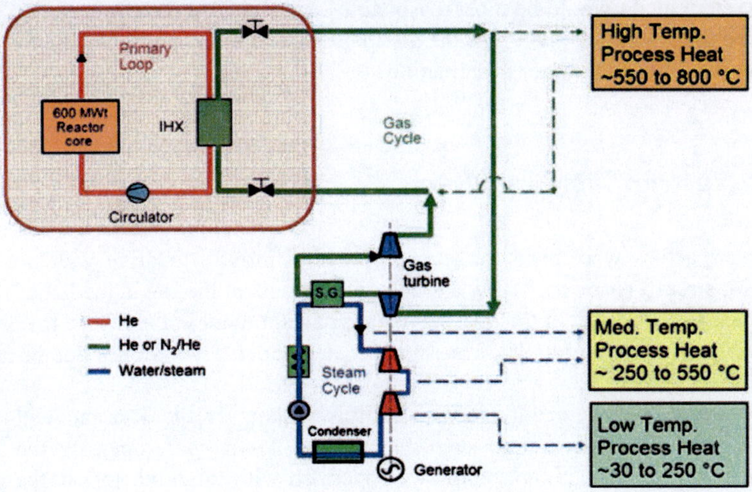

Fig. 1.3 Gas cooled nuclear power plant producing process heat and electricity [5]

incentive of finding substitutes for oil and natural gas as primary sources of energy for producing electricity and for transport purposes. In principle, the waste heat of nuclear power plants, as of coal fired power plants, can be exploited to supply district heat to cities or industrial regions. Optimum utilization of nuclear heat will be achieved in dual purpose nuclear power plants, in which the steam generated will first be partly expanded in turbines for electricity generation and then extracted either partly or entirely from the final turbine stage for purposes of heat supply (back pressure turbine process). The combined generation of power and heat in dual purpose nuclear power plants not only offers advantages in terms of energy (overall thermal efficiency, 75–85%), but is especially attractive also from an economic point of view. Figure 1.3 shows the basic flow sheet of a dual purpose gas cooled nuclear power plant producing process heat of different temperature levels and electricity.

Since nuclear power plants are built in large units for economic reasons, and since heat cannot be transported over long distances economically (the costs of the necessary distribution system decisively influence the costs of district heat), the use of nuclear power in combined dual purpose plants offers economic advantages only in areas of high and concentrated heat requirements [6].

A large part of industrial process heat is generated in the range of temperatures between 200 and 400°C, especially in chemical industries. For this application nuclear power from dual purpose power plants would constitute a solution. A special area of application of nuclear process heat is the generation of fresh water by sea water distillation. The world's fresh water requirement increases very much like the energy requirement. In many developing countries, the supply of fresh water by this technique will become a vital necessity. Since, for technical reasons, a steam quality not exceeding 150°C is sufficient for the desalination process, a combined dual purpose system again would be most reasonable economically. The high capital outlay required for the power plant plus the distillation plant today still hampers the use of nuclear power for sea water desalination.

1.3.2 Nuclear Ship Propulsion

Nuclear reactors were used for ship propulsion initially to drive warships (submarines, aircraft carriers). Today's PWRs, which rank at the top in the list of reactor concepts, are a product of this marine reactor development in the fifties. In commercial shipping, nuclear propulsion has not been developed beyond a few demonstration projects.

The vessel to be mentioned first in this respect is the Russian icebreaker, N.S. "Lenin", which serves to keep the Western Arctic route open for the Soviet marine in winter. The N.S. "Lenin" was operated with three reactors in the period 1959–1966 and has been run on two improved reactors since 1970.

Over a period of eight years of operation, between 1962 and 1970, the American N.S. "Savannah" accumulated the necessary operating experience and was then decommissioned for cost reasons.

The N.S. "Otto Hahn". the German nuclear power research vessel, was operated between 1968 and early 1979 and, like the N.S. "Savannah", has produced excellent operating results. The Japanese N.S. "MUTSU" performed a similar trial phase of operation.

The market penetration of commercial vessels with nuclear propulsion systems is now mainly an economic question. In addition, the future of commercial nuclear shipping depends very much on the establishment of international agreements about port entry permits for the most important commercial ports in the world.

1.3.3 Nuclear High Temperature Process Heat

High temperature gas cooled reactors (HTGRs) can greatly expand the use of nuclear process heat. They attain coolant outlet temperatures of 700–950°C. In that range of temperature, especially processes of direct nuclear coal gasification are of interest in which the process heat required for conversion is supplied as nuclear heat from an HTGR in a temperature range up to 950°C.

1.3.4 Nuclear Power for Hydrogen Generation

Hydrogen as a future secondary fuel can make major contributions to the supply of energy in all areas of consumption of final energy. Nuclear power plants can produce hydrogen both by electrolysis and by thermo-chemical water splitting processes at high temperatures. Hydrogen can be carried to load centres over long distances in pipelines of the type successfully applied today in the chemical industry. According to experience gathered in Europe, hydrogen can be stored, e.g., in underground cavities.

For water electrolysis, large electrolytic plants of several 100 MW(e) power will have to be developed in the future with low capital costs and high efficiencies for hydrogen generation. One main incentive in electrolysis processes may lie in the utilization of off-peak electricity, which means that the surplus electricity generated in nuclear power plants outside peak load times can be used to produce hydrogen.

The problems inherent in thermo-chemical water splitting processes today still lie in the choice and demonstration of economically viable processes. They can work successfully only in a range of temperatures offered at present by high temperature gas cooled reactors.

1.4 Economic Aspects of Nuclear Energy

The future of nuclear power decisively depends on its economic prospects. Economics in this case not only implies the operation of nuclear power plants, but also the facilities going with them to supply nuclear fuel and dispose of nuclear waste.

The economic advantage of nuclear power lies in its relatively low fuel costs, which means that, e.g., changes in the uranium price will only have moderate effects on the overall electricity generating costs of a nuclear power plant. This characteristic has a stabilizing influence on the energy market. On the other hand, nuclear power plants have capital costs clearly higher than those of fossil fired power plants. However, nuclear power plants have a much more pronounced cost degression for larger units.

1.4.1 Electricity Generating Costs by Nuclear Power Reactors

The costs of generating electricity are determined as average cost values over the full life time of the nuclear plant and over the fuel cycle time periods of the fuel. They can be divided into three main components [7]:

- capital investment (construction costs and interest during the construction time),
- operation and maintenance cost (materials, man power and services, insurances, safety inspections and safeguards etc.),
- fuel cycle costs (uranium acquisition, conversion, enrichment, fuel fabrication, reprocessing and radioactive waste disposal).

Figure 1.4 shows the cost structure of nuclear electricity generating costs. Nuclear power reactors typically have high investment costs of about 60%, maintenance cost

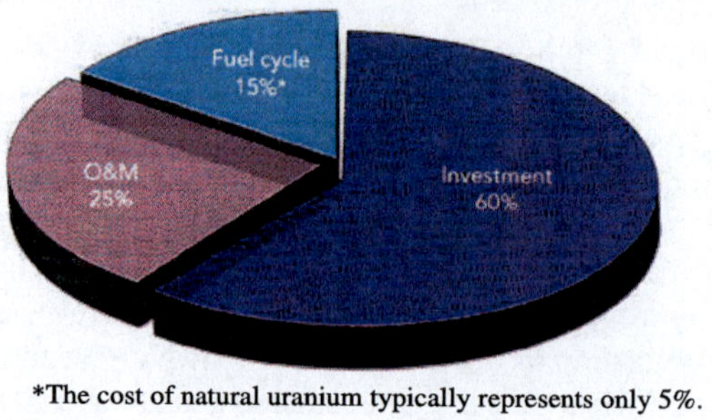

Fig. 1.4 Cost structure of nuclear electricity generation [2]. *Source* NEA and IAEA (2005)

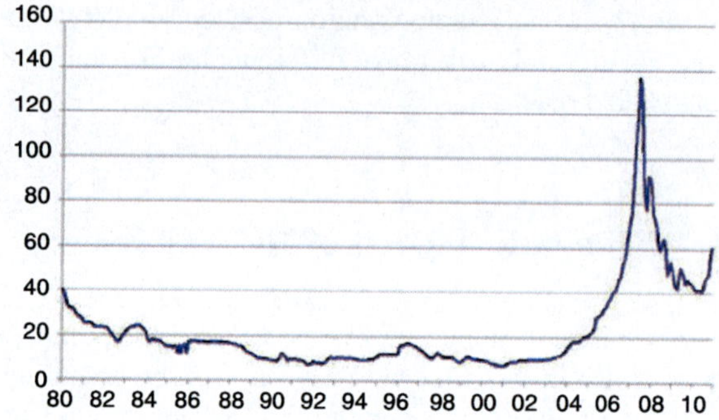

Fig. 1.5 Uranium price development between 1976 and 2010 for uranium coming from different regions of the world [8]

of about 25% and fuel cycle costs of only about 15%. The costs of the uranium typically represent only about 5% of the total electricity generation costs [2].

Figure 1.5 shows the uranium market price development between the years 1976 and 2010. The uranium price development depends on a number of factors. The predominant factors are the uranium production and supply situation by the uranium mining industry and the demand of uranium by the nuclear power reactor industry.

1.4.2 Example for Projected Nuclear Electricity Generating Costs of a 1.6 GW(e) PWR

The OECD/NEA reported the following cost structure for a 1.6 GW(e) PWR in 2007 (UK Department of Trade and Industry and Direction Générale de l'énergie et des matières premières of the (DTI) French Ministry of Economy) [2]:

Construction cost of the plant	2,500 US $/kW(e)
Construction period	6 years
Plant availability factor	90%
Operation life time	60 years
Discount rate	10%

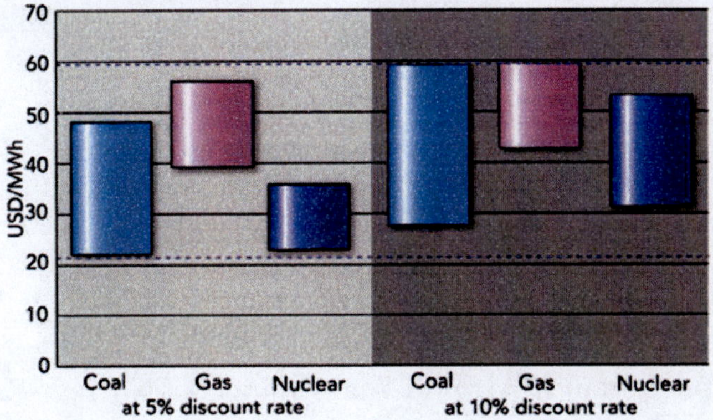

Fig. 1.6 Comparison of electricity generating costs for coal, gas and nuclear energy (for 5 and 10% discount rate) [2]

This leads to nuclear electricity generating costs of 6.2 US ¢/kWh. If a 30% cost overrun of construction cost would be assumed (DTI) these nuclear electricity generating costs could rise up to 8.8 US ¢/kWh. A lower discount rate of 5% instead of 10% would lower the nuclear electricity generating cost by about 30–35% [2]. The dependency on the discount rate is a characteristic of capital intensive technologies, such as nuclear power plants.

1.4.3 Cost Comparison for Nuclear Energy, Coal and Gas

Structures for construction costs, operation and maintenance, interest rates, taxes etc. are different in different countries. The OECD/NEA published in 2005 extensive studies and comparisons for electricity generation in different countries for the three main primary energies (coal, gas, nuclear). Figure 1.6 shows the results of this sensitivity study. The total electricity generation costs of three electricity generating base load plants (coal, gas and nuclear) with a 5 and 10% discount rate are compared. (The ranges shown do not contain the lowest and highest values obtained in these studies).

The nuclear energy cost ranges are in most cases lower than the competing plants using coal and gas. However, there are also cases for which coal can be as competitive as nuclear (5% discount rate case).

The cost projections given in Sect. 1.4.2 are somewhat higher, since they are based on higher construction cost.

Table 1.3 Average CO_2 emissions by energy source (kg CO_2/kWh) [2, 9, 10]

Energy chain	Average CO_2 emissions
Lignite	1.2
Hard coal	1.07
Oil	0.9
Natural gas (combined cycle)	0.4
Solar PV	0.060
Wind (offshore)	0.014
Wind (onshore)	0.011
Nuclear	0.008
Hydro	0.005

1.5 Comparison of Environmental Impacts of Nuclear Energy

1.5.1 Carbon Dioxide Emissions

Nuclear power avoids to a large extent the emission of CO_2 during electricity generation. Table 1.3 shows the CO_2 emissions in kg CO_2 per kWh electricity generation for different energy sources. The data were determined as average values on a full life-cycle basis (construction, operation, decommissioning) of the power plants [2, 9, 10]. Each of the time periods for construction, operation and decommissioning produces it own CO_2 emissions since their realization consumes energy.

Lignite and hard coal have the highest CO_2 emissions per kWh during operation of these plants for energy production followed by oil and gas. Plans for coal fired plants to sequester the CO_2 from the flue gas and store it underground can lower these CO_2 emissions by about 80%.

Nuclear power together with hydro power, wind and solar power have the lowest CO_2 emissions. CO_2 emissions are only produced during their construction and decommissioning phases. According to OECD/NEA nuclear power has already avoided between 1971 and 2004 more than 20% of the CO_2 which would have been emitted otherwise [2].

1.5.2 Particulate Emissions

Particulate emissions refer to aerosol particles (solid or liquid) in the atmosphere. Power stations and other industrial processes or diesel engines are the main sources of particulate emissions. Particles smaller than 10 microns (PM_{10}) are responsible for health damage (lung diseases). Figure 1.7 shows the life-cycle releases of particulates (PM_{10}) in kg/kWh for selected energy chains. Lignite, hard coal and oil are mainly responsible for the highest particulate emissions followed by Diesel engines and wood burning.

Natural gas, wind, solar, hydro and nuclear power show the best performance for the emissions of particulate matter (PM_{10}).

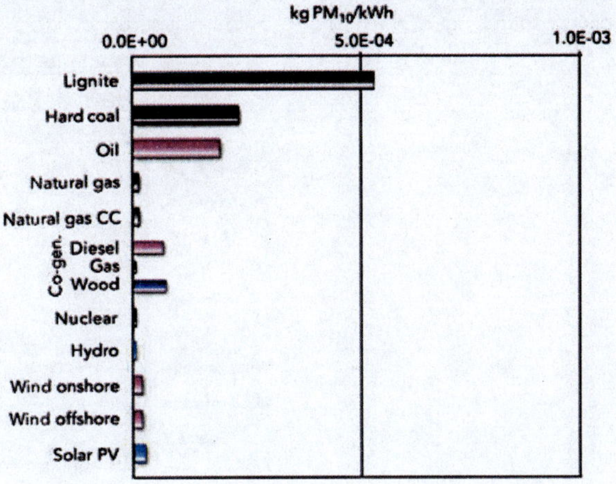

Fig. 1.7 Particulate emissions by energy sources (kg PM_{10}/kWh) (OECD-NEA) [2, 9, 10]

1.5.3 SO_2 Emissions

For fossil fuel (lignite, hard coal, oil) power plants the SO_2 emissions depend on the sulphur content of the fuel. Figure 1.8 shows the SO_2 emissions for different energy chains on a life-cycle basis in kg SO_2/kWh. Lignite, hard coal and heavy oil have the highest SO_2 emission. Again natural gas, solar, wind, hydro and nuclear power have the lowest SO_2 emissions.

1.5.4 NO_x Emissions

Figure 1.9 shows the NO_x emissions on a life-cycle basis for the different energy chains. The high temperature of the combustion process with air for the fossil fuel is mainly responsible for the high NO_x emissions of lignite, hard coal, oil and natural gas. Again hydro, wind and nuclear power have the lowest NO_x emissions.

1.5.5 Radioactive Gas and Liquid Emission

The above mentioned emissions do not cover the emission of radioactive substances, e.g. hard coal contains uranium and thorium and their daughter products on a ppm level.

In case of nuclear energy the whole nuclear fuel cycle starting from uranium mining and conversion to fuel fabrication, nuclear energy production in fission reactors, fuel reprocessing and refabrication as well fuel waste disposal must be considered. During uranium mining and milling, radium and radon will be released or

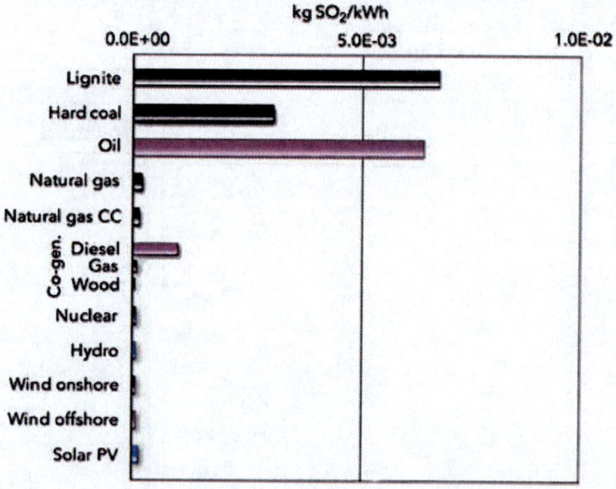

Fig. 1.8 SO_2 emissions by energy source (kg SO_2/kWh) (OECD-NEA) [2, 9, 10]

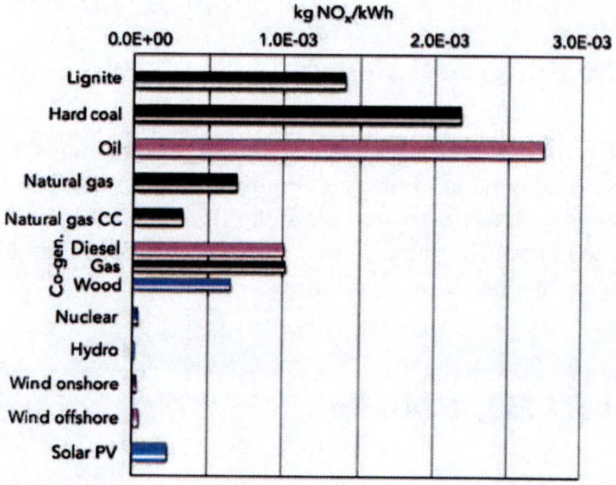

Fig. 1.9 NO_x-emissions by energy source (kg NO_x/kWh) (OECD-NEA) [2, 9, 10]

emitted. The tails from mining and milling must be covered by a layer of several meters of sand to prevent radon emission. Uranium mining and milling cause similar collective global radioactivity exposure as nuclear fission reactors (Chap. 10) [2].

Uranium enrichment and fuel fabrication plants have much lower radioactivity releases to the environment than nuclear fission reactors (Chap. 10). The radioactivity release of nuclear fission reactors and of spent fuel reprocessing plants are given in Chap. 10. These radioactive substances released into the air or water typically lead to radioactivity exposures of the public which are about 1% of the natural background radiation levels (Chap. 10) [2].

Both the operational and potential accidental radioactivity exposures to the environment will be discussed in detail in Chaps. 10 and 11.

1.6 Sustainability of Nuclear Energy

The definition of sustainability of energy-systems was published by the Brundtland Commission of the World Commission for Environment and Development of the United Nations [11]. This can be cast into the following requirements for energy [12]:

- No short time depletion of resources.
  This can assured by the large scale introduction of the FBR technology and the utilization of U-238 and Th-232 ores for many thousands of years as mentioned in Sects. 1.2 and 2.2.1 and presented in detail in Chap. 6.
- Extremely low emission of noxious or radioactive substances.
  This was shown in Sect. 1.5 and will be presented in more detail in Chap. 10.
- Extremely low risk for the population and environment.
  The extremely low risk of nuclear energy technology in comparison to other energy technologies will be presented and discussed in Chap. 11. The transformation of the very long term risk of nuclear waste disposal into a few thousands years problem by partitioning and transmutation or destruction of long-lived radiotoxic nuclides will be discussed in Chaps. 7 and 9.
- The economical competitiveness of nuclear energy was shown in Sect. 1.4.

It will be shown in the subsequent sections that nuclear energy can satisfy the requirements for sustainability of the Brundtland Commission of the United Nations.

References

1. American Nuclear Society (2011) Nuclear news—a publication of the American Nuclear Society, Mar 2011. http://www.ans.org
2. Nuclear Energy Outlook 2008 (2008) OECD-NEA No. 6348
3. Uranium 2007—Resources, Production and Demand (2008) OECD-NEA-IAEA No. 6345
4. IAEA (2008) Power reactor information system of the International Atomic Energy Agency. http://www.allcountries.org/rankings/nuclear_share_electriticy_generation_by_country_2009.html
5. AREVA (2006) AREVA HTR: a process heat source to power many industrial applications. http://www.areva-np.com/common/liblocal/docs/Brochure/VHTRadd_A_2006.pdf
6. Kessler G (1983) Nuclear fission reactors. Springer, Vienna
7. Owen P, Omberg R et al (1981) Economic analysis of nuclear reactors. In: Waltar A, Reynolds A (eds) Fast breeder reactors. Pergamon Press, New York
8. http://www.mongabay.com/commodities/price-charts/price-of-uranium.html
9. Hirschberg S et al (2004) Sustainability of electricity supply technologies under German conditions. PSI report No. 04–15, PSI, Villingen
10. Dones R et al (2005) Externalities of energy: extension of accounting framework and policy applications. Final report on work package 6: new energy technologies. http://www.externe.info/expolwp6.pdf
11. Report of the World Commission on Environment and Development (1987) Our common future. Oxford University Press, Oxford
12. Voss A (1999) Nachhaltige Entwicklung ohne Kernenergie, Deutsches Atomforum. Inforum, Adelaide

Chapter 2
Nuclear Fuel Supply

Abstract According to IAEA and OECD/NEA the reasonably assured and inferred natural uranium resources amounted in 2007 to 5.47 million tonnes in the world for the cost category up to 130 US $/kg. The reasonably assured resources are based on high confidence estimates compatible with decision making standards for mining. Inferred resources still require additional measurements, before making decision for mining. Prognosticated resources amounted to another 2.8 million tonnes and speculative natural uranium resources were estimated to about 7.5 million tonnes. Together with so-called unconventional natural uranium resources the global amount of uranium is estimated to about 22 million tonnes. For a 400 GW(e) scenario the present LWRs operating in a once through fuel cycle mode would consume about 5.46 million tonnes of natural uranium over a time period of about 80 years. Plutonium recycling reactors, e.g. LWRs and FBRs, operating in a partially or fully closed fuel cycle would consume a reduced amount of natural uranium over a very long period of time.

2.1 The Nuclear Fuel Cycle

The exploitation of nuclear power in power plants begins with the only fissile isotope U-235 occurring naturally. This isotope is contained in natural uranium in an abundance of 0.72%, the balance being 99.2745% of U-238 and about 0.0055% U-234. Natural uranium can be found in uranium ores in varying concentrations ranging from fractions of a percent up to several percent. It must be extracted as uranium oxide (mainly U_3O_8) by open pit mining or underground mining and subsequent ore dressing (Fig. 2.1). Only a few reactor types (see Chap. 8) can operate with the natural U-235 content in uranium. Since most reactor types require U-235 fuel with a low enrichment, uranium oxide is converted first into gaseous uranium hexafluoride (UF_6) and raised to the desired enrichment level of about 3.3–4.5% in isotope enrichment plants. This produces depleted uranium with a U-235 content of approximately

G. Kessler, *Sustainable and Safe Nuclear Fission Energy*, Power Systems,
DOI: 10.1007/978-3-642-11990-3_2,

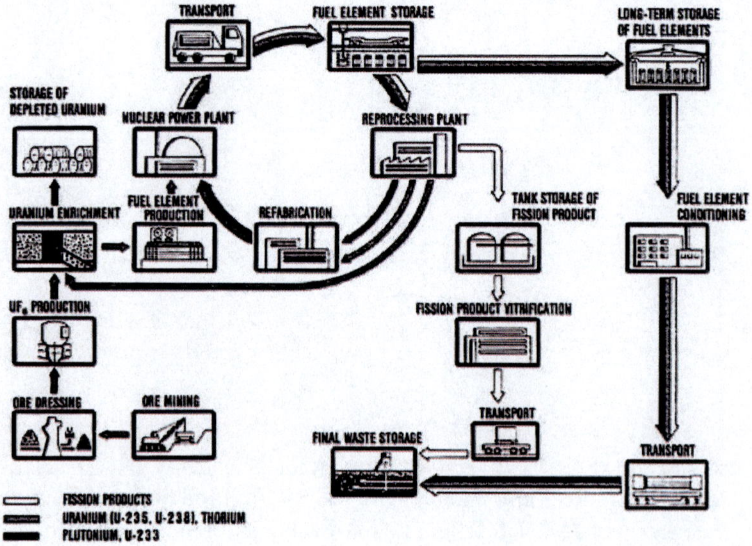

Fig. 2.1 Nuclear fuel cycle options [1]

0.2%, which is first stored and can later be used as a fuel, e.g. in fast neutron breeder reactors. Enriched UF_6, is reconverted into UO_2, in a chemical conversion process, fabricated into cylindrical pellets, which are stacked in zircaloy or steel tubes and then assembled into fuel elements. These fuel elements are loaded in the core of the reactor plant, where they generate nuclear power through fission processes. In this process, the enrichment in the U-235 isotope is reduced continuously, while radioactive fission products and transuranium isotopes are produced. After having generated energy in the core, the fuel elements are unloaded and, after short interim storage in the reactor plant, shipped to an intermediate storage facility for spent fuel.

After several years of storage of the spent fuel elements, e.g. 7–10 years, two basic options are open for further treatment:

- The fuel elements can be directly put into a repository in deep geological strata after prolonged temporary storage (more than 35 years) and special conditioning (see Chap. 7).
- In a chemical reprocessing plant, the fission products and higher actinides can be separated from uranium and plutonium in the spent fuel elements. After special waste treatment, the fission products and higher actinides can be shipped to a repository and stored in deep geological strata. The valuable uranium and plutonium can be recycled and reused as new fuel for energy production in the cores of nuclear reactors (see Chaps. 8 and 9).
- In addition also the actinides neptunium and americium can be separated and incinerated. The remaining fission products and losses of actinides during the chemical separation process can also be stored in deep geological strata (Chap. 9).

- Thorium-232 is a fertile material. It can be mixed with either U-235 enriched uranium or plutonium. Neutron capture in Th-232 will lead to U-233. In a chemical reprocessing plant the fission products and actinides can be separated. U-233 and plutonium can be recycled. In such a way also a U-233/Th-232 fuel cycle can be established.

In the following sections of this chapter, the front end of the fuel cycle will be treated, while the different reactor lines will be described in Chap. 9. The remaining part of the fuel cycle (the back end) will be outlined in Chaps. 7 and 9.

2.2 Uranium Resources and Requirements

2.2.1 Uranium Consumption in Various Reactor Systems

Projections of the future contribution of nuclear power towards meeting the world energy requirement invoke the question of the consumption of natural uranium, and the worldwide geological resources and recoverable reserves, respectively, of natural uranium and thorium. As U-235 is the only natural heavy atomic nucleus which can be split by thermal neutrons, nuclear reactors in the initial phase of a nuclear power economy must be operated with a nuclear fuel containing the U-235 uranium isotope in its natural, or a higher, enrichment (isotope enrichment). It is only by nuclear reactions during reactor operation that, following neutron capture, U-238 and Th-232 (fertile materials) are converted into artificial fissile isotopes, such as Pu-239 and U-233:

$$^{238}_{92}\text{U} \xrightarrow{(n,\gamma)} {}^{239}_{92}\text{U} \xrightarrow[23.5\,\text{min}]{\beta^-} {}^{239}_{93}\text{Np} \xrightarrow[2.35\,\text{d}]{\beta^-} {}^{239}_{94}\text{Pu}$$

$$^{232}_{90}\text{Th} \xrightarrow{(n,\gamma)} {}^{233}_{90}\text{Th} \xrightarrow[22.1\,\text{min}]{\beta^-} {}^{233}_{91}\text{Pa} \xrightarrow[27.0\,\text{d}]{\beta^-} {}^{233}_{92}\text{U}$$

The artificial fissile isotopes, U-233 and plutonium can be chemically separated from the spent fuel (reprocessed) and returned (recycled) to the nuclear reactors either together with or instead of U-235.

In principle, any type of reactor can be operated on U-235 fuel. However, the development of nuclear power so far has shown that the initial phase was dominated by three different reactor lines. LWRs have gained a significant lead (82%) over HWRs (10%) in the nuclear power market, and gas-graphite reactors (GGRs), AGRs (4%), originally the market leaders, now only play a minor role. Among the advanced reactors it is the Pu recycling converter (PWR-Pu) and the FBR which can reuse the plutonium generated in today's "conventional" reactors that have received the greatest attention. In principle, however, plutonium can also be recycled in HWRs or AGRs. Within the development of HWRs and HTGRs thorium has been considered as a fertile material. The artificial fissile U-233 produced in this case can be used in

Table 2.1 Natural uranium consumption (t) for a 400 GW(e) scenario over 40–80 years with different reactor types

Time period (year)	LWR-once through[a] (t)	HWR once through[b] (t)	LWR U/Pu-recycle[b] (t)
40	2,732,800	1,715,200	1,768,000
60	4,099,200	2,572,800	2,652,000
80	5,465,600	3,430,000	3,56,000

[a]Load factor 0.93
[b]Load factor 0.85

advanced HWRs and high temperature thorium reactors (HTR-Th), which achieve good fuel economics.

Natural uranium consumption in various reactor types is covered in detail in Chap. 8. At this point, only a few figures will be singled out to explain the orders of magnitude involved [1].

The consumption of natural uranium in the present LWR design is still relatively high: 171 or 156 t/GW(e) over one year of operation at an average load factor of 0.93 or 0.85, respectively. The natural uranium consumption in the AGR is slightly lower, amounting to 140 t/GW(e). The HWR needs 151 t/GW(e) of natural uranium per year of operation. Fuel recycling reduces the natural uranium consumption of LWRs to levels around 110 t/GW(e) over one year of operation with a load factor of 0.85.

Advanced reactors can have even lower natural uranium consumption levels. They attain those levels with optimized reactor cores and by recycling the artificial fissile fuels (for details, see Chap. 8).

FBRs do not need natural or enriched uranium when started up with plutonium, for instance from LWR spent fuel. In that case, FBRs only need some 1.5 t of U-238 per GW(e) and year. This U-238 is available in the large quantities of depleted uranium accumulating as waste from the enrichment plants. However, also for FBR fuel reprocessing and recycling is an absolute necessity.

On the basis of the aforementioned data the consumption of natural uranium can be determined for various nuclear energy scenarios and reactor types.

Instead of using nuclear energy scenarios and forecasts a simple consideration can be made assuming a constant 400 GW(e) installed for a time period of 60–80 years. This assumption roughly corresponds to IAEA low in Fig. 1.1. Table 2.1 shows the natural uranium consumption of the above mentioned reactor types for the time period up to 80 years. Higher nuclear energy projections (see NEA high in Fig. 1.1) would lead to a natural uranium consumption of about 5.3 million tonnes already up to the year 2050 [2].

Reprocessing and recycling of the fuel can help to curb natural uranium consumption. However, a major cutback in the consumption of natural uranium can be achieved only by the introduction of the FBR. For the time afterwards, where breeder reactors would dominate the nuclear energy scenario the depleted uranium left over from uranium enrichment will be sufficient to produce fission energy for several thousands of years (see Chap. 6).

There are uncertainties about the time when LWRs ought to be supplemented or substituted by FBRs. World scenario studies show that FRs should be deployed on a large scale around the mid of the twenty-first century [3].

2.3 Worldwide Available Uranium Reserves

In the previous section, the quantities of natural uranium required for the different types of nuclear power reactors have been discussed; now it is necessary to contrast the worldwide available uranium reserves with those findings.

Uranium was discovered first in Czechoslovakia in the pitchblende mineral and was found also in Africa and north-western Canada. Pitchblende is a dense, impure form of uranium oxide, which occurs mixed with ores of copper, tantalum or other minerals.

Natural uranium ores are also found in sandstone deposits. The most important deposits of this type occur in the USA, Australia, Gabon and Niger. The uranium content of the mineralized rock is commonly between 0.04 and 0.25% of U_3O_8. The size of the deposits ranges between a few hundred tonnes and over 60,000 tonnes of uranium. At Elliot Lake in Canada and the gold–uranium deposits of Witwatersrand in South Africa, uranium can also be found in quartz pebble conglomerates. Their uranium content varies between 0.03 and 0.13%. Deposits may be as large as 75,000 tonnes of uranium. Another type found at Lake Athabasca, Canada, and at Alligator River of Australia is called proterozoic unconformity. These deposits range up to 150,000 tonnes of uranium with contents up to several percent of uranium. Other deposits are associated with igneous and metamorphic rocks, such as granite-pegmatites or carbonatites. Such deposits are known to be located in Namibia, Greenland, Alaska and Brazil. Calcrete deposits have been discovered in Namibia and Australia. They contain up to 40,000 tonnes of uranium at contents of 0.13%.

Uranium reserves are commonly classified by the cost of uranium recovery, which includes direct costs of mining, processing and cost of the production unit. It is commonly accepted to use cost categories, e.g. up to $80/kg U or up to $130/kg U. However, it should be noted that these cost categories do not reflect the prices at which uranium will be available to the user. Uranium reserves are also classified by the extent of geologic knowledge and the confidence in the estimates. The two categories commonly applied to this classification are "reasonably assured" and inferred resources.

Reasonable assured resources (RAR) are based on high confidence estimates which are compatible with decision-making standards for mining. Inferred resources are defined on a similar basis. However, additional measurements are required before making decisions for mining. Undiscovered resources (prognosticated and speculative) are expected to exist based on geology or previously discovered resources. Both prognosticated and speculative resources still require significant efforts for exploration.

Table 2.2 Known recoverable resources of uranium: reasonably assured resources plus inferred resources up to US $130/kg U (IAEA 2007) [4]

Country	Tonnes U	Percentage of World (%)
Australia	1,243,000	23
Kazakhstan	817,000	15
Russia	546,000	10
South Africa	435,000	8
Canada	423,000	8
USA	342,000	6
Brazil	278,000	5
Nambia	275,000	5
Niger	274,000	5
Ukraine	200,000	4
Others	636,000	11
Total	**5,469,000**	**100**

The OECD nuclear energy agency (NEA) and the IAEA collect the resource data of the different countries in the world on a yearly basis [2, 4].

Table 2.2 shows the uranium resources (RAR) and inferred for the cost category up to 130 US $/kg known to exist in the world in 2007. They amount to a total of 5.469 million tonnes of natural uranium. Most of these resources are located in Australia (23%), Kazakhstan (10%), Russia (10%), South Africa (8%) and Canada (8%).

Prognosticated resources are estimated by OECD/NEA and IAEA (2007) to total about 2.8 million tonnes of uranium recoverable at <US $130/kg U. The total reported speculative uranium resources are estimated to amount to about 7.5 million tonnes of uranium [2, 4].

2.3.1 Unconventional Uranium Resources

Unconventional resources are resources from which uranium is only recoverable as a by-product. These are uranium resources associated with phosphate rocks, non-ferrous ores, carbonatite etc. The total of unconventional resources, dominated by phosphorite deposits amounts to about 7.3–7.6 million tonnes of uranium—as reported by OECD/NEA and IAEA in 2007 [2, 4].

Other estimates of uranium resources with marine and organic phosphorite deposits point to the existence of almost 9 million tonnes of uranium in four countries alone: Jordan, Mexico, Marocco and the United States. The global total amount of uranium was estimated to 22 million tonnes. Seawater contains uranium in concentrations of 3–4 parts per billion. Research is carried out on extracting uranium from sea water. But the recovery costs are considered to be too high at present [4].

Table 2.3 World production of uranium in 2006 [2, 4]

Country	Tonnes U	Percentage of World (%)
Canada	9,862	25
Australia	7,593	19
Kazakhstan	5,279	13
Niger	3,434	9
Russia	3,262	8
Namibia	3,067	8
Uzbekistan	2,260	6
USA	1,672	4
Others	3,000	4
Total	**39,429**	**100**

2.3.2 Uranium Production

On the basis of the uranium reserves described above, it is mainly the existing yearly production capacity of uranium mines, which controls the availability of uranium on the world market. The main uranium producers of the world are shown by Table 2.3. Canada, Australia and Kazakhstan produced 57% of the world uranium in 2006.

Future increases in uranium production will depend on a variety of parameters. Capital must be available for exploration as well as uranium mine and mill constructions. Markets and prices must be attractive for investors. In the USA, Canada, Australia, and Africa, expansions of uranium mines and mills as well as the opening of new uranium mines are planned or underway.

2.4 Worldwide Available Thorium Reserves

Thorium contains only fertile material, no fissile isotopes. This assigns thorium reserves a quality different from that of natural uranium reserves. Thorium can be compared with U-238. However, the limits to the reserves of natural uranium when used in the present line of converter reactors are imposed by U-235.

Thorium can be recovered as a by-product from minerals mined for the extraction of titanium, tin and zirconium. Monazite is the main thorium-bearing mineral ($ThSiO_4$). Most of the reasonably assured reserves are located in India, China, Brazil and the USA, with large estimated additional reserves in Canada, Egypt, Australia and the USA.

Monazite sands in India, Brazil, Australia and Egypt contain 4.6–7% thorium. In the United States and Canada, thorium reserves are also found in vein deposits. Table 2.4 indicates the major thorium depositions and resources in the world. Worldwide thorium resources are estimated to total about 6.078 million tonnes of thorium including undiscovered resources. Opportunities exist to expand known reserves and

Table 2.4 Major thorium deposit types and resources [2, 4]

Deposit type	Resources (1000 t Th)
Carbonatite	1,900
Placer (glacial, alluvial)	1,500
Vein-type	1,300
Alkaline rocks	1,120
Other	258
Total	**6,078**

discover new areas. However, due to limited markets for thorium there has been little exploration to discover new reserves.

In addition to monazite sand, thorium also occurs in uranium ores processed for uranium production. The thorium recoverable from such uranium mines in Canada has reached a production level of several thousand tonnes per annum. Further increases should be attainable from uranium deposits specially mined for their thorium value.

2.5 Concentration, Purification and Conversion of Uranium

Uranium ores usually contain only a few tenths of a percent of uranium and must be separated from by-products to reduce the weight for subsequent shipment. This concentration stage is predominately performed by leaching processes. However, physical concentration methods are also applied including crushing and sizing, gravity, magnetic, electrostatic and flotation types of separation. Roasting may be employed to improve the solubility of the uranium. The type of leaching agent used depends on the uranium-bearing mineral and can be either sulfuric, nitric, hydrochloric acids or alkaline carbonates depending on the chemistry of the uranium ore and the host rock. If acid solutions are employed, solvent extraction or ion exchange is the preferred method.

Figure 2.2 shows a uranium ore processing flow sheet for acid leaching. The ores arriving from the mine are crushed and then ground to the consistency of fine sand. If wet grinding is used, the resulting slurry is fed to a leaching circuit where acid is added. With many ores, an oxidant must be added to convert the uranium to the hexavalent state, which is readily soluble. After leaching, the solids and liquids are separated. The solids are washed to recover the adhering leach solution. Uranium is extracted from the leach solution by solvent extraction or ion exchange. In the solvent extraction process, the active agent is usually an organic amine salt diluted in kerosene that can selectively extract the uranium ions into an organic complex insoluble in water. This organic phase is separated from the organic complex by contacting it with an inorganic salt solution. The uranium is then precipitated from the strip solution and the resultant concentrate (yellow cake) is dried and packaged for shipment to a refinement plant.

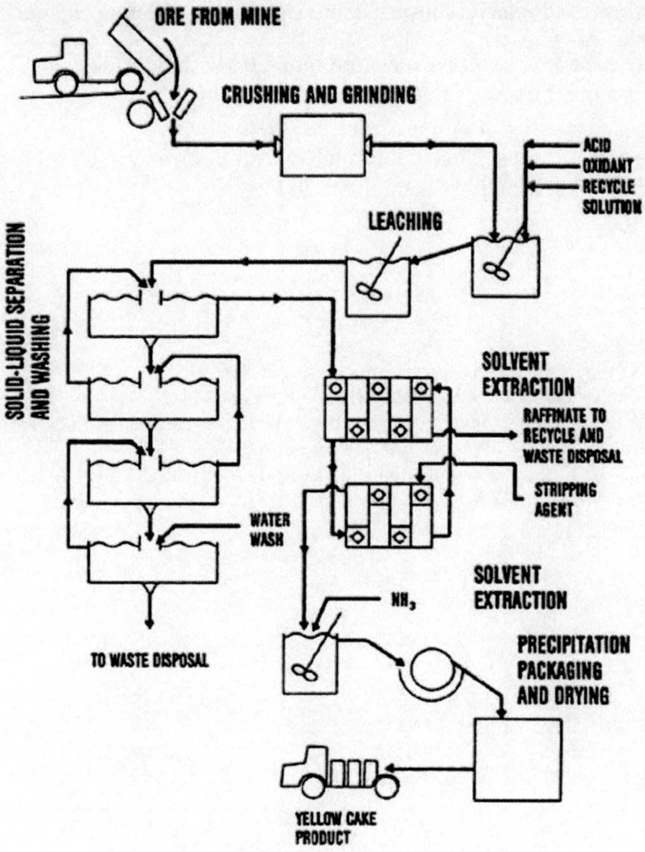

Fig. 2.2 Flow sheet of uranium ore processing [5]

For its application as a reactor fuel, these natural uranium concentrates must then be purified. For purification, uranium concentrates are dissolved in nitric acid. The resulting uranyl nitrate is then extracted in a solution of tributyl phosphate in kerosene. This purification step removes to a level of a few ppm such neutron absorbing elements as boron, cadmium and the rare earths. In addition, the contents of many other elements must be below rather low levels. The product of purification is usually one of the oxides of uranium, UO_2, UO_3 or U_3O_8.

These uranium oxides are converted into gaseous uranium hexafluoride, UF_6, for the necessary enrichment steps in isotope separation plants. Uranium hexafluoride, UF_6, is the most volatile compound of uranium, with a boiling point (sublimation point) of 56.5°C at 0.1 MPa.

The uranium oxide can be converted into UF_6 in two different chemical processes:

- In one chemical process it is converted into UF_4 with fluoride acid. Then the UF_4 is converted into UF_6 with fluorine gas.
- In the dry process the uranium oxides are first reduced to UO_2 with a mixture of hydrogen and nitrogen. This is fluorinated with fluorine gas HF to form uranium tetrafluoride UF_4. Finally, it is fluorized to UF_6.

References

1. Kessler G (1983) Nuclear fission reactors. Springer, Vienna
2. Nuclear Energy Outlook 2008 (2008) OECD-NEA no. 6348
3. Carré F et al (2009) Overview on the French nuclear fuel cycle strategy and transition scenario studies. In: proceedings of global 2009, Paper no. 9439, Paris
4. Uranium 2007: Resources, Production and Demand (2008) OECD-NEA-IAEA, no. 6345
5. Seidel DC (1981) Extracting uranium from its ores. Int At Energy Agency Bull 23(2):24–28

Chapter 3
Some Basic Physics of Converters and Breeder Reactors

Abstract The most important reactor physics characteristics needed for the understanding of the design and operation of nuclear reactors and of their fuel cycle are presented. This comprises the criticality factor, the neutron and temperature distributions in the reactor core and reactivity effects to be controlled by the safety systems. The evolution of the isotopic composition during burnup, i.e., the buildup of fission products and actinides in the reactor fuel, and the importance of conversion and breeding ratios are discussed together with the fuel utilization. Inherent safety characteristics like the negative fuel Doppler coefficient and the negative coolant temperature coefficient are essential for the safe operation and control of nuclear reactors.

3.1 Basic Nuclear and Reactor Physics

The important nuclear reactions in the cores of fission reactors are primarily caused by neutrons interacting with atomic nuclei of the fuel, the coolant, the structural materials and absorber materials. Four main neutron interactions have to be considered:

Elastic scattering,
Inelastic scattering,
Neutron capture,
Nuclear fission.

3.1.1 Elastic Scattering

Neutrons can be scattered elastically by atomic nuclei. In such scattering processes, neutrons will change their flight path and velocity (kinetic energy). Neglecting the effects of chemical bonding and the influence of crystalline materials observed at fairly low neutron energies allows such nuclear reactions to be treated by the collision

G. Kessler, *Sustainable and Safe Nuclear Fission Energy*, Power Systems,
DOI: 10.1007/978-3-642-11990-3_3,

laws of classical mechanics. According to these laws, neutrons generated by nuclear fission (fast neutrons) lose energy with each elastic collision they suffer until their energy becomes comparable to that of the thermal motion of the nuclei. The average loss of energy per collision varies inversely with the atomic weight A of the nucleus involved. For atomic nuclei with A = 100, for instance, it is only approximately 2%, but in an elastic collision with a hydrogen nucleus with A = 1, it is 50%. Multiple elastic collisions finally slow down high velocity neutrons to kinetic energies, where they are in thermal equilibrium with the atomic nuclei. Their velocity spectrum then follows approximately a Maxwellian distribution with an average kinetic energy of 0.0253 eV at a temperature of 20°C in a weakly absorbing, predominantly scattering material.[1]

The average number of collisions needed to slow down (moderate) fission neutrons to thermal energies of about 0.0253 eV is about 16 for moderation in light water, 28 in heavy water, and 91 in graphite [1–9].

3.1.2 Inelastic Scattering

In inelastic scattering interactions, the incident neutron is briefly absorbed by the atomic nucleus. The resulting compound nucleus will be in a highly excited state, which is partially relieved by emitting a neutron of a lower kinetic energy. The remaining excited nucleus usually returns to its original ground state by emitting γ-radiation. Inelastic scattering is a threshold reaction, i.e., it can occur only if the energy of the incident neutron exceeds a certain minimum kinetic energy. This energy threshold is around a few MeV for light atomic nuclei and decreases to a few keV for intermediate and heavy atomic nuclei [1–9].

3.1.3 Neutron Capture

Another possible neutron interaction with matter is the so-called capture process where the nucleus involved and the captured neutron form a compound nucleus, which usually is in an excited state so that it may

(a) change into its stable state after the emission of γ-radiation ((n, γ)-reaction),
(b) become unstable and emit an electron (β^--decay), a proton ((n, p)-reaction) or an alpha particle ((n, α)-reaction).

Occasionally, two or more neutrons may be emitted when a nucleus is struck by a high energy neutron ((n, 2n)- or (n, 3n)-reactions).

[1] 1 eV = 1.602×10^{-19} J is the kinetic energy acquired by an electron passing through a potential gradient of 1 V. 1 keV is equal to 10^3 eV and 1 MeV is equal to 10^6 eV. The energy of 0.0253 eV corresponds to a neutron velocity of 2,200 m/s.

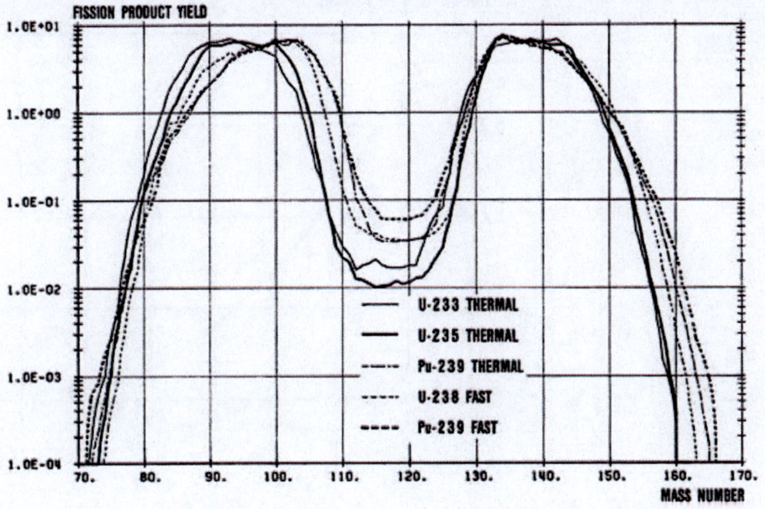

Fig. 3.1 Fission product yield (%) for fission reaction by thermal (T) and fast (F; E>1.2 MeV) neutrons [11]

As an example of neutron capture followed by β^--decay, the conversion or breeding process in the U-238/Pu-239 cycle and in the Th-232/U-233 cycle is shown below:

$$^{238}_{92}\mathrm{U} \xrightarrow{(n,\gamma)} {}^{239}_{92}\mathrm{U} \xrightarrow[23.5\,\mathrm{min}]{\beta^-} {}^{239}_{93}\mathrm{Np} \xrightarrow[2.35\,\mathrm{d}]{\beta^-} {}^{239}_{94}\mathrm{Pu}$$

$$^{232}_{90}\mathrm{Th} \xrightarrow{(n,\gamma)} {}^{233}_{90}\mathrm{Th} \xrightarrow[22.1\,\mathrm{min}]{\beta^-} {}^{233}_{91}\mathrm{Pa} \xrightarrow[27.0\,\mathrm{d}]{\beta^-} {}^{233}_{92}\mathrm{U}$$

Neutron capture in ^{10}B leads to (n, α) reaction:

$$^{10}_{5}\mathrm{B} + \mathrm{n} \rightarrow {}^{7}_{3}\mathrm{Li} + {}^{4}_{2}\mathrm{He}$$

3.1.4 Nuclear Fission

If a neutron is absorbed by a heavy nucleus, e.g., U-235, the resulting compound nucleus may become highly unstable and split into two (sometimes even three) fragments. The two fragments (fission products) generated in nuclear fission are unlikely to be identical, but are produced in accordance with a certain probability distribution. Figure 3.1 shows this probability distribution as a double-humped yield curve. In addition to the fission products, 2–3 neutrons are produced immediately after the fission event. The kinetic energy of these prompt fission neutrons also follows a certain distribution curve. This fission neutron spectrum (Fig. 3.2) and the correlated dependence on mass number of the fission product yields depend on the

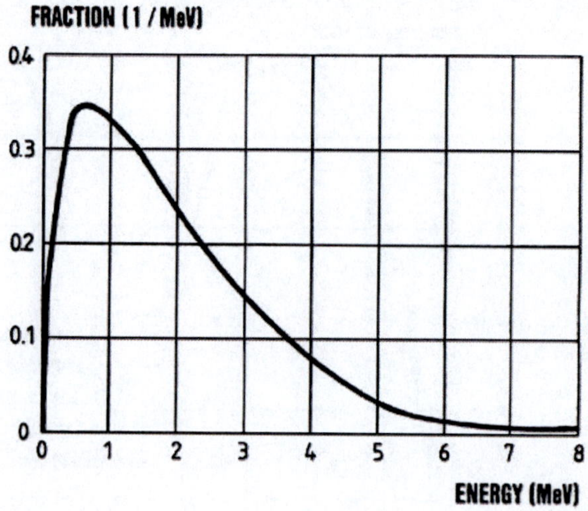

Fig. 3.2 Fission neutron energy distribution for fission of U-235 [12]

kinetic energy of the incident neutron (fast or thermal) initiating fission and on the heavy atomic nucleus undergoing fission [10].

In certain heavy nuclei, nuclear fission can only be initiated by incident neutrons above a minimum energy (threshold):

Atomic nucleus	Th-232	U-233	U-234	U-235	U-238	Pu-239
Minimum neutron energy required (MeV)	$\gtrsim 1.3$	0	$\gtrsim 0.4$	0	$\gtrsim 1.1$	0

Fission products may be gaseous, volatile or solid. Differences in fission product mass yields for different fuel isotopes, e.g., U-235 or Pu-239, have to be considered in the safety analysis of fission reactors and can be used for detection purposes.

3.1.5 Energy Release in Nuclear Fission

The energy in nuclear fission, Q_{tot}, is released as the kinetic energy, E_f, of the fission products, as the kinetic energy, E_n, of the prompt fission neutrons, as β^--radiation, E_β, and γ-radiation, E_γ. In addition, a small amount of energy, E_ν, is emitted as neutrino radiation which does not produce any heat in the reactor. The thermal energy, Q_{th} released per fission of different fissile nuclei and the different component energies are listed in Table 3.1. For U-235 about

$$194\,\text{MeV/fission or } 3.11 \times 10^{-11}\,\text{J/fission}$$

Table 3.1 Total energy release per fission and component energies (in MeV) [10]

Target nucleus	Incident neutron energy	E_f	E_n	E_β	E_γ	E_ν	Q_{tot}	Q_{th}
Th-232	3.35	161.79	4.70	8.09	14.01	10.87	196.11	185.24
U-233	thermal	168.92	4.90	5.08	12.53	6.82	198.25	191.43
	0.5	169.37	4.9	5.05	12.52	6.79	198.13	191.34
U-235	thermal	169.75	4.79	6.41	13.19	8.62	202.76	194.14
	0.5	169.85	4.8	6.38	13.17	8.58	202.28	193.7
U-238	3.10	170.29	5.51	8.21	14.29	11.04	206.24	195.2
Pu-239	thermal	176.07	5.90	5.27	12.91	7.09	207.24	200.15
	0.5	176.09	5.9	5.24	12.88	7.05	206.66	199.61
Pu-240	2.39	175.98	6.18	5.74	12.09	7.72	206.68	198.96
Pu-241	thermal	175.36	5.99	6.54	14.22	8.80	210.91	202.11
	0.5	175.62	6.0	6.49	14.19	8.73	210.53	201.8
Pu-242	2.32	176.79	4.59	6.62	12.98	8.90	206.15	197.25

are released as thermal energy, Q_{th} for U-235. Since 1 g of U-235 metal contains $6.025 \times 10^{23}/235$ = about 2.56×10^{21} atoms, complete fission of 1 g of U-235 results in

$$7.96 \times 10^{10}\,\text{J or } 2.21 \times 10^{4}\,\text{kWh}$$

of thermal energy, which corresponds to

$$0.92\,\text{MWd(th)}.$$

1 g of Pu-239 contains $6.025 \times 10^{23}/239$ = about 2.52×10^{21} atoms. As the fission of one Pu-239 atom yields about 200 MeV a complete fission of 1 g of Pu-239 yields 8.07×10^{10} J or 0.93 MWd(th) [10].

For other fissile materials, e.g., U-233 or Pu-239, the result is roughly the same. In reactor technology it is usually assumed that the mass of 1 g of fissile materials produces a 1 MWd/d(th). The energy produced by the fuel in the reactor core is indicated in MWd(th) per t_{HM} of fuel. Therefore this figure roughly corresponds to the mass (in grams) of heavy atoms fissioned in one tonne of fuel.

3.1.6 Decay Constant and Half-Life

Radioactive decay changes the number of isotopes, N(t), existing per cm^3 as a function of time, (t). This change can be described by the exponential law of

$$N(t) = N_0 \cdot \exp(-\lambda\, t) \tag{3.1}$$

where λ is the decay constant and N_0 the number of atomic nuclei per cm^3 at the time $t=0$. Instead of the decay constant, λ, one can also use the half-life, $T_{1/2}=(\ln 2)/\lambda$, which is the time by which half of the nuclei existing at $t=0$ have decayed. The decay rate, λ N(t), is called the activity of a specimen of radioactive material. This activity is measured in units of Curie or Becquerel [2, 3].

One Becquerel, denoted Bq, is defined as one disintegration per second. One Curie, denoted Ci, is defined as 3.7×10^{10} disintegrations per second, which is approximately the activity of 1 g of radium. Low activities are also measured in $mCi = 10^{-3}$ Ci or $\mu Ci = 10^{-6}$ Ci.

3.1.7 Prompt and Delayed Neutrons

More than 99% of the neutrons generated by fission appear within some 10^{-14} s of the splitting of the atomic nuclei (prompt neutrons). The fission products (fragments) generated, however, will be in a highly excited state and some of them may emit a so-called delayed neutron with delay times on the order of seconds. Thus, a small fraction of less than 1% of the total fission neutron yield is produced by disintegrating fission fragments (precursors of delayed neutrons), a process following the exponential law mentioned above. These delayed neutrons are generally combined in six groups according to the disintegration characteristics of their precursors (parent nuclei). The half-lives, $T_{1/2}$, of these six groups of delayed neutron precursors differ between 56 and about 0.2 s (Table 3.2). Their average kinetic energies are between 400 and 500 keV, which is below those of prompt fission neutrons (see Fig. 3.2). The delayed neutrons allow a reactor core to be controlled fairly easily by relatively slow movements of the absorber rods (see Sect. 3.8.3) [12–14].

The average kinetic energies of delayed neutrons and the disintegration characteristics of the precursors are roughly in the same ranges of energy and time for the most important fuel isotopes, i.e., Th-232, U-233, U-235, U-238, Pu-239, Pu-240, Pu-241, Pu-242. In addition, they are nearly independent of the energy of the neutron inducing fission. However, the fraction of delayed neutrons occurring per fission differs, as indicated in Table 3.3, and depends on the energy of the neutron inducing fission.

3.1.8 Afterheat of the Reactor Core

The gradually decaying fission products generate heat in the reactor core, even if the neutron fission chain reaction has been interrupted (after shutdown of the reactor core). This afterheat, or decay heat, is composed of the contributions by the decay chains of the fission products and of contributions by U-239, Np-239, and the higher actinides, which are unstable. It is a function of the power history of the reactor core before shutdown and is thus strongly influenced by the burnup of the fuel. Figure 3.3

Table 3.2 Delayed neutron data for six groups of delayed neutron precursors (fission by fast neutrons) [12, 13]

Group i	Half-life $T_{1/2}$(s)			Decay constant λ (1/s)		
	U-235	U-238	Pu-239	U-235	U-238	Pu-239
1	55.72	52.38	54.28	0.0124	0.0132	0.0128
2	22.72	21.58	23.04	0.0305	0.0321	0.0301
3	6.22	5.00	5.60	0.111	0.139	0.124
4	2.30	1.93	2.13	0.301	0.325	0.325
5	0.610	0.49	0.618	1.14	1.41	1.12
6	0.230	0.172	0.257	3.01	4.02	2.69

Table 3.3 Total fractions of delayed neutrons for different fuel isotopes (in %) [12, 13]

Fissionable isotopes	Th-232	U-233	U-235	U-238	Pu-239	Pu-240
Fraction of delayed neutrons						
Thermal fission	0	0.266	0.641	0	0.204	0
Fast fission	2.03	0.267	0.65	1.48	0.212	0.266

by way of example shows the relationship between the power of the fuel elements in the reactor core of a PWR after shutdown, P(t), and the power during operation, P_0. This afterheat, P(t), drops very sharply as a function of time. Immediately after shutdown it is slightly below 10%, after 10 s it is still 7%, after one hour 1.5%, after one month about 0.15%, and after one year it is 0.03% of the nominal reactor power, P_0, during operation [2, 3, 15–17].

3.2 Neutron Flux and Reaction Rates

The neutrons produced in nuclear fission have a certain velocity (kinetic energy) and direction of flight. During their lifetime they may be scattered elastically or inelastically or absorbed by atomic nuclei. In some cases, they may induce nuclear fissions so that successive generations of fission neutrons are produced and a fission chain reaction is established. The number of neutron-nuclear reactions per cm^3 at the point $\vec{r}$ of the reactor core is calculated according to the following considerations [1–9].

Let $n(\vec{r}, v, \Omega)$ be the number of neutrons at point $\vec{r}$ with the velocity, v, and the direction of flight, Ω. Within a volume element, dV, they can react with N$\cdot$dV atomic nuclei, N being the number of atomic nuclei per cm^3 of reactor volume. The number of reactions per second and cm^3 between neutrons and atomic nuclei is proportional to $v \cdot \sigma\, n(\vec{r}, v, \Omega)$ and $N \cdot dV$, the proportionality factor, $\sigma(v)$, being a measure of the probability of the nuclear reaction. $\sigma(v)$ is called the microscopic cross section of the nucleus for the respective reaction and is indicated in units of 10^{-24} $cm^2 \triangleq 1$ barn. It

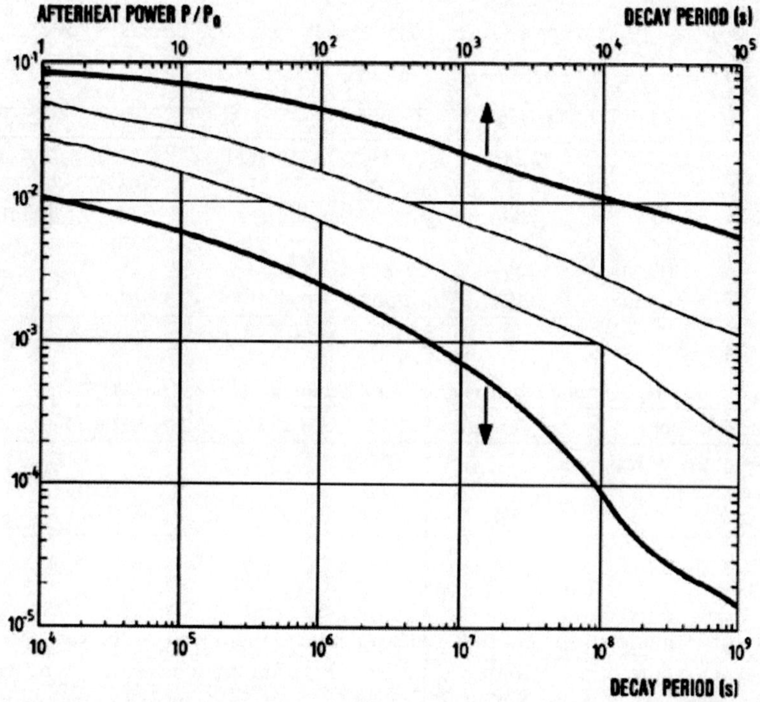

Fig. 3.3 Post-shutdown afterheat of a PWR core as a function of time (initial enrichment 3.2% U-235, burnup 32,000 MWd(th)/t) [11]

is a function of the velocity or kinetic energy, E, of the neutron, the type of reaction, and differs for every type of atomic nucleus. For scattering processes, it also depends on the angle between the direction of the incident and scattered neutron but, except for very special cases, not on Ω.

Integration over all flight directions furnishes the nuclear reactions of all neutrons with the velocity v. The reaction rate per cm^3 of reactor volume at point $\vec{r}$ then turns out to be:

Reaction rate:

$$R(\vec{r}, v) = \sigma(v) \cdot N(\vec{r}) \cdot v \cdot n(\vec{r}, v) = \Sigma(\vec{r}, v) \cdot \phi(\vec{r}, v) \tag{3.2}$$

The quantity

$$\Sigma(\vec{r}, v) = N(\vec{r})[cm^{-3}] \cdot \sigma(v)[cm^2] \tag{3.3}$$

is called the macroscopic cross section. The quantity

$$\phi(\vec{r}, v) = v \cdot n(\vec{r}, v)[n/cm^2 s] \tag{3.4}$$

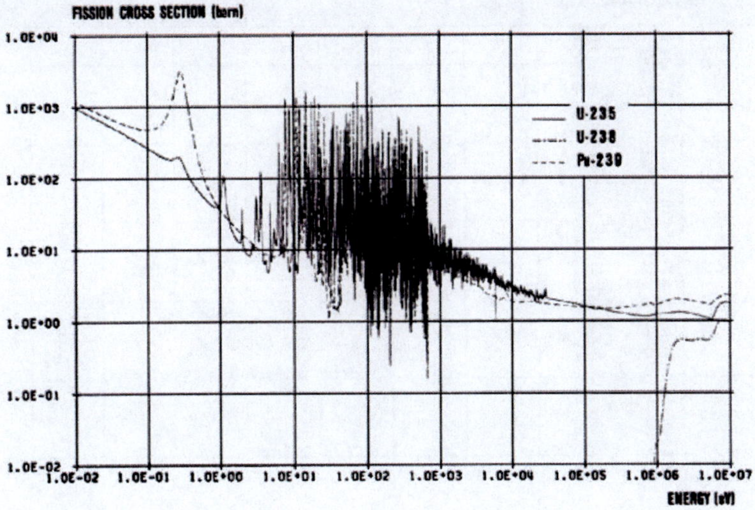

Fig. 3.4 Fission cross sections of U-235, U-238 and Pu-239 as a function of kinetic energy of the incident neutron [11]

is usually called the neutron flux or, sometimes, more appropriately labeled the neutron flux density.

Generally, the microscopic cross section, σ(E), decreases with increasing neutron velocity or kinetic energy, E, of the neutron. Exceptions to this rule are the resonance reactions and threshold reactions. As an example, Fig. 3.4 shows the fission cross sections for U-235, U-238 and Pu-239. Microscopic cross sections are collected, e.g., in JEFF [18], ENDF/B [19] or JENDL [20].

3.3 Spatial Distribution of the Neutron Flux in the Reactor Core

The number of neutrons of a certain velocity or energy and direction of flight is determined as a function of space and time from the solution of an integrodifferential equation with boundary conditions. This Boltzmann neutron transport equation is essentially a balance equation counting the number of neutrons gained and lost via different reaction processes, such as scattering, fission, capture and spatial neutron migration (for time dependent problems, also the delayed neutrons and the change in time of the number of neutrons need to be considered). For many practical applications the migration of the neutrons may be described by the neutron diffusion equation, which is an approximation to the Boltzmann neutron transport equation. The neutron diffusion equation is derived as a coupled system of G partial differential equations for G neutron energy groups. The microscopic cross sections are averaged over each neutron energy group [21, 22]. Usually the neutron diffusion equation can

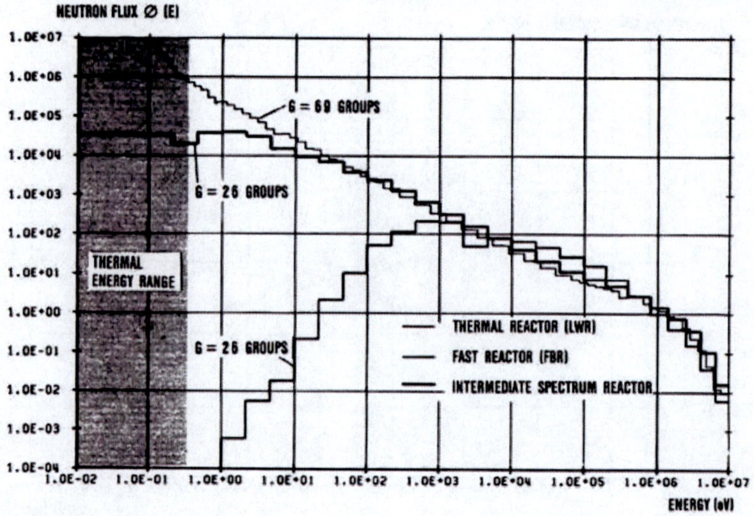

Fig. 3.5 Neutron flux, (E), as a function of neutron energy for an LWR core, an FBR core and an intermediate spectrum reactor core [11]

not be applied near local neutron sources, external boundaries and internal interfaces between material regions with different nuclear properties and within strongly neutron absorbing materials. In these cases, the Boltzmann neutron transport equation must be applied [23–26]. Figure 3.5 shows the neutron energy spectra in a thermal reactor core (LWR), a fast reactor core (FBR), and an intermediate spectrum reactor core. In this figure, the neutron flux spectra are normalized and plotted as a function of the neutron kinetic energy, E [1, 7–9].

Solutions for the spatial distribution of the neutron flux, $\phi_g(\vec{r})$, of a specific energy group, g, are usually determined numerically by means of computer programs run on digital computers [23–27]. Figure 3.6 shows the spatial distribution of the thermal neutron flux, $\phi_{th}(\vec{r})$, in a PWR core with the control rods partly inserted. The thermal neutron flux is tilted somewhat in the vicinity of the neutron absorbing control rods and decreases rapidly at the outer core boundaries.

The ratio between the number of neutrons absorbed in the reactor core and escaping from the reactor and the number of neutrons newly generated characterizes the so-called criticality parameter or effective multiplication factor, k_{eff}. For $k_{eff} = 1$, the reactor core is just critical and can be operated in a steady state condition. In this case, the statistical average of the number of neutrons appearing in successive neutron generations does not change with time [1–9].

At $k_{eff} < 1$, the reactor core remains subcritical. With $k_{eff} > 1$, more neutrons are produced than are consumed, i.e., the number of neutrons increases steadily with time. However, in the case that k_{eff} is not exceeding 1 plus the delayed neutron fraction, the increase in the number of neutrons is determined by the properties of

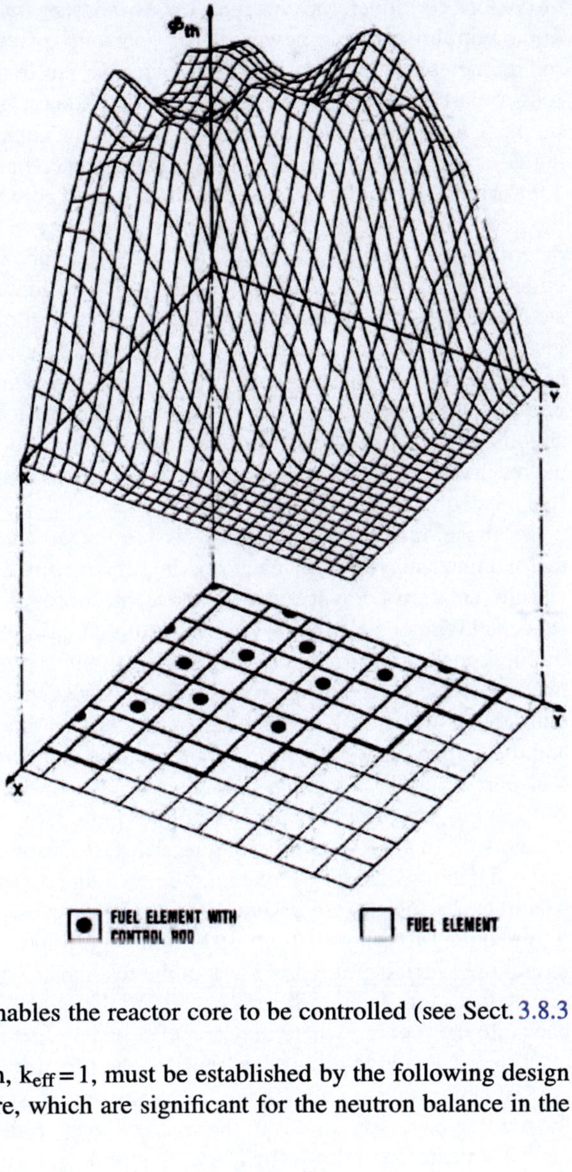

Fig. 3.6 Spatial distribution of the thermal neutron flux, $\phi_{th}(\vec{r})$, in a PWR core with partially inserted control rods [28]

the delayed neutrons. This enables the reactor core to be controlled (see Sect. 3.8.3 for more details).

The steady state condition, $k_{eff} = 1$, must be established by the following design parameters of the reactor core, which are significant for the neutron balance in the reactor core:

- geometric dimensions of the fuel rods and of the entire reactor core,
- volume fractions and types of coolant and/or moderator, structural and absorber materials in the reactor core and their geometrical arrangement,
- choice of enrichment, i.e., the isotopic composition of U-235, U-238, Pu-239, Pu-240, etc., or Th-232, U-233, U-234, etc. in the fuel rods.

Usually the choice of the coolant and/or moderator, the structural material and the type of fuel with its fissile isotopes (U-235, U-233, Pu-239) is made first. The type of coolant or moderator and fuel, by its inherent materials properties, determines the maximum possible power density per cm^3 of core volume. For a given fuel rod diameter this leads to a certain power per cm length of fuel rod. Usually these rods are arranged in regular arrays forming square or triangular lattices as shown in Fig. 3.7, the hexagonal lattice with a rod at its center is equivalent to a triangular lattice at each corner. Assuming a certain power output of the reactor core (e.g., 3,000 MW(th)), one then calculates the required core volume (radius and height) or the necessary number of fuel rods of a given length. Thermo-hydraulic conditions determine the coolant mass flow through the reactor core at specific coolant inlet and outlet temperatures (see Chaps. 5 and 6). Given these design parameters, the next step is to calculate the necessary enrichment in fissile isotopes. This is done for the given reactor geometry by solving numerically the steady state multigroup diffusion equation for its smallest eigenvalue, $1/k_{eff}$, and the neutron flux distribution in energy group g and space, $\phi_g(\vec{r})$. The fissile fuel enrichment is determined such that k_{eff} is slightly above 1, to allow for control throughout the whole period of operation of the reactor core. Enrichment and overall core size furnish the total quantity of fissile fuel needed within the core.

In those reactor cores where fission events are predominantly caused by thermal neutrons (thermal reactors), the substructure of the neutron flux in fuel rods and the ambient coolant must be taken into account by subdividing the core into subcells (Wigner-Seitz cells) (Fig. 3.7). In that case, the neutron flux is determined by superposition of the micro distribution in the subcell and the macro distribution over the entire core. The micro distribution in the subcell is obtained by solving the multigroup diffusion equations for characteristic subcells consisting of a fuel rod and the respective moderator. In more precise calculations, the Boltzmann neutron transport equation is applied [24]. For the boundary conditions of the subcells the simplifying assumption is made that a periodic lattice system of infinite extension is involved. In a two-energy group model, fast fission neutrons are generated in the fuel rod, from where they move into the surrounding moderator. Although they may penetrate the fuel region several times, they are slowed down to thermal energies, thus avoiding to a large extent undesired capture processes of epithermal neutrons in the fuel. Such capture processes in the resonance energy region otherwise would reduce the multiplication factor, k_{eff}. From the moderator, thermal neutrons diffuse back into the fuel rod where they are absorbed by fuel nuclei, thus producing fission reactions. The above discussion explains the distributions shown in Fig. 3.7, where the fast neutron flux is high in the fuel region and the thermal neutron flux is high in the moderator region. In a well thermalized reactor, the life of a neutron from birth (as a fast neutron) to death (by absorption or leakage out of the reactor) can also be described by the so-called four-factor formula for k_{eff}. Knowing the micro distribution of the neutron flux, this allows average material constants of the subcell to be determined by some kind of homogenization procedure. These average parameters are subsequently used in the calculation of the macroflux distribution.

To determine the macrobehavior of the neutron flux in the reactor core, averaged materials constants are established by detailed calculations of the fine structures of materials and the neutron fluxes in the subcells. Afterwards, the multigroup diffusion

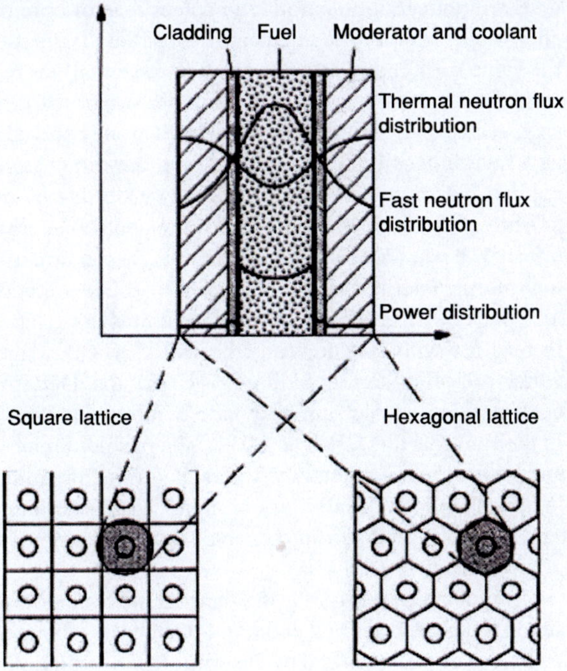

Fig. 3.7 Sub-cells in thermal reactor cores (Wigner-Seitz cells) [29]

equations are solved with the macrostructure of the reactor core and its boundary conditions taken into account.

The local power distribution, $P(\vec{r})$, in the reactor core is determined from the fission reaction rate as

$$P(\vec{r}) = 3.11 \times 10^{-11} \sum_{g=1}^{G} \Sigma_f(\vec{r})_g \phi_g(\vec{r}) [\mathrm{W/cm^3}] \tag{3.5}$$

where $\Sigma_f(\vec{r})_g$ is the macroscopic fission cross section and $\phi_g(\vec{r})$ is the neutron flux in energy group g at location $(\vec{r})$. The factor 3.11×10^{-11} corresponds to 3.11×10^{-11} J/fission for U-235 (Sect. 3.1.5). For, e.g., Pu-239 or U-233 this factor changes slightly.

Thermal reactor cores, such as PWRs, typically have a U-235 enrichment of about 3–5%. As the microscopic fission cross sections in the thermal energy range (0.025 eV) are relatively high (about 580 barns) (see Fig. 3.4), the corresponding average thermal neutron flux, ϕ_{th} is in the range of 3×10^{13} n/cm^2 s, to achieve power densities of about 100 kW(th)/l of core volume. However, fast reactor cores with average neutron energies of about 100 keV require a fissile enrichment of about 15–25%. As the microscopic fission cross sections in the 100 keV range are rather low (about 1.8 barns (see Fig. 3.4)), a much higher average neutron flux of about 3×10^{15} n/cm^2 s is required to produce an average power density of about 300 kW(th)/l [30].

For practical applications in nuclear reactor core design, predictions of the local energy spectrum of the neutrons, the spatial distributions of neutron flux and power, the value for k_{eff} and other parameters essentially rely on two and three-dimensional numerical methods involving computer programs run on fast digital computers. In special cases, the neutron transport equation is solved by numerical methods in one- and two-dimensional geometries or three-dimensional Monte Carlo methods are applied. The number of neutron energy groups to be used depends upon the problem to be treated. Frequently, instead of the neutron energy, E, a corresponding variable, lethargy, $u = \ln(E_o/E)$, is used where E_o is the upper limit of the energy scale. This logarithmic energy scale is suggested by the fact that the average logarithmic energy loss per collision of a neutron with a nucleus is an energy independent constant. In reactors with fast neutrons, basically 20–30 neutron energy groups are treated, which are often condensed into 6–12 groups. Determining the fine structure of the neutron energy spectrum may take hundreds or even thousands of neutron groups. In thermal reactors, such as LWRs, the problem may well be reduced to two or four energy groups. In reactors with fast neutrons, the microstructure of the neutron flux in the subcells is usually less pronounced than in thermal reactors; in most cases it can therefore be taken into account by a heterogeneity correction of the so-called group constants [21].

As indicated in Fig. 3.7, the thermal power is low in the cladding of the fuel rod and in the moderator and coolant. It is mainly generated there by elastic and inelastic collisions of neutrons and by absorption of γ-radiation. The radial temperature distribution within the fuel rod and the axial temperature increment in the cooling channel surrounding the fuel rod is indicated by Fig. 3.8. During normal operation, the fuel must not reach its melting temperature and the maximum cladding temperature must not exceed certain limits specific to materials (see Chaps. 5 and 6).

3.4 Fuel Burnup, Fission Product and Actinide Buildup

The fission of atomic nuclei and generation of fission products, neutron capture and radioactive α- and β-decay change the concentration of various isotopes in the reactor core during reactor operation. The change in concentration, $N^i(\vec{r}, t)$, of the isotope i as a function of time at location $\vec{r}$ per cm^3 of reactor volume can be written as a balance equation between the production rate, $R^i_p(\vec{r}, t)$, and the loss rate, $R^i_l(\vec{r}, t)$, taking into account all neutron reaction processes as well as radioactive decay processes of the nuclei. This leads to an ordinary differential equation for each isotope i. Integration over a given time period furnishes the concentration of the isotope i at location $\vec{r}$ and time t [2, 3, 6, 23, 24, 28, 30–36].

The possible chains of production and decay of fuel isotopes or higher actinides in the uranium/plutonium cycle and the thorium/uranium cycle are presented in Fig. 3.9. Neutron capture of a fuel isotope increases the mass number of the isotope by 1 (vertical step). The radioactive β-decay, with the mass number unchanged, means

Fig. 3.8 Radial and axial temperature distributions in a fuel rod and the surrounding coolant channel [29]

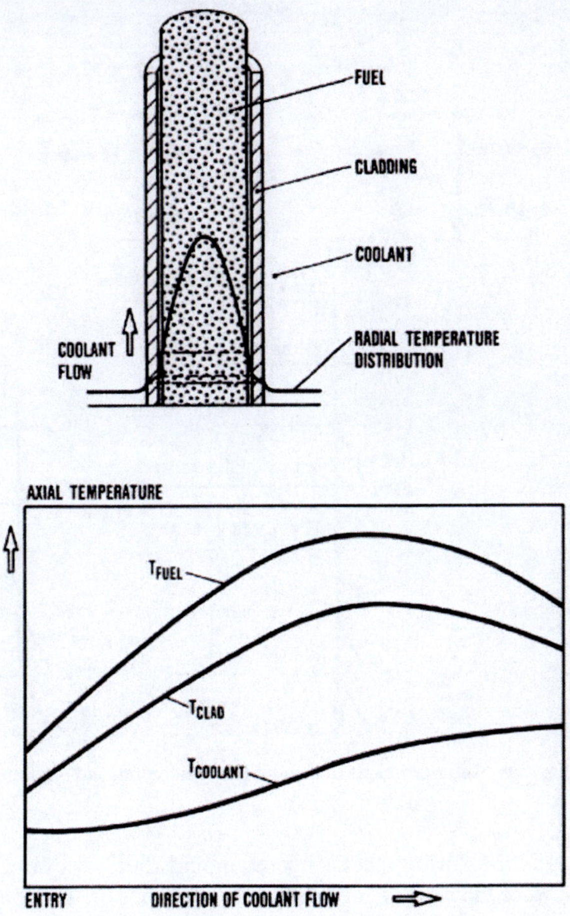

an increase by 1 in the atomic number (protons) (horizontal step). The radioactive α-decay implies a reduction by 4 in the mass number and by 2 in the atomic number.

In nuclear reactors with U-235/U-238 fuel, isotopes of the elements uranium, neptunium, plutonium, americium and curium are built up over the period of operation. Reactors containing fertile thorium in addition to U-235/U-238 fuel (HTGR, CANDU) build up possible isotopes of thorium and protactinium and their decay chains, while reactors operated only in the Th-232/U-233 fuel cycle build up almost no plutonium, americium and curium isotopes, but correspondingly larger amounts of uranium, thorium and protactinium isotopes.

In thermal reactors, the concentrations of Xe-135 and Sm-149 fission products and their precursors I-135, Nd-149 and Pm-149, respectively, must be observed carefully because of their high absorption cross sections. Their variation as a function of space

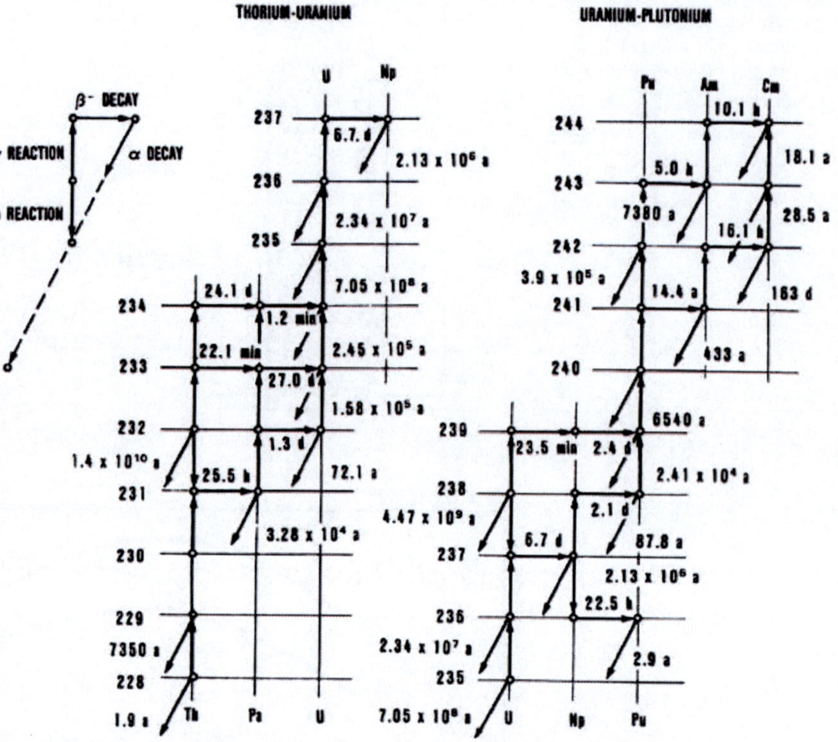

Fig. 3.9 Actinide chains in the uranium–plutonium and uranium–thorium nuclear fuel cycles [11]

and time during reactor operation and shutdown affect considerably the criticality, k_{eff}, of the reactor core and the spatial neutron distribution.

Figures 3.10a, b show the decrease of fissile U-235 nuclei (burnup), the buildup of U-236 and of the different plutonium isotopes, as well as the buildup of Am-243, Cm-244, Sr-90 and Cs-137 in one tonne of LWR fuel over a period of about five years of operation. It is seen that the isotopic distribution of plutonium (Pu-239 : Pu-240 : Pu-241 : Pu-242) also changes as a function of burnup and time. While the enrichment of U-235 has dropped to roughly 0.8% or 8 kg in 1 t of fuel by the end of the period of operation, the plutonium content (all plutonium isotopes) has risen to almost 0.9% over the same period. The plutonium buildup is due to the continuous in situ conversion of U-238 and Pu-240 by neutron capture. In this way, the fissile isotopes, Pu-239 and Pu-241, produced during reactor operation make major contributions to the fission reaction rate and, hence, to power and heat production. A corresponding process of conversion into fissile U-233 is possible as a result of the capture of neutrons in Th-232.

To determine the change in time of the isotopic composition of the fuel, including actinides and fission products, ordinary differential equations (Bateman equations) must be solved for all relevant isotopes in the actinide and fission product chains. This is done by taking into account the initial concentrations for each isotope for a sufficiently large number of space points and material zones in the reactor core. After preparation of the corresponding average one-group cross sections, programs are run on digital computers to follow these concentrations in time. In this way, the isotopic concentrations in the reactor core and their repercussions upon k_{eff} are determined for various time steps of a power cycle. The burnup of fissile isotopes and the buildup of neutron absorbing fission products and actinides mostly cause k_{eff} to decrease over a given operating cycle of a reactor core [32–35].

3.5 Conversion Ratio and Breeding Ratio

The ratio between the production rate of new fissile material (U-233, Pu-239, Pu-241) continuously generated from fertile Th-232, U-238 and Pu-240 and the continuous destruction rate of fissile U-233, Pu-239 and Pu-241 atomic nuclei is called the conversion ratio (CR). It is called breeding ratio (BR), if it attains values ≥ 1 [30, 31].

Most thermal reactors (LWR, CANDU, HTGR) have conversion ratios between 0.5–0.8 and thus are net consumers of fissile material. Because of their relatively low conversion ratios they are called converter reactors. If CR always equals 1, the fissile core inventory does not change during reactor operation. Breeding ratios close to 1 can be attained by near breeder reactors. Breeding ratios of approximately 1.15–1.30 are only reached in breeder reactors with a fast neutron spectrum (FBR).

The conversion ratio or breeding ratio of a reactor core can be described by the relation

$$\left.\begin{matrix} \mathrm{CR} \\ \mathrm{BR} \end{matrix}\right\} = \bar{\eta} - 1 - \bar{a} - \bar{\ell} + \bar{f} \tag{3.6}$$

where $\bar{\eta}$ is the neutron yield, i.e., the total number of fission neutrons generated per neutron absorbed averaged over all fissile isotopes, the neutron spectrum and the whole reactor. The quantities $\bar{a}$, $\bar{\ell}$ and $\bar{f}$ are equivalent corresponding averages; $\bar{a}$, the so-called parasitic absorption, describes the neutron loss fraction arising from absorption in coolant, structural and control materials; $\bar{\ell}$ is the neutron leakage from the reactor core (or, in FBRs, from the surrounding breeding blanket); $\bar{f}$ is the fractional contribution of U-238 or Pu-240 fissions (fast fission effect) (see Sect. 3.1.4).

The dominating quantity in the above relationship is the neutron yield: Fig. 3.11 shows the individual neutron yields for the fissile fuel isotopes, U-233, U-235, Pu-239, and Pu-241. It is seen that Pu-239 assumes the highest possible η-values in the range of neutron energies >100 keV and that U-233 has higher values than the other fissile isotopes in the thermal energy range. With U-235 as the fissile material, CR reaches relatively low values. In thermal reactors using previously bred U-233 fissile material, it is claimed that CR can attain a maximum possible value of

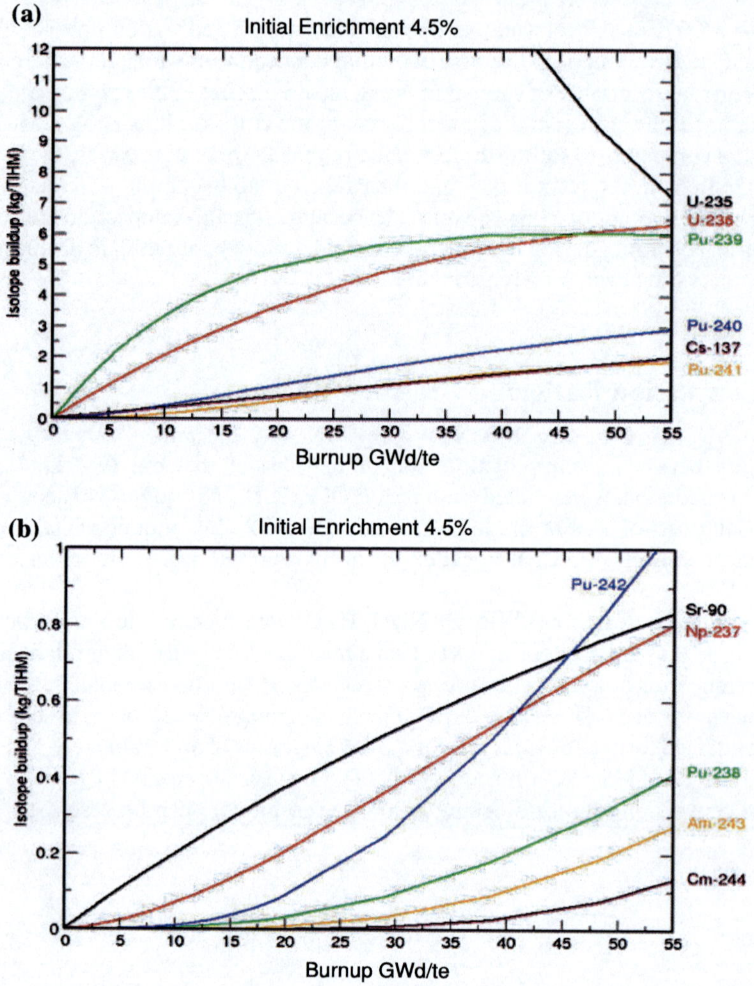

Fig. 3.10 **a** Isotope concentrations for U-235, U-236, Pu-239, Pu-240, Pu-241, Cs-137 as a function of burnup (GWd/t) of LWR fuel. Initial fuel enrichment 4.5% U-235 [11]. **b** Isotope concentrations for Pu-242, Sr-90, Np-237, Pu-238, Am-243 and Cm-244 as a function of burnup (GWd/t) of LWR fuel. Initial fuel enrichment 4.5% U-235 [11]

approximately 0.9–1.03 at $\eta = 2.28$. The highest values of CR are obtained with plutonium as a fissile material in reactors having neutron spectra in the range of several hundred keV. This is achieved by employing in the cores of those reactors relatively small fractions of coolants and structural materials with relatively high atomic mass numbers and low capture cross sections so that neutron slowing-down (moderation) is diminished and parasitic neutron absorption remains low. These reactors can clearly attain BR > 1, which is why they are called breeder reactors. As

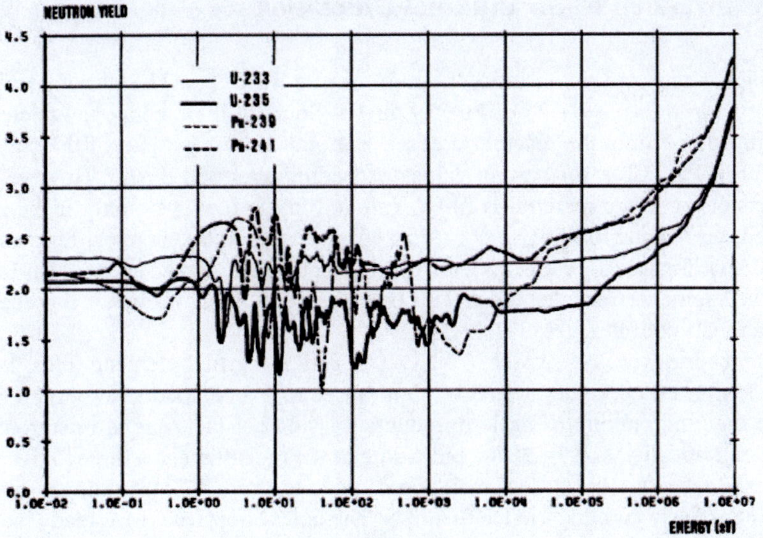

Fig. 3.11 Neutron yield as a function of neutron energy for various fissile fuel isotopes [11]

Table 3.4 Conversion/breeding ratios for different reactors

Reactor type	Initial fuel	Fuel cycle	Conversion/breeding ratio
PWR	3–5% U-235	U/Pu	0.55–0.60
BWR	3–5% U-235	U/Pu	0.55–0.60
CANDU/PHWR	0.72–2% U-235	U/Pu	0.7–0.8
HTGR/HTR	8% U-235	U/Pu	0.5–0.7
PWR/BWR	5% Pu	U/Pu	0.71–0.77
PWR	4% U-233	U/Th	0.78
LMFBR	15–25% Pu	U/Pu	1.2–1.4
LMFBR	12–20% U-233	Th/U-233	1.03–1.15

will be explained in Chap. 6 the net production of new fissile material (Pu) occurs essentially in axial and radial blankets surrounding the core. Because of their neutron spectrum (see Fig. 2.5) with predominantly fast neutrons they are usually called fast breeder reactors (FBRs). FBRs with plutonium/uranium mixed oxide fuel, steel as a structural material and sodium as a coolant sometimes, also called Liquid Metal cooled Fast Breeder Reactors (LMFBRs), reach BR values of 1.15–1.30.

The fast fission effect $\bar{f}$, i.e., the fraction of fissions of fertile atoms by fast neutrons, is only around 0.01–0.03 in thermal reactors. In FBR cores, the fast fission effect may well reach levels of 0.1–0.15.

Table 3.4 indicates conversion and breeding ratios achieved by different types of reactors.

3.6 Conversion Ratio and Fuel Utilization

The higher the conversion ratio, the more fertile Th-232 or U-238 nuclei will be converted into fissile U-233 or Pu-239 nuclei. Some of these converted nuclei will be utilized in situ in the reactor core, see, e.g., the almost flat part of the curve for Pu-239 in Fig. 3.10a for burnup values exceeding about 30 GWd/t. However, this conversion of fertile material is fully exploited in the fuel cycle only if the newly formed, man-made fissile U-233 or Pu-239 is recovered by chemical reprocessing after unloading the fuel elements from the reactor core and is then used for fabricating new fuel elements (recycle mode). This involves small losses in chemical processing and fabrication of the fuel [37].

In converter reactors (LWR, CANDU, HTGR), the fuel elements must be unloaded, after an operating period of about three to five years due to burnup of the fissile fuel and buildup of fission products. Fission products capture neutrons and decrease criticality, k_{eff}. In FBRs, unloading of the fuel after some three to five years is mainly necessary because of radiation damage to structural materials and buildup of gaseous fission products in the fuel rods. The latter effect leads to a steady increase in pressure that the clad of the fuel pin has to withstand.

Fuel utilization is the fraction of the original nuclear fuel that can be ultimately converted into thermal energy by nuclear fission after having passed through the whole fuel cycle once or several times, taking into account possible losses during reprocessing and refabrication of the fuel.

Fuel utilization depends on:

- the tails assay and the enrichment of natural uranium,
- the neutronic properties of the reactor core, which determine the conversion or breeding ratios. These factors, above all, are the design parameters of the reactor core (fractions and types of coolant, structural material, absorber), the choice of fissile material (U-233, U-235, Pu-239) and fertile material (U-238, Th-232), and the neutron spectrum present in the reactor core (thermal neutrons, fast neutrons).

If the reactor operates in the recycle mode, fuel utilization also depends on:

- the burnup of the fuel when unloaded from the reactor core, which determines the frequency of reprocessing and refabrication cycles (the lower the fuel burnup, the less energy will be generated while the fuel remains in the reactor core and the more frequently the fuel will pass through the fuel cycle when recycled),
- the fuel losses occurring during reprocessing and refabrication of the fuel.

Fuel utilization can be determined by analyzing these functional relationships and listing in a balance sheet the amount of fuel which can be utilized (fissioned) in each operating cycle of the reactor core. This leads to Fig. 3.12, which represents fuel utilization as a function of the conversion ratio or breeding ratio for various types of reactors. Reactors not operated in the recycle mode, such as present LWRs, only have conversion ratios of about 0.55, attaining fuel utilizations as low as 0.6%. CANDUs and AGRs reach conversion ratios around 0.68 and, correspondingly, slightly better fuel utilizations.

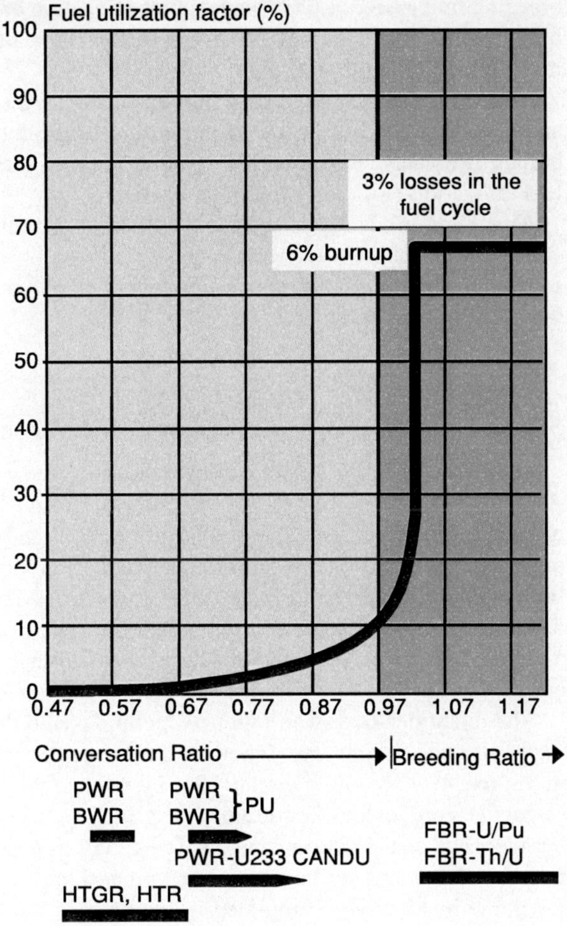

Fig. 3.12 Utilization of uranium or thorium as a function of the conversion or breeding ratios [37]

Reactors operating in a recycle mode and having appropriately optimized core neutron physics properties may reach conversion ratios of 0.9 or close to 1.0. In that case, fuel utilization may rise to several percent.

Fuel utilization shows a steep increase in the range of low breeding ratios (BR approximately 1.03), attaining a constant level at a certain limit. At this limit of the breeding ratio, the reactor system is just able to make up for the fuel cycle losses by breeding sufficient fissile material. Below this limit of the breeding ratio, fuel utilization is severely affected by relatively slight changes in the design and operating parameters of the reactor core and the fuel cycle (reprocessing and refabrication).

FBRs employing the uranium/plutonium cycle operate beyond these limits. In this region, fuel utilization is a function only of the average fuel burnup (averaged over the reactor core and the blanket) and of the fuel losses in the fuel cycle. FBRs with breeding ratios BR > 1.03 achieve a fuel utilization of roughly 60% and above. This is approximately a factor of 100 above the level reached by present LWRs operating in the once-through fuel cycle.

The value 60% is based on a burnup of 60 GWd/t (averaged over the FBR core and blankets) and a 3% loss of plutonium in the fuel cycle. With higher average burnup (less recycles) and a loss of plutonium of less than 1% in the fuel cycle, a fuel utilization close to 80% can be attained.

This high utilization of the fuel in FBRs is the ultimate reason for the low fuel consumption of breeder reactors. FBRs started on a core containing plutonium/U-238 fuel will need only small amounts of U-238 or depleted uranium for further operation (Chap. 6).

3.7 Radioactive Inventories in Fission Reactors

As outlined in Sects. 3.1 and 3.4 the processes of nuclear fission and the capture of neutrons in the fuel, coolant, structural material and associated radioactive decay processes generate chains of radioactive fission products and higher actinides. Radioactivity is measured in Becquerel or Curie. The units of Bq or Ci only indicate the number of disintegrations, not the type of radioactive decay (β^--, γ- or α-decays). Particle and radiation energy and the range of penetration through matter can be very different for the various decay paths.

The quantities of fission products, actinides and the radioactivities of structural materials and coolants are a function of the time for which the fuel elements are kept in the reactor core and generate heat by nuclear fission. They also depend on the respective core loading conditions (fresh fuel elements or fuel elements with higher burnups) and on the kind and purity of materials used as the fuel, coolant or structural material. Table 3.5 lists the most important fission products and actinides responsible for the radioactivity accumulated in the core of a PWR. These radioactivity data are calculated for 1 t of UO_2 fuel at its target burnup of 50,000 MWd(th)/t. Only one fourth of the fuel elements in a PWR core reach this maximum burnup at the end of each cycle, because the reactor core of a PWR is loaded in such a way that, in the course of a year, one fourth of the fuel elements increase their burnups from 0 to 12,500 MWd(th)/t, another one fourth from 12,500 to 25,000, another one fourth from 25,000 to 37,500 MWd(th)/t and the last one fourth from 37,500 to 50,000 MWd(th)/t. At the end of the burnup cycle, one fourth of the fuel reaching its maximum burnup is unloaded and replaced by fresh fuel elements. Accordingly, the maximum radioactivity has accumulated at the end of the burnup cycle prior to unloading. PWRs contain roughly 80 tonnes of fuel per GW(e) and, at the end of the burnup cycle, they contain some 10^{10} Ci or 3.7×10^{20} Bq or somewhat more of radioactivity, mainly in the fission products. The bulk of the fission products disintegrate relatively quickly, while most actinides are extremely long-lived. The time dependence of the activities of the different fission products and actinides plays a major role in the nuclear fuel cycle. Fission products and actinides are separated

Table 3.5 Radioactivity of particular fission products and actinides of PWR spent fuel after a burnup of 50,000 MWd(th)/t. The specific radioactivity is given for the time of unloading of the spent fuel and for several years cooling time [11]

Isotope	Half-life	Specific radioactivity (Ci/t)				
		Discharge	1 year	3 years	5 years	7 years
Fission products						
H-3	12.349 years	8.246E+02	7.796E+02	6.968E+02	6.228E+02	5.567E+02
KR-85	10.720 years	1.507E+04	1.413E+04	1.242E+04	1.091E+04	9.589E+03
SR-90	29.121 years	1.027E+05	1.003E+05	9.566E+04	9.121E+04	8.697E+04
Y-90	2.6667 days	1.031E+05	1.003E+05	9.568E+04	9.123E+04	8.699E+04
ZR-95	63.981 days	1.788E+06	3.422E+04	1.254E+01	4.592E-03	1.682E-06
NB-95	35.150 days	1.821E+06	7.703E+04	2.783E+01	1.019E-02	3.734E-06
RU-106	1.0080 years	7.838E+05	3.941E+05	9.965E+04	2.520E+04	6.371E+03
RH-106	2.2 hours	8.686E+05	3.941E+05	9.965E+04	2.520E+04	6.371E+03
CS-134	2.0619 years	3.096E+05	2.212E+05	1.130E+05	5.768E+04	2.945E+04
CS-137	29.999 years	1.547E+05	1.512E+05	1.444E+05	1.379E+05	1.316E+05
BA-137M	2.5517 min	1.469E+05	1.430E+05	1.366E+05	1.304E+05	1.245E+05
CE-144	284.26 days	1.406E+06	5.772E+05	9.726E+04	1.639E+04	2.761E+03
PR-144	17.283 min	1.418E+06	5.772E+05	9.726E+04	1.639E+04	2.761E+03
PM-147	2.6235 years	2.059E+05	1.643E+05	9.689E+04	5.713E+04	3.368E+04
EU-154	8.6001 years	1.201E+04	1.108E+04	9.428E+03	8.025E+03	6.830E+03
Actinides						
U-234	2.45E05 years	2.525E-02	4.187E-02	7.587E-02	1.095E-01	1.426E-01
U-235	7.04E08 years	0.900E-02	1.900E-02	1.900E-02	1.900E-02	1.900E-02
U-236	2.34E07 years	0.875E-01	3.875E-01	3.876E-01	3.876E-01	3.876E-01
U-238	4.5E09 years	3.097E-01	3.097E-01	3.097E-01	3.097E-01	3.097E-01
NP-237	2.14E06 years	5.066E-01	5.175E-01	5.180E-01	5.188E-01	5.199E-01
NP-239	2.3553 days	2.468E+07	4.064E+01	4.063E+01	4.062E+01	4.062E+01
PU-238	87.744 years	5.617E+03	5.978E+03	5.976E+03	5.886E+03	5.794E+03
PU-239	24,064 years	3.791E+02	3.858E+02	3.858E+02	3.858E+02	3.857E+02
PU-240	6537.3 years	6.067E+02	6.074E+02	6.086E+02	6.097E+02	6.108E+02
PU-241	14.399 years	1.871E+05	1.783E+05	1.619E+05	1.471E+05	1.336E+05
PU-242	3.87E05 years	3.234E+00	3.234E+00	3.254E+00	3.254E+00	3.254E+00
AM-241	432.23 years	1.647E+02	4.570E+02	9.997E+02	1.491E+03	1.935E+03
AM-243	7380.2 years	4.059E+01	4.064E+01	4.063E+01	4.062E+01	4.062E+01
CM-242	163.19 days	8.949E+04	1.908E+04	8.581E+02	3.864E+01	1.793E+00
CM-244	18.110 years	6.879E+03	6.638E+03	6.149E+03	5.696E+03	5.276E+03

chemically during reprocessing and then stored in waste repositories after having been conditioned. Some of the fission products, such as tritium and certainly krypton and xenon, appear in gaseous form, while iodine and cesium are highly volatile. All other fission products and actinides are non-volatile (for further details, see Chap. 7).

3.8 Inherent Safety Characteristics of Converter and Breeder Reactor Cores

3.8.1 Reactivity and Non-Steady State Conditions

It has been explained in Sect. 3.3 that $k_{eff} = 1$ corresponds to the steady state condition of the reactor core, in which case the production of fission neutrons is in a state of equilibrium with the number of neutrons absorbed and the number of neutrons escaping from the reactor core. For $k_{eff} \neq 1$, either the production or the loss term becomes dominant, i.e., the number of neutrons varies as a function of time. The neutron transport equation or, by way of approximation, the multigroup diffusion equation for calculation of the time dependent neutron flux must then be solved for the non-steady state case. However, in most cases it is a sufficiently good approximation to solve the so-called point kinetics equations in connection with equations describing the temperature field and its impacts on k_{eff} [2, 5, 12–14, 31].

Axial movements of the absorber rods in the core change the loss term for neutrons and influence k_{eff}. The relative change as a function of time of $k_{eff}(t)$ is called reactivity:

Reactivity

$$\rho\,(t) = \frac{k_{eff}(t) - 1}{k_{eff}(0)} = \frac{\Delta k_{eff}(t)}{k_{eff}(0)} \tag{3.7}$$

The point kinetics equation describing the reactor power as a function of time can be derived from the time dependent multigroup diffusion equation under the assumption that the space distributions of the neutron flux or power and of the concentrations of delayed neutron precursors always equal those in steady state conditions. The point kinetics equations read

$$\frac{dP(t)}{dt} = \left[\frac{\rho(t) - \beta_{eff}}{l_{eff}}\right] \cdot P(t) + \lambda_{eff} \cdot C(t) \tag{3.8}$$

$$\frac{dC(t)}{dt} = -\lambda_{eff} \cdot C(t) + \frac{\beta_{eff}}{l_{eff}} \cdot P(t) \tag{3.9}$$

with the initial conditions of

$P(t = 0) = P_0$ (steady state reactor power)
$C(t = 0) = C_o$ (steady state concentration of the parent nuclei (precursors, see Sect. 3.1.7) for all delayed neutrons combined in one group).

In these equations, P(t) is the reactor power, C(t) describes the averaged concentration of parent nuclei (precursors) of the delayed neutrons, λ_{eff} is the average decay constant of all parent nuclei of delayed neutrons, β_{eff}, the effective fraction of all delayed neutrons (integrated over all fissile isotopes and averaged over the reactor), l_{eff} the lifetime of the prompt neutrons in the reactor, i.e., the average time required

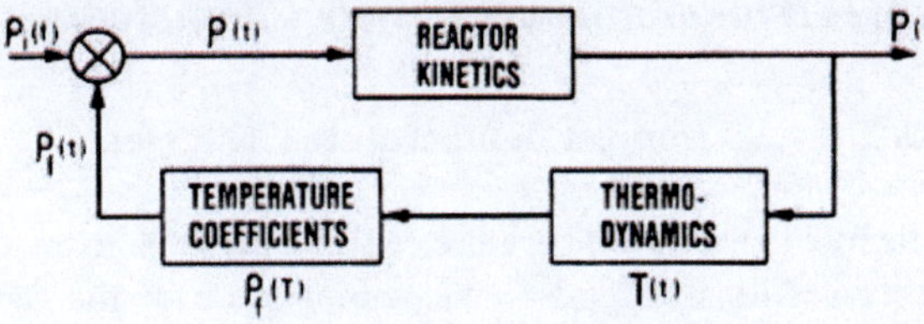

Fig. 3.13 Schematic of the reactor dynamics feedback loop

for a prompt fission neutron to induce a new fission process. Values of β_{eff} and λ_{eff} are calculated from those indicated in Sect. 3.1.7. Values of 1_{eff} are in the range of 10^{-3}–10^{-5} s for thermal reactors and around 4×10^{-7} s for FBRs. The average decay constant for a one-group treatment is in the range of about $\lambda_{eff} = 0.08\text{s}^{-1}$ for all fissile fuel isotopes.

The reactivity, $\rho(t)$, is composed of the superimposed or initiating reactivity, $\rho_i(t)$, which can be caused, e.g., by movements of absorber rods or fuel, and the feedback reactivity, $\rho_f(t)$, which takes into account all repercussions of temperature changes in the reactor core,

$$\rho(t) = \rho_i(t) + \rho_f(t) \tag{3.10}$$

Movements of absorbers produce local changes in the macroscopic cross sections and the neutron flux in certain material zones in the reactor core and, accordingly, also in $k_{eff}(t)$ and $\rho_i(t)$ (superimposed or initiating perturbations). The resultant change as a function of time of the neutron field and the power level alters the temperatures in the reactor core. Temperature changes provoke changes in material densities (expansion and displacement) and microscopic cross sections by the Doppler broadening of resonances (see Sect. 3.8.2.1). Also the neutron flux spectrum can be shifted by changing the moderation of the neutrons. Moreover, the dimensions of the reactor core and its components are changed by thermal expansion inducing, e.g., relative movements of control rods with respect to the upper core boundary. All these feedback reactivities, $\rho_f(t)$, resulting from changes in power and temperature together with external perturbation reactivities constitute a feedback loop (Fig. 3.13).

For numerical treatment, the feedback reactivity, ρ_f, is split up into individual contributions by different temperature effects

$$d\rho_f = \operatorname{Sum}_{m=1}^{M} \frac{\partial k_{eff}}{\partial T_m} \tag{3.11}$$

where ∂T_m are the average changes in the temperatures of the fuel, moderator, coolant, structural or absorber materials. In LWRs, the coolant is also the moderator. In other types of reactors (HTGR), the moderator (graphite) is distinct from the coolant (gas).

3.8.2 Temperature Reactivity Coefficients

3.8.2.1 Fuel Doppler Temperature Coefficient

The fuel Doppler temperature coefficient is due to the fact that the neutron resonance cross sections depend on the temperature of the fuel and the relative velocities, respectively, of neutrons and atomic nuclei [12–14, 30, 31, 38, 39].

The resonance cross sections, σ(E,T), for U-238, Th-232 and U-233, U-235, Pu-239, etc. show very pronounced peaks at certain neutron kinetic energies (see Fig. 3.4). An increase in fuel temperature, T_f, broadens this shape of the resonance curve which, in turn, results in a change in fine structure of the neutron flux spectrum in these ranges of resonance energy. The reaction rates are changed as a consequence. Above all, the resonance absorption for U-238 increases as a result of rising fuel temperatures, while the effect of a temperature change in the resonance cross sections of the fissile materials, U-235 and Pu-239, is so small that it can generally be neglected if the fuel enrichment is not extremely high. The increases in fission and capture reactions in U-235 and Pu-239 partly compensate each other. For these reasons, temperature increases in the fuel result in a negative temperature feedback effect (Doppler effect) brought about by the increase in neutron absorption in U-238. For Th-232, the effects are similar. The Doppler effect is somewhat less pronounced at very high fuel temperatures because adjacent resonances will overlap more and more. The resonance structure then is no longer as pronounced as at low temperatures, which leads to a reduction of the negative Doppler effect.

Due to the specific energy distribution of the neutron spectrum (see Fig. 3.5), the Doppler effect in thermal reactors follows

$$\frac{1}{k_{eff}}\frac{\partial k_D}{\partial T_f} \sim \frac{-1}{\sqrt{T_f}} \tag{3.12}$$

whereas in FBRs it follows the relation

$$\frac{1}{k_{eff}}\frac{\partial k_D}{\partial T_f} \sim -\frac{1}{T_F^x} \text{ with } \frac{1}{2} \leqq x \leqq 3/2 \tag{3.13}$$

The Doppler coefficient is always negative in power reactor cores because, given the relatively low enrichment in U-235 and Pu-239, respectively, the resonance absorption of U-238 will always dominate. It is an instantaneous negative feedback coefficient of reactivity, which immediately counteracts increases in power and fuel temperature. For the sake of completeness it should be mentioned that structural material, in particular the isotope Fe-56, and such strong resonance absorbers like Tc-99 or erbium may contribute small fractions to the total Doppler reactivity feedback.

Besides the Doppler coefficient, the fuel expansion coefficient also leads to a negative feedback coefficient of reactivity. Especially in fast breeder cores it may well attain an importance equal to that of the Doppler coefficient.

3.8.2.2 Coefficients of Moderator or Coolant Temperatures

The main contributions to the coefficients of moderator or coolant temperatures stem from changes in the densities of the moderator or coolant and from resultant shifts in the neutron spectrum. Temperature rises decrease the density of the coolant and accordingly reduce the moderation of neutrons. The neutron spectrum is shifted towards higher energies. As a result of the lower moderator density and the correspondingly higher transparency to neutrons of the core it is also possible that far more neutrons escape from the reactor core and neutron absorption will be reduced [38].

For thermal reactors, typically LWRs, the sign of the moderator temperature coefficient depends on the degree of moderation, i.e., whether undermoderated or overmoderated. Furthermore, in particular for other reactor types it is influenced by the kind of fuel, e.g., UOX or MOX and its burnup (due to the influence of fission products and the change in fuel isotopic composition).

Moreover it is important if coolant and moderator are different materials (e.g., H_2O and graphite as in the Russian RBMKs where the neutron absorption is of dominant importance) or if coolant and moderator (e.g., H_2O and D_2O) belong to separate coolant circuits with associated different temperatures as e.g., in CANDU reactors.

For the present line of PWRs, the total sum of the individual contributions to changes in various energy ranges finally leads to a negative coefficient of the moderator temperature which, however, also depends an the concentration of boric acid dissolved in the coolant and the burnup condition of the reactor core. In large graphite moderated HTGRs containing U-233, the moderator temperature coefficient is usually positive. In small HTRs the moderator coefficient is negative.

Also in sodium cooled FBRs with core sizes in excess of about 100–150 MW(e), the coolant temperature coefficient is positive because the neutron spectrum is shifted towards higher energies as a consequence of the reduced moderation. The resultant increased contribution by the fast fission effect of U-238 as well as the higher η-values (see Fig. 3.11) add to the reactivity. These positive reactivity contributions cannot compensate all negative contributions coming from an increase in the leakage rate of neutrons escaping from the core (which is the dominating effect in small sodium cooled FBRs with power levels of less than approximately 100–150 MW(e)).

3.8.2.3 Structural Material Temperature Coefficient

Especially in FBRs, the structural material temperature coefficient also plays an important role. Increasing temperatures cause the core structure to expand radially and axially and, in this way, result both in indirect changes in material densities and in changes of size of the reactor core and, as a consequence, of neutron leakage. The structural material temperature coefficient must be determined by detailed analyses of all expansion and bowing effects for given core and fuel element structures also taking into account the core restraint (clamping) system. For FBRs, the structural

Table 3.6 Typical temperature coefficients of reactivity for various reactor lines

Temperature coefficient Δk/K	PWR	BWR	Large HTGR[a]	LMFBR
Moderator or coolant (fresh fuel)	-2×10^{-4}	-1.3×10^{-3}[b]	+0.5 to 1.7×10^{-5}	$+5 \times 10^{-6}$
Doppler coefficient (500–2800°C)	−1.7 to -2.7×10^{-5}	-1.3×10^{-5} to −2.5	−4 to -2×10^{-5}	-1.1×10^{-5} to -2.8×10^{-6}

[a]End of operating cycle
[b]Coolant void coefficient

material coefficient is also negative. This is accomplished by the specific design of the core support plate and the core restraint system.

For analysis of the control behavior of a reactor core and its behavior under accident conditions, the non-steady state neutron flux, power, temperature and all feedback reactivities must be considered in detail. Negative feedback reactivities or temperature coefficients always counteract increases in power and temperature. Positive coefficients of moderator or coolant can be tolerated as long as all the other temperature coefficients, above all the sufficiently fast prompt Doppler coefficient, are negative and larger in magnitude than the positive coefficients of moderator or coolant. Table 3.6 shows typical temperature coefficients of reactivity for various types of reactors.

PWRs or BWRs have highly negative coolant or moderator temperature coefficients, whereas HTGRs at the end of their operating cycle and also large sodium cooled FBRs throughout their whole operating cycle have positive coolant temperature coefficients. Thermal reactors, such as PWRs, BWRs and HTGRs, have more negative Doppler coefficients than FBRs.

3.8.3 Reactor Control and Safety Analysis

3.8.3.1 Reactivity Changes During Start-Up and Full Power Operation

As the reactor core is slowly being started up from almost zero power to full power, the temperatures of coolant and core structure rise by several 100°C. At the same time, the fuel temperature increases by more than 1,000°C. This causes a negative reactivity effect, which must be overcome by moving absorber (control) rods out of the reactor core. In LWRs, this reactivity span is in the range of several percent. In sodium cooled FBRs, it is smaller mainly because of the lower value of the negative Doppler coefficient [2, 9, 24, 28, 30].

The buildup of fission products and actinides as well as the burnup of fissionable isotopes leads to a reactivity loss of up to 12% in LWRs and about 3% in sodium cooled FBRs. Sufficient excess reactivity, i.e., $k_{eff} > 1$, therefore must be provided in a core with fresh (non-irradiated) fuel and zero power at the beginning of an operating cycle. At this time, the excess reactivity is counterbalanced by the insertion into the core of such absorber materials as boron, cadmium, gadolinium, or indium, which provide a sufficient reactivity span for reactor control. Due to the burnup effects as well as fission product buildup mentioned in Sect. 3.4, the negative reactivity must be reduced during the operating cycle. This is accomplished by several methods, e.g., withdrawing absorber rods, reducing the concentration of soluble poisons, such as boric acid, and by the diminishing absorption effect of burnable poisons, such as cadmium or gadolinium contained in fixed rods.

The fraction of excess reactivity for fissile isotope burnup and fission product buildup designed into the fresh core determines the length of operation of a core (operating cycle). This length of the operating cycle is usually chosen in the light of an optimization of core physics properties and fuel cycle economics. The reactor core is shut down by moving into the core absorber rods with sufficient negative reactivity. In this case, the reactivity span from full power (high temperature) to zero power (low temperature) has to be overcome. In addition, the reactor core must be held subcritical, which means that it has to attain and maintain a k_{eff} well below 1.

3.8.3.2 Qualitative Description of a Reactor Core Under Transient Power Conditions

Reactivity changes of the reactor core lead to power changes. They can be approximately described by the point kinetics equations (3.8), (3.9), which must be solved in conjunction with the respective equations for temperature and feedback reactivity (3.10), (3.11) by means of programs run on digital computers [2, 3, 9, 40].

Several important reactivity and transient power ranges can be distinguished in discussing the point kinetics equations:

(a) $\rho < 0$
If the reactivity step, ρ, is negative. The reactor power, P(t), will decrease exponentially. For a sufficiently large negative reactivity step the power decreases rapidly and remains at the level of the afterheat (reactor shut down).

(b) $0 < \rho < \beta_{eff}$
For a positive reactivity where $0 < \rho < \beta_{eff}$, the reactor core is called delayed critical. The power increase as a function of time is mainly determined by the delayed neutrons. After a small rapid increase in the reactor power, its time behavior increases exponentially. The time behavior of the power and the time period T can be described by Eq. 3.14

$$P/(t)/P_o = \exp(t/T) \text{ with } T \approx \frac{\beta_{eff}}{\lambda_{eff} \cdot \rho} \tag{3.14}$$

As long as ρ does not exceed the value of 0.5 β_{eff}, the time period T (Eq. 3.14) is larger than $1/\lambda_{eff} = 1/0.08$ (s^{-1}) (Sect. 3.8.1), which is in the range of more than 12 s. This allows easy and reliable control of the reactor core by slowly moving absorbers. As the decay constants for delayed neutron precursors are very similar for all fissile isotopes as well as for thermal and fast neutron spectra (see Table 3.2), the design data of control systems of thermal reactors, e.g., LWRs and FBRs, are very similar.

(c) For $\rho = \beta_{eff}$, the reactor core is called prompt critical; for $\rho > \beta_{eff}$, it is called superprompt critical. The reactor power, P(t), in that case is dominated entirely by the prompt neutrons. The time period T is determined in this case by Eq. 3.15

$$P(t)/P_o = \exp(t/T) \text{ with } T = \frac{l_{eff}}{\rho - \beta_{eff}} \tag{3.15}$$

As l_{eff} is small, the time period becomes short and the power would increase very rapidly as long as the effective reactivity input is not reduced by counteracting inherent negative feedback effects or negative reactivity effects of the shutdown system.

Figure 3.14 shows the time period, T, as a function of positive reactivity steps, ρ, and for different prompt neutron lifetimes, l_{eff}, as a parameter. Instead of absolute values, the reactivity is usually indicated in units β_{eff}. The quantity β_{eff} is historically defined as a reactivity unit known as a dollar [1, 3, 9, 40]

$$1\text{dollar}(1\$) = \beta_{eff}$$

The hundredth part of a dollar is called a cent. These reactivity units are normally written behind the amount. In case of U-235 1 $=0.0064.

A more detailed description on the treatment of time-dependence and the role of β_{eff} in reactor kinetics and dynamics may e.g., be found in [40].

In power reactors, counteracting feedback reactivity effects are always involved. This is especially valid for case (c), where the core may become prompt critical or superprompt critical. Therefore, in quantitative analyses of the dynamic behavior of reactor cores, the point reactor kinetic equations are solved by means of sophisticated computer codes also describing the temperature and feedback mechanisms. If, in the course of such analysis, the core becomes prompt critical and the power increases very rapidly, also the fuel temperature increases and a negative Doppler reactivity accumulates. This Doppler reactivity decreases the total reactivity below prompt critical and the reactor power, after having attained a certain peak level, rapidly drops again. If the reactor core were not shut down shortly after this power peak and if the initiating reactivity increased further, an oscillating behavior would result. If sufficient energy had been accumulated during such a postulated excursion accident, the core temperatures would become so high that the fuel and the structural material would melt and material relocations would result. In the latter case, the core would be destroyed and the coolant vaporized and expelled from the core. The reactor core would then shut itself down by material relocations and would turn subcritical.

For extremely rapid power changes and variations of the neutron flux distributions during the power transient the approximation by point kinetics are no longer

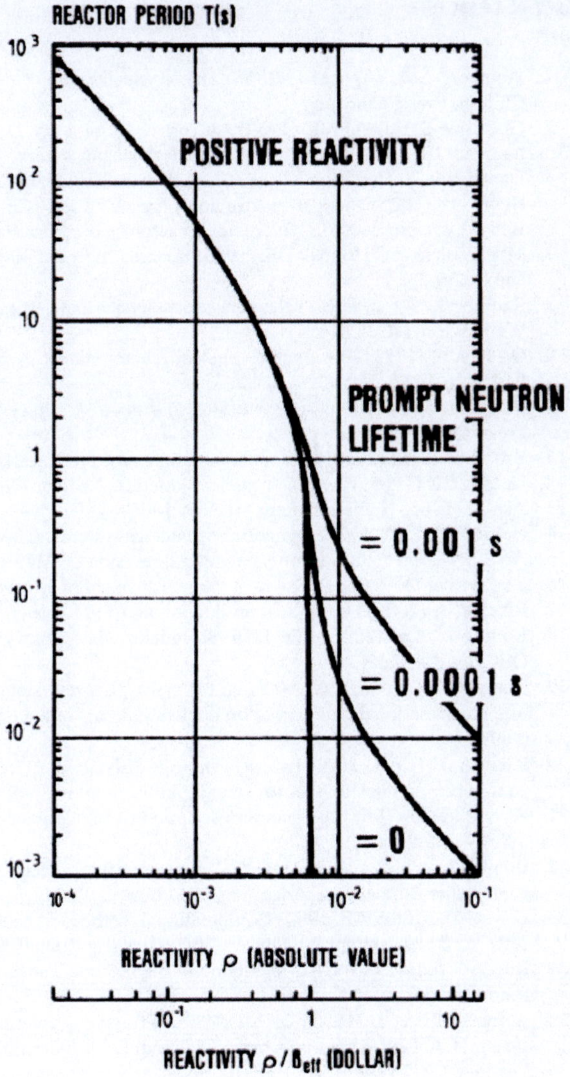

Fig. 3.14 Reactor period as a function of positive reactivity steps for a U-235 fueled reactor [3]

valid and a more sophisticated treatment has to be applied for the analysis of such phenomena [9].

Such prompt critical conditions of a reactor core could only be envisaged if, in the course of an accident sequence, sufficient positive reactivity could be generated and if the diverse and redundant control and shutdown systems of a reactor core failed completely on demand. Such questions, among a number of others, are generally analyzed for each type of reactor in detailed safety analyses (see Chaps. 11 and 12).

References

1. Weinberg AM, Wigner EP (1958) The physical theory of neutron chain reactors. University of Chicago Press, Chicago
2. Glasstone S, Edlund MC (1952) Nuclear reactor theory. D. Van Nostrand, Princeton
3. Lamarsh JR (1983) Introduction to nuclear reactor theory, 2nd edn. Addison-Wesley, Reading
4. Duderstadt JJ, Hamilton LJ (1976) Nuclear reactor analysis. Wiley, New York
5. Henry AF (1975) Nuclear-reactor analysis. MIT Press, Cambridge
6. Bell GI, Glasstone S (1970) Nuclear reactor theory. Van Nostrand Reinhold, New York
7. Meghreblian RV, Holmes DK (1960) Reactor analysis. McGraw-Hill, New York, pp 160–267 and 626–747
8. Radkowsky A (ed) (1964) Naval reactors physics handbook. U.S. Atomic Energy Commission, Washington, DC (Chap. 5)
9. Ott K et al (1983) Introductory nuclear reactor statics. American Nuclear Society, LaGrange Park
10. Michaudon A (1981) Nuclear fission and neutron induced fission cross sections. Pergamon Press, Oxford
11. Broeders CHM (2010) Personal communication, KIT Karlsruhe
12. Keepin GR (1965) Physics of nuclear kinetics. Addison-Wesley, Reading
13. Ash M (1965) Nuclear reactor kinetics. McGraw-Hill, New York
14. Hetrick DL (ed) (1972) Dynamics of nuclear systems. University of Arizona Press, Tucson
15. ANS-5.1-1994 (1985) Element standard, revision of ANSE/ANS-51-1979; R 1985
16. Nusbaumer O (2006) Decay heat in nuclear reactors. http://decay-heat.tripod.com/
17. Rineiski A (2008) Decay heat production and TRU burner. Prog Nucl Energy 50:377–381
18. Koning A et al (2006) The JEFF-3.1 nuclear data library, JEFF report 21, NEA No. 6190, OECD/NEA, Paris
19. Roussin RW, Young PG, McKnight R (1994) Current status of ENDF/B-VI. In: Proceedings of the international conference on nuclear data for science and technology, vol 2. Gatlinburg, p 692
20. Kikuchi Y (1994) JENDL-3, Revision 2: JENDL 3-2. In: Proceedings of the international conference on nuclear data for science and technology, vol. 2. Gatlinburg, p 685
21. Askew J et al (1966) A general description of the lattice code WIMS. J Br Nucl Energy Soc 5:564
22. Bondarenko I et al (1964) Group constants for nuclear reactor calculations. Translation–consultants Bureau Enterprice Inc., New York
23. Lewis EE, Miller WF (1993) Computational methods of neutron transport. Wiley-Interscience, New York (1984); reprinted by American Nuclear Society, LaGrange Park
24. Ronen Y (ed) (1986) CRC handbook of nuclear reactors calculations, vol I. CRC Press, Boca Raton
25. Alcouffe RE et al (1995) DANTSYS: a diffusion accelerated neutral particle transport code system, LA-12969-M. Los Alamos National Laboratory, Los Alamos
26. Lawrence RD (1983) The DIF3D nodal neutronics option for two- and three-dimensional diffusion theory calculations in hexagonal geometry, ANL-83-1. Argonne National Laboratory, Argonne
27. Briesmeister JF (ed) (2000) MCNP–a general Monte Carlo N-particle transport code, version 4C. Technical report, LA-13709-M. Los Alamos National Laboratory, USA
28. Oldekop W (1975) Einführung in die Kernreaktor- und Kernkraftwerkstechnik, Teil I. Karl Thiemig, München
29. Kessler G (1983) Nuclear fission reactors. Springer, Vienna
30. Waltar A et al (1981) Fast breeder reactors. Pergamon Press, New York
31. Stacey W (2007) Nuclear reactor physics. Wiley, New York
32. Wiese HW, Fischer U (1981) KORIGEN—Ein Programm zur Bestimmung des nuklearen Inventars von Reaktorbrennstoffen im Brennstoffkreislauf. Kernforschungszentrum Karlsruhe, KfK-3014

33. Haeck W et al (2007) An optimum approach to Monte Carlo burnup. Nucl Sci Eng 156:180–196
34. Fission Product Nuclear Data (FPND)—1977 (1978) In: Proceedings of the second advisory group meeting on fission product nuclear data, Energy Centrum Netherlands, Petten, 5–9 September 1977. International Atomic Energy Agency, IAEA-213, Vienna
35. ORNL (2005) SCALE a modular code system for performing standardized computer analyses for licensing evaluations, ORNL/TM-2005/39, version 5, vols I–III
36. Poston DI et al (1999) Development of a fully-automated Monte Carlo burnup code MONTEBURNS, LA-UR-99-42
37. Heusener G (1980) Personal communication, KfK Karlsruhe
38. Hummel H et al (1970) Reactivity coefficients in large fast power reactors. American Nuclear Society, LaGrange Park
39. Nicholson R et al (1968) The doppler effect in fast reactors, advances in nuclear science and technology, vol 4. Academic Press, New York, p 109
40. Ott K et al (1985) Nuclear reactor dynamics. American Nuclear Society, LaGrange Park

Chapter 4
Uranium Enrichment

Abstract Only gas graphite reactors and heavy water reactors can operate with natural uranium (~0.7% U-235). However, the burnup of their fuel is limited. Present light water reactors operating with a fuel burnup of about 55–60 GW_{th}/t need their uranium fuel enriched to 4–5% U-235 content. Uranium enrichment is performed almost exclusively by the gaseous diffusion and gas centrifuge process. The gaseous diffusion enrichment plants in the USA and France provide about 42% of the world-wide enrichment capacity. Gaseous diffusion plants will phase out in the near future as more economic gas centrifuge plants will be built which provide already about 58% of the world wide enrichment capacity. The separative work unit (SWU) which is a measure of the amount of energy necessary to produce a certain unit (amount) of enriched uranium is by an order of magnitude lower (in e.g. SWU/kg U) for centrifuge enrichment than for gaseous diffusion enrichment. Laser isotope separation, chemical isotope separation and plasma isotope separation were scientifically studied. Only one laser isotope separation (SILEX) plant is being built in the USA.

4.1 Introduction

For neutron physics reasons, the use of natural uranium as a reactor fuel requires the presence of weakly neutron absorbing materials as moderators, coolants and structural materials. Such materials, for instance, are heavy water, graphite, gaseous media and low neutron absorbing metals. Conversely, if steel is used as a structural material, or normal light water as a moderator and coolant, it is not possible to build a nuclear reactor with natural uranium with a U-235 content of about 0.72% as a nuclear fuel.

For practical purposes, this means that heavy water reactors and gas-graphite reactors (e.g., those of the MAGNOX type) are the only reactor lines to be run on natural uranium. All other types of reactors, especially LWRs and AGRs need the uranium fuel enriched in its U-235 content.

G. Kessler, *Sustainable and Safe Nuclear Fission Energy*, Power Systems,
DOI: 10.1007/978-3-642-11990-3_4, © Springer-Verlag Berlin Heidelberg 2012

Fig. 4.1 UF_6 transport container going to enrichment plant (URENCO)

AGRs and LWRs require low enriched uranium (LEU) with an enrichment in the range of about 2–5% of uranium-235. Advanced reactor types, e.g. HTGRs, need fuel with 8–20% U-235 enrichment. Reactor types with fuel enrichment of more than 20% U-235 enrichment are technically feasible, but are excluded here for proliferation reasons [1].

Natural uranium is delivered as UF_6 in big containers to the enrichment plant. It has a sublimation point of 56.5°C at 0.1 MPa. At the enrichment plant these containers are heated in autoclaves to obtain the UF_6 in gaseous form (Fig. 4.1).

4.2 Elements of Enrichment Plants

Isotope separation plants contain as an elementary unit the "separating element," in which the feed material is fractionated into a "head fraction" enriched in the desired isotope and a "tails fraction" depleted in the same isotope (Fig. 4.2a).

One or more separating elements connected in parallel are called a "stage" (Fig. 4.2b). In all elements of one stage, the feed, head, and tails have the same fractions of isotopic composition. The desired isotope concentration (enrichment) can be achieved by connecting many stages in series. This stage arrangement is known as a "cascade." In isotope separation plants, a counter current flow cascade is generally adopted in which the tails fraction of each stage is subjected to further fractionation in the next lower stage (Fig. 4.2c) [1–3].

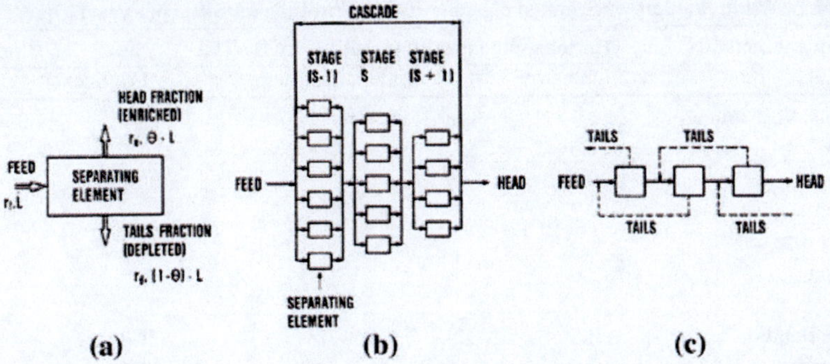

Fig. 4.2 Separating element (**a**), stage and cascade arrangement (**b**) and counter current symmetric cascade (**c**) in enrichment plants [2]

The effectiveness of the enrichment process is expressed in terms of the separation factor. If r designates the ratio of the numbers of U-235 atoms to U-238 atoms (abundance ratio) in the uranium (UF_6) gas, and if r_e defines this ratio at the head (enriched) side, r_d at the tails (depleted) side and r_f, at the feed side of the separation unit or stage, the separation factor is given by

stage separation factor $\alpha = r_e/r_d$
and the heads separation or enrichment factor $\beta = r_e/r_f$

The feed stream, L, is divided into the head stream, $\theta \cdot L$, and the tails stream, $(1 - \theta) \cdot L$. The ratio of head to feed streams is known as the cut, θ. In the case of natural uranium as feed, the abundance ratio starts at $r_f = 0.0072$ (0.72% U-235) and a certain number of stages of cascade arrangement are needed to achieve the desired enrichment [2–4].

The tails assay, r_d, of present enrichment plants is 0.2–0.3%. Lower tails assays down to the range of 0.1% are under discussion for future enrichment activities and would lead to lower natural uranium requirements [1].

The separative work is a measure of the amount of work necessary in the enrichment plant to produce a certain amount of enriched uranium. The Separative Work Unit, SWU, has the dimension of mass and is indicated in kg SWU or tonne SWU. Similarly, the separative capacity of an enrichment plant is generally given in tonnes SWU/a. The energy consumption for a separative work unit is given in kWh/kg SWU [2–4].

4.3 World Uranium Enrichment Plant Capacities

Uranium enrichment is performed almost exclusively by the gaseous diffusion and gas centrifuge process. The gaseous diffusion process was originally developed in the United States of America within the framework of the Manhattan Engineer

Table 4.1 World uranium enrichment capacity in 2008 (million kg SWU per year) [5]

Enrichment method	Enrichment capacity in million kg SWU/a		
	Operating	Under construction	Pre-licensing planned
Gaseous diffusion			
USA	11.3		
France	10.8		
China	0.2		
Gas centrifuge			
Russia	2.0		
UK	3.7		
Netherlands	3.5		1.0
Germany	1.8		2.7
USA		3.0	6.8
France			7.5
Japan	0.3		1.2
China	1.0		0.5
Brazil		0.13	
Iran			0.25
India	0.01		
Pakistan	0.02		
Laser (SILEX)			
USA			3.5–6.0
Total	52.63	3.13	23.45–25.95

District project. It still provides about 42% of the world wide enrichment capacities. But it will be phased out in future as more gas centrifuge plants will be built. The gas centrifuge process was developed to the stage of commercialization in Russia, Europe, the USA and Japan. This technique is much more economic than the gaseous diffusion method. The nozzle separation (Germany) or the Helikon method (South Africa), reached the technical demonstration but not the commercialization stage. Laser isotope separation, chemical isotope separation and plasma isotope separation were scientifically studied. Only a laser Isotope separation (SILEX) plant is being built in USA from 2010 on. This technique has still to be demonstrated on large scale [5, 6].

Table 4.1 shows the existing and planned enrichment capacities in the world. In 2008, the world uranium enrichment capacity was approximately 53 million kg SWU per year. About 42% or 22 million kg SWU were installed in gaseous diffusion plants (USA, France) and 31 million kg SWU in gas centrifuge plants. An additional capacity of 23 million kg SWU was under construction or in the planning phase, almost entirely using the centrifuge technology. Most of this gas centrifuge capacity is located in Russia, Europe and the USA. Laser isotope separation technology may become an additional economical, promising uranium enrichment technology.

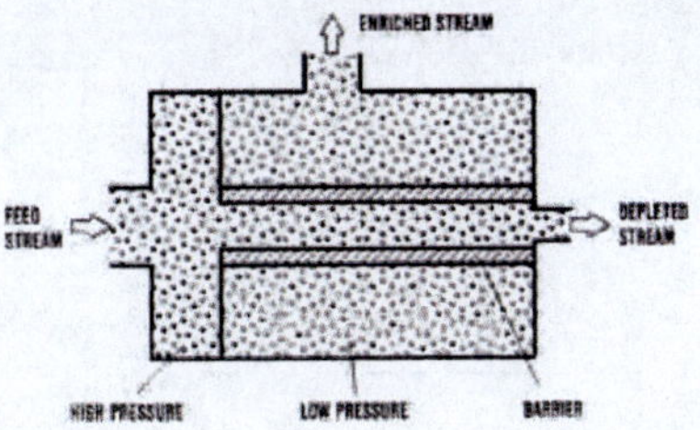

Fig. 4.3 Schematic of a gaseous diffusion cell [7]

4.4 Uranium Enrichment by Gaseous Diffusion

The principle of enrichment by gaseous diffusion is the phenomenon of molecular diffusion through the micro pores in a membrane (barrier). In a closed cell in thermal equilibrium all molecules of a gas mixture have the same average kinetic energy. Hence, the lighter molecules of 235-UF_6 travel faster on the average and strike the cell wall more frequently than the heavier 238-UF_6 molecules. Micropores in the cell wall will therefore preferably allow the passage (diffusion) of the lighter U-235-UF_6 molecules. The remaining nondiffused mixture is then depleted in the lighter isotope [1–3, 5, 6].

The elementary unit of the gaseous diffusion process is a diffusion cell (diffusor) divided into two compartments by a porous barrier. A compressor maintains a steady pressure at the cell inlet. A heat exchanger removes the heat of compression of the uranium hexafluoride gas. On the depleted gas side a regulation valve controls the flow rate of the process gas (Fig. 4.3).

The elementary enrichment factor of a gaseous diffusion cell is low (ideal enrichment factor $\alpha = 1.0043$). Gaseous diffusion plants therefore typically need 1,000–1,400 stages to produce low enriched uranium for LWRs. Several thousand stages are needed in series to produce highly enriched uranium (Fig. 4.4).

Figure 4.5 shows the different components of a large stage of a US gaseous diffusion plant. Multistage axial compressors are used. Protection of the process gas, UF_6, against leakage from the compressor or against the entrance of air into the compressor is achieved by special seals using nitrogen under overpressure. The barrier holding diffusor resembles the design of a heat exchanger. The cooler is integrated in the diffusor design.

The barrier design in diffusion cells must satisfy a number of conflicting requirements. It must be chemically resistant against the highly reactive uranium hexafluoride. It must be as thin as possible, have pores with very small radii of 10^{-6} cm or less and, at the same time, have a high porosity (more than 10^9 pores/cm^2). In addition, it must have sufficient mechanical strength at operating temperatures around 80°C and a differential pressure on both sides of about 1 bar. Two kinds of barrier design are feasible:

- film type barriers, where pores are generated by leaching a certain constituent of a well-dispersed alloy or by electrolytic etching of thin aluminum foils,

Fig. 4.4 Schematic of a gaseous diffusion cascade [7]

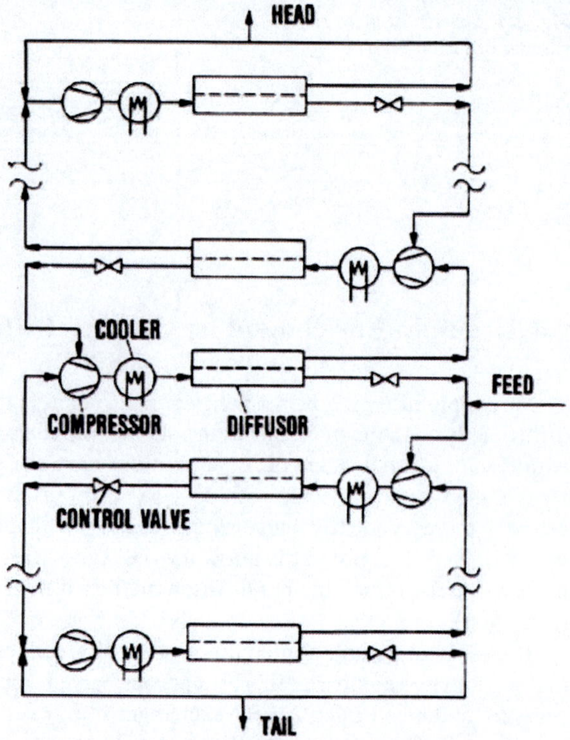

- aggregate type barriers, where pores are generated by sintering powders (e.g., nickel or alumina powders).

The barriers of present US gaseous diffusion plants are sintered nickel tubes assembled in tube bundles housed in cylindrical diffusors (see Fig. 4.4).

The gaseous diffusion plant in operation in the USA at Paducah represents the greatest part of separative work capacity available on the market together with the gaseous diffusion plant at Tricastin in France (see Table 4.2).

The specific power consumption of gaseous diffusion plants is on the order of 2,400 kWh/kg SWU; this is relatively high, compared to the centrifuge process (cf. next Section). Gaseous diffusion, on the other hand, requires lower specific capital costs [2, 6].

4.5 Gas Centrifuge Process

The gas centrifuge process is based on the separation effect of UF_6, isotopes in a strong centrifugal field, suitably combined with the effect of a counter current axial flow circulation within the centrifuge. This process has been developed in several countries since the mid-1950s and 1960s.

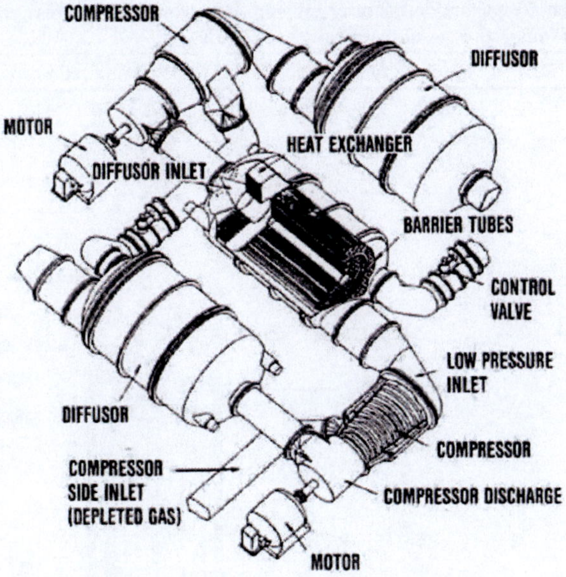

Fig. 4.5 Components of a large stage in a US enrichment plant [2, 6]

Table 4.2 Comparison of technical data of commercial size gaseous diffusion and gas ultracentrifuge plants [1]

Process		Gaseous diffusion	Gas ultracentrifuge
Plant name		EURODIF/Tricastin	URENCO/Almelo-Capenhurst
Capacity	1,000 t SWU/a	10.8	1
Enriched product	%	3.25/2.6/1.9/1.5	3
Feed material	%	0.72	0.72
Feed rate	te	14,050	1,272
Tails assay	%	0.2	0.2
Tails production rate	t U/a	11,260	1,040
Number of stages in series		1,400	12
Number of elementary separations (units)		14,000	Tens of thousands
Power requirements	MW(e)	3,000	20

Within the high speed centrifuge, the centrifugal force causes the heavier isotopes, 238-UF_6, to move closer to the wall of the centrifuge than the somewhat lighter isotopes, 235-UF_6. This produces partial separation of the isotopes in the radial direction. Depending upon the peripheral speed (400–700 m/s), very high pressure ratios between the centrifuge axis and centrifuge wall are attained (Table 4.3). This generates vacuum conditions at the centrifuge axis.

Table 4.3 Elementary separation factor of uranium isotopes and UF_6 pressure ratio for various peripheral speeds for a given gas ultracentrifuge (T=310 K) [2, 7–9]

Peripheral speed (m/s)	Elementary separation factor	Pressure ratio between axis and wall
400	1.0975	5.5×10^4
500	1.156	2.5×10^7
600	1.233	4.6×10^{10}
700	1.329	

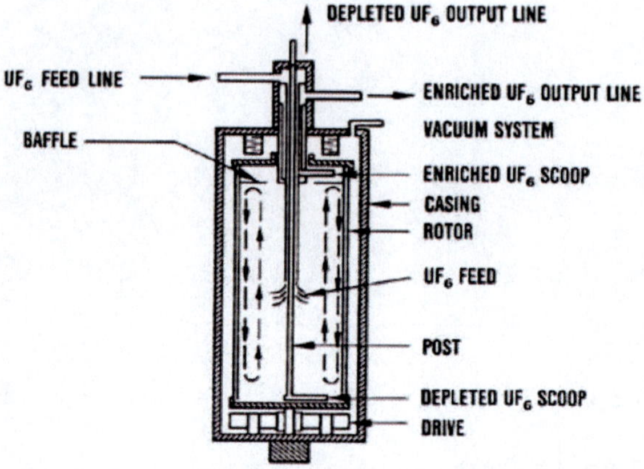

Fig. 4.6 Schematic representation of a counter current gas ultracentrifuge [2]

The elementary separation factor indicated in Table 4.3 not only depends upon the peripheral speed, but also upon the axial temperature field. Centrifuges of present enrichment plants are operated with an internal countercurrent flow. Superposition of the axial counter current flow on the wall flow considerably increases the elementary separation effect of a centrifuge. In this way, the stage (centrifuge) separation factor attains a range of 1.2–1.5. For the production of low enriched uranium of about 3% U-235 enrichment, therefore, only 12 stages are needed. A schematic representation of an ultracentrifuge with an internal counter current flow is shown in Fig. 4.6.

The gaseous uranium hexafluoride, UF_6, feed is introduced into the centrifuge through a tube on the axis. Enriched UF_6 is extracted from the upper part near the wall through a scoop resembling a small Pitot tube arranged perpendicular to the rotor axis. On the enriched side, this scoop is enclosed in a small chamber, which permits the extraction of the enriched gas through a screen while preventing any interaction of the scoop with the internal gas flow. The depleted UF_6 gas is extracted by a similar scoop at the lower end of the rotor. This scoop not only removes the depleted UF_6, but also interacts with the spinning gas. This interaction generates the counter current flow, which ascends along the inner axis of the centrifuge and descends along the wall. A similar effect can also be generated by a linear temperature profile along the axial solid boundary of the centrifuge (slight heating of the bottom combined with some slight cooling of the top). The rotor of such high speed centrifuges requires

Fig. 4.7 Modern centrifuge enrichment plant (URENCO)

a material of high strength and low specific weight. Aluminum, which allows a maximum peripheral speed of about 400 m/s to be used, is more and more being replaced by other materials like titanium, special steels and composite materials, such as glass and carbon fibers as well as aromatic polyamides of the nylon family (peripheral speed, 600–700 m/s) (Fig. 4.7).

While the investment costs of centrifuge enrichment plants are higher, the specific energy consumption is in the range of 100 kWh/kg SWU, i.e., an order of magnitude lower than in gaseous diffusion plants.

4.6 Aerodynamic Methods

Two aerodynamic separation techniques were developed in Germany (separation nozzle) and South Africa (advanced vortex tube). The separation nozzle process is based on the centrifugal force in a fast curved jet flow (Fig. 4.8). The process gas is a mixture of about 96% hydrogen or helium and about 4% UF_6. The principle of the advanced vortex tube process is based on a vortex system for the separation of uranium isotopes contained in a hydrogen carrier gas employing the Helikon cascade technique [7, 10].

Both aerodynamic processes did not reach the stage of commercial application.

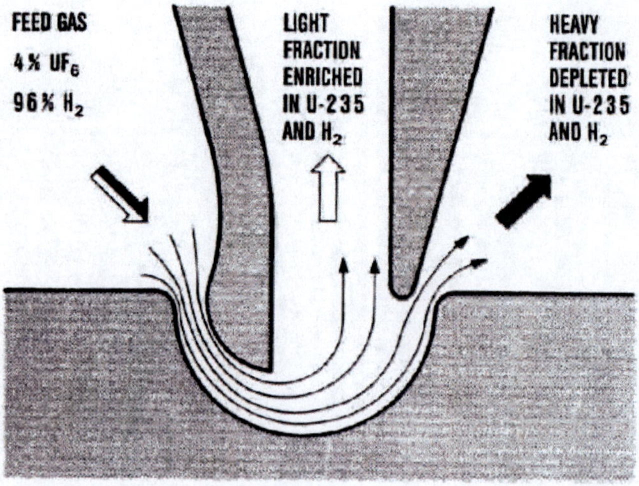

Fig. 4.8 Schematic of the separation nozzle process [7, 10]

4.7 Advanced Separation Processes

Laser separation processes based on selective excitation with subsequent ionization, dissociation or chemical combination of uranium atoms or uranium hexafluoride molecules by laser beams were developed in several countries.

The atomic vapour laser isotope separation (AVLIS) technology was suspended by the USA in 1994. A variation of the molecular laser isotopic separation (MLIS) technology called SILEX was developed in Australia. It will be applied in a large scale SILEX uranium enrichment plant constructed in the USA [5, 6].

4.8 Effects of Tails Assay and Economic Optimum

A 1 GW(e) LWR requires about 84 t of enriched uranium as the initial load. This means that assuming a tails assay of 0.2% of U-235, about 367 t of natural uranium and 257 t SWU are needed. Including all reloads at 3.15% of U-235 enrichment over a life of 30 years of the reactor plant, a total of 4,220 t of natural uranium and 3,320 t SWU are necessary. If the tails assay were decreased, the required amount of natural uranium would be reduced, but the amount of separative work would increase.

Both natural uranium savings as a function of decreasing tails assay and increased separative work lead to cost variations with opposed functional dependencies. Thus, there is an optimum tails assay with minimum costs, which is only dependent on the costs of separative work and the cost of natural uranium including uranium conversion to UF_6. For given costs of natural uranium of US $100/kg and costs of $100/kg SWU,

Fig. 4.9 UF_6 container for low enriched uranium leaving of enrichment plant (URENCO)

the optimum tails assay is determined as 0.218%. Present enrichment plants are operated around this optimum tails assay. Tails assays below 0.20% are technically feasible but, under these conditions, would not guarantee economic operation of enrichment plants [1].

4.9 Fuel Fabrication

Uranium is now used in the reactor core almost exclusively in the form of UO_2 pellets (LWR, HWR, AGR). Only in very few cases are uranium carbide particles used (HTGR and HTR). The source material for the production of enriched UO_2 pellets is uranium hexafluoride gas UF_6, which is supplied by enrichment plants in 1.5 t cylinders (Fig. 4.9). Conversion into uranium oxide can follow a number of procedures.

One process used is the ammonium uranyl carbonate AUC process (Fig. 4.10). It involves mixing of the UF_6 gas with water in a reaction column, thus producing, uranyl fluoride UO_2F_2. Further mixing with ammonia, NH_3, and carbon dioxide, CO_2, generates ammonium uranyl carbonate, which is precipitated from the solution. The suspension is passed through an rotary filter, washed and then put into a fluidized bed furnace where ammonia and carbon dioxide is split off by thermal decomposition. This produces uranium trioxide, UO_3, which is reduced to uranium dioxide, UO_2, by means of hydrogen at temperatures around 500°C. Fluoride residues in the uranium dioxide powder are decreased to 110 ppm by water vapor at 650°C [11–14].

The production of UO_2 pellets follows powder metallurgical processes. First the UO_2 powder is homogenized and a suitable grain size distribution is achieved by

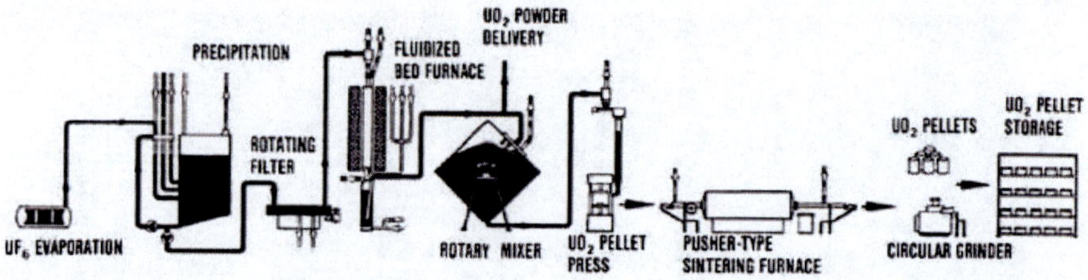

Fig. 4.10 Flowsheet of UO_2 pellet production [13, 14]

crushing and screening. After adding binders and lubricants, pellets of about 10 mm diameter and height are pressed to a density of approximately 5.5 g/cm^3. In a pusher-type sintering furnace, the binders and lubricants as well as fluorine are first expelled at 500–1,000°C and hyper stoichiometric uranium oxide is reduced to UO_2 by hydrogen. This is followed by the sintering process, which takes 2–3 h at 1,600–1,750°C, achieving a density of the UO_2 pellets of 10.3–10.5 g/cm^3. Afterwards, the pellets are ground on centerless circular grinders to the required geometric dimensions and tolerances.

The UO_2 pellets then pass quality control, where the geometric dimensions, density, surface quality, moisture content and the O/U stoichiometry are checked. After this quality control, the UO_2 pellets are stored ready for fuel fabrication. In PWR, BWR and CANDU fuel elements, zircaloy tubes are filled with the UO_2 pellets and welded tight. These fuel rods are then assembled into fuel elements (see e.g. Figs. 5.3 and 5.4) [15, 16].

The fabrication of uranium and thorium particles with ceramic coatings for HTGRs will be described in Sect. 5.2. The design of prismatic or spherical fuel elements with such dispersed fuel particles will be addressed in Sect. 5.2.2.1. Similarly, the fabrication of PuO_2/UO_2 pellets or particles for Plutonium recycling and LMFBR fuel elements are described in Chaps. 7 and 8.

References

1. International Nuclear Fuel Cycle Evaluation (1980) Enrichment availability. Report of INFCE working group 2. International Atomic Energy Agency, Vienna
2. Villani S (ed) (1979) Uranium Enrichment. Topics in applied physics, vol 35. Springer, Berlin
3. Benedict M et al (1981) Nuclear chemical engineering. McGraw-Hill, New York
4. Cohen K (1951) The theory of isotope separation as applied to the large-scale production of U-235. McGraw-Hill, New York
5. Laughter M (2007) Profile of world uranium enrichment programs—2007, ORNL/TM-2007/193
6. Rahn FJ et al (1984) A guide to nuclear power technology. Wiley, New York
7. Ehrfeld W, Ehrfeld U (1980) Anreicherung von U-235. In: Gmelin Handbuch der Anorganischen Chemie, Uran, Ergänzungsband 2A, Isotope. Springer, Berlin

8. Wood M (2008) Effects of separation processes on minor uranium isotopes in enrichment cascades. Sci Glob Secur 16:26–36
9. Bukharin O (2004) Russia's gas centrifuge technology and uranium enrichment complex. Science and Global Security, Princeton University, Princeton
10. Becker EW et al (1981) Uranium enrichment by the separation nozzle method within the framework of German/Brazilian cooperation. Nucl Technol 52:105–114
11. Brandberg SG (1973) The conversion of uranium hexafluoride to uranium Dioxide. Nucl Technol 18:177–184
12. Hackstein KG, Plöger F (1967) Neue Anlage zur Erzeugung von UO_2-Pulver aus UF_6. Atomwirtschaft/Atomtechnik 12:524–526
13. Hardy CJ (1978) The chemistry of uranium milling. Radiochimica Acta 25:121–134
14. Keller C, Möllinger H (eds) (1978) Kernbrennstoffkreislauf, Band I. Dr. Alfred Hüthig, Heidelberg
15. Schneider VW, Plöger F (1978) Herstellung von Brennelementen. In: Baumgärtner F (ed) Chemie der nuklearen Entsorgung, Teil I. Karl Thiemig, München, pp 115–138
16. Seidel DC (1981) Extracting uranium from its ores. Int At Energy Agency Bull 23(2):24–28

Chapter 5
Converter Reactors With a Thermal Neutron Spectrum

Abstract Nuclear power generation is currently mainly based on light water reactors, designed as pressurized water reactors and boiling water reactors. These are built by a number of manufacturers in various countries of the world. In this chapter, the standard German PWR of 1,300 MW(e) and the European Pressurized Water Reactor (EPR) will be described. In addition, the chapter deals with the German Standard BWR of 1,280 MW(e) and the newer design SWR-1,000 (KERENA). Gas cooled and graphite moderated commercial reactors with natural uranium were developed in the United Kingdom and in France and built in the 1950s and 1960s (MAGNOX reactors). Advanced gas cooled reactors (AGCRs) with graphite as moderator and carbon dioxide as coolant gas have been built in unit sizes up to 620 MW(e). High temperature gas cooled reactors with gas outlet temperature of 700–740°C use helium as a coolant gas. Their fuel elements have been developed as prismatic or spherical pebble fuel elements. High temperature gas cooled reactors with medium enriched uranium are now designed mainly as small modular reactors for safety reasons. Power reactors with heavy water as the moderator and heavy water or light water as coolant have been developed in Canada, Europe and Japan up to unit sizes of 630 MW(e). The advanced CANDU reactor (ACR) is developed currently to a unit size of up to 1,000 MW(e). Homogeneous core thermal breeders with molten salt and light water breeder reactors together with accelerator driven subcritical reactor cores are still in the design or development phase.

5.1 Light Water Reactors

Nuclear power generation is currently mainly based on light water reactors (LWRs) designed as pressurized water reactors (PWRs) or boiling water reactors (BWRs) (see Table 1.1, Sect. 1.2). LWRs use low enriched uranium fuel, which makes for greater flexibility in the choice of reactor core materials, especially allowing normal (light) water to be used as a coolant and moderator. PWRs deliver the heat generated in

G. Kessler, *Sustainable and Safe Nuclear Fission Energy*, Power Systems,
DOI: 10.1007/978-3-642-11990-3_5,

Table 5.1 Design characteristics of the large PWR power plants of Kraftwerk Union and AREVA [1, 2]

	Kraftwerk-Union PWR-1,300	AREVA EPR-1,600
Reactor power		
Thermal (MW(th))	3,780	4,500
Electrical (MW(e))	1,300	1,600
Plant efficiency (%)	32.8	35.6
Reactor core		
Equivalent core diameter (m)	3.6	3.77
Active core height (m)	3.9	4.2
Specific core power (kW(th)/l)	95	96
Density (kW(th)/kg U)	36.5	28.4
Number of fuel elements	193	241
Total amount of fuel (kg U)	125,000	153,000
Fuel element and control element		
Fuel	UO_2	UO_2
U-235 fuel enrichment (wt%)	3.5–4.5	4–5
Cladding material	Zircaloy-4	Zircaloy-M5
Cladding outer diameter (mm)	10.75	9.50
Cladding, thickness (mm)	0.725	0.57
Fuel rod spacing (mm)	14.3	12.6
Av. specific fuel rod power (W/cm)	208	156
Fuel assembly array	16 × 16	17 × 17
Control/absorber rods		
Absorber material	20 rods inserted in fuel element Ag, In, Cd, B_4C	24 rods inserted in fuel element Ag, In, Cd and B_4C
Number of fuel rods per fuel element	236	265
Burnable poison	Boron acid	Gd and boron acid
Number of control elements	61	89
Heat transfer system		
Primary system		
Total coolant flow		
Core flow rate (t/s)	18.8	21.4
Coolant pressure (MPa)	15.8	15.5
Coolant inlet temp. (°C)	292	295.6
Coolant outlet temp. (°C)	326	328.2
Steam supply system		
Steam pressure (MPa)	6.8	7.8
Steam temperature (°C)	285	293
Fuel cycle		
Average fuel burn up (MWd/t)	55,000	70,000
Refuelling sequence	1/3 to 1/4 per year	1/4 to 1/5 per year
Average fissile fraction in spent fuel		
U-235 (wt%)	0.8	0.8
Plutonium (wt%)	0.7	0.7

their reactor core to water circulating under high pressure in primary coolant circuits. From here the heat is transferred to a secondary coolant system via a steam generator to produce steam driving a turbo-generator system. In BWRs with direct cycle, the steam for the turbo-generator system is generated right in the reactor core and sent directly to the turbo-generator. PWRs and BWRs have been advanced to a high level of technical maturity. They are built in unit sizes up to 1,250 and 1,600 MW(e).

5.1.1 Pressurized Water Reactors

In the 1950s PWRs were developed in the USA particularly as power plants for nuclear submarines ("N.S. Nautilus" 1954). The successful application of the PWR concept then resulted in the construction of the first non-military experimental nuclear power plants in Russia and the USA.

Present PWRs are built basically according to the same technical principles by a number of manufacturers in various countries of the world. In this chapter, the standard PWR of 1,300 MW(e) will be described as built in the 1990s by Kraftwerk Union in Germany [1]. In addition, the 1,600 MW(e) European Pressurized Water Reactor (EPR) will be described as built by AREVA [2]. PWR designs by other manufacturers have small technical differences relative to this concept but these are not relevant to understanding the basic principles of design [3–6].

Figure 5.1 shows the main design principles of a PWR. The heat generated by nuclear fission in the reactor core is transferred from the fuel elements to the coolant in the primary system. The highly pressurized water (15.5 or 15.8 MPa) is circulated by coolant pumps and heated in the reactor core from 292 or 296°C (inlet) to 326 or 328°C (outlet) (Table 5.1). It flows to four steam generators, where it transfers its heat to the secondary steam system. In the secondary system, steam of 6.8 or 7.8 MPa and 285 or 293°C is generated. The steam drives the turbine and generator. The steam exhausted by the turbine is precipitated in the condenser, and the condensate water is pumped back into the steam generators. Waste heat delivered to the condenser is discharged into the environment either to river, lake or sea water or through a cooling tower.

5.1.1.1 Core

The core initially contains fuel elements with three or four different levels of U-235 enrichment. In case of EPR some fuel elements also contain gadolinium as burnable poison. The higher enriched fuel elements are arranged at the core periphery, the less enriched fuel elements are distributed throughout the interior of the core (Fig. 5.2). This provides for a relatively flat power distribution over the core and adapts to later so-called equilibrium core loadings. In later core reloadings, fuel elements with different burnups and different enrichments are also arranged in similar patterns. The specific power density in the core is about 95 or 96 kW(th)/l. Around 2008 the

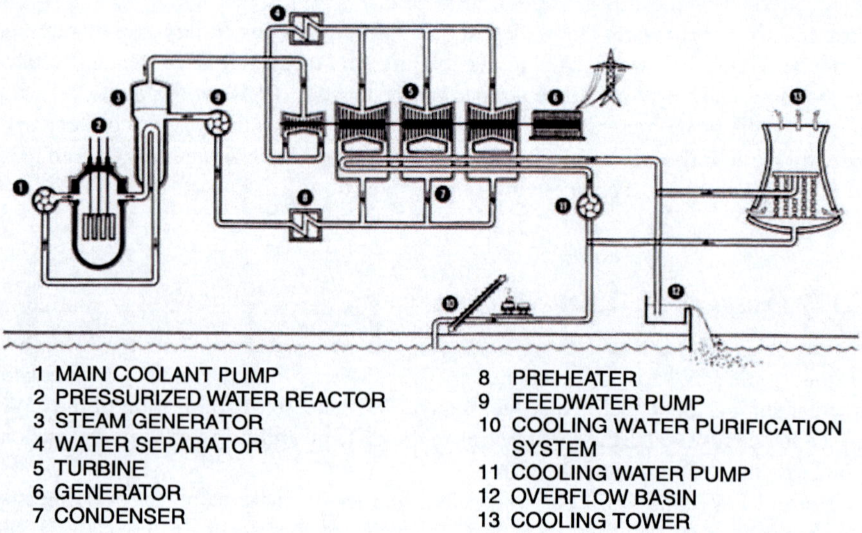

Fig. 5.1 Functional design diagram of a pressurized water reactor power plant (Kraftwerk Union) [1]

Fig. 5.2 EPR core with typical initial fuel element enrichments (AREVA) [2]

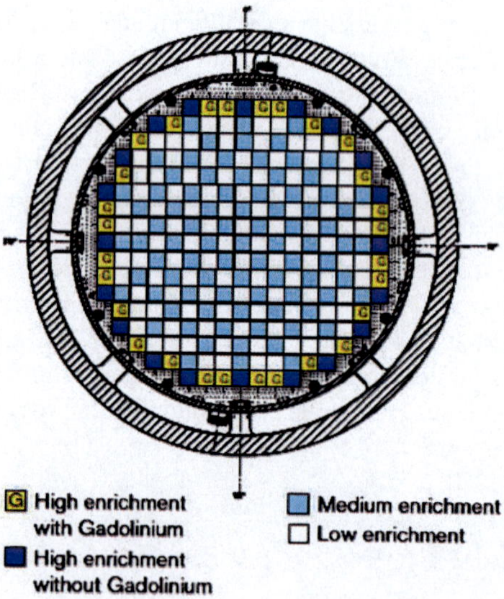

average maximum burnup of the fuel was about 55,000 MWd/t over an irradiation period in the core of about four years. Around 2010–2020, a maximum burnup to 70,000 MWd/t is strived for. A one year reloading cycle can be reached by unloading either one fourth or one fifth of the fuel elements at maximum burn up [7–10].

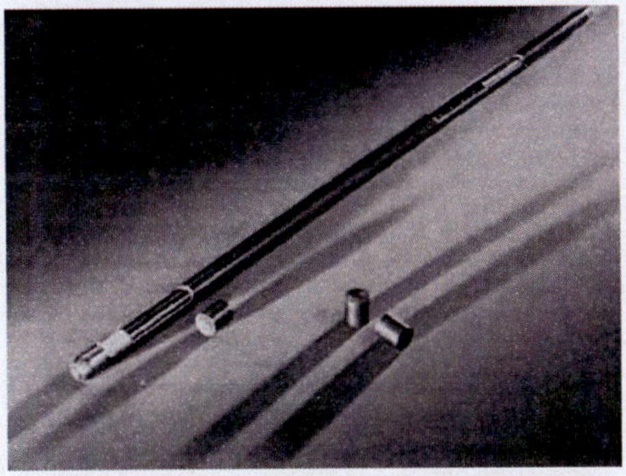

Fig. 5.3 PWR fuel rod (Kraftwerk Union) [1]

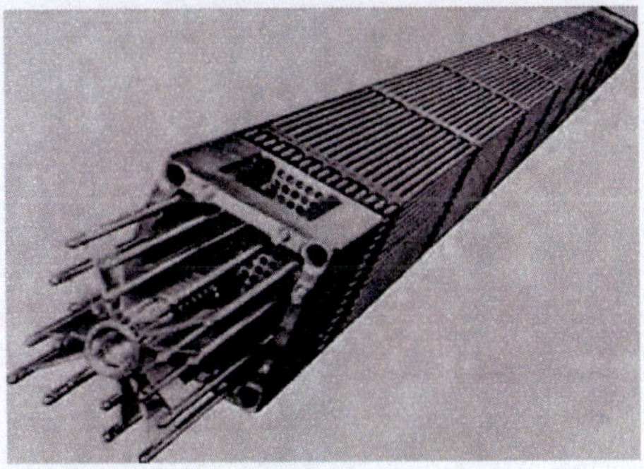

Fig. 5.4 Fuel element and control element of a 1,300 MW(e) PWR (Kraftwerk Union) [1]

Uranium dioxide (UO_2) is used as a core fuel. The UO_2 powder is pressed and sintered into pellets of about 10 mm height and diameter with an average density of 10.40 g/cm^3. The pellets along with pressurized helium are placed into tubes (cladding) made of Zircaloy-4 or Zircaloy-M5 with an active core length of 3.9 or 4.2 m. This cladding material is chosen for its low neutron absorption as well as good mechanical and corrosion properties. The tubes are welded and assembled into fuel elements (Fig. 5.3).

A fuel element of a 1,300 or 1,600 MW(e) reactor contains 236 or 265 fuel rods, respectively. The core has 193 or 241 fuel elements, respectively, with a total uranium mass of 125 t in case of the PWR of Kraftwerk-Union PWR or 153 t in case of EPR of AREVA. Some of the fuel elements contain control elements with 20 or 24 axially movable absorber rods (Fig. 5.4) [11, 12]. These absorber rods are filled with boron carbide or silver-indium-cadmium as neutron absorbing materials.

5.1.1.2 Reactor Pressure Vessel

The fuel elements, control elements and core monitoring instruments are contained in a large reactor pressure vessel (RPV) designed to withstand the operating pressures at operating temperatures (Fig. 5.5). The vessel has an inner diameter of about 5 m,

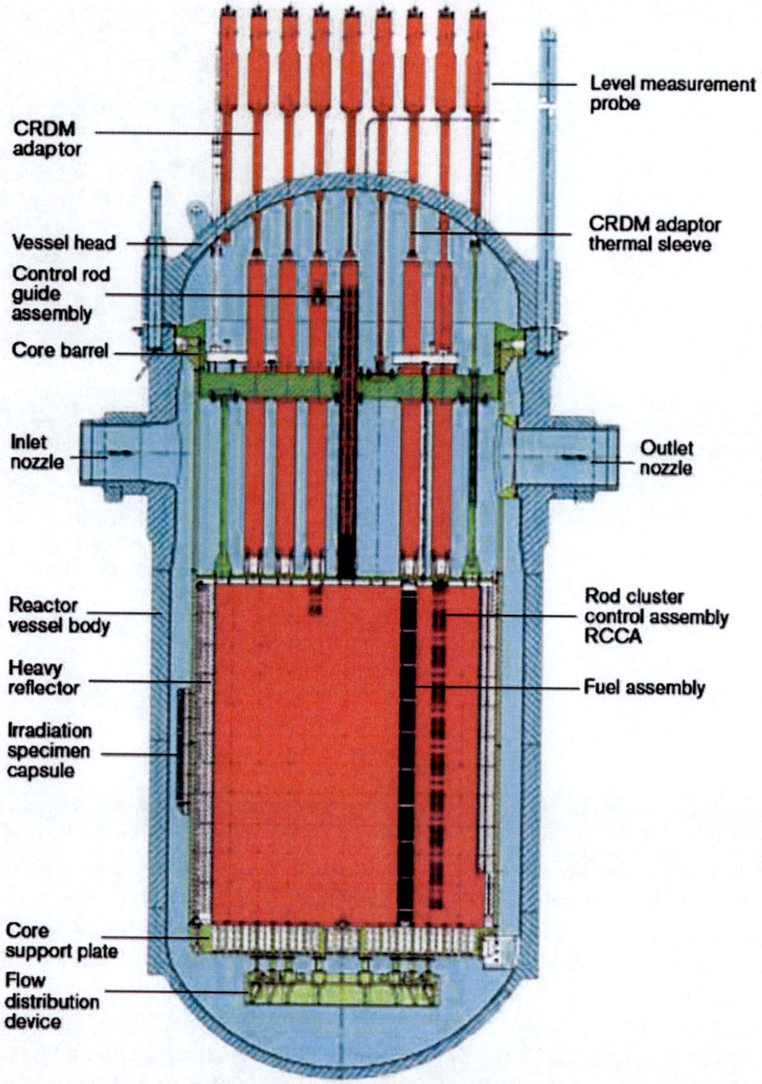

Fig. 5.5 Reactor pressure vessel with core and internal components of EPR (AREVA) [2]

a thickness of the cylindrical wall of 250 mm and a height of approximately 13 m. The cover head of the vessel, which holds all the control rods and control rod drive mechanisms (CRDM), can be removed for refueling. The water flowing at a rate of 18.8 and 21.4 t/s, respectively, enters the reactor vessel through four inlet nozzles close to the top and flows downward through the annulus between the vessel and the thermal shield and neutron reflector to the core inlet near the bottom. It then returns upward through the core leaving the four outlet nozzles of the pressure vessel. The reactor vessel is made of 22NiMoCr37 steel, in case of Kraftwerk Union PWRs and of 16 MND5 for EPR. Its inner surface is plated with austenitic steel.

5.1.1.3 Coolant System

The water leaving the outlet nozzles of the pressure vessel transports the heat generated in the reactor core through four identical primary coolant circuits to four steam generators and is then recirculated to the pressure vessel. The inner diameter of the primary system pipes is 750 mm for the Kraftwerk Union or 780 mm for EPR. Each primary coolant pump has a pressure head of 0.8 or 1.0 MPa and consumes 5.4 or 9 MW(e) of power, respectively. The whole primary system is also plated with austenitic steel. The pressure of 15.8 or 15.5 MPa in the primary coolant system is maintained by a pressurizer filled partly with water and partly with steam. It has heaters for boiling the water and sprayers for condensing the steam to keep the pressure within specified operating limits.

In the steam generators heat is transferred through a large number of tubes (Fig. 5.6). Feed water at a pressure of about 7 MPa is evaporated and steam is passed through separators to remove water droplets and attain a moisture content of less than 0.25 or 0.015%. In the steam generator, the primary coolant system is separated from the secondary coolant system by the tubes. The EPR steam generator (Fig. 5.6) with economizer makes it possible to reach a saturation pressure of 7.8 MPa by special feed water flow guidance. This leads to a thermal efficiency of about 36%.

The steam flows through the turbine valves into the high pressure section and after reheating into the low pressure section of the turbine. The expanded steam is precipitated in the condenser and pumped back as feed water by the main condensate pumps into the feed water tank. The main feed water pumps move the pressurized water from the feed water tank through four main feed water pipes into the four steam generators.

In case the turbine would have to be suddenly disconnected from the grid as a result of some fault condition, steam can be directly passed into the condenser by means of bypass valves. If the condenser should not be available due to some failure, the steam can be blown off to the atmosphere by means of blow down valves and safety valves.

A number of supporting systems are required for operation of the coolant circuit systems. The volume control system offsets changes in volume of the reactor cooling system resulting from temperature and operational influences. It is controlled by the water level in the pressurizer (Fig. 5.7). Part of the cooling water is extracted continuously and purified in ion exchangers. At the same time, corrosion products and radioactive products are removed.

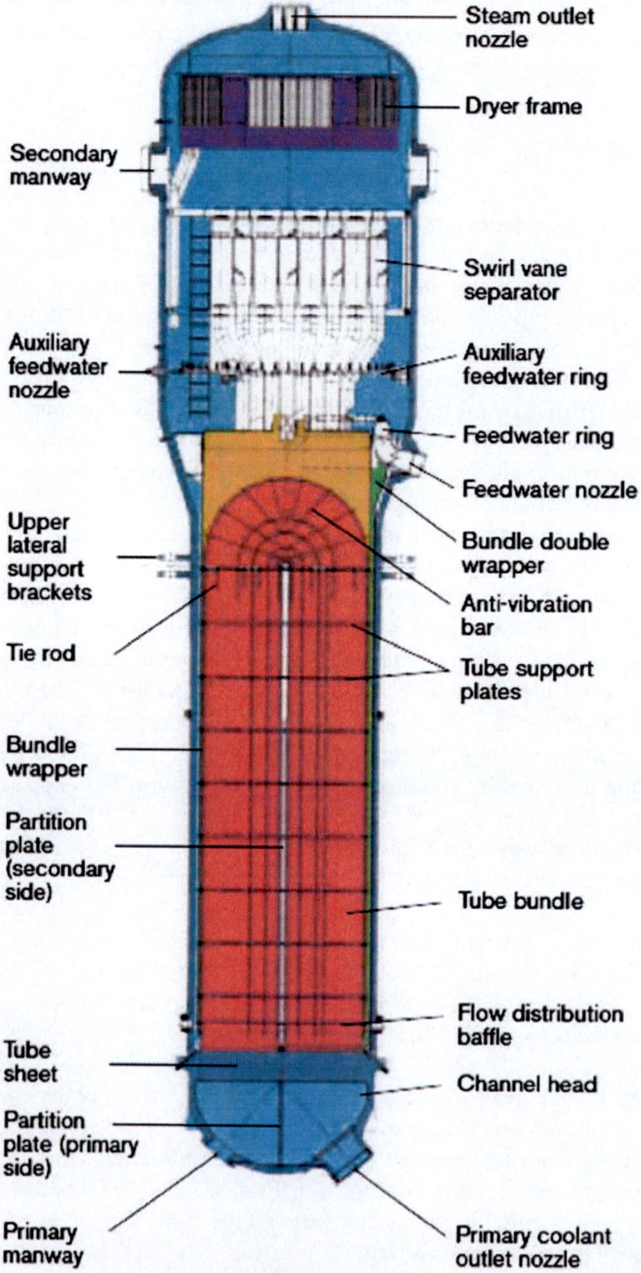

Fig. 5.6 Cutaway of steam generator for EPR (AREVA) [2]

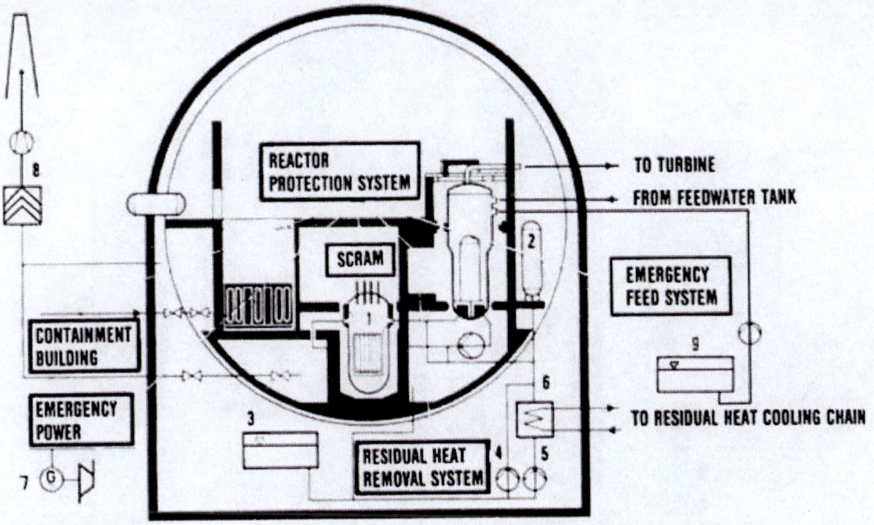

1 SCRAM SYSTEM
2 PRESSURIZER
3 FLOODING TANK
4 SAFETY FEED PUMP
5 RESIDUAL COOLING PUMP
6 RESIDUAL HEAT COOLER
7 EMERGENCY POWER SYSTEM
8 VENTILATION SYSTEMS
9 EMERGENCY FEEDWATER SYSTEM

Fig. 5.7 Reactor protection system of a PWR (Kraftwerk Union) [1]

5.1.1.4 Containment

In case of the Kraftwerk Union PWR, the reactor pressure vessel, the coolant pumps, steam generators, emergency and afterheat cooling systems as well as the vault for fresh and spent fuel elements are arranged within the reactor building, which is enclosed in a spherical double containment (Fig. 5.7). The double containment is made up of the inner steel containment and an outer concrete shield which is 1.80 m thick. The inner steel containment is held at a lower pressure than the atmospheric pressure. In this way only leakage from the outside to the interior of the containment is possible. The spherical inner steel containment has a diameter of approximately 56 m and is designed to an internal pressure of about 0.5 MPa. Penetrations of the piping through the containment are equipped with vented double bellows and can be checked for leaks. The outer steel reinforced concrete shield protects the reactor against external impacts and shields the environment against radiation exposure in case of accidents. External impacts which are considered as a design basis for the containment are earthquakes, floods, storms, airplane crashes, and pressure waves generated by chemical explosions.

The EPR containment is a cylindrical double containment of prestressed concrete. The outer concrete shield is up to 2 m thick and has a diameter of 75 m and height of 60 m (Fig. 5.8). It protects the inner containment against external events as in case

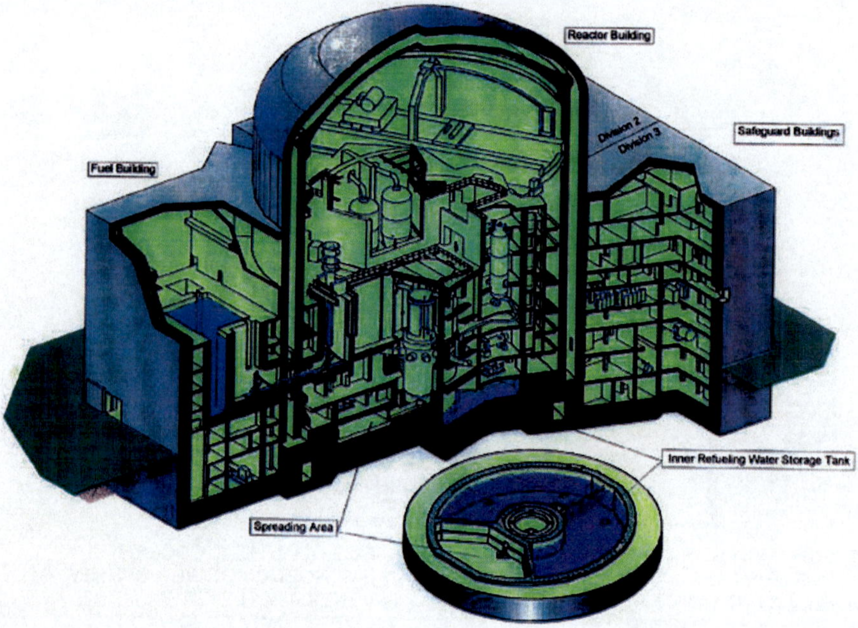

Fig. 5.8 EPR double containment with RPV, cooling systems and molten core spreading area (AREVA) [2]

of the spherical containment of the Kraftwerk Union PWR. The upper part of the inner containment houses sprinkler systems supplied with water by the containment heat removal system (CHRS) for ultimate heat removal in case of a severe accident. Below the reactor pressure vessel, a so-called molten core spreading area with special cooling systems is located which can cool the hot core masses in case of a core melt accident (Fig. 5.9). Leak tightness of the inner containment and filter systems guarantee extremely low releases of radioactivity to the environment, even in case of severe accidents. Relocation or evacuation of the population can be avoided even in case of severe accidents (Chap. 11).

The inner containment of modern European PWRs can also withstand the pressure resulting from large scale hydrogen combustion processes which might occur under severe accident conditions. Hydrogen can result from the chemical reaction of the fuel cladding material zircaloy with hot steam during a severe core melt accident. Also so-called hydrogen recombiners are distributed all over the containment to decrease the hydrogen concentrations in the inner containment in case of slowly developing core melt accidents. In Chap. 11 additional aspects for more rapid accident progression with faster hydrogen generation will be discussed.

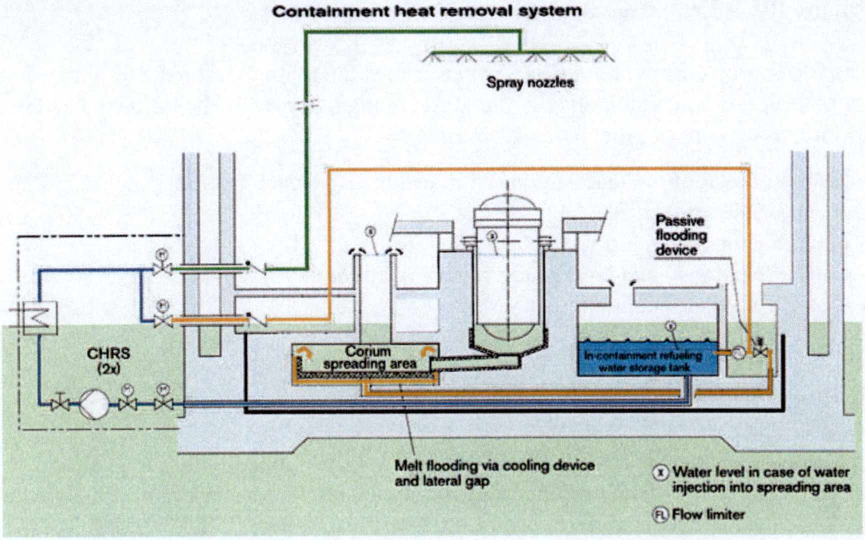

Fig. 5.9 Containment heat removal system in case of core melt for EPR (AREVA) [2]

5.1.1.5 Control Systems

At nominal power PWRs have negative reactivity coefficients of coolant temperature and power (see Sect. 3.8.2.2). Any reduction in core coolant temperature therefore will result in an increase in reactivity and power. If higher loads are demanded by the generator and the turbine, more heat must be extracted from the primary cooling system through the steam generators. This is done by opening the turbine governor valve which causes the primary coolant temperature in the reactor core to drop and subsequently the reactor power to rise; the power automatically balances out at a slightly higher level. However, in order to prevent the steam temperature and the steam pressure from dropping too far, the control elements are also moved at the same time.

Reactivity changes in the core are balanced by axial movements of the control elements: slow changes of the kind brought about by fuel burnup and fission product build up are controlled by changing the boric acid concentration of the primary coolant. Also gadolinium is mixed with the UO_2 fuel as a burnable poison if very high burn up of the fuel shall be attained.

The coolant pressure is kept at the systems pressure by the pressurizer (Fig. 5.7). The water level in the pressurizer is controlled by the volume control system. The addition of feed water to the steam generator must be matched by the feed water control system as a function of the amount of steam extracted for the turbine.

5.1.1.6 PWR Protection System

The PWR protection system (Fig. 5.7) processes the main measured data important for plant safety, and automatically initiates, among other steps, the following actions as soon as certain set points are exceeded:

- fast shutdown of the reactor core with turbine trip (reactor scram) and separation from the electrical grid,
- emergency power supply,
- afterheat removal and emergency water injection.
- closure of reactor containment.

Reactor Scram

In a reactor scram, the absorber rods of the PWR are dropped into the reactor core by gravity (Fig. 5.7). The reactor goes subcritical and the reactor power drops to the level of afterheat generation (Sects. 11.6.6 and 11.6.7).

Emergency Power Supply

During normal reactor operation, the plant is connected to the public grid system. In the case of a breakdown of the public grid, the plant is disconnected from the grid and the output of the turbogenerator and the power generation of the reactor is reduced to the level of the plant requirement. The plant generates its own supply in isolated operation and is maintained operational. If also the isolated mode of operation fails, only the safety systems will be supplied by emergency power. Emergency power is supplied in a redundant layout by diesel generators and battery systems.

Emergency Feedwater System

The emergency feed water system (Fig. 5.7) supplies feed water to the steam generators, if the main feed water pumps can no longer do so. They are supplied by emergency power in case the main power supply were to fail. The emergency feed water system is equipped with fourfold redundancy, having borated water reserves (feed water tank and demineralized water tank) which allow the removal of afterheat to be maintained for many hours. Removal of the afterheat from the reactor core is ensured first by natural convection of the primary cooling water flowing through the core and steam generators. Natural convection is enforced by installing the steam generators at a higher level than the reactor core. After the coolant pressure in the reactor cooling system has dropped to a sufficiently low level the afterheat removal system takes over this function.

Emergency Cooling and Afterheat Removal Systems

The emergency cooling system consists of the high pressure safety feed system, the flooding tank accumulators (Kraftwerk Union PWRs) and the low pressure safety feed system. The low pressure safety feed system is combined with the residual heat or afterheat system which serves for both operational and safety related purposes (see Fig. 5.7):

- When the power of the PWR is shutdown by the protection system, the emergency cooling and afterheat removal system is started up automatically, after the pressure and the temperature in the primary systems have dropped to sufficiently low levels. The system then removes the afterheat and continues to cool the reactor core and coolant circuits.
- In a loss-of-coolant accident the emergency cooling and afterheat removal system has to maintain the coolant level in the reactor pressure vessel and ensure cooling of the reactor core. The emergency cooling and afterheat removal system has fourfold redundancy and is supplied by emergency power. It can feed water both into the cold (inlet) and the hot (outlet) main coolant lines by means of a high pressure feed system (11 MPa in case of Kraftwerk Union PWRs and 9.2 MPa in case of EPR) and a low pressure feed system <1 MPa. Major leaks would cause the pressure in the reactor cooling system to drop so quickly that the high pressure feed system would not be started up. In that case, borated water will be injected into the primary main cooling system directly from the flooding tank accumulators at a 2.5 MPa (Kraftwerk Union PWRs) and through the low pressure injection systems at <1 MPa from the flooding tanks. If the flooding tanks or accumulators have run dry, the low pressure injection system feeds water from the reactor building sump into the primary systems. The building sump can collect leakage water from the primary system (sump recirculation operation). In case of EPR, water discharged inside the containment is collected in the in-containment refuelling water storage tank (IRWST) located at the bottom of the containment.

Smaller leaks cause the pressure to drop only gradually, so that initially only the high pressure feed system will start to function. However, the pressure and temperature drop in the main coolant system is supported by a temperature drop of 100°C/h on the secondary side. This is done automatically. After the coolant pressure and temperature have dropped sufficiently below 1 MPa, the low pressure feed systems will be started up (Chap. 11).

The afterheat is discharged into river, lake or sea water, a cooling pond or special cooling towers by way of intermediate heat exchangers (secondary parts of the emergency cooling and afterheat removal system), which are also fourfold redundant and driven by emergency power.

5.1.1.7 Supercritical Water Cooled Reactor

The SCWR shall operate at a water pressure above its critical point (22.1 MPa, 374°C) to achieve a thermal efficiency of more than 40%. The SCWR design takes

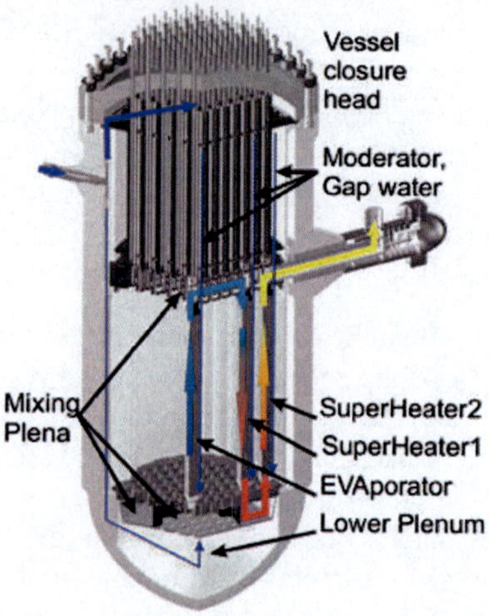

Fig. 5.10 Three pass core of a SCWR with flow direction of moderator water in part of fuel elements as well as central core evaporator region, superheater 1 region (annular core assemblies) and superheater 2 region (peripheral assemblies) [17]

advantage from experience already available from coal fired plants with supercritical water cycle.

In the reactor core, water at a pressure of 25 MPa is heated up from 310 to 500°C [13, 14]. This leads to a ten fold higher enthalpy increase compared to modern PWRs operating at 15.5 MPa. Consequently, a ten fold lower water flow rate is required to produce the same power output. This allows a reduction of component sizes (turbine, condenser, pipes etc.). Similar to BWRs, steam generators—as in PWRs—are not necessary.

A three pass coolant concept was proposed and designed (Fig. 5.10) [15–19]. One of the objectives is a peak cladding temperature below 630°C. This peak cladding temperature is not exceeded in the core, despite of a core outlet temperature of 500°C at 25 MPa. The water density changes on its way through SCWR core by a factor of about 7. For good neutron thermalization the fuel elements, therefore, have square inner water channels (Fig. 5.12) similar as BWR fuel elements. A part of the supercritical feed water with 280°C, is guided to the top structures of the SCWR core (Fig. 5.10) and flows down through the square channels of all fuel elements and is mixing in the lower plenum with the remaining feed water of 280°C. It enters the core subassembly inlet with a temperature of 310°C and flows first upwards through a central part of assemblies (evaporator), then downwards in surrounding assemblies (superheater 1) and then again upwards in outer peripheral assemblies (superheater 2) (Fig. 5.11). Between each of these heat up steps the supercritical coolant with slightly different temperatures from different assemblies is mixed in lower and upper mixing chambers to compensate temperature variations. The thermal efficiency is 43.7%. The temperatures after each heat up step are given in Fig. 5.11 [18].

The fuel element of the three pass SCWR is similar to a modern BWR fuel element with an inner 26.9 mm^2 water-filled zone (Sect. 5.1.2.1). The SCWR assembly contains 40 fuel rods with an outer cladding diameter of 8 mm at a pitch of 9.44 mm

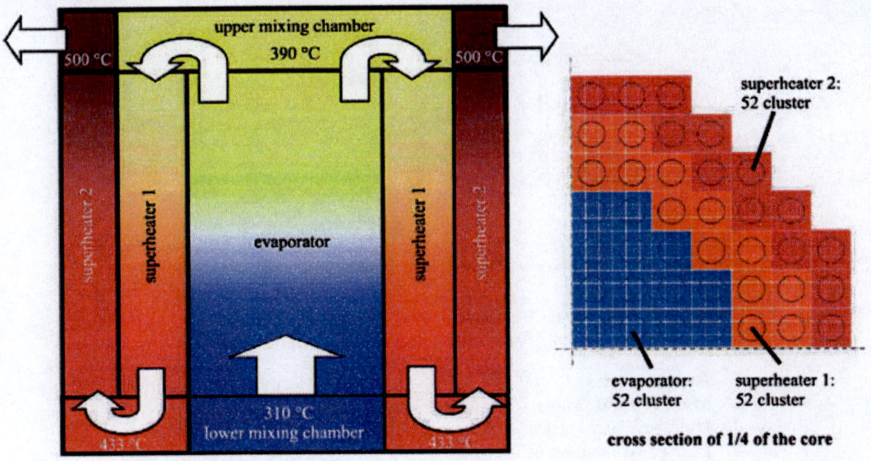

Fig. 5.11 Temperatures of supercritical water in evaporator and superheater regions in the SCWR core [18]

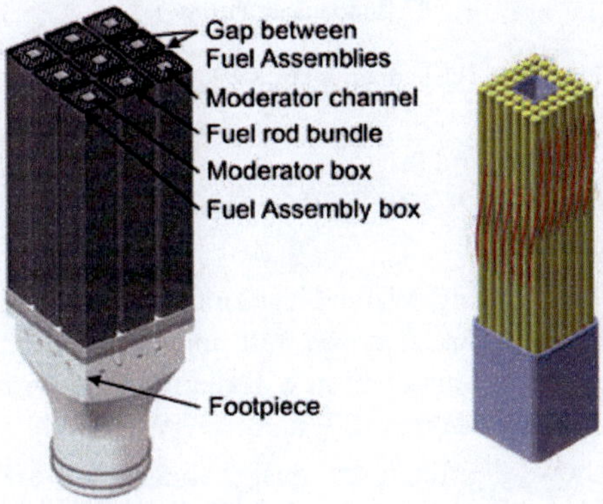

Fig. 5.12 Fuel element cluster and single fuel element with square moderator channel [18]

(Fig. 5.12). Austenitic stainless steels and oxide dispersed steels (ODS) are under consideration as cladding material. The fuel rods are housed in a square stainless steel box of 2.5 mm wall thickness and 72.5 mm outer size. A single wire of 1.34 mm diameter is wrapped around each fuel rod. Nine fuel elements with common foot and head pieces are connected to a cluster (Fig. 5.12). The active core height is 4.2 m [14–19].

Cross shaped absorber rods are inserted and move within the square moderator water tubes of the nine fuel assembly cluster. The reactor pressure vessel is designed for a pressure of 28.7 MPa and a temperature of 350°C. The inner diameter of the pressure vessel is about 4.5 m and the overall height 14.3 m. The cylindrical wall thickness is 350 mm.

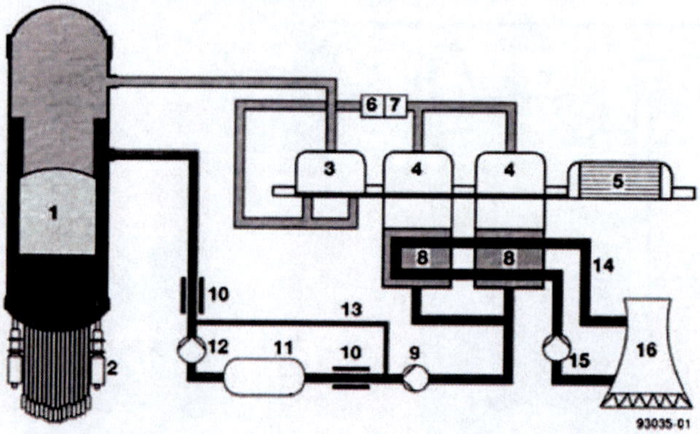

1 Reactor core
2 Main coolant pump
3 High pressure part of turbine
4 Low pressure part of turb ine
5 Generator
6 Water separator
7 Superheater
8 Condenser
9 Condensate Pump
10 Pre-heater
11 Feedwater tank
12 Feedwater pump
13 Interconnecting pipe
14 Cooling water
15 Cooling-water pump
16 Cooling tower

Fig. 5.13 Basic diagram of SWR 1,000 [21]

5.1.2 Boiling Water Reactors

The Boiling Water Reactors (BWRs) built today by a number of manufacturers in the USA, Europe and Japan are characterized by almost identical technical designs. This Section will mainly deal with the standard BWR of Kraftwerk Union of 1,280 MW(e) [20] and the SWR-1,000 as designed by AREVA [21]. Figure 5.13 shows the functional design diagram of a BWR (SWR $\triangleq$ Siedewasserreaktor is the German expression for BWR).

5.1.2.1 Core, Pressure Vessel and Cooling System

The reactor core consists of a square array of 784 or 664 so-called ATRIUM fuel elements, about 3.74 or 3.0 m long, respectively (Table 5.2). The fuel elements each contain either 10 × 10 (ATRIUM 10) or 12 × 12 (ATRIUM 12) fuel rods with outer diameters of 10.28 mm in a closed square box of 131 × 131 or 155 × 155 mm, called a fuel channel (Fig. 5.14). The ATRIUM fuel elements have a 3 × 3 or 4 × 4 cm central water channel to achieve a relatively flat power profile across the fuel element [11, 12, 22, 23]. For moderation of the neutrons and cooling of the core, water flows through the core and is allowed to boil in the upper part of the core. Cruciform absorber plates (Fig. 5.15), containing boron carbide or hafnium as absorber material,

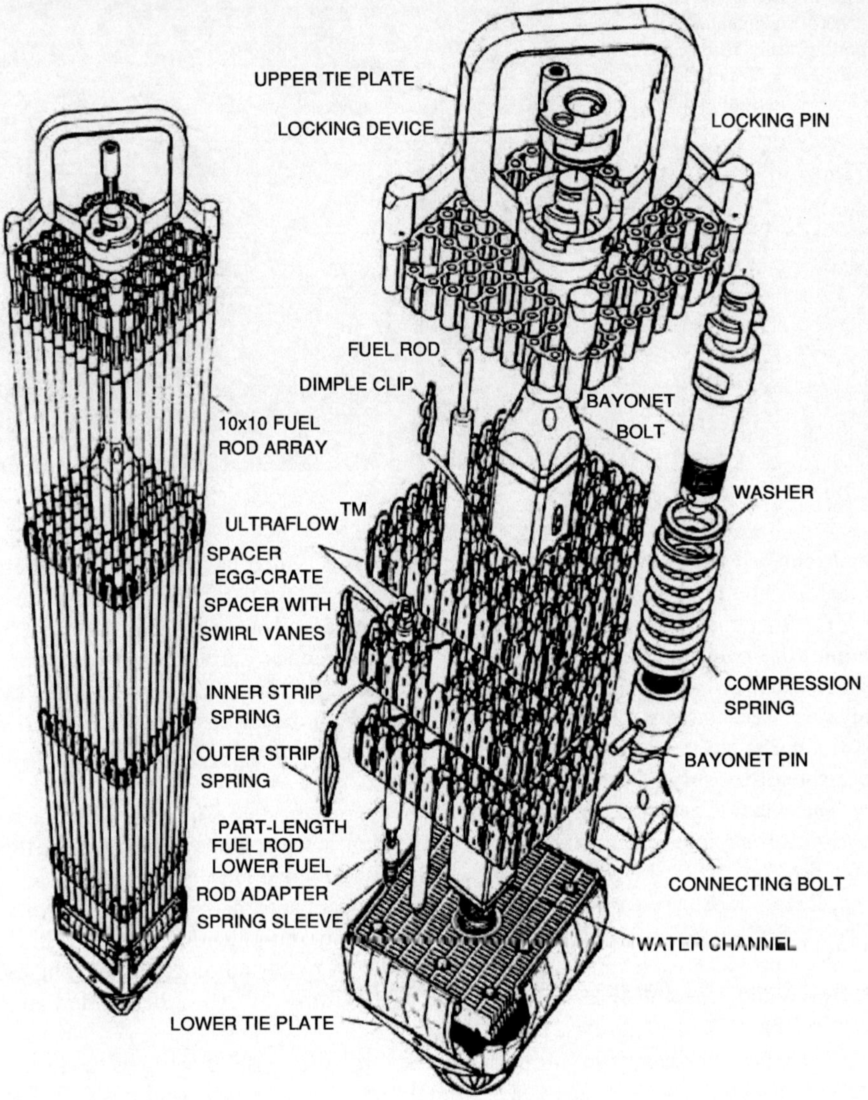

Fig. 5.14 ATRIUM-10 boiling water reactor fuel element [11]

are installed in between a set of four square fuel elements. The absorber plates are moved hydraulically into and out of the reactor core from below.

The fuel rods have claddings of Zircaloy and contain UO_2 pellets with enrichments of about 4.5 or 5.4% U-235. The fuel is unloaded after a maximum burnup of 55,000 or 65,000 MWd/t. Roughly one quarter of the fuel elements are unloaded after 18 months

Fig. 5.15 Cruciform absorber element of a boiling water reactor core with a 7 × 7 fuel element (KKW-Gundremmingen) [20]

and replaced by fresh fuel elements. Fuel elements which have not attained their maximum burnups at that time are reshuffled.

The fuel rods contain gadolinium as a burnable poison to compensate for the burnup of fissile material and the build up of fission products during reactor operation. The average power density in the core is 57 or 51 kW(th)/l and 28.2 or 24.7 kW(th)/kg uranium, respectively. The water inlet temperature in the core is 216 or 220°C, the steam outlet temperature is 286 or 289°C, which corresponds to a saturation steam pressure of roughly 7.0 or 7.5 MPa (Table 5.2).

The steam is generated by water boiling in the reactor core. To provide sufficient core flow for ample heat transfer, BWRs employ internal jet pumps which increase the core flow rate. The core with the absorber plates is contained in a large steel pressure vessel (Fig. 5.16). Above the core, there are the steam separators and steam dryers. The reactor vessel head can be removed for loading and unloading fuel elements. A BWR pressure vessel has a diameter of 6.6 or 7.1 m, a wall thickness of roughly 160 mm, and a height of 23 m. It is made of 22NiMoCr37 steel, the inside being plated with austenitic stainless steel.

The saturated steam flows from the reactor pressure vessel directly to the turbo-generator system and is pumped back from the condenser to the pressure vessel. The condenser is cooled by water from a cooling tower or from a river (see Fig. 5.13).

A question of particular interest in BWRs with direct cooling systems is the radioactivity of the cooling water. The amount of radioactivity is determined by the impurities contained in the water and by an (n, p)-reaction of oxygen producing nitrogen, N-16, with a half life of 7.2 s. Many years of experience in operating boiling water reactors have shown that, because of the short half-life of the N-16, maintenance work on the turbine, the condenser and the feed water pumps is not impaired by radioactivity.

Table 5.2 Design characteristics of the BWR power plants BWR-1,300 of Kraftwerk Union and SWR-1,000 of AREVA [1, 2]

	BWR-1,300 (Kraftwerk Union)	SWR-1,000 (KERENA) (AREVA)
Reactor power		
Thermal (MW(th))	3,840	3,370
Electrical net (MW(e))	1,284	1,254
Plant efficiency (net) (%)	35	37.2
Reactor core		
Equivalent core diameter (m)	4.8	5.3
Active core height (m)	3.74	3.0
Specific core power (kW(th)/l)	56.8	51
Number of fuel elements	784	664
Total amount of fuel (kg U)	136,000	136,300
Fuel element and control element		
Fuel	UO_2	UO_2
U-235 fuel enrichment average (w/%)	4	5.4
Cladding material	Zircaloy	Zircaloy (Fe combined)
Cladding outer diameter (mm)	10.28	10.28
Cladding thickness (mm)	0.62	0.62
Av. specific fuel rod power (W/cm)	129	116
Fuel assembly array	10 × 10 ATRIUM	12 × 12 ATRIUM
Control/absorber rod type	Cruciform control element inserted from the bottom between a set of 4 fuel assemblies	Cruciform control element inserted from the bottom between a set of 4 fuel assemblies
Number of control elements	193	157
Heat transfer system		
Primary system		
Total coolant flow rate (core flow) (t/s)	14.3	13.2
Coolant pressure (MPa)	7.0	7.5
Coolant inlet temp. (°C)	216	220
Steam outlet temp. (°C)	286	289
Steam supply system		
Steam generation (t/s)	1.94	1.85
Steam pressure (MPa)	7.0	7.3
Steam temperature (°C)	286	289
Fuel cycle		
Average fuel burn up (MWd/t)	56,000	65,000
Refuelling sequence	1/4 per 18 months	1/4 per 18 months
Average fissile fraction in spent fuel		
U-235 (%)	0.8	0.8
Pu-fiss (%)	0.7	0.7

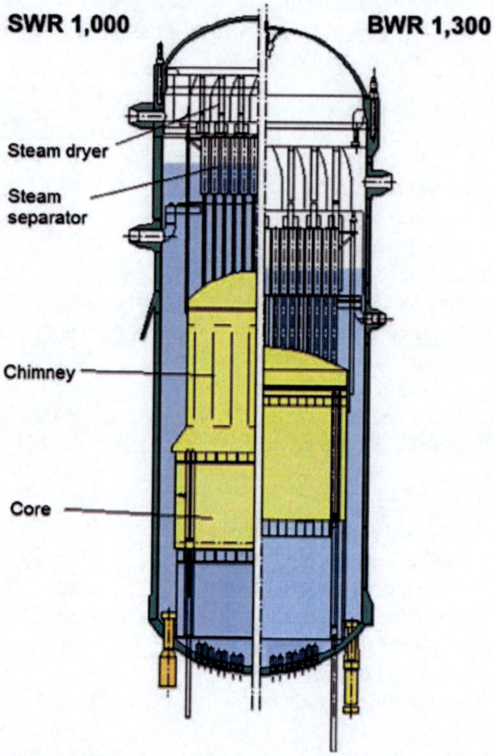

Fig. 5.16 Reactor pressure vessel of a SWR-1,000 or a BWR-1,300 [20, 21, 23]

The water circulation in the reactor pressure vessel and through the core can be used for changing the reactor power. Reduction of water flow through the core will result in a higher evaporation rate and in a larger volume of bubble formation. Increasing the volume of steam in the core reduces the moderation of the neutrons, and as a consequence, the reactivity and the reactor power will be reduced. In this way, changes in the water flow can be used to control the reactor power in a broad range without movement of control rods. BWRs can automatically follow the load requirements of the turbine, by sensing pressure disturbances at the turbine, transmitting these signals to the recirculation flow control valve and regulating core flow and therefore reactor power.

In order to ensure high quality of the reactor feed water, all the feed water recirculated from the turbine condenser is pumped through filters (demineralizer units) and cleared of any corrosion products and other impurities.

5.1.2.2 Safety Systems

If there are reactivity perturbations or losses of coolant flow, the reactor is shut down in a short time by rapid insertion of the absorber rods. As a backup shutdown system,

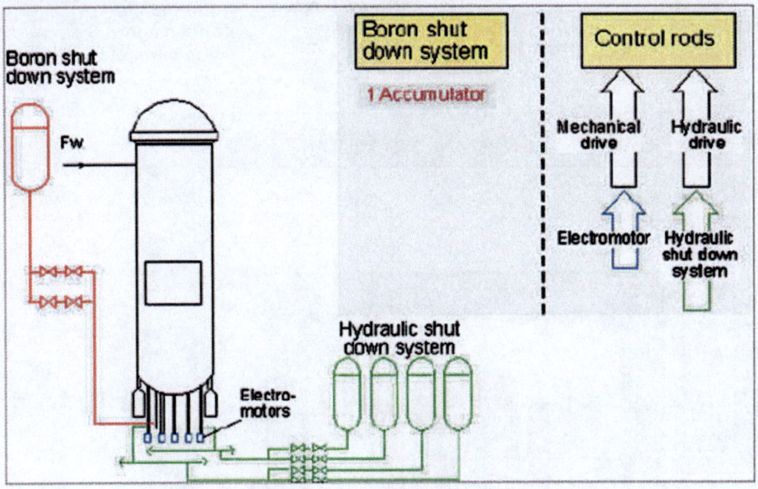

Fig. 5.17 Shutdown systems of BWR-1,300 and SWR-1,000 [21, 23]

the BWR can poison the coolant (moderator) with a neutron absorbing boron acid solution and, in this way, also quench the nuclear reaction and shut down the reactor (Fig. 5.17).

Ruptures in any pipe in the primary cooling system (e.g., the recirculation system) cause the main steam pipes and the feed water pipes to be blocked by two series-connected isolation valves and by two series-connected non-return valves, respectively. This action isolates the reactor pressure vessel within the containment from the outer turbine and condenser cycles. When these isolation valves are closed, or if there is overpressure in the reactor pressure vessel, BWR safety/relief valves are automatically actuated, allowing a path for steam to be discharged from the reactor vessel. In this case, the steam is discharged into a large water pool inside the pressure suppression containment.

In case of the 1,284 MW(e) BWR of Kraftwerk Union steam is discharged into a large pressure suppression pool shown in Fig. 5.18. In case of SWR-1,000 of AREVA more passive safety systems are installed and the steam is discharged into four core flooding pools. In the following part both designs are described separately.

Pressure Suppression System of Kraftwerk Union BWR

Figure 5.18 shows the pressure suppression containment system of the Kraftwerk Union BWR-1,300. The reactor pressure vessel, the recirculation system and the pressure relief valves of the main steam pipes are accommodated in the inner containment, also called drywell, which isolates them from the rest of the reactor containment. The drywell consists of a reinforced concrete structure with horizontal

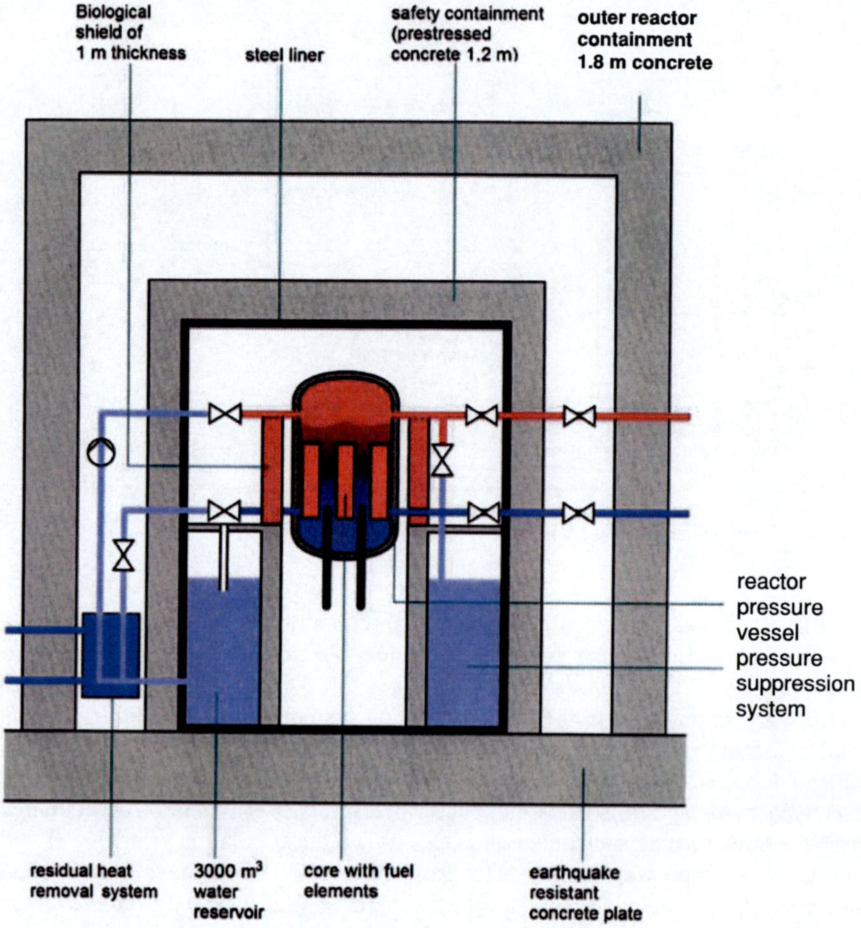

Fig. 5.18 Inner containment of BWR-1,300 (Kraftwerk Union) [20]

vent openings which allow communication between the pool and the drywell. The drywell is kept dry.

If primary coolant is released through a leak in the primary system pipes, steam will enter into the drywell and will be channelled into the pressure suppression pool, where it will condense. A steel containment surrounds the drywell pressure suppression system and all reactor equipment. It is designed to withstand temperatures and pressures that could be caused by a loss-of-coolant accident and is designed to retain non-gaseous fission products which could potentially be released from the reactor system during an accident.

Emergency Cooling and Afterheat Removal Systems

If the water level in the reactor pressure vessel drops, or if there is a leak (loss-of-coolant accident), water is automatically added to the reactor pressure vessel by the following systems):

- a high pressure core water injection system.
- a low pressure coolant injection system and
- the residual core heat removal system.

The pressure pump of the high pressure water injection system initially takes water from tanks that store condensate water or, if necessary, takes water from the pressure suppression pool. The water is then directly pumped into the pressure vessel. Its function is to supply large quantities of water to the core in the case of a loss of coolant accident while the reactor is still in a high pressure condition. It prevents fuel cladding damage in the event the core becomes uncovered.

In case it becomes necessary to use the low pressure system, the reactor pressure vessel can be depressurized. This is accomplished by the opening safety relief valves and by discharging steam to the pressure suppression pool for condensation. After this is done, the low pressure water injection system can be used by taking water from the suppression pool and feeding water directly into the inside of the reactor vessel.

The low pressure coolant injection system is actually one mode of the residual heat removal system. Two heat exchanger loops can be used to cool the suppression pool or the heat exchangers can be bypassed and water can be injected at low pressure directly into the pressure vessel to cool the core.

An additional function of the residual heat removal system is to remove and condense any steam bypassing the drywell. Water is taken from the pressure suppression pool and sprayed into the inner containment free volume. Thus over-pressurization of the containment can be prevented.

SWR-1,000 Containment and Passive Cooling Systems

The reactor pressure vessel is housed in an inner prestressed concrete containment with a steel liner (Fig. 5.19). This inner containment is subdivided into a pressure suppression chamber, the drywell and four hydraulically connected large core flooding pools. The core flooding pools serve as a heat sink for passive heat removal from the reactor pressure vessel by emergency condensers and the safety relief valves. The drywell houses the reactor pressure vessel with control rod drive and shutdown system, the three steam lines and two feed water lines, as well as the safety relief valves, the containment isolation valves and the pressure pulse transmitters. In addition the high pressure part of the water cleanup system and the flooding lines for passive flooding of the reactor pressure vessel and flooding of the drywell in case of drastic loss of coolant are located in the drywell. During operation the inner containment is inertized by nitrogen, to prevent hydrogen combustion in case of a loss of coolant

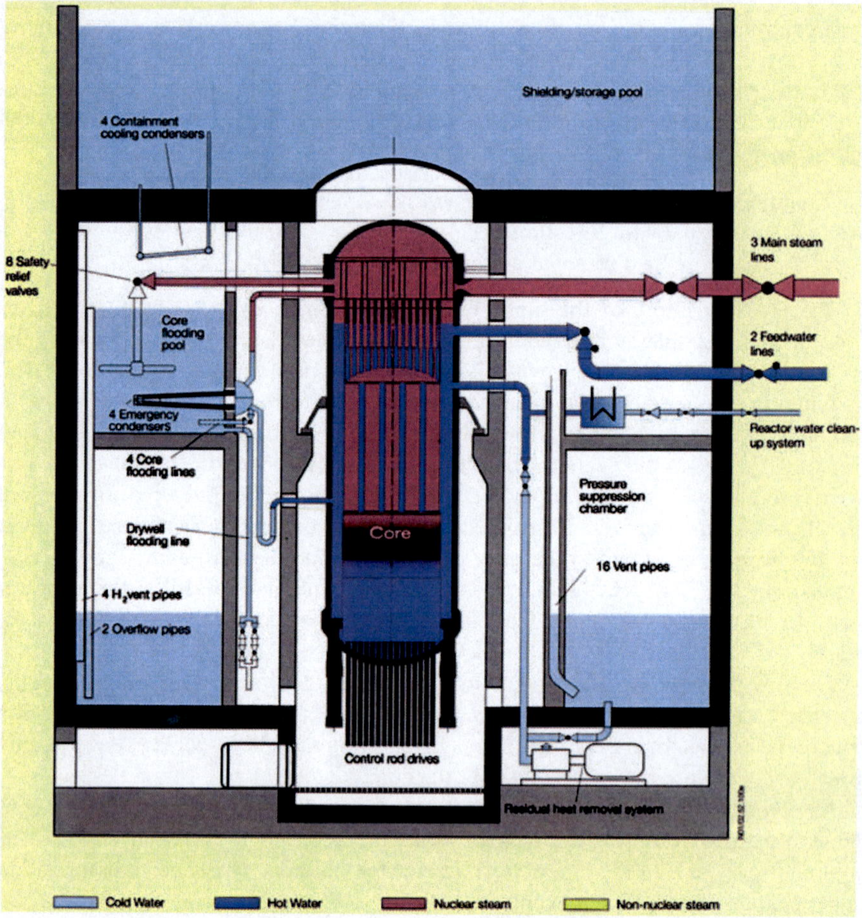

Fig. 5.19 Inner containment with safety systems of SWR-1,000 [21, 23]

accident. Above the inner containment the shielding pool and the fuel element storage pool are located. In case of refuelling the reactor pressure vessel cover is opened and the fuel elements are loaded or unloaded through the shielding water pool and transferred to the fuel element storage pool.

The main steam lines and the feed water lines are equipped with two containment isolation valves, one located inside and one outside of the inner containment wall. The residual heat removal systems are located underneath the pressure suppression chamber in non-inertized compartments which are accessible for inspection and repair from the outside.

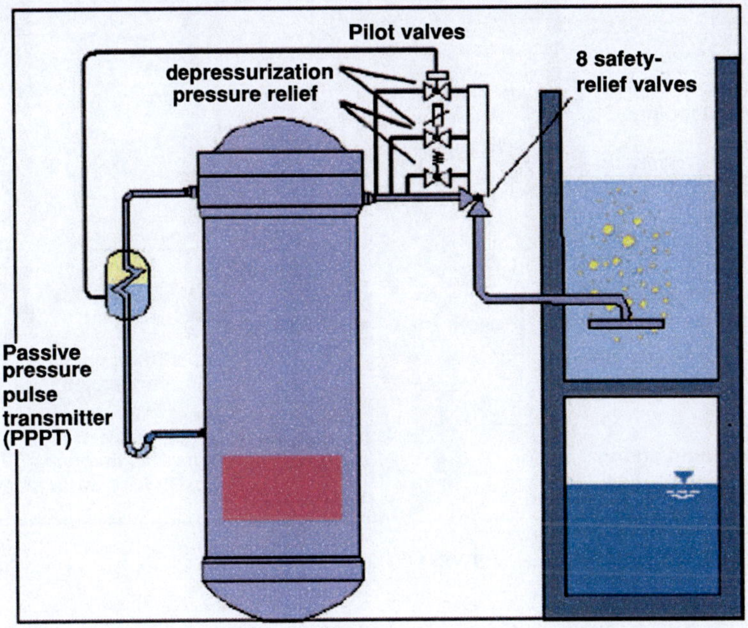

Fig. 5.20 RPV safety relief valve system of SWR-1,000 [21, 23]

Safety Relief Valve System

The safety relief valve system (Fig. 5.20) consists of the safety relief valves, the relief lines and steam quenchers which are located in the core flooding pools.

They act in case of:

- overpressure to protect the pressure boundaries (pressure relief)
- the water level in the reactor pressure vessel falling below specified limits (loss of coolant accident, automatic depressurization)
- turbine trip (short term removal of excess steam)
- severe accident mitigation (depressurization).

The eight main safety relief valves are actuated by solenoid pilot valves or by diaphragm pilot valves via passive pressure pulse transmitters. Standard spring loaded valves are also used to initiate the pressure relief function.

Emergency Condensers

Each of the four emergency condensers (Fig. 5.21) consists of a steam line (top connection), the heat exchanger tubes and a condensate return line (lower connection). Each return line is equipped with an anti-circulation loop and a passive outflow re-

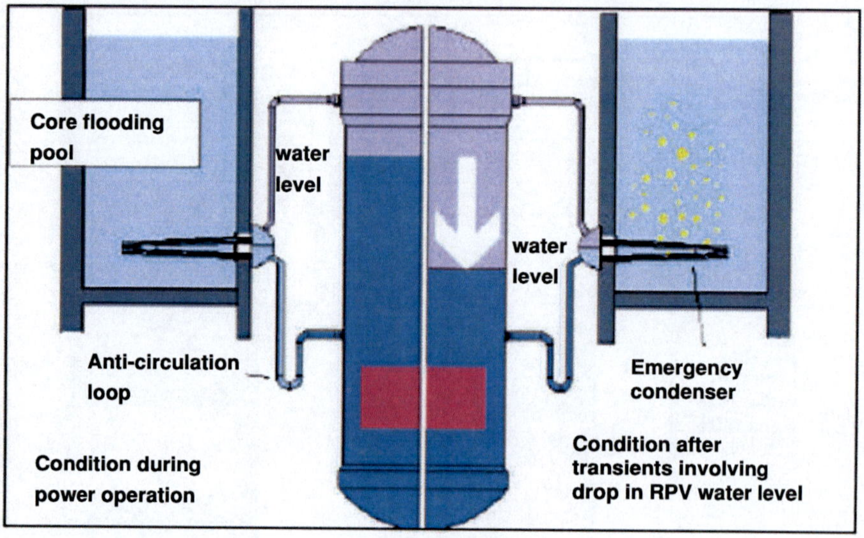

Fig. 5.21 Emergency condenser of SWR-1,000 [21, 23]

ducer. The emergency condensers are actuated passively when the water level in the reactor pressure vessel drops to a certain level. In this case the emergency condenser tubes fill with steam which condensates. The condensate returns back to the reactor pressure vessel.

Containment Cooling Condensers

The containment cooling condensers (Fig. 5.22) remove residual heat passively from the containment atmosphere to the shielding storage pool located above the inner containment. Each containment condenser consists of tubes arranged at a slight angle to horizontal. The tubes penetrate the upper wall of the inner containment. They are open to the shielding storage pool. The containment cooling condensers are passive cooling systems. In case of high temperatures of the upper containment atmosphere the water in the tubes warms up or evaporates and the heat is transferred to the shielding/storage pool.

Residual Heat Removal and Core Flooding Systems

Two active low pressure core flooding and residual heat removal systems perform the following tasks:

- reactor core cooling during operational shut down as well as under accident conditions (loss of coolant)

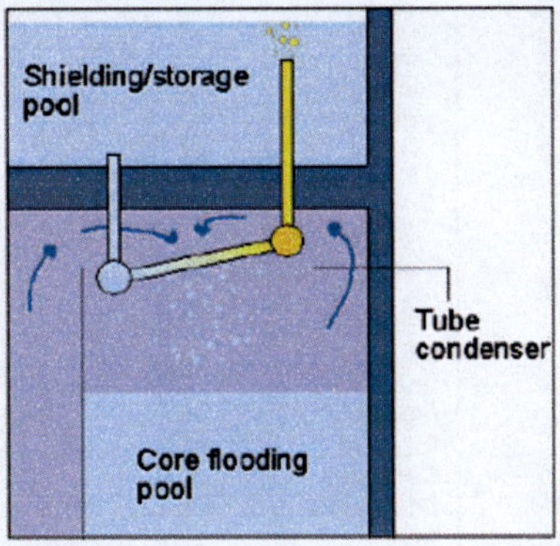

Fig. 5.22 Containment cooling condenser of SWR-1,000 [21, 23]

- heat removal from the core flooding pools and the pressure suppression water in case of pressure relief conditions or loss of the main heat sink
- water transfer operations prior and after core refuelling operations.

Systems for Control of Severe Accidents

Because of the above described passive cooling systems a high pressure water injection system as in other BWRs is no more needed for the SWR-1,000.

Severe core melt accidents could occur if all active and passive injection functions would fail. This is extremely improbable. In this case a core melt under high pressure in the reactor pressure vessel can be ruled out by the design of the depressurization system. If the reactor core would melt at low pressure it could be kept in the lower part of the pressure vessel and be cooled by water from the outside by thermal conduction through the remaining steel structures in the lower part of the reactor pressure vessel (Fig. 5.23). A special flooding system can flood water from the core flooding pools to the lower part of the drywell. A water pool would surround the lower part of the pressure vessel. Steam arising from cooling of the reactor pressure vessel would be condensed at the containment cooling condensers. The heat would be transferred to the shielding/storage pool above the inner containment (Chap. 11).

Auxiliary Cooling Systems

The water pool for spent fuel elements is cooled by four heat exchangers operating in the natural convection mode. Reactor water and spent fuel pool cleanup systems with filters operate to keep the cooling water always at specified conditions.

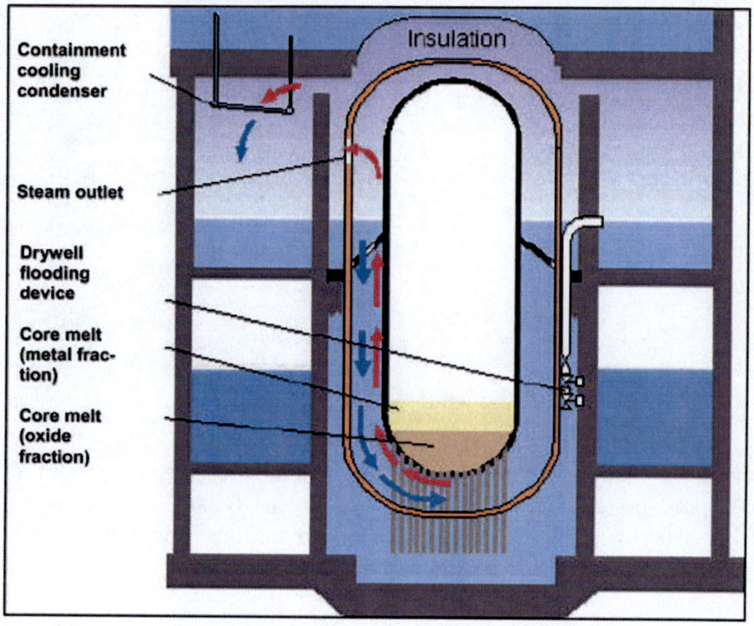

Fig. 5.23 Cooling of RPV exterior in the event of core melt of SWR-1,000 [21, 23]

Emergency Power Supply

An external grid supplies emergency power for all electrical loads which have to remain available even in the event of a loss of the normal auxiliary power supply grid. Emergency diesel generators can take over in case and provide an independent power supply.

Containment System

A reinforced concrete containment (Fig. 5.18) encloses and protects the inner containment against such external events as they were described for the PWR in Sect. 5.1.1.3 above. The annulus between the outer concrete shield building and the containment is maintained at a pressure lower than atmospheric so that any radioactive gases leaking into this annulus can be filtered prior to the release to the environment.

In case of the SWR-1,000, radioactivity releases to the outer environment would be extremely low. Even in case of severe core melt down accidents, evacuation or relocation of the population would not be necessary.

5.2 Gas Cooled Thermal Reactors

Graphite is a good neutron moderator with a relatively low absorption rate of neutrons. Thermal reactors with graphite moderators and gas (carbon dioxide or helium) as a coolant can therefore be operated on natural uranium. Because of the low concentration of fissile material, the attainable burn up of the fuel is low. For this reason, advanced gas cooled reactors use U-235 enrichment of the fuel. The lower capacity for moderating fission neutrons to thermal neutron energies results in a relatively large moderator volume, if graphite is used, which makes for a large reactor core and very low power density. From the point of view of thermal reactor physics this is obvious when comparing the so-called mean square distance (Fermi age) a neutron migrates in slowing down from fission energy to thermal energy of e.g. 1.46 eV for graphite as moderator. If compared to light water, this characteristic is by about a factor of 25 higher. For heavy water this factor is about 2.5.

Gas cooled and graphite moderated natural uranium commercial reactors were developed in the United Kingdom and in France and built in the 1950s and 1960s (MAGNOX reactors). The world's first nuclear power plant, Calder Hall (4×40 MW(e), commissioned in 1956) belonged to this category of reactors. MAGNOX reactors are no longer built.

In Britain, further technical development then led to the advanced gas-cooled reactor (AGR). In the USA, Japan and in Germany, the high temperature gas cooled reactor line (HTGR and HTR) has been developed. These advanced gas cooled reactors are attractive, above all, because of their high gas outlet temperatures and the resultant high thermal efficiencies. The high gas outlet temperatures, moreover, allow such plants to be used as sources of industrial high temperature process heat (see Sect. 1.3.1).

5.2.1 Advanced Gas Cooled Reactors

The AGR line has so far been built in unit sizes up to 620 MW(e). It also uses graphite as a moderator and carbon dioxide (CO_2) as a coolant gas. The primary cooling system operates at a gas pressure of 4.2 MPa and coolant gas outlet temperatures of 645°C. This allows steam temperatures to be reached in the secondary cooling system of 540°C at a steam pressure of 16 MPa. The overall plant efficiency of 41% (with sea water cooling) is correspondingly high. The whole primary cooling system, i.e., the reactor core, the gas circulators and the steam generators, are arranged in a prestressed concrete reactor pressure vessel. Gas circulators ensure circulation of the coolant gas between the core and steam generators. The fuel element consists of 36 rods with UO_2 pellets enclosed in steel claddings. The UO_2 pellets contain slightly enriched uranium fuel (2% U-235). The fuel elements are arranged in vertical cooling channels in a graphite structure acting as a moderator. On-load fuelling is possible. The main data of an AGR with 620 MW(e) power are listed in Table 5.3.

5.2.2 High Temperature Gas Cooled Reactors

High temperature gas cooled reactors (HTGRs) also use graphite as a moderator, but helium as a coolant. Helium, being an inert gas, will not react with graphite at high temperatures and, consequently, allows even higher coolant outlet temperatures to be reached than in the AGR line. Two high temperature gas cooled reactors will be briefly described below, which mainly differ in the shapes of their fuel elements and, accordingly, in the arrangements of their reactor cores, namely the high temperature gas cooled reactor (HTGR) with prismatic fuel elements developed in the USA and pursued in France and Japan and the high temperature reactor (HTR) with spherical fuel elements developed in the Federal Republic of Germany and now pursued in China.

So-called very high temperature reactors (VHTRs) are designed to attain gas outlet temperatures of 900°C or more.

5.2.2.1 HTGR With Prismatic Fuel Elements

The Fort St. Vrain HTGR prototype reactor was built in the USA by General Atomic in a unit size of 300 MW(e) after successful operation of the smaller prototype reactors, Peach Bottom in the USA and Dragon in the UK. Larger plants of 1,160 MW(e) were planned in the early 1970s, but were not built. Table 5.3 includes the main design characteristics of such an 1,160 MW(e) HTGR power plant [24–26].

The HTGR was designed to operate in the uranium/thorium fuel cycle. The fuel consists of enriched uranium particles (fissile) and thorium particles (fertile) with ceramic coatings (Fig. 5.24). The fissile particles have diameters of 200–800 μm and contain UO_2 either enriched U-235 or recycled U-233. These fuel elements, due to their high enrichment and coated fuel particles, can attain a relatively high burn-up of 100,000 MWd/t. But reprocessing of such fuel elements, to make use of the U-233, would be very difficult.

The fuel particles are coated with pyrolytic carbon and silicon carbide layers of 150–200 μm thickness. The fertile particles contain Th-232 as ThO_2 and are coated with pyrolytic carbon only. The particles are dispersed in a graphite matrix to form a fuel rod. Figure 5.24 shows the fuel elements of an HTGR core. The fuel rods are again incorporated in a hexagonal graphite block to form a hexagonal fuel element. These hexagonal fuel elements are arranged in groups of seven elements to make up the core block. The core block is composed of several hundred hexagonal graphite elements, each consisting of three bottom reflector graphite blocks, eight fuel element blocks, and three upper reflector graphite blocks. The helium coolant gas flows downward through vertical holes in the hexagonal fuel elements.

HTGRs were studied and designed up to an electrical output power of 1,160 MW(e) in the 1980s (Table 5.3). Later, design tendencies shifted to small modular type HTGRs.

Table 5.3 Design characteristics of large gas cooled thermal power reactor plants (UKAEA, General Atomic, AREVA)

	AGR (Hinkley Point B)	HTGR	HTR modular reactor
Reactor power			
Thermal (MW(th))	1,493	3,000	200
Electrical net (MW(e))	620	1,160	–
Plant efficiency (net) (%)	41.6	38.6	40
Reactor core			
Equivalent core diameter m	9.1	8.4	3
Active core height m	8.3	6.3	9.6
Specific core power kW(th)/l	2.76	8.6	3
Power density kW(th)/kg	13.1	76.5	83.5
Number of fuel elements	308	3,944	360,000
Total amount of fuel kg	114,000 U	1,725 U + 37,500 Th	2,396
Fuel element	UO_2 fuel (2.0–2.55% enriched) hollow pellets 5.1 mm i.d. 14.5 mm o.d. Cladding: stainless steel	Th/U-235 (93% enriched) coated particles in cylindrical rods, 15.6 mm diameter	U oxide kernels (8% enriched), 0.5 mm diameter, coated with pyrolytic carbon in a spherical graphite element of 6 cm diameter
Control Absorber rod	44 control rods (boron inserts in stain-less steel claddings) 21 override regulating rods	73 pairs, hollow cylindrical B_4C graphite elements	6 Control rods operating in holes in the reflector around the core 18 absorber ball systems
Heat transfer system			
Primary system Coolant	CO_2	He	He
Total coolant flow rate (core flow) (t/h)	13,250	5,080	274
Coolant pressure (MPa)	4.2	5.1	6
Coolant inlet temp. (°C)	292	316	250
Coolant outlet temp. (°C)	645	740	700
Steam supply system			
Steam generation (t/h)	2,200	3,900	288
Steam pressure (MPa)	16	16.9	19
Steam temperature (°C)	540	510	530
Fuel cycle			
Average fuel burnup (MWd/t)	18,000	100,000	80,000
Refuelling sequence	On-load, continuously 3 channels per week	Off-load	On-load, multipass

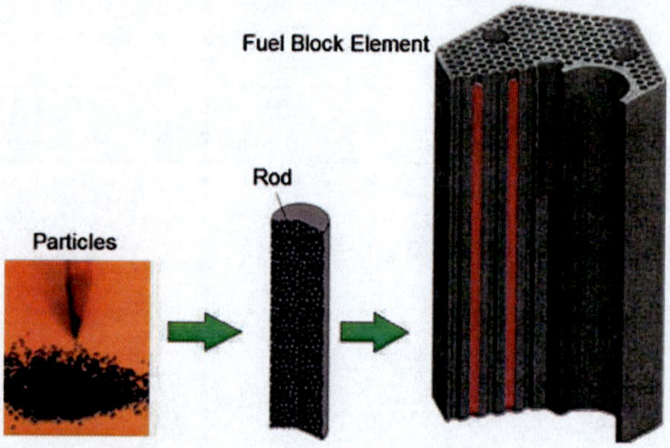

Fig. 5.24 Fuel particles, fuel rod and fuel element of an HTGR [25]

The reactor core of an 1,160 MW(e) HTGR plant would have a diameter of 8.4 m and a height of 6.3 m. The core power density is 8.6 kW(th)/l, which is considerably lower than in LWRs. The helium transmits its heat to the secondary system while flowing upward in the steam generators. The core, the steam generators and the circulators of the primary cooling system are contained in cavities of a prestressed concrete reactor vessel (PCRV). These cavities are steel lined for sealing the high pressure coolant system. A thermal barrier protects the prestressed concrete from the high temperature. The helium coolant gas is at a pressure of 5.1 MPa and leaves the core at a temperature of about 740°C. This leads to an overall efficiency of the plant of about 39%.

In Japan, the High Temperature Engineering Test Reactor (HTTR) [26] with a thermal power output of 30 MW(th) was constructed. It achieved a helium outlet temperature of 950°C in 2004. It is equipped with prismatic block type fuel elements and ceramic heat exchangers aiming at the demonstration of process heat application for hydrogen production.

5.2.2.2 HTR with Spherical Fuel Elements

A small experimental test reactor (AVR, 6 MW(e)) was taken into operation in Germany in 1966. This reactor was successfully operated over 20 years with helium gas outlet temperatures of 950°C. The thorium high temperature reactor, THTR 300 with 300 MW(e) power was taken into operation in Germany in 1986. This prototype test reactor was shut down in 1989 [27–29].

A similar prototype test reactor, HTR-10, was taken into operation in China around 2003 [6]. Between 2000 and 2010, small HTR module reactors were proposed for construction in Germany, South Africa and in France [30].

The HTR reactor core consists of several 100,000 spherical fuel elements of 6 cm diameter each (Fig. 5.25). These spherical fuel elements contain the fuel as UO_2 and ThO_2 particles which, as in the HTGR fuel, are coated with pyrolytic carbon

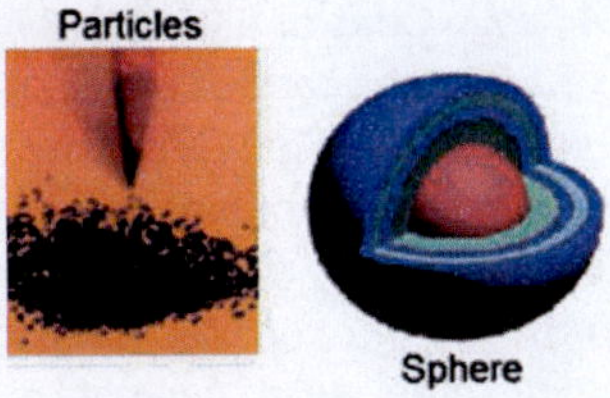

Fig. 5.25 Fuel particles and spherical fuel element of an HTR [31, 32]

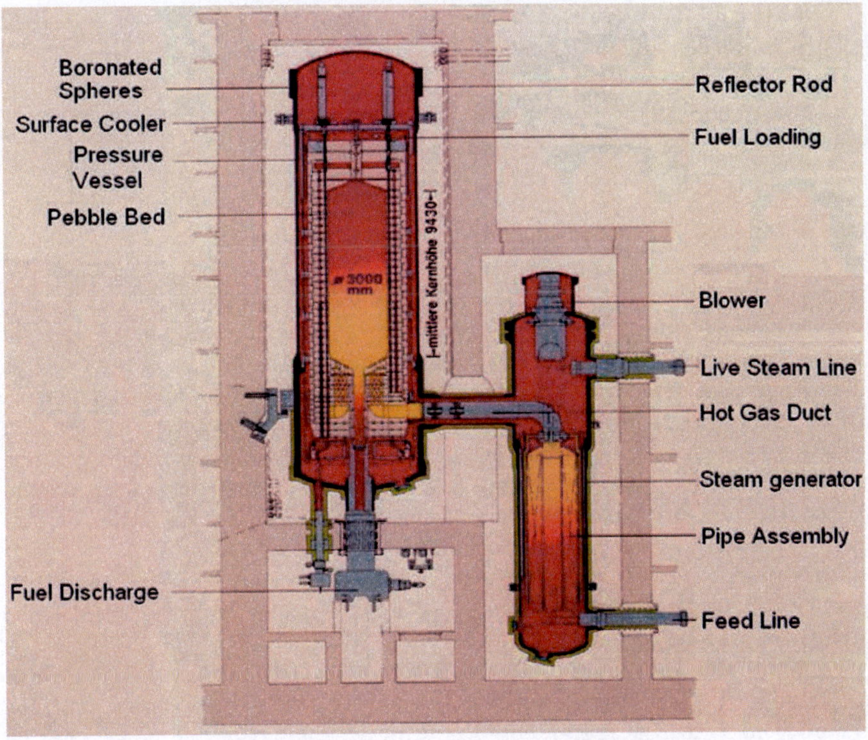

Fig. 5.26 Scheme of modular HTR [31, 32]

and a SiC layer [33]. The pebble bed is enclosed in a graphite structure (Fig. 5.26). The bottom reflector is conical, terminating in a discharge tube for the spheres. The fuel elements are loaded continuously during operation into the core through refuelling tubes located above the core. After passing through the core, they are removed continuously during operation. The spherical fuel elements are passed through the reactor core from top to bottom about six or seven times until they have reached their maximum burn up. The average core power density is around 3 kW(th)/l. The whole primary system with the coolant circulators and steam generators is integrated in a prestressed concrete or steel reactor vessel in case of THTR-300. The modular type

HTRs have steel vessels [31, 32]. Control and absorber rods are moved vertically within the graphite reflector for reactor control and shutdown. The coolant pressure in the primary system is about 6 MPa. The core structure is surrounded by steam generators with coolant circulators. The coolant passes through the pebble bed in a downward flow and is heated to a core outlet temperature of 700°C. It then passes through holes in the bottom reflector into the plenum below the core and is circulated from there through hot gas ducts to the steam generators. In the steam generators the heat is transferred to a secondary steam-water system with a turbo-generator unit. The overall efficiency of the plant is 40%.

Some small HTR reactors designs, e.g. the South Africa PBMR design use the hot helium gas directly for driving a gas turbine in the Brayton cycle [34].

In France, the modular high temperature gas cooled reactor design, called ANTARES, aims at an indirect cycle to provide a flexible heat source either for heat supply or cogeneration [30].

5.2.2.3 General Safety Considerations of HTGRs and HTRs

Since the cores of HTGRs and HTRs are basically made up of graphite and ceramic fuel, the melting and sublimation points of the fuel elements are very high. The solid core (power density of 3–8 kW(th)/l) has a relatively high heat capacity, which is important in case of failure of the core heat removal system. In normal operation the reactor shows high stability and very good self regulating properties. The high heat capacity of the graphite moderator and the comparatively low power density retard and mitigate all temperature transients. Even considerable increases in temperature do not result in abrupt or irreversible changes of the physical properties of core components, such as melting or evaporation [31, 32, 35].

The helium coolant is chemically inert. Because of possible chemical reactions of graphite with water and air after defects in a steam generator or air ingress, detailed analyses of graphite corrosion is necessary.

For small modular type HTGR and HTR systems about 200–300 MWth, the loss of coolant following a pipe break can be counteracted by the self-regulating properties of the reactor core. The reactor core reacts by self-shut-down and the heat can be dissipated by thermal conduction and radiation. This was demonstrated by the AVR and by the HTR-10 reactor [30–32, 35].

Control and Shutdown Systems

HTGRs with cylindrical fuel elements use rod type absorbers. The central fuel element of each group of seven fuel elements has two adjacent absorber rod channels. In these two channels a pair of absorber rods driven by a common drive can be inserted. An 1,160 MW(e) HTGR plant would have 73 such rod pairs. There is a third channel adjacent to the two absorber rod guide channels. This third channel enables small boron carbide granules, which are stored in hoppers between the control rod drives and the upper thermal shield, to be introduced in the core. These hoppers can be pressurized to break their rupture disks, thus allowing the absorber balls to fall through a guide tube into their respective channels in the core. This absorber system acts as a standby or secondary shutdown system.

In the HTR pebble bed core, scram and hot shutdown is achieved by a number of absorber rods. These absorber rods are inserted in bores in the radial reflector. Reactor control and secondary shutdown is provided by a bank of rods freely inserted under pressure into the pebble bed. For large pebble bed reactors rotating rods and helical absorbers are proposed as an alternative design. In addition, small neutron poison granules penetrating into the gaps between the pebbles are suggested as a standby shutdown system [31, 32].

Afterheat Removal and Emergency Cooling

Under normal operating conditions, the afterheat of the reactor is accommodated by the primary systems with steam generators. Auxiliary cooling loops are designed in large scale HTGRs or HTRs to remove the afterheat in case of failure of the main coolant loops.

Design Base Accidents

In the safety analyses of high temperature gas cooled reactor plants a number of accidents have to be considered, e.g., uncontrolled control rod withdrawal, steam/water leaks into the primary system, loss of forced circulation of the helium coolant, and primary system depressurization into the reactor building, with a potential for air ingress [36].

To limit graphite corrosion in case of a steam generator tube rupture, the amount of water entering the core must be limited and the reactor core temperature must be decreased to a level below 700°C. After steam or water has been detected in the coolant system, the defective loop is isolated and the reactor is shut down. Flow restrictors at penetrations through the reactor vessel reduce the helium loss, if the vessel integrity is violated. In case of a leak in the penetration of the pressure vessel the reactor must be shut down and long term cooling of the afterheat of the reactor core must be assured.

The purpose of the outer containment is to protect the environment against severe internal accidents with radioactivity releases and to protect the reactor against external events (earthquakes, airplane crashes, gas cloud explosions).

5.3 Heavy Water Reactors

Heavy water (D_2O) is an excellent moderator, absorbing far fewer neutrons than light water (H_2O) and less than pure graphite. Reactors using heavy water as a moderator and coolant can therefore be run on natural uranium fuel. However, this requires a ratio of the D_2O/UO_2 volumes of almost 20, which means relatively wide fuel rod lattices or relatively low power densities. Variants of the heavy water reactor (HWR) design also use light water or CO_2 as coolants besides heavy water as a moderator. In that case, slight enrichment of the fuel is required [37–41].

Power reactors with heavy water as the moderator have been developed in Canada, Europe and Japan. The steam generating heavy water reactor (SGHWR) developed in the United Kingdom uses light water in pressure tubes surrounded by the D_2O moderator. The coolant can boil in these pressure tubes.

In the Atucha heavy water reactor, built by Siemens and operated in Argentina, on the basis of the prototype MZFR built and operated at Forschungszentrum Karlsruhe, Germany, the heavy water coolant in the cooling channels and the surrounding heavy water moderator are at the same pressure level. The D_2O coolant under high pressure will not reach the boiling point. In France, the CO_2 cooled, D_2O moderated pressure tube reactor was developed.

In this Section, first the 630 MW(e) standard CANDU reactor type developed in Canada will be explained as an example. These CANDU reactors are being redesigned and developed up to a power level of 1,000 MW(e).

Table 5.4 shows some characteristic design data of a 600 MW(e) CANDU nuclear power plant.

5.3.1 CANDU Pressurized Heavy Water Reactor

In the CANDU PHWR (Canada deuterium uranium pressurized heavy water reactor) D_2O is used as the coolant and moderator (Fig. 5.27). So far, this line has only been built in unit sizes up to 750 MW(e) [37–39]. Design studies of 1,000 MW(e) have been completed. In the primary system, the heavy water coolant flows through individual pressure tubes of the reactor core, thus cooling the fuel bundles. The coolant is kept at a pressure of roughly 10 MPa and at temperatures of 267°C at the inlet and 310°C at the outlet of the cooling channel. For thermal insulation, the pressure tubes are surrounded by another gas filled annular gap. Some 380 of these tubes are arranged horizontally in the reactor vessel filled with D_2O, which is kept at a pressure close to atmospheric (Fig. 5.28). Each pressure tube is connected via distribution headers to the pumps and steam generators of the primary system. A pressurizer in the primary system designed similar to the device used in PWRs maintains the coolant pressure. The heavy water moderator in the calandria has its own cooling system and is kept at a temperature of approximately 70°C and a pressure close to atmospheric. The calandria in a 680 MW(e) plant has a diameter of 7.6 m and an inside (core) length of about 6 m. Because of the low internal pressure, its wall thickness is only 29 mm. The reactor power is removed by two cooling circuits with pumps and steam generators. In the secondary system, light water serves as the coolant to generate steam to drive a turbogenerator system. Due to the low temperature level the overall thermal efficiency of the CANDU reactor line is 29.4%. Figure 5.28 shows an overall

Table 5.4 Design characteristics of large HWR CANDU power plants (AECL) [37–44]

Reactor power	
Thermal (MW(th))	2,156
Electrical gross (MW(e))	680
Electrical net (MW(e))	633
Plant efficiency (net)	29.4
Reactor core	
Equivalent core diameter (m)	6.28
Active core height (m)	5.94
Specific core power (kW(th)/l)	11
Density (kW(th)/kg U)	24
Number of fuel elements	380
Total amount of fuel (kg U)	86,000
Fuel element and control rod	
Fuel	UO_2
U-235 fuel enrichment (%)	
Initial	0.72 (natural)
Reloadings	0.72 (natural)
Cladding material	Zircaloy
Cladding outer diameter (mm)	13.1
Cladding thickness (mm)	0.38
Fuel channel spacing (cm)	28.6
Nominal fuel element power	
Outer ring (W/cm)	508
Intermediate ring (W/cm)	417
Inner ring (W/cm)	365
Reactivity devices	
Control system	
Light water compartments	14
Cd absorber rods	4
Stainless steel adjuster rods	21
Safety systems	
Cd shutoff units	28
Gd injection nozzles	6
Heat transfer system	
Primary system	
Total coolant flow rate (core flow) (t/s)	7.6
Coolant pressure (MPa)	10
Coolant inlet temperature (°C)	267
Coolant outlet temperature (°C)	310
Steam supply system	
Steam generation (t/s)	1.05
Steam pressure (MPa)	4.7
Steam temperature (°C)	260

(continued)

Table 5.4 (continued)

Fuel cycle	
Average fuel burn up (MWd/t)	7,000
Refuelling sequence	Continuously, on-load
Average fissile fraction in spent fuel	
U-235 (wt%)	0.2
Pu-fiss. (wt%)	0.3

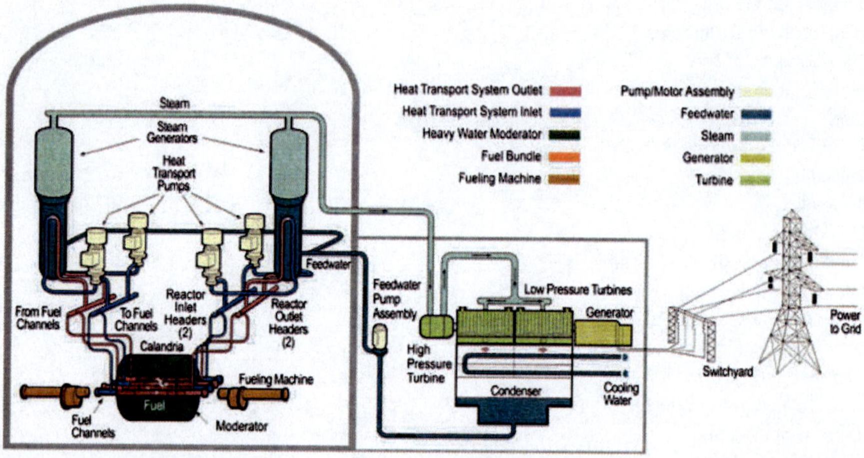

Fig. 5.27 Functional design diagram of a CANDU reactor [37]

view of the calandria and the primary systems. The reactor building is a prestressed concrete structure [40–42].

5.3.1.1 Fuel Elements

The CANDU reactor uses natural uranium (0.72% U-235) as a fuel in UO_2 pellets welded into Zircaloy tubes of 13.1 mm outer diameter. A fuel bundle contains 37 of these fuel elements combined into a cylindrical cluster (Fig. 5.29). Twelve of these fuel bundles, about 49 cm long, are loaded in the pressure tube in series.

The reactor core contains 4,560 fuel bundles and a total of 96 t of UO_2 fuel. It has a diameter 6.3 m and a length of 5.9 m, attaining a power density of 11 kW(th)/l. In CANDU reactors, the fuel bundles can be exchanged during operation without requiring the reactor to be shut down. During about 200 full load days, the fuel attains a burn up of 7,000 MWd/t, still containing some 0.2% U-235 and 0.3% plutonium when unloaded.

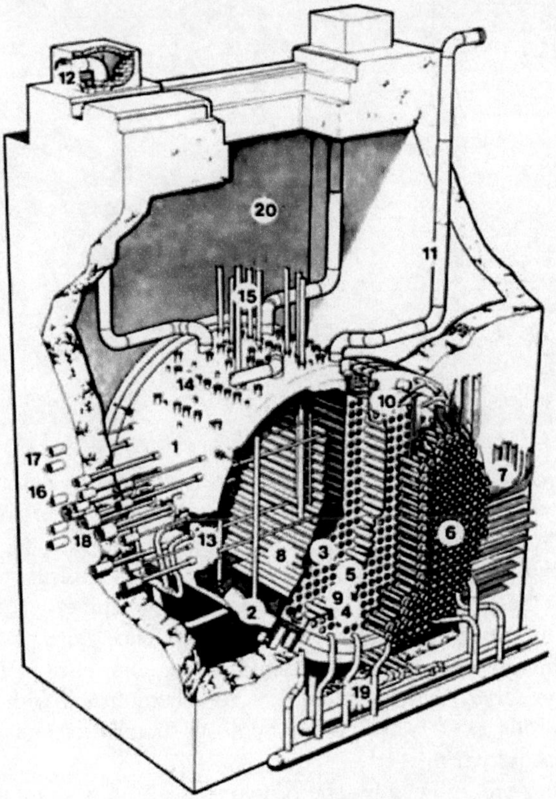

1 CALANDRIA
2 CALANDRIA SHELL
3 CALANDRIA SIDE TUBE SHEET
4 FUELING MACHINE SIDE TUBE SHEET
5 LATTICE TUBES
6 END FITTINGS
7 FEEDERS
8 CALANDRIA TUBES
9 STEEL BALL SHIELDING
10 ANNULAR SHIELDING SLAB
11 PRESSURE RELIEF PIPES
12 RUPTURE DISC
13 MODERATOR INLETS (4 EACH SIDE)
14 REACTIVITY CONTROL NOZZLES
15 REACTIVITY CONTROL DEVICES
16 HORIZONTAL FLUX DETECTORS (9)
17 POISON INJECTOR NOZZLES (6)
18 ION CHAMBER HOUSING (3 EACH SIDE)
19 END SHIELD COOLING PIPING
20 CALANDRIA VAULT (LIGHT WATER SHIELD)

Fig. 5.28 Cutaway view of a 600 MW(e) CANDU reactor core [37]

Refuelling is done by two refuelling machines. Fuel bundles are pushed into the reactor channel by a remotely operated fuelling machine. Spent fuel bundles are discharged simultaneously into another fuelling machine at the opposite end of the reactor. The fuel is then transferred to a water filled storage bay.

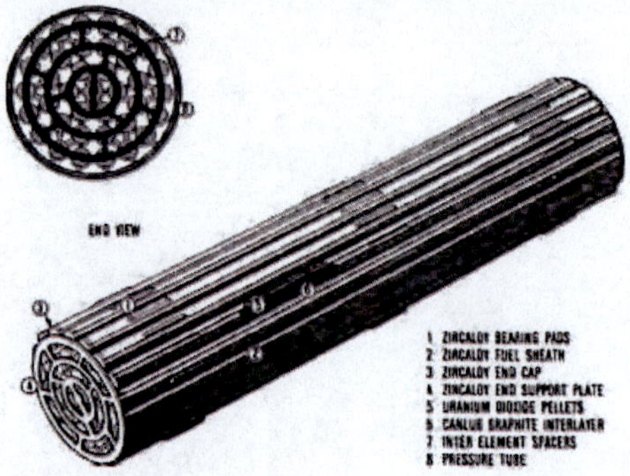

Fig. 5.29 37-element fuel bundle of a 600 MW(e) CANDU reactor [37]

5.3.1.2 Reactivity Control

Long term control of burn up and fission products is achieved by on-power refuelling. Minor reactivity variations are compensated by a liquid zone control system. This consists of 14 compartments of variable amounts of light water (acting as neutron absorbers). Bulk reactivity control is achieved by variation of the average amount of light water in these compartments, whereas spatial (stability) control is achieved by differential filling or draining of these compartments. Core power distribution flattening is achieved with a set of 21 stainless steel adjuster rods. Fast, controlled (as distinct from scram) power reductions are accomplished with four cadmium loaded absorber rods. Dissolved moderator poisons (both boron and gadolinium) are also used for reactivity trim.

Reactor scrams are initiated by one of two special safety shutdown systems, as discussed in Sect. 5.3.1.4.

5.3.1.3 Shutdown Cooling System

A shutdown cooling system is provided in the CANDU PHWR to cool the fuel and the primary heat transport system after a reactor shutdown. The system consists essentially of a pump and a heat exchanger at each end of the reactor, connected between the inlet and outlet headers of both heat transport circuits.

5.3.1.4 Safety Systems

CANDU PHWRs have four special safety systems:

- shutdown system 1,
- shutdown system 2,
- emergency core cooling system,
- containment system.

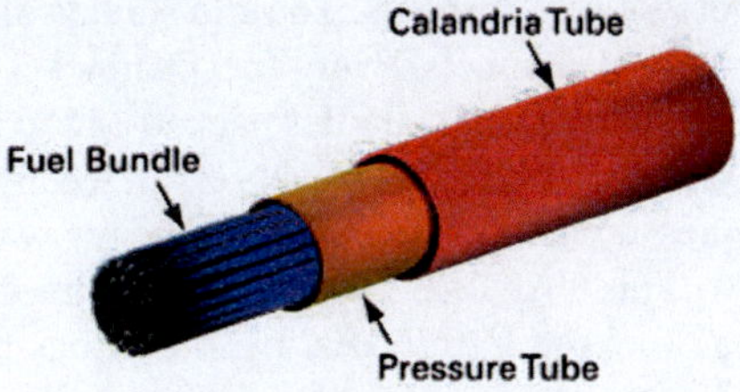

Fig. 5.30 CANFLEX fuel element with pressure tube and calandria tube [37]

These systems are independent of each other and of any of the process or regulating systems (Fig. 5.28).

Shutdown system 1 makes use of 28 cadmium loaded shutoff units. These enter the core vertically.

Shutdown system 2 injects a liquid neutron poison (gadolinium nitrate) into the moderator. This is done through the use of six horizontally placed, perforated injection nozzles, each connected to its own poison tank.

The emergency core cooling system would inject cooling water into the core after a loss-of-coolant accident. The system has a high pressure, intermediate pressure, and low pressure stage. During the high pressure stage water is injected from a high pressure, external tank. The intermediate stage takes water from a tank located under the dome of the reactor building. The low pressure stage recirculates water collected in the reactor building sump. In the CANDU PHWR design, failure of the emergency core cooling system can be tolerated because in such cases the heavy water moderator, which is independent of the primary coolant becomes the heat sink. Subsequent failure of this moderator heat sink, however, would lead to core melt down.

The containment system in single-unit stations consists of a concrete structure with a plastic liner. Reactor building air coolers and a dousing tank act as energy suppression systems during loss of coolant accidents. Multi-unit stations use a common vacuum building to contain potential radioactivity releases. In these stations each unit reactor building is connected via a duct to this vacuum building.

5.3.1.5 Advanced CANDU Reactors

Based on the experience with CANDU PHWRs an advanced CANDU reactor ACR-700 with 700 MW(e) power output was developed in Canada [43]. This new design shall be extended to 1,000 MW(e). The ACR-700 differs from the CANDU PHWR by restricting the use of heavy water to its function as a low pressure moderator in the reactor core. Low enriched uranium fuel (about 2% U-235 enrichment) is used as fuel. Light water as primary coolant allows the core to be designed more compact. This reduces the total required heavy water inventory by a factor 4 [43].

The ACR-700 reactor core is also housed in a horizontal, cylindrical tank (calandria). The calandria with 5.2 m diameter is filled with heavy water. The core consists of 284 fuel channels. The end shields of the cylindrical calandria support the ends of the fuel channels. Every fuel channel contains a pressure tube holding 12 cylindrical CANFLEX fuel bundles packed end to end (Fig. 5.30). The fuel channels isolate the pressure tubes and their pressurized light water coolant from the surrounding low pressure and low temperature heavy water moderator.

The CANFLEX fuel bundle with 43 fuel rods has evolved from the standard 37 fuel rod CANFLEX bundle. It has the same external dimension as the standard 37 fuel rod bundles with 10.3 cm diameter and 49.5 cm length. The foreseen burn up is 20,000 MWd/t which is about three times higher as in the CANDU PHWR. The smaller spacing of the ACR-700 CANFLEX fuel bundle guarantees a negative coolant void reactivity coefficient (CANDU PHWRs, have a positive coolant reactivity coefficient) [43, 44].

On-power refuelling of the ACR-700 as in all other CANDU PHWRs is done remotely by two refuelling machines acting in tandem on both sides of the reactor core or calandria. Fresh fuel bundles are inserted into the inlet end of the channel while spent fuel bundles are removed at the opposite outlet end.

The operating temperatures and coolant pressures in the coolant system of the ACR-700 are increased in comparison to the CANDU PHWR. The core outlet temperature is 325°C which leads to a steam pressure of 6.5 MPa. This increases the thermal efficiency.

5.3.1.6 Control and Shut Down Systems

Due to remote refuelling at full power the CANDU-PHWRs and the ACR-700 do not need reactivity control mechanisms for balancing reactivity changes during burn up of the fuel elements.

Nine vertical control assemblies are arranged in three symmetrical rows of three assemblies each. They are capable to provide control for the Xenon buildup after shutdown. They also can reduce the reactor power from 100 to 75%. Four additional vertical control assemblies can be dropped into the core by gravity for fast power reductions.

A first shut down system consisting of 20 vertical mechanical absorber rods can shut off the reactor by dropping into the core. A second shut down system consists of eight liquid poison injection tubes. Gadolinium nitrate solution is injected in the tubes to shut down the reactor in all possible accident situations.

5.3.1.7 Emergency Core Cooling System

An emergency core cooling system injects high pressure light water into the reactor cooling system in the event of failure of the coolant piping system.

5.3.1.8 Containment System

The containment system includes a steel lined prestressed concrete reactor building as in case of the CANDU PHWR.

5.4 Near Breeder and Thermal Breeder Reactors

In a thermal neutron spectrum, the neutron yield, η, of U-233 is considerably higher than that of Pu-239 or U-235. This value of η for U-233 opens up a potential of high conversion ratios or even a breeding ratio around 1.0 (see Sect. 3.5). Reactors able to attain breeding ratios of 1–1.03 with U-233/Th fuel and a thermal neutron spectrum are called "thermal breeders". Converter reactors attaining conversion ratios between 0.94 and 1.0 are sometimes called "near breeders". In such reactor cores, parasitic neutron absorption and neutron leakage losses must be kept extremely low. Fission product poisoning and removal of Pa-233 with a high neutron absorption cross section and a half-life of 27.4 days are then two of the key problems. Enhanced fission product removal necessitates either frequent refuelling or the introduction of the homogeneous reactor concept with continuous fission product removal. Pa-233 poisoning can be avoided by separating the fissile U-233 from the fertile Th-232 fuel containing Pa-233 in two separate circuits.

5.4.1 Homogeneous Core Thermal Breeders

A reactor concept of the homogeneous core type is the molten salt breeder reactor (MSBR), a small test reactor developed, built and operated for some time at Oak Ridge, USA, in the 1960s [45–48].

The MSBR (Fig. 5.31) is fueled with a homogeneous fluid salt containing both the fissile uranium and the fertile thorium fuels. The fuel carrier salt is a mixture of the fluorides of lithium, beryllium and thorium (LiF-BeF_2-ThF_4). The fissile material is contained as UF_4. Either U-235 and U-233 or plutonium can be used. The fuel carrier salt is pumped through a core structure of bare graphite. Pa-233 and some of the neutron absorbing fission products are removed continuously by a purification system and on-site reprocessing.

The heat of the core fuel salt is transferred in a heat exchanger to a secondary coolant (molten salt), which then passes through steam generators to generate steam. All metal surfaces contacting fuel salt are made of Hastelloy-N, a nickel based alloy.

5.4.2 Light Water Breeder Reactors (LWBRs)

In principle, thermal converter reactors can attain high conversion ratios 0.9 in the Th/U-233 cycle (see Sect. 3.6). Further improvement in the neutron economy of the

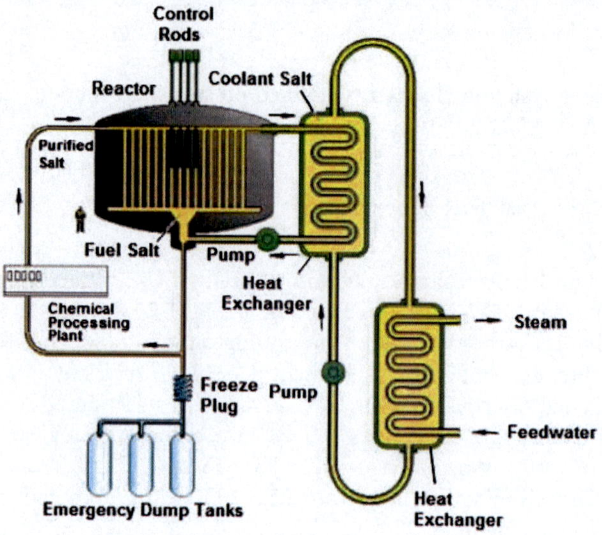

Fig. 5.31 Molten salt near breeder reactor [48]

core and short time refuelling can raise the conversion ratio even further. However, short time refuelling adds to the fuel cycle costs, because the fuel remains in the core for power production only for short periods of time and must pass through reprocessing and re-fabrication more often [49].

One particularly interesting variant of a high converting thermal reactor is the so-called seed- and blanket concept in a light water reactor. In the seed and blanket reactor concept, the core is subdivided into a number of modules each containing a seed and a blanket region. The reactor core is water cooled. Seed and blanket have Th/U-233 fuel of different enrichments and different neutron spectra. The seed region contains mainly fissile material, whereas the blanket region is made mainly of fertile material. Because of the concentration of reactivity in the seed region, the reactor can be controlled by axial movement of the seed regions. Axial movement of the seed changes the leakage of neutrons from the fissile region (seed) to the fertile region (blanket). Moreover, the fuel and moderator volume fractions are chosen in such a way that a thermal spectrum dominates in the seed zone. Consequently, the advantages of the high thermal η-value of U-233 can be utilized. The blanket region contains less moderator. As a consequence, the neutron spectrum has a high average neutron energy, which increases the fast fission effect in Th-232 and the fraction of (n, 2n)-reactions. All these measures support the breeding effect. It was claimed that this reactor could attain a conversion/breeding ratio close to 1 [48, 50].

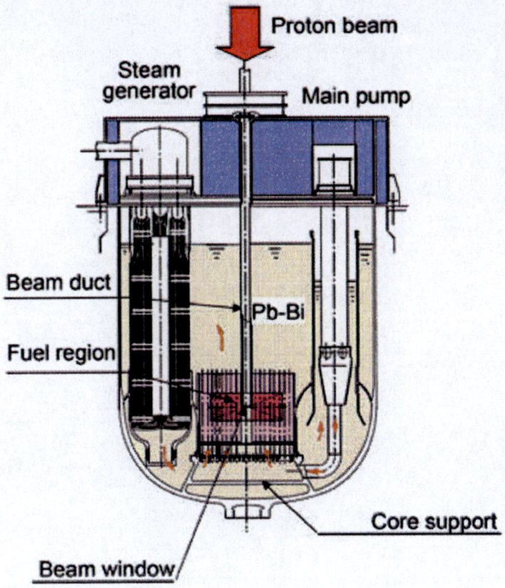

Fig. 5.32 Simplified design scheme of an accelerator driven system (ADS) [53]

5.5 Accelerator Driven Systems

5.5.1 Spallation Process, Breeding and Transmutation

ADSs are operating on the basis of the spallation process in which high energy protons with energies of about 1 GeV collide with a heavy nucleus, e.g. lead, bismuth, tungsten or uranium. During spallation, intranuclear cascade and evaporation processes occur which produce spallation fragments and a high number of neutrons with relatively high kinetic energies.

Calculations with the code LAHET [51] show that a proton beam with an energy of 1 GeV interacting with a lead target of about 10 cm radius can produce about 22 neutrons per proton from the intranuclear cascade process and from the subsequent evaporation process of the excited compound nucleus [52]. The kinetic energy of the neutrons ranges between about 0.1–100 MeV. In addition to the neutrons, a large number of different spallation products is produced. They remain in the lead target. Figure 5.32 shows the simplified design scheme of an accelerator driven system based on the spallation process.

The spallation process was originally proposed in the 1950s by E.O. Lawrence (USA) and W.B. Lewis (Canada) for breeding of Pu-239 or U-233 from U-238 or Th-232, respectively. However, these early projects were abandoned for economical reasons. They were proposed again during the 1980–1990s by Takahashi (BNL) [54], Bowman (LASL) [55] and Rubbia (CERN) [56] for breeding, but also for transmutation and incineration of minor actinides (neptunium, americium) as well as long-lived fission products. The new idea was to drive a subcritical fission reactor core by a central spallation neutron source. In this way the neutron multiplication of

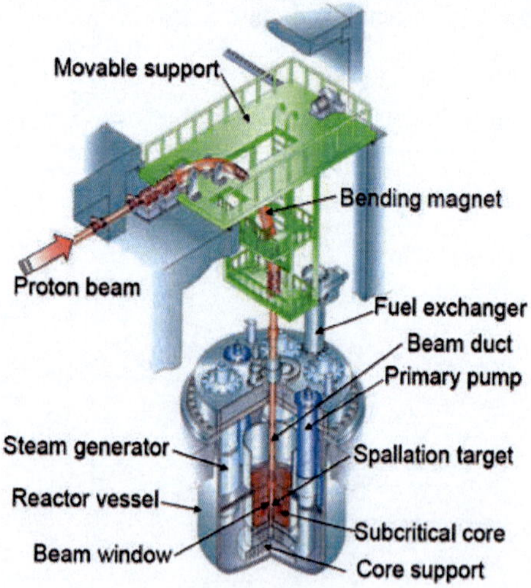

Fig. 5.33 800 MW(th) ADS design, lead-bismuth cooled (JAERI) [53]

the subcritical fission reactor core can be applied and the necessary proton current of the accelerator can be drastically reduced. The neutron multiplication M of a subcritical core is given approximately by [57]

$$M = \frac{1}{1 - k_{eff}}$$

where k_{eff} is the effective multiplication factor for the subcritical core surrounding the spallation target. For $k_{eff} = 0.95$ the neutron multiplication factor is 20.

5.5.2 Design Concept for ADS

Design proposals show that a proton accelerator can drive a subcritical reactor system of 1,500 MW(th) by a 1 GeV proton beam with a current beam of about 20 mA (dependent on the subcriticality multiplication). This appears to be technically feasible [56].

In this technical configuration ADSs offer the potential of loading the subcritical fission reactor core with a higher percentage of minor actinides or long-lived fission products than this would be possible in conventional critical fission reactors with a thermal or fast neutron spectrum.

The essential reactivity coefficients guaranteeing the safety of fission reactors would be deteriotated too much in critical fission reactor cores by the presence of a higher percentage of minor actinides or long-lived fission products. This does not play a role for accelerator driven subcritical fission reactor systems.

ADSs can operate in different neutron spectrum modes. In thermal neutron spectrum mode the inventories of fissile materials can be reduced due to the high thermal cross sections. Fluorides of lithium, beryllium and fissile-fertile materials can be used as coolant and fuel in that case (similar as in the case of thermal neutron spectrum breeders). Fast neutron spectrum ADSs have advantages due to better neutron economy. Sodium, lead or lead-bismuth are suggested as coolant in such cases [53, 55–60].

Various technical options of ADSs for actinide incineration and power production were proposed in different countries. Figure 5.33 shows one example of a 800 MW(th) ADS design of JAERI. Lead-Bismuth Eutectic (LBE) is used as a coolant [58].

The proton beam enters the reactor core through a beam window into a beam duct. The spallation target is arranged in a beam duct in the center of the subcritical core. The reactor vessel houses all thermal heat transfer circuit components, e.g. pumps, steam generators etc.. The control and shut down systems can be simpler compared to fission reactors as the core is operating in a subcritical mode. In some designs they have been completely omitted relying completely on beam shut off or beam interruption.

References

1. Druckwasserreaktor (1981) Kraftwerk Union AG, K/10567, Erlangen
2. EPR—European Pressurized Water Reactor, the 1600 MWe Reactor. http://www.areva-np.com/common/liblocal/docs/Brochure/EPR_US_%20May%202005.pdf
3. Oldekop W (ed) (1974) Druckwasserreaktoren für Kernkraftwerke. Karl Thiemig, München
4. International Atomic Energy Agency (1981) Technical evaluation of bids for nuclear power plants—a guide book. International Atomic Energy Agency, TR 204, Vienna
5. Lunin G et al (2002) The Russian advanced VVER reactor designs. Nucl News 45:28–37
6. Nero AV (1979) A guidebook to nuclear reactors. University of California Press, Berkeley
7. Märkl H (1976) Core engineering and performance of KWU pressurized water reactors. Kraftwerk Union AG, Erlangen
8. Lamarsh JR (1975) Introduction to nuclear engineering. Addison-Wesley, Reading
9. Lewis EE (1977) Nuclear power reactor safety. Wiley, New York
10. Smidt D (1979) Reaktor-Sicherheitstechnik. Springer, Berlin-Heidelberg-New York
11. Strasser A et al (2005) Fuel fabrication process handbook. Advanced nuclear technology international, Surahammer, Sweden
12. Güldner R et al (2009) Contribution of advanced fuel technologies to improved nuclear power plant operation. In: The uranium institute 24th annual symposium, London
13. Oka Y et al (1993) Concept and design of a supercritical-pressure, direct-cycle light water reactor. Nucl Technol 103:295–302
14. Monti L (2009) Multi-scale, coupled reactor physics/thermo-hydraulics systems and applications to the HPLWR3 pass core. Dissertation, Fakultät für Maschinenbau, Universität Karlsruhe (TH)
15. Hofmeister J et al (2005) Optimization of a fuel assembly for a HPLWR. In: Proceedings of JCAPP05, Seoul, Korea
16. Ehrlich K et al (2004) Materials for high performance light water reactors. J Nucl Mater 327:140–147
17. Schulenberg T et al (2007) Core design concepts for high performance light water reactors. Nucl Eng Technol 39(4):249–256

18. Fischer K et al (2009) Design of supercritical water cooled reactor with three pass arrangement. Nucl Eng Des 239:800–812
19. Starflinger J et al (2008) Results of mid term assessment of the high performance light water reactor phase 3 project. In: Proceedings of ICAPP09, Tokyo
20. Kernkraftwerk Gundremmingen 1280 MWe BWR (2000). www.kkw.gundremmingen.de
21. SWR-1000, An advanced boiling water reactor with passive safety features (2007). www.areva.com
22. Sauer A (1969) Siedewasserreaktoren für Kernkraftwerke, 10th edn. AEG Telefunken, Handbücher Band, Berlin
23. Stosic Z et al (2008) Boiling water reactor with innovative safety concept, the generation III + SWR-1000. Nucl Eng Des 238:1863–1901
24. Boyer VS et al (1976) High temperature reactors. In: Zaleski P (ed) Nuclear energy maturity, proceedings of the European nuclear conference, Paris, vol 2. Pergamon Press, Oxford, 21–25 April 1975, pp 191–217
25. Dahlberg RC et al (1974) HTGR fuel and fuel cycle summary description. General Atomic Company, GA-A 12801, San Diego
26. Electric Power Research Institute (1976) Development status and operational features of the high temperature gas cooled reactor. Electric Power Research Institute, EPRI NP-142, Palo Alto
27. Harder H et al (1971) Das 300 MWe Thorium-Hochtemperatur-Kernkraftwerk (THTR). Atomwirtschaft/Atomtechnik 16:238–245
28. Schulten R et al (1974) The Pebble-bed high temperature reactor as a source of nuclear process heat. Kernforschungsanlage Jülich. Jül 1115-RG
29. Zhang Z et al (2006) Design aspects of the Chinese modular high temperatur gas cooled reactor HTR-PM. Nucl Eng Des 236:485–490
30. Bogusch E et al (2009) Programs and projects for high temperature reactor development. Atomwirtschaft 54(2):84–88
31. Reutler H (1988) Plant design and safety concept of the HTR module reactor. Nucl Eng Des 109:335–340
32. Reutler H, Lohnert G (1984) Advantages of going modular in HTRs. Nucl Eng Design 78:129–136
33. DelCal G et al (2002) TRISO-coated fuel processing to support high temperature gas cooled reactors, ORNL/TM-2002/166
34. Koster A et al (2003) PBMR design for the future. Nucl Eng Design 222:231–245
35. Bernnat W et al (2001) Models for reactor physics calculations for HTR pebble bed modular reactors. Nucl Eng Des 222:331–347
36. Wessmann GL, Mofette TR (1973) Safety design bases for HTGR. Nucl Saf 14:618–634
37. Canada enters the nuclear age, published by AECL (1997) McGill-Queens University Press, Montreal-Kingston
38. Foster JS et al (1975) The status of the Canadian nuclear program and possible future strategies. Ann Nucl Energy 2:689–703
39. McIntyre HC (1975) Natural uranium heavy water reactors. Sci Am 233(4):17–27
40. Smith HA (1976) A review of the development status of CANDU nuclear power plants. In: Zaleski P (ed) Nuclear energy maturity, Proceedings of the European nuclear conference, Paris, vol 2. Pergamon Press, Oxford, 21–25 April 1975, pp 38–44
41. Electric Power Research Institute (1977) Study of the development status and operational features of heavy water reactors, Palo Alto, CA. EPRI NP-365
42. Woodhead LW et al (1976) Performance of Canadian commercial nuclear units and heavy water plants. In: Zaleski P (ed) Nuclear energy maturity, proceedings of the European nuclear conference, Paris, vol 2. Pergamon Press, Oxford, 21–25 April 1975, pp 160–162
43. Torgerson D (2002) The ACR-700-Raising the bar for reactor safety performance, economics and constructability. Nucl News 4(11):24–32
44. Luxat J (2002) Thermal-Hydraulic aspects of progression to severe accidents in CANDU reactors. Nucl Technol 167:187–210

45. Oak Ridge National Laboratory (1971) Conceptual design study of a single fluid molten salt breeder reactor. ORNL-4541
46. Oak Ridge National Laboratory (1972) The development status of molten salt breeder reactors. ORNL-4812
47. Perry AM, Weinberg AM (1972) Thermal breeder reactors. Ann Rev Nucl Sci 22:317–354
48. International Atomic Energy Agency (1979) Status and prospects of thermal breeders and their effect on fuel utilization. IAEA Technical Report Series No. 195, International Atomic Energy Agency, Vienna
49. Nuttin A et al (2005) Potential of thorium molten salt reactors: detailed calculations and concept evolution with a view to large scale energy production. Prog Nucl Energy 46(1):77–99
50. Radkowsky A et al (1998) The nonproliferative light water thorium reactor: a new approach to light water reactor core technology. Nucl Technol 124:215–222
51. Prael RE et al (1989) User Guide to LCS: The LAHET Code System, LA-UR-89-3014
52. Broeders CHM et al (1997) Neutronenphysikalische Analysen von beschleunigergetriebenen unterkritischen Anordnungen. Nachrichten—Forschungszentrum Karlsruhe 29:411–420
53. Saito S et al (2006) Design optimization of ADS plant proposed by JAERI. Nucl Instrum Methods Phys Res A 562:646–649
54. Rief H, Takahashi H (1994) Some physics considerations in actinide and fission product transmutation. In: Proceedings of the international conference on reactor physics and reactor computations, Tel-Aviv, Ben-Gurion University of the Negev Press, 23–26 Jan 1994, p 108
55. Bowman CD et al (1992) Nuclear energy generation and waste transmutation using an accelerator-driven intense thermal neutron source. Nucl Instrum Methods A 320:336–367
56. Rubbia C et al (1995) Conceptual design of a fast neutron operated high power energy amplifier, CERN/AT/95-44, Geneva
57. Weinberg AM, Wigner EP (1958) The physical theory of neutron chain reactors. University of Chicago Press, Chicago
58. Ahlström PE (2004) Partitioning and transmutation. Current developments—2004, Technical Report TR-04-15. Svensk Kärnbränslehantering AB, Stockholm
59. Maschek W et al (2008) Accelerator driven systems for transmutation: fuel development, design and safety. Prog Nucl Energy 50:333–340
60. Wade D (2000) ATW neutronic design studies. International seminar on advanced nuclear energy systems towards zero release of radioactive waste, Susono, Japan

Chapter 6
Breeder Reactors With a Fast Neutron Spectrum

Abstract Breeder reactors with a fast neutron spectrum have a sufficiently high breeding ratio to attain a fuel utilization of more than 60% which is almost by a factor of 100 higher than that of present light water reactors. They can operate on the U-238/Pu-239 or on the Th-232/U-233 breeding process and utilize depleted or natural uranium or thorium. In this way they can open up an energy potential with the existing uranium and thorium reserves which can last for many thousand years. Construction of breeder reactors began in the USA, the UK and the former Soviet Union already before 1960. Their development started with small test reactors and continued with the construction and operation of prototype power reactors of unit sizes of 300 MW(e) up to 1,250 MW(e) in the USA, Europe, Russia, India and Japan. This proved their technical feasibility. Fast breeder reactors with a fast neutron spectrum use sodium or in more recent designs lead or a lead-bismuth-eutecticum (LBE) as a coolant. Plutonium-uranium mixed oxide fuel, but also metallic alloys and nitride fuel were developed for the fuel of fast breeder reactors. At present the small test reactors JOYO (Japan) and BOR 60 (Russia) and the fast breeder reactor BN 600 in Russia are operating since several decades whereas MONJU (Japan) is close to full power operation and BN 800 as well as SVBR/75/100 in Russia and PFBR (India) are under construction.

6.1 The Potential Role of Breeder Reactors With a Fast Neutron Spectrum

Breeder reactors with a fast neutron spectrum have a sufficiently high breeding ratio to allow independence of any supply of enriched uranium in practical operation. The U-238/Pu-239 or the Th-232/U-233 conversion breeding processes enable this type of reactor to utilize depleted uranium or thorium theoretically with 100% efficiency and practically, including losses in the fuel cycle, with an efficiency in excess of 60% (see Sect. 3.6). This is about a factor of 100 higher than the fuel utilization in

G. Kessler, *Sustainable and Safe Nuclear Fission Energy*, Power Systems,
DOI: 10.1007/978-3-642-11990-3_6, © Springer-Verlag Berlin Heidelberg 2012

present LWRs (once through cycle) and approximately a factor of 50 higher than in converter reactors operating with high conversion ratios and in the recycling mode (see Sect. 3.6).

The initial plutonium or U-233 core inventory of FBRs must come from reprocessing of spent fuel of e.g. LWRs or other converter reactors including fast reactors operating with U-235/U-238 or U-235/Th-232 fuel.

The specific energy release per gram of uranium or plutonium completely undergoing fission is roughly 1 MWd(th) (see Sect. 3.1.5). As the fuel utilization in breeder reactors with fast neutron spectrum is about 60% (Sect. 3.6) including losses in the fuel cycle, this results in:

$1 \times 0.6 = 0.6$ MWd(th) of energy extracted from 1 g of uranium (U-235 or U-238 being converted in Pu-239). This allows the uranium consumption of a fast breeder reactor (FBR) to be determined as

$$\frac{1}{0.6} \cdot 365 \frac{1}{0.40} \left[\frac{\text{gU}}{\text{MWd(th)}} \cdot \frac{\text{d}}{\text{y}} \cdot \frac{\text{MW(th)}}{\text{MW(e)}} \right] = 1,521(\text{gU}/(\text{MW(e)} \cdot \text{y}))$$

According to this formula, FBRs annually consume about 1.52 tons of natural or depleted uranium (the contribution of 0.72% U-235 being small) fuel per 1,000 MW(e) (at 40% thermal efficiency and 100% plant load factor). Because of their low uranium consumption, FBRs are quite insensitive to the price of uranium.

If the estimated world nuclear energy requirement and uranium resources listed in Chaps. 1 and 2 are included in these assessments, breeder reactors with a fast neutron spectrum are found to open up an energy potential with the existing uranium reserves, which can be good for several thousand years. FBRs can utilize for nuclear fission not only fertile U-238, but also fertile Th-232. This would further add to the energy potential referred to above. Accordingly, with the use of FBRs, the supplies of nuclear fuel can be considered inexhaustible far beyond any time scale of conceivable planning interest. This is comparable with the energy potential that is hoped to be tapped by fusion reactors operating on the D-T cycle with lithium as the breeding material [1, 2]. However, fusion reactors are still in their infancy of development compared to FBRs.

6.2 Brief History of the Development of Fast Breeder Reactors

The principle of breeding had been recognized at the very beginning of the development of nuclear fission reactors [2, 3]. Construction of the first reactors with a fast neutron spectrum was begun in the USA before 1950. In the UK and the former Soviet Union, this development started in the early fifties. These first-generation breeder reactors, however, mainly served for studies of fast neutron physics (CLEMENTINE, EBR-I, BR-1, BR-2). In addition, they were to demonstrate the feasibility of the technical solutions adopted (EBR-II, EFFBR, DFR). Consequently, some of them had rather low thermal power levels. In accordance with the state of the art at that time, they were equipped with fuel elements of enriched uranium or plutonium metals. Because of the high power per unit volume in the core, liquid metals such as mercury, sodium-potassium or sodium were used as coolants. They were called liquid metal cooled fast breeder reactors, (LMFBRs) [3].

Along with the development of LWRs it became apparent around 1960 that LMFBR cores would be able to attain the required relatively low fuel cycle costs only at a high burnup of 100,000 MWd/t or more if fueled with ceramic fuel (PuO_2/UO_2) (based on experience with UO_2 fuel in LWRs). The use of PuO_2/UO_2 mixed oxide fuels and higher volume fractions of the sodium coolant in large sodium cooled fast reactor (SFR) cores led to higher neutron moderation than in the initial small LMFBR cores. As a consequence, reactivity coefficients, such as the instantaneous Doppler coefficient and the sodium void coefficient, became subjects of detailed investigations (see Sect. 3.8.2). SFBR test reactors up to powers of 60 MW(th) were therefore built in an interim phase between 1960 and 1970, which mainly served to demonstrate properties of ceramic fuels up to high burnups and the safe operation of this type of reactor (SEFOR, BR-5, BOR-60, RAPSODIE). All of these reactors had PuO_2/UO_2 mixed oxide fuels and were cooled with sodium. Coolant temperatures were chosen to allow steam conditions for high thermal plant efficiencies [3].

The proven good predictability of the physics and safety characteristics and the good operating experience accumulated in this first generation of test reactors have been the basis for construction of the second generation of sodium cooled fast breeder reactor (SFBR) plants with electric powers around 250–300 MW(e) (Table 6.1). The technical data and the design features of these second generation reactors are already geared to the characteristics of commercial size SFBRs. They have PuO_2/UO_2, mixed oxide (MOX) fuels with target burnups of 100,000–150,000 MWd/t and sodium as a coolant. The core outlet temperature of the coolant is approximately 550°C, permitting steam conditions to be reached in the turbogenerator system with thermal efficiencies around 40%. The breeding ratios of these SFBRs are around 1.15–1.20. Regarding the primary cooling systems, two design alternatives are applied: the loop and the pool type systems (see Sect. 6.4).

Three prototype reactors in this power category have already accumulated many years of operating experience. The French PHENIX prototype [4] SFR has been in operation from 1973 until 2009. The Soviet BN 350 prototype SFR [5] was first connected to the grid in 1973 and shut down 1999. In addition to electrical energy production it provided also thermal energy for desalination. The British PFR [6] prototype fast reactor reached criticality in 1974 and has delivered power into the public grid system from 1975 until 1994. For all three prototype plants, the original design characteristics were confirmed in terms of fast reactor core physics, control and safety engineering parameters, and performance of the primary cooling system. The control and safety performance of these prototypes has been tested not only in normal operation, but also under simulated emergency cooling conditions, and has fully met expectations. Initial difficulties in running the large sodium components (pumps, heat exchangers and steam generators) were overcome.

Table 6.1 Fast breeder prototype and demonstration SFBRs [3]

	France	
	PHENIX	SUPERPHENIX
Reactor power		
Thermal (MW(th))	568	3,000
Electrical net (MW(e))	250	1,200
Primary circuit	Pool	Pool
Number of primary circuits	3	4
Primary/secondary coolant	Na/Na	Na/Na
Coolant temperature at		
Core inlet (°C)	385	395
Core outlet (°C)	552	545
Steam conditions turbine inlet		
Pressure (bar)	168	177
Temperature (°C)	510	487
Diameter of reactor vessel (m)	11.8	21
Core dimension		
Eq. diameter (cm)	139	366
Height (cm)	85	100
Fuel	UO_2/PuO_2	UO_2/PuO_2
Cladding material	316 SS	316 SS
Pin diameter (mm)	6.6	8.5
Number of pins per fuel element	271	271
Core power density (kW(th)/l)	406	280
Max. linear pin power (W/cm)	450	450
Mass of core fuel PuO_2/UO_2 (t)	4,52	36,9
Mass of blanket fuel UO_2 (t)	19	74
Fuel burnup		
Average (MWd/t)	40,000	70,000
Maximum (MWd/t)	72,000	100,000
Breeding ratio	1.16	1.18

During the operation of PHENIX, a breeding ratio of 1.16 was demonstrated, while a large proportion of the irradiated fuel assemblies have already been reprocessed.

The FFTF reactor [7] in the USA, a large sodium cooled test reactor of 400 MW(th), began operation in 1980 after extensive pretest programs. It was mainly used for fuels and materials testing until 1992.

The Russian BN 600 [8], the first LMFBR with an electric power of 600 MW(e), went into full operation in 1982 and was still operating by 2010. It represents an intermediate step between the prototypes of the 250–300 MW(e) class and commercial size LMFBRs.

In the Federal Republic of Germany, the small sodium cooled test reactor, KNK-II, 20 MW(e), was in operation until 1989 and a 300 MW(e) prototype SFBR, SNR 300 [3], reached its final construction stage in 1991, but was not taken in operation for political reasons.

In Japan, a 280 MW(e) prototype LMFBR, MONJU [3] reached first criticality in 1994 based on sufficient operating experience with the experimental fast reactor JOYO. During preparations for full-power operation, however, the operation had to be suspended due to sodium leaks from the secondary system in 1995. After overall plant checking and function tests MONJU was restarted in 2010.

The phase of commercial size demonstration power plants was begun in France in 1977 with the construction of SUPERPHENIX [9]. The plant had a net electric power of 1,200 MW(e), a thermal efficiency of 40% and, like PHENIX, was a pool type SFBR. It was operated between 1983 and 1998 and then shut down for political reasons.

In Russia, the 800 MW(e) BN 800 is expected to go into operation by 2012. In India the 500 MW(e) PFBR is expected for begin of operation in 2011. Studies on commercial size SFBRs are underway in Japan and in France [3, 10, 11].

6.3 The Physics of SFBR Cores

6.3.1 SFBR Core Design

SFBR cores mainly consist of a cylindrical arrangement of hexagonal fuel elements surrounded radially by hexagonal blanket elements. The hexagonal fuel elements are designed in such a way that the inner core is also axially surrounded by an upper and a lower breeding blanket (Fig. 6.1). The blanket elements initially contain depleted uranium as uranium dioxide (UO_2). By contrast, the fuel elements of the core are filled with PuO_2/UO_2 MOX fuel as the fissile material. Most cores contain two or three radial zones of different enrichments in order to make the radial power distributions as flat as possible.

In an SFBR core, such as SUPERPHENIX with a net power generation of 1,200 MW(e), the inner radial core zone contained 196 hexagonal fuel elements with 14% Pu enrichment, and the outer radial core zone comprises 171 hexagonal fuel elements with 18% Pu enrichment. The core was surrounded by 233 hexagonal blanket elements. It also incorporated 24 positions for hexagonal absorber elements containing boron carbide (B_4C) as an absorber. Insertion or withdrawal of these absorber elements regulated the criticality and power of the reactor and guaranteed safe control and shutdown conditions. The core had a diameter of 3.66 m, a height of 1 m and a volume of 10.8 m^3. This corresponded to an average power per unit volume of the core of 280 kW(th)/l and a maximum power per unit volume of 435 kW(th)/l. The radial blanket had a thickness of 50 cm, the axial upper and lower blankets a height of 30 cm each.

The fuel assembly was made up of 271 fuel pins held in position by helical wire wrapped spacers. The fuel pins had outside diameters of 8.5 mm and were filled in the core region with cylindrical pellets made of PuO_2/UO_2 mixed oxide. The fuel rod cladding was made of Ti-stabilized austenitic steel with a wall thickness of

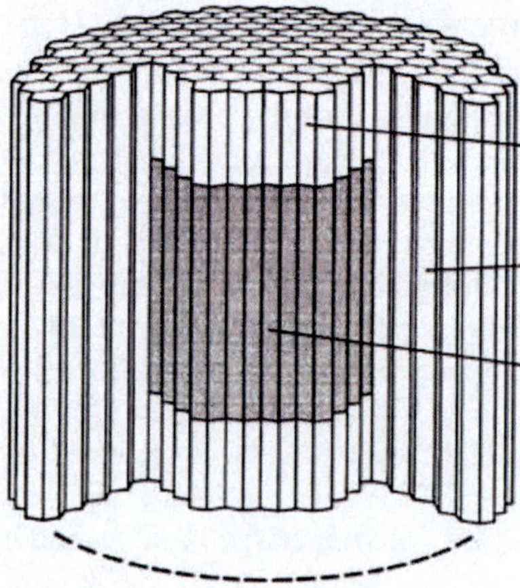

Fig. 6.1 Schematic of a fast breeder reactor core [2]

approximately 0.7 mm. After fuel pin fabrication, the pins were filled with helium and welded tight at the ends. The fissile zone of a fuel pin was 1 m long. It was followed on both ends by a fertile zone each of 30 cm length. Below the lower axial blanket there was a gas plenum 850 mm long, in which gaseous fission products would accumulate. Sodium flowed through the fuel elements at a rate of approximately 3–6 m/s. On its way through the fuel elements, it was heated from 395 to 545°C. The fuel elements were surrounded by a hexagonal can made of austenitic or ferritic steel with a wall thickness of 4.6 mm. The complete fuel element had a length of 5.4 m and was 16.6 cm across the hexagonal flats. The blanket rods had diameters of 16.3 mm, which were larger than the diameters of the fuel rods; the blanket elements accordingly only contained 91 rods. The maximum fuel pin power in the core was roughly 450 W/cm of fuel pin length. A 1,200 MW(e) core contains some 4.6 t of Pu-239 and Pu-241. This corresponded to 3.7 t of (Pu-239 and Pu-241)/GW(e). The core contained a total of 37 t of PuO_2/UO_2, the axial breeding blanket had 22 t of UO_2, and the radial blanket has 52 t of UO_2. Each absorber element consisted of 18 absorber rods with an outside diameter of 18.8 mm, filled with B_4C pellets. B_4C is enriched to about 93% in its B-10 isotope [10].

6.3.2 *Energy Spectrum and Neutron Flux Distribution*

The cores of LMFBRs in the 1,200 MW(e) power category contain some 34–37% of PuO_2/UO_2 mixed oxide fuel, some 39–41% of sodium, and 22–27% of austenitic steel [12, 13]. This results in a neutron energy spectrum (see Fig. 3.5), whose mean energy is in the range of 200 keV. The low energy part of the neutron energy spectrum extends into the resonance ranges of the fission cross sections of the fuel and the capture cross sections of the fuel, steel and sodium. In this region, the resonances of these cross sections in part overlap very strongly. This must be taken into acount

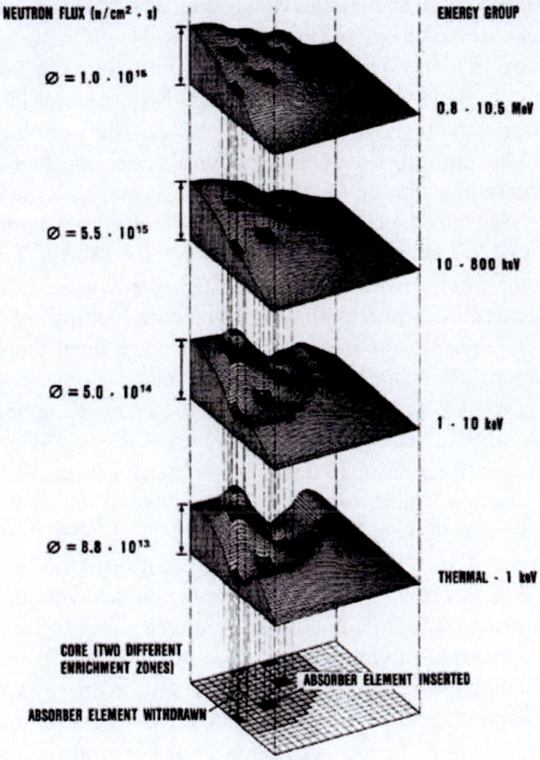

Fig. 6.2 Neutron flux distribution in a fast reactor core (KfK) [2]

in determining the temperature coefficient of reactivity (Doppler coefficient caused by temperature-induced broadening of resonance shapes). The high energy part of the neutron energy spectrum extends into the region of the fission cross sections of U-238, where it contributes to a pronounced fast fission effect (see Sect. 3.5).

Figure 6.2 shows the spatial distribution of the neutron flux in such an SFBR core for a rough subdivision into only four groups of the neutron energy spectrum. The neutron flux drops markedly towards the edge of the core. Neutron flux and power distribution attain their peaks in the center of the core. At maximum power generation, 450 W/cm, the neutron flux in the center of the core reaches approximately 7×10^{15} n/cm^2 s, integrated over the entire energy spectrum. Neutrons leaving the core are captured in the U-238 of the radial and axial blankets eventually generating Pu-239 there.

6.3.3 Breeding Ratio

The total breeding ratio of a 1,200 MW(e) core, such as SUPERPHENIX, was calculated to be 1.18 [3, 10, 12, 13]. It was composed of contributions by the core of

0.8 and by the axial and radial blankets of 0.38. The annual excess production in that case was 83 kg of (Pu-239 and Pu-241)/GW(e)·y at a load factor of 0.75 and 2% losses in the fuel cycle. This excess plutonium can be saved, e.g. for construction of another large SFBR core. For an SFBR such as SUPERPHENIX, this would have taken some forty years (doubling time). This doubling time was relatively long.

The doubling time was considered very important during the early phase of SFBR development in the 1960s. However, it plays no major role during the future commercial introduction phase of SFBRs since the plutonium is predominantly made available externally by operating LWRs. About 2,500 tonnes of plutonium were already existing in spent LWR fuel elements in 2010 and can be made available by chemical reprocessing for initial core loadings of SFBRs.

The physics of large SFBR cores have been thoroughly investigated in numerous critical assemblies, such as ZPR and ZPPR in the USA, ZEBRA in the UK, MASURCA in France, SNEAK in the Federal Republic of Germany, FCA in Japan, and BFS in Russia. Microscopic cross section data libraries, e.g., the Evaluated Nuclear Data File (ENDF), allow special computer programs to be used to prepare group constants as inputs for extended multigroup calculations (see Sect. 3.3). Examples of such programs are the ETOE-2/MC^2-2/SDX program package developed at ANL (USA), the GALAXY program of the UKAEA or the MIGROS program of KfK (Germany). Computation techniques were developed to account for energy resonance self-shielding in the resonance range of the cross sections and for heterogeneity effects in fast reactor cores. A few hundred up to 2,000 neutron energy group calculations are run to calculate the neutron energy spectra within the LMFBR core. Condensed group constant sets are developed and finally used for multidimensional transport or diffusion calculations. Further group collapsing is usually applied so that only very few energy groups (most times even a one group approach) are used for burnup calculations (Chap. 3).

6.3.4 Reactivity Coefficients and Control Stability

Changes in k_{eff} resulting from the buildup of fission products and higher actinides and the changes in k_{eff} during startup and shutdown of the reactor determine the design of control and shutdown systems [14]. Because of the relatively small cross sections for fission and capture in the important range of a fast reactor neutron energy spectrum, particularly resulting in a lower fraction of parasitic absorption, e.g. in fission products, and a better value of neutron production per neutron absorbed in fissile material, these changes in k_{eff} are smaller in SFBRs than in thermal converter reactors (see Sects. 3.2 and 3.8). In a 1,200 MW(e) core, such as SUPERPHENIX, the buildup of fission products and actinides causes only a 3% change of k_{eff}, over a power cycle of roughly one year. The feedback reactivities (Doppler coefficient, density changes and core expansion) to be covered as a result of temperature changes during startup from zero power to full power only amount to roughly a 1.5% change of k_{eff}. As a consequence, sufficient absorber material for a control system and two shutdown systems can be accommodated in about 24 positions of SFBR cores. The aggregate negative reactivity of these control and shutdown systems is about −13% of k_{eff}. Since the capture cross sections for absorbers are relatively small in the important range of the fast reactor neutron energy spectrum, the absorber material B_4C has to be enriched in its B-10 content in order to provide the desired amount of negative reactivity within a given volume available for absorber elements.

For SUPERPHENIX, the reactivity coefficient for increasing power (power coefficient) was -0.19×10^{-5}/MW(th), at reference power level. This means that increases in power are counteracted by strong negative feedback reactivities. Accordingly, large LMFBR cores can be controlled in an inherently stable mode.

6.3.5 The Doppler Coefficient

The negative Doppler coefficient makes the greatest contribution to the negative power coefficient [3, 10, 12, 13, 15]. It instantaneously generates a negative reactivity effect whenever the power and as a consequence the fuel temperature increase. It results from the Doppler broadening of resonances of the capture and fission cross sections. In large cores, its main contribution comes from the range of the neutron energy spectrum below 75 keV. Unlike thermal reactors, where the main contributions to the Doppler coefficient arise from resolved resonances of the cross sections, some 50% of the contributions in fast reactors result from the range of unresolved resonances. The negative contributions by the capture resonance cross sections of mainly U-238 and Pu-239 by far outweigh the small positive contributions by fission resonance cross sections of Pu-239 (see Sect. 3.8.2).

The Doppler coefficient in fast reactor cores was measured in a number of critical assemblies, such as ZPPR, ZEBRA, SNEAK etc., and in the SEFOR reactor. In large SFBR cores, there is at present only an uncertainty of about ±15% between theoretical calculations of the Doppler coefficient and its experimental verification. The Doppler coefficient, i.e. the change of reactivity per degree of changing average fuel temperature, under conditions of rising average fuel temperature, T_f , in large SFBR cores is in the range of

$$\frac{1}{k_{eff}} \cdot \frac{\partial k_{eff}}{\partial T_f} = -\frac{0.008}{T_f^x}$$

For normal operating conditions, the exponent, x, is between 0.8 and 1.0. The Doppler coefficient can be influenced by the composition of the core:

- Within the composition of fuel isotopes, Pu-241 raises the Doppler coefficient, whereas Pu-240 lowers it.
- The fractions of sodium and steel in the core influence the contribution of the neutron energy spectrum below 75 keV. Voiding of the sodium from the core (boiling) may reduce the Doppler effect by up to 50%.

- Other constituents such as Fe-56 in steel or resonance absorbers like Tc-99 in sufficient amounts may also contribute to the Doppler coefficient.
- Adding solid moderator, such as BeO, increases the fraction of the neutron spectrum below 75 keV and raises the Doppler coefficient.

6.3.6 The Coolant Temperature Coefficient

Increases in temperature of the sodium coolant decrease its density [3, 10, 12–14]. Increases in temperature of the cladding and fuel result in radial expansion of the fuel rod and, at the same time, expulsion of sodium from the core. This leads to a reduction of the sodium density per cm^3 of core volume. Three different individual effects must be distinguished. The reduction in sodium density

- increases neutron leakage. This effect is negative (reducing k_{eff}), dominating mainly in the outer core regions at high spatial gradients of the neutron flux. It depends on the size of the reactor core;
- changes moderation. The neutron energy spectrum is shifted towards higher energies. This effect is usually positive (increasing k_{eff}), because the shift towards higher energies slightly increases the average η-value of fissile isotopes and also increases fast fission of U-238;
- reduces neutron absorption. This usually relatively small reactivity effect is positive throughout the core (adding to k_{eff}).

While the leakage term is influenced by the size and shape of the reactor core, the spectral term depends on the plutonium enrichment, the content of higher Pu isotopes, and the composition of the core, e.g., the solid moderator fraction. Only in small size SFBRs of about 50–100 MW(e) the overall sodium temperature coefficient can be negative. In a cylindrical two zone SFBR core with 1,200 MW(e) power, the overall coolant temperature coefficient of reactivity is positive and about 5×10^{-6} per °C of coolant temperature increase. Special core designs can reduce this coolant temperature coefficient. For instance, for a so-called heterogeneous core (see Sect. 6.7) containing also blanket elements in the fuel zone, it can be reduced to about 2×10^{-6} per °C of coolant temperature increase. However, in this case of heterogeneous cores also the negative Doppler coefficient is reduced, due to the higher fissile enrichment of the core fuel elements, and its feedback effect is reduced due to a time-lag in heating the in-core blanket fuel elements during a power transient [12]. Other design possibilities to reduce the positive sodium temperature coefficient are: flat cores or the addition of BeO or ZrH_x ($x \approx 1.7$) as solid moderator materials for mitigating the effect of spectral shift to higher neutron energies.

The positive sodium temperature or void coefficient is dominated in the normal operating power range of LMFBR cores by the quantitatively higher negative contributions of the Doppler coefficient, the axial fuel expansion and by radial structural expansion of the core as well as by axial expansion of the control rod drive structures. The isothermal temperature coefficient for large SFBRs, e.g. SUPERPHENIX, was

-2.7×10^{-8}/°C. The power coefficient of large SFBRs, e.g. SUPERPHENIX, was -0.19×10^{-5}/MW(th).

Only, if in extremely improbable accident situations the central regions of the core were to be voided suddenly, e.g. by ingress of a large gas or vapor bubble, the positive sodium void coefficient could become dominant.

6.3.7 Fuel and Structural Temperature Coefficients

Temperature increases of the fuel cause the fuel rod to expand axially and decrease its density, thus reducing k_{eff} [3, 12–14]. In a large SFBR core, the fuel temperature coefficient of reactivity due to axial expansion is in the range -2×10^{-6} per °C of fuel temperature increase.

The radial expansion of the core structure resulting from temperature increases in the steel causes the core to expand and thereby reduces the fuel density. This again results in a negative temperature coefficient of reactivity (structural expansion coefficient). In a large SFBR core the structural expansion coefficient is determined by two effects. These are the expansion of the core grid plate and bowing effects of core fuel elements caused by radial temperature gradients. Radial clamping and the structural design of the core prevent bowing fuel elements from leading to significant overall positive bowing reactivity contributions. Early difficulties experienced with a positive bowing coefficient in the EBR-I are avoided in today's SFBRs. The expansion coefficient of the grid plate of large LMFBR cores is in the range of -10^{-5} per °C of coolant inlet temperature variation. The bowing coefficient caused by radial temperature gradients is in the range of -5×10^{-7} per °C of temperature difference between core inlet and outlet.

6.3.8 Delayed Neutron Characteristics and Prompt Neutron Lifetime

In addition to the feedback reactivity coefficients, the following set of characteristics govern the dynamics of SFBR cores, as with any other fission reactor cores (see Sect. 3.1.7) [2, 3, 12–14, 16]:

- the effective fraction of delayed neutrons, β_{eff},
- the average decay constant of delayed neutron precursors, λ,
- the lifetime of prompt neutrons, l_{eff}

These characteristics are compared in Table 6.2 with those of a PWR representing the class of thermal converter reactors. This qualitative comparison is made to show differences between thermal converter reactors and SFBR cores and to discuss whether SFBR cores show specific peculiarities in their dynamic behavior compared

Table 6.2 Comparison of design characteristics (approximate values) of control and shutdown systems of PWRs and SFBRs (KfK) [1, 2]

	PWR (1,300 MW(e))	Fast breeder SNR 300
Fuel	U-235	Pu-239
Prompt neutron lifetime (l_{eff} (s))	2.5×10^{-5}	4.5×10^{-7}
Effective fraction of (β_{eff}) delayed neutrons	0.005–0.0065	0.0035
Average decay constant (λ (1/s)) of delayed neutrons	0.077	0.065
Speed of control-rod (mm/s)	1	1.2
movements (10^{-2} ($/s))	2.5	$\leqq 4$
Speed of shutdown (cm/s) rod movements	156	85–190
Delay time prior to reaction (s) of shut-down system	0.2	0.2
Time span for full insertion (s) of shut-down rods in core	2.5	0.6
Reactivity of shutdown (Δk ($)) system	11	10
Reactivity of shimrod (Δk ($)) system (burnup reactivity)	19	8

to thermal converter reactors. As can be seen from Table 6.2, the average decay constants of the precursors of delayed neutrons are similar for U-235 fueled thermal spectrum PWRs and Pu-239 fueled fast spectrum SFBRs. The effective delayed neutron fraction, β_{eff}, of a plutonium fueled SFBR is smaller by almost a factor of 2 than β_{eff} in PWRs.

In the delayed critical regime, i.e., as long as reactivity increases, ρ, remain essentially below β_{eff}, the dynamic behavior of any reactor core is determined by the average decay constant, λ, of the precursors of the delayed neutrons. As can be seen from Table 6.2, the decay constants of plutonium fueled SFBR cores and U-235 fueled PWRs are very similar. Consequently, the control and stability behavior of SFBR cores in the delayed critical range for control and power rise or shutdown operations is not essentially different from that of PWRs or other thermal converter reactors. The designs of control and shutdown systems are very similar to those of PWRs or other thermal converter reactors. This is apparent from Table 6.2.

Under prompt critical and prompt supercritical conditions, i.e., when $\rho \geq \beta_{eff}$ the short neutron lifetime is not a fundamental problem as long as the power coefficient of reactivity or the dominant fuel temperature coefficient of reactivity is negative. With a negative power coefficient, as represented by the negative Doppler coefficient, SFBR cores are subject to sharply limited narrow power bursts. Such prompt supercritical power bursts were intentionally produced within the SEFOR experimental program, as is shown in Fig. 6.3. The SEFOR core was safely shut down even after such a

Fig. 6.3 SEFOR core I super-prompt power transient (KfK) [2]

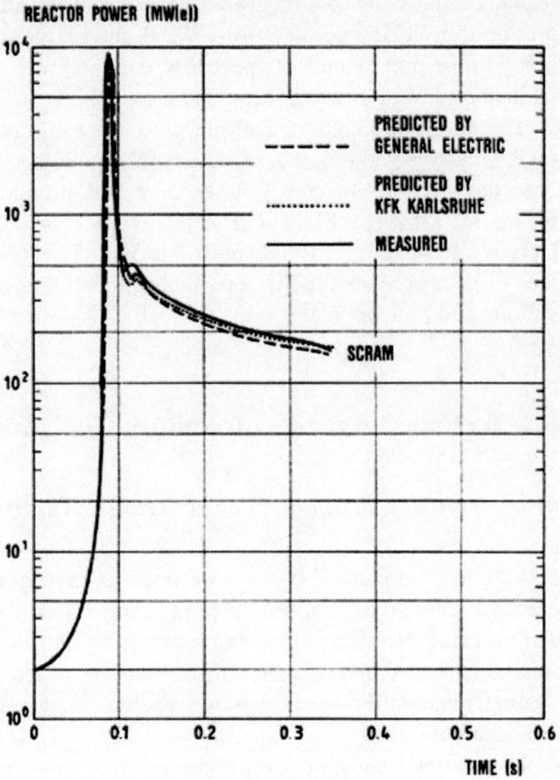

prompt supercritical power burst. The energy released during such a power burst even decreases with decreasing neutron lifetime in the presence of a highly negative Doppler coefficient.

In conclusion, it can be stated that the differences in β_{eff} and l_{eff} of SFBR cores do not lead to significant differences in control behavior. Therefore, there is no need for electromechanical designs of control and shutdown (scram) devices to be significantly different from those in thermal converters.

6.3.9 Masses of Fuel, Fission Products and Actinides

SFBR plants produce roughly the same amounts of fission products per GW(th)·y as thermal converter reactors do. As a consequence of the higher thermal efficiency (40%), this quantity of fission products, relative to the electricity generated, is approximately as high as in AGRs and HTRs and slightly lower than in LWRs and HWRs. Although plutonium fission by fast neutrons clearly dominates, the fission product yield distribution is rather similar to that encountered in thermal converter reactors usually fueled with uranium (see Chap. 3). The quantities of higher actinides produced in SFBRs per GW(th)·y are slightly larger than in thermal converter reactors. On the other side fast spectrum reactors present excellent capabilities for the incineration of higher actinides (see Chap. 9).

The absolute quantity of fissile plutonium differs relative to U-235 and U-233 fueled reactors and is still by a factor of 2 higher than in LWRs recycling plutonium. Also the total fuel inventory in the core and the blankets, which was 111 t of fuel for the 1,200 MW(e) SUPERPHENIX (Table 6.1), can be compared, e.g. with 122 t UO_2 in a 1,240 MW(e) PWR (see Table 5.1). However, in LMFBRs, roughly 90% of the power is produced in the core, which constitutes only about one third of the fuel volume and fuel weight.

6.4 Technical Aspects of Sodium Cooled FBRs

6.4.1 Sodium Properties and Design Requirements

Sodium as a coolant of fast reactor cores at present clearly dominates fast reactor development and demonstration programs all over the world [3, 10, 11, 17–19]. In addition lead-bismuth cooled fast reactors are pursued by Russian industry primarily as small modular FRs. Helium cooled fast reactor concepts were studied as an alternative coolant concept, but no helium cooled fast test or demonstration reactor has been built so far.

The design concept of liquid sodium cooled fast breeder reactors (SFRs) is mainly determined by the thermal and nuclear properties of the coolant: good heat transfer, small moderating effect and low neutron capture cross section. Capture processes in sodium lead—after (n, 2n) processes—to the formation of radioactive Na-22 with a half-life of 2.6 years. This activates the primary sodium flowing through the core. Sodium has a high melting point (98°C) and a high boiling point (892°C at atmospheric pressure, sodium boiling starts at 900–1,000°C at coolant pressures within the core). It has a high specific heat and very good thermal conductivity.

The high melting point of sodium requires preheating of pipes and components of the cooling system before sodium filling and after certain maintenance periods. The high boiling point allows high coolant temperature conditions to be maintained at very low system pressures (0.6 MPa). This results in high thermal efficiencies around 40% for the whole SFBR plant. The relatively high specific heat permits moderate coolant flow rates of 2–6 m/s in the fuel elements and low pumping powers, while the good thermal conductivity, coupled with other thermal properties, leads to very good natural convection conditions in the core and the cooling system in case of a pump failure. However, these excellent thermal properties also give rise to special design and operating requirements. Thermoshock problems at components in the reactor vessel must be avoided during short term power reductions or reactor scram conditions.

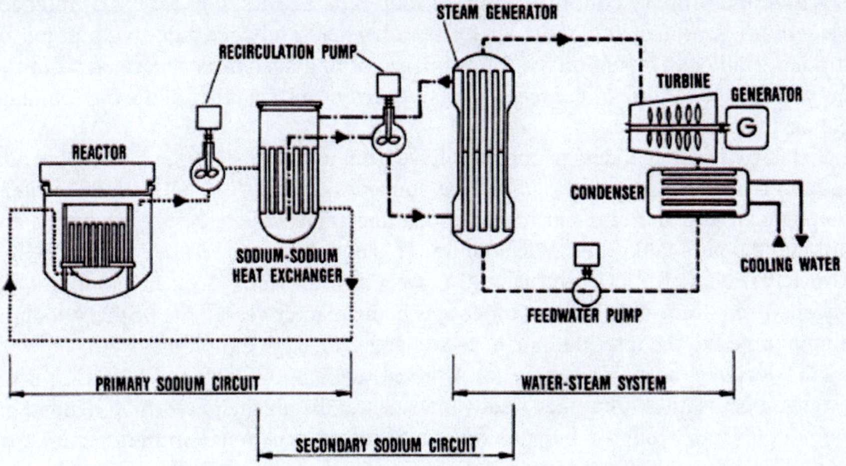

Fig. 6.4 Heat transfer system of a loop type sodium cooled fast breeder reactor (KfK) [2]

Although sodium is not very corrosive to stainless steel, its impurities, mainly oxygen and carbon, must be held at acceptably low contents (5–10 ppm for O_2, <50 ppm for carbon). High impurity contents cause radioactive corrosion products to be detached from the surfaces of fuel claddings and then transported to low temperature parts of the primary coolant system (heat exchangers). Such undesired concentrations of corrosion products must be avoided because of possible maintenance and repair difficulties, which could arise after several years of plant operation. It was observed that Mn-54 produced by the (n, p)-reaction of Fe-54 is rapidly transferred from the hot core regions to low temperature areas.

The opacity of sodium affects the design of the refuelling systems and requires ultrasonic devices to be used for supervision of refuelling and repair processes.

6.4.2 Sodium Cooling Circuits and Components

Major design consequences arise from the potential of sodium to enter into chemical reactions with water and air. This property, together with the fact that sodium becomes radioactive under neutron irradiation in the core, leads to a plant design with

- a primary cooling system containing the radioactive sodium heated up in the core;
- a secondary cooling system coupled with the primary system by intermediate heat exchangers;
- a tertiary water system producing steam for electricity generation in the turbogenerator system. Figure 6.4 shows a loop type SFBR with primary, secondary and tertiary coolant system.

Within the primary cooling system of loop type SFBRs, e.g. MONJU, radioactive sodium is protected against air by steel barriers and cells filled with argon or nitrogen. Radioactive sodium of the primary cooling system is separated from the non-radioactive sodium of the secondary system by the steel tubes of the intermediate heat exchangers.

So far, two principal design concepts have been used for SFBRs, i.e., the pool and the loop type concepts. In the pool type concept (see Fig. 6.7), all primary system components with the core, primary pumps and intermediate heat exchangers are built into the pool tank filled with sodium. This concept was used in PFR, PHENIX, SUPERPHENIX, BN 600 and BN 800 (see Fig. 6.8, Sect. 6.5.2). In the loop type concept by contrast, only the reactor core is in the reactor vessel, the primary sodium being pumped to the intermediate heat exchanger through a piping system.

This design concept was chosen in BN 350, SNR 300, FFTF and MONJU. Both design concepts have a number of advantages and disadvantages, which depend on design details and roughly balance out. Only future experience in the operation of large SFBRs can tell whether one of these design concepts must be preferred to the other.

Much care is required in the design and construction of sodium heated steam generators. FBR steam generators contain non-radioactive secondary-system sodium and water or steam, both separated from each other only by the steel tube walls. As is well known, sodium and water develop a violent chemical reaction when brought into contact, due to their different physical and chemical properties. Many design aspects, therefore, have to be taken into account, such as fabrication, operational availability, leak detection, inspectability, corrosion effects, repair, etc.. Consequently, much research within SFBR programs has been devoted to steam generator development and testing in full scale test rigs and reactor plants. Figure 6.5 shows the helical tube steam generator of MONJU. A number of other design concepts, e.g. straight tube steam generators are employed in other fast reactors. Chemical sodium-water interactions in steam generators and design optimization of pressure relief systems for SFBRs have been the subject of intensive research. There is no doubt that large SFBR steam generators can be built upon a sound and safe sodium technological concept. In Sect. 6.5.3 a double walled steam generator concept will be described. This design concept avoids sodium water reactions. The aspect of sodium fires will be discussed in Sect. 6.4.5 and Chap. 12.

6.4.3 Control and Shutdown Systems

SFBR cores are controlled by means of absorber rods containing B_4C as an absorber material [3, 21]. Also shutdown can be brought about by means of B_4C absorbers. After being released, the absorber rods can be introduced into the core by dropping under gravity within 0.7–0.8 s. To add to the reliability of this shutdown concept, SFBRs have two completely independent, diverse shutdown systems. In this way, a failure threshold for the shutdown systems of $<10^{-6}$ failures/a can be attained. By

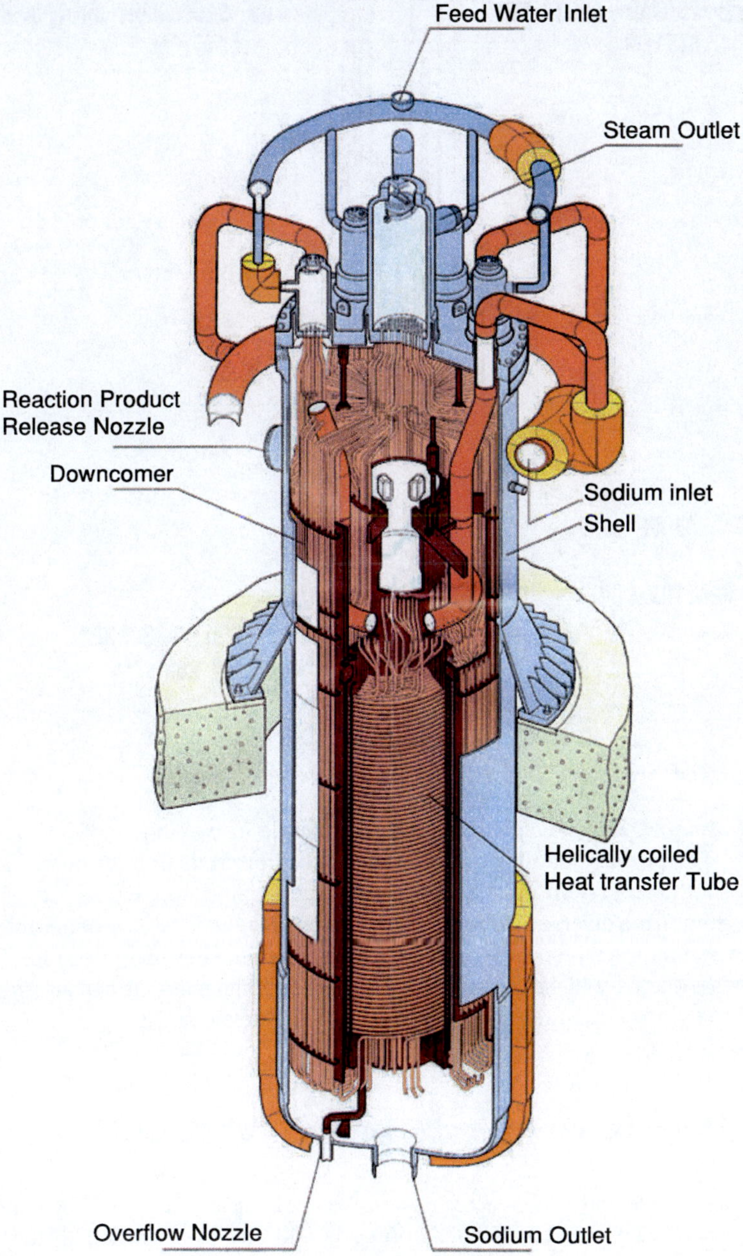

Fig. 6.5 Steam generators of the MONJU FBR demonstration plant (JAEA) [20]

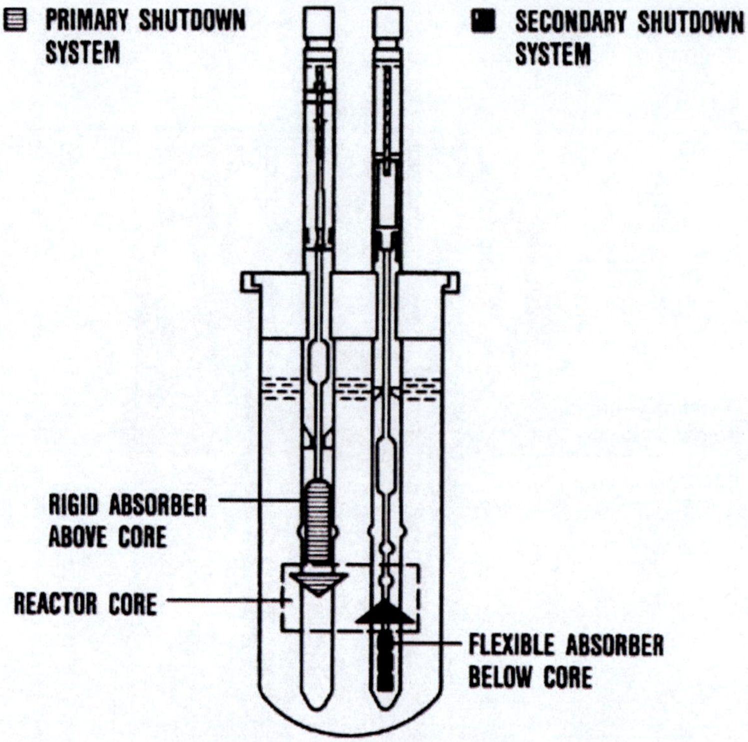

Fig. 6.6 Redundant and diverse shutdown systems (KfK) [2, 3]

way of example, Fig. 6.6 shows the design principle of two independent shutdown systems. The primary shutdown system drops absorber rods into the core. The secondary shutdown system pulls a flexible absorber chain into the core from below. Both systems have diverse electronic channels. The magnetic release of the absorbers is direct and indirect, respectively. In Chap. 11 additional design options for diverse shutdown systems will be presented based on the axial thermal expansion of the control rod guide structures and diverse de-latching mechanisms.

6.4.4 Afterheat Removal and Emergency Cooling of SFBR Cores

Even after shutdown of the reactor power, the decay heat of an LMFBR core must still be removed safely [3]. Under normal conditions, it is carried to the steam generators by the main heat transfer systems in such a way that the pumps are driven by pony motors at low speed. However, sodium has such excellent natural circulation characteristics that the afterheat can also be dissipated from the reactor core through the main systems by way of natural circulation. This has been proven

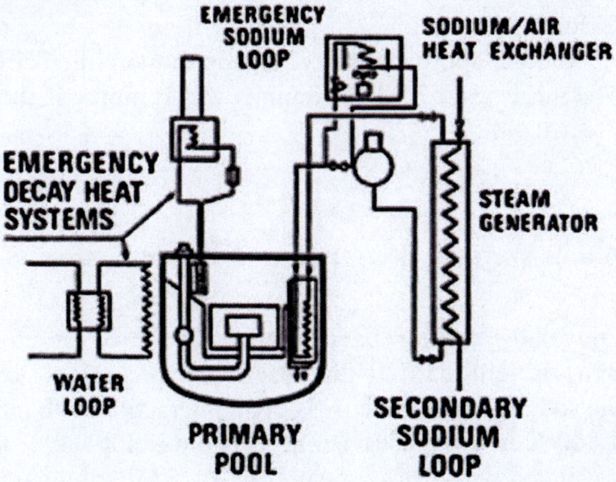

Fig. 6.7 SUPERPHENIX decay heat removal and emergency system [3]

experimentally both for pool type reactors, such as PHENIX, PFR and Superphenix, and for loop type reactors in SEFOR and FFTF. In case the main heat transfer systems were not available, immersion coolers arranged in the reactor tank below the emergency sodium level transfer the decay heat to sodium-air coolers. Even if all active components of the emergency cooling systems were to fail, enough heat would still be delivered through the surfaces of the pipes of the primary system to leave the maximum temperature in the reactor tank at $<700°C$. This thermal conduction effect can also be used to dissipate the afterheat through the wall of the reactor double tank to outside water filled pipings on surface of the tank. This design possibility had been applied to PHENIX and Superphenix (Fig. 6.7).

Unlike water or gas cooled thermal converter reactors, which need active emergency cooling systems, LMFBRs thus have the potential to remove decay heat by natural convection without any active systems.

6.4.5 Sodium Fires

Sodium fires in the primary cooling systems containing radioactive sodium are prevented by enclosing those systems in cells or double walled pipings filled with nitrogen or argon [3]. Free sodium surfaces in tanks are covered with argon. Sodium leak detection systems survey the leaktightness of the cooling system. The secondary non-radioactive systems are usually surrounded by air.

Sodium leaking out of the cooling systems will be directly collected in special catch pans and containers underneath the pipes and pumps. Access of oxygen to the hot sodium will largely be prevented in those installations. Fire extinguishing systems are also available. Extensive experimental data on sodium burning rates, burning temperatures and sodium aerosol formation are at hand from large out-of-pile test rigs in the USA and Europe. In addition, extensive operating experience with demonstration SFBRs has proved the high standard of experience of sodium technology.

Another approach is to equip the sodium filled circuit with double walled piping. The inner space of those double walled piping is then filled with argon or nitrogen (Sect. 6.5.3).

6.4.6 Sodium-Water Reactions in Steam Generators

Only in the steam generators non-radioactive sodium could contact water as a result of a tube failure [3]. In that case, a high pressure water steam jet would penetrate into the sodium and a violent Na-H_2O interaction with maximum temperatures of nearly 1,300°C in the reaction zone would develop. Hydrogen and sodium hydroxide are generated with peak pressures of 9–13 MPa in the reaction zone. Steam generators are therefore fitted with special pressure relief systems. A rupture disk would break on the sodium side, and the hydrogen, sodium and sodium hydroxide would pass through a release pipe into a reaction tank where the hydrogen would be separated from the sodium and sodium hydroxide. Hydrogen would be vented to the air where it would ignite spontaneously and burn. Hydrogen monitors and other detectors in the pressure relief system produce clear signals of such an accident, and the steam generator would then be shut down, taken out of operation and repaired.

Extensive experimental data on temperature and pressure buildup and pressure wave propagation through the steam generator bundle are available in the US, Europe, Japan and Russia. Hydrogen detectors detecting small leaks over heating tubes have been developed and are being used. They allow timely detection of small leaks before these can develop with tube failure propagation or overheating tube rupture.

Another approach is to equip the steam generators with double walled steam generator tubes in order to avoid sodium water reactions and mitigate their consequences. This design approach is being pursued for JSFR in Japan (Sect. 6.5.3).

6.5 BN 800: A Near Commercial Size Demonstration SFBR

BN 800 has a thermal power of 2,100 MW(th) and an electric net power of 800 MW(e) with a net plant efficiency of 41% [3, 10, 11]. The reactor core is cooled by sodium. The design of the primary heat transfer systems is based on the pool principle, i.e., the primary coolant systems are contained in the pool tank. The technical data of both BN 600 and BN 800 are summarized in Table 6.3.

Table 6.3 Design characteristics of BN 600 and BN 800 (pool type SFBRs) [3, 10]

	BN 600	BN 800
Reactor Power		
Thermal (MW(th))	1,400	2,100
Electrical net (MW(e))	600	870
Plant efficiency (%)	41	41
Reactor Core		
Fuel	UO_2 and PuO_2/UO_2	PuO_2/UO_2
Core outer diameter (cm)	205	256
Core height (cm)	103	88
Pu eq. enrichment		
Inner core (2 zones)	17/26	19/22[a]
Outer core zone (%)	21	24[a]
U-235 or Pu eq. mass (t)	2.6	3.5
Total UO_2/PuO_2 mass in core (t)	12	16
Fuel rod outer diameter (mm)	6.9	6.6
Length of fuel pin (mm)	2445	2445
Core power density		
Average (kW(th)/l)	445	407
Maximum (kW(th)/l)	603	635
Residence time of fuel (d)	420	480
Max. fuel rod power (W/cm)	480	480
Max. burnup (MWd(th)/t)	110,000	110,000
Blankets		
Fuel	UO_2	UO_2
Axial length (cm)	30 (upper)	35 (lower)
Radial thickness (cm)	47	10
Total UO_2 mass (t)	15	25
Fertile rod outer diameter (cm)	14.0	14.0
Total breeding ratio	0.85–1.0	1.0
Fissile Fuel Bundles		
Number of bundles	369	565
Number of pins per bundle	127	127
Pin total length (m)	2.4	2.0
Bundle total length (m)	3.5	3.5
Cladding material	Stainless steel	Stainless steel
Cladding maximum rated temperature (°C)	695	695
Fertile Fuel Bundles		
Number of bundles	362	90
Number of pins per bundle	37	37
Pin total length (m)	1.84	1.98
Bundle total length (m)	3.5	3.5
Cladding material	Stainless steel	Stainless steel

(continued)

Table 6.3 Continued

	BN 600	BN 800
Control Bundles		
Main shutdown system:		
Number of bundles	14	15
Number of absorber elements		
Per bundle	7/31/8	7
Pin length (m)	1.1	1.3
Primary System		
Coolant	Sodium	Sodium
Primary Na mass (t)	770	820
Rated flow (t/s)	6	8.60
Core sodium inlet temperature (°C)	365	354
Core sodium outlet temperature (°C)	535	547
IHX sodium inlet temperature (°C)	537	545
IHX sodium outlet temperature (°C)	362	351
Secondary System		
Coolant	Sodium	Sodium
Secondary Na mass (t)	830	1,100
Rated flow (t/s)	6.1	8.4
SG sodium outlet temperature (°C)	315	309
IHX sodium inlet temperature (°C)	315	309
IHX sodium outlet temperature (°C)	510	505
SG sodium inlet temperature (°C)	510	505
Water-Steam System		
SG water inlet temperature (°C)	240	210
Turbine steam inlet temperature (°C)	502	487
Turbine steam inlet pressure (MPa)	13.2	13.7

[a] Pu_{tot}

6.5.1 Reactor Core and Blankets

The BN 800 core contains PuO_2/UO_2 fuel elements. The operational cycle is approximately 160 full power days. The fuel elements remain for 480 days in the core. They attain a maximum burnup of 110,000 MWd/t. The blanket elements have a longer residence time, the inner rows of elements being unloaded earlier and more frequently than the outer ones. Fuel rod claddings attain maximum temperatures at their inner surfaces of about 695°C and, towards the end of their burnup periods in the core, must sustain internal pressures of about 5 MPa. This internal pressure results from the gaseous fission products building up in the fuel rod. At high burnup and high specific power, the PuO_2/UO_2 mixed oxide fuel considerably changes its physico-mechanical and chemical behavior. The fuel rod cladding must withstand these loads and, in addition, considerable thermal stresses and still maintain its good mechanical properties. Fuel rod claddings and fuel element cans are also exposed to high temperatures and radiation damage brought about by fast neutrons. The maximum neutron fluence reaches approximately 3×10^{23} n/cm^2. As a consequence, volume swelling, creep effects and high temperature embrittlement occur in the steel of the fuel rod claddings and the fuel element cans. Austenitic steel, such as Ti-stabilized alloys, and ferritic steel might be replaced by oxide dispersion strengthened (ODS) steel under these severe conditions in future.

6.5.2 Reactor Tank and Primary Coolant Circuits

The core and the radial blanket are supported on a core diagrid plate resting in a support structure in the pool tank (Fig. 6.8) [3, 10]. Sodium enters this double bottom core support plate at a temperature of 354°C from the primary sodium pumps and flows into the fuel elements through an orifice in the fuel element feet. It flows through the core and radial blanket from bottom to top and is heated to a temperature of 547°C. The core and the radial blanket are surrounded by a radial steel reflector and the neutron shield. The hot sodium now flows upward within the inner vessel structure into the large sodium pool. It moves on into the inlet openings of the three intermediate heat exchangers (IHX) arranged radially around the core inside the pool tank. Sodium passes through these IHXs from top to bottom and is cooled to 351°C by secondary sodium moving in a countercurrent flow. The cooled sodium is now taken in by three primary sodium pumps and forced back into the core diagrid plate.

The internal tank structure serves to separate the hot sodium leaving the core at 547°C from the sodium cooled to 351°C.

The sodium contained in the pool tank is kept at atmospheric pressure. The primary pumps only generate a fairly low pressure, which is mainly necessary to overcome the gravity and drag forces in the core and in the IHXs. The large mass of sodium in the pool tank ensures that the whole primary cooling system will react only slowly to increases in power or in the outlet temperature of the core. The radial neutron shield around the core prevents activation of the secondary sodium in the IHXs.

The pool tank is a double walled structure, the space between the primary main vessel and the secondary safety vessel being filled with argon. The pool tank is made of stainless steel welded on site. It is closed at the top by a roof slab which also carries the two rotating plugs with the fuel element transfer and loading machine and the core cover plug with the control and shutdown systems and the instrumentation. On its outer circumference, the roof slab also supports the three IHXs and the three primary sodium pumps. The space between the open primary sodium level in the pool tank and the roof slab is filled with argon as a cover gas.

Two excentric rotating plugs allow the positions of each fuel element, blanket element or radial shielding element to be reached precisely for the loading and unloading procedures. The fuel elements and blanket elements are removed from the core or the radial blanket by means of the fuel element transfer machine and brought into a transfer cask in a loading position. During the loading and unloading processes, the absorber rod drives are uncoupled. All absorber rods are in the core. The fuel element transfer machine can move freely above the core.

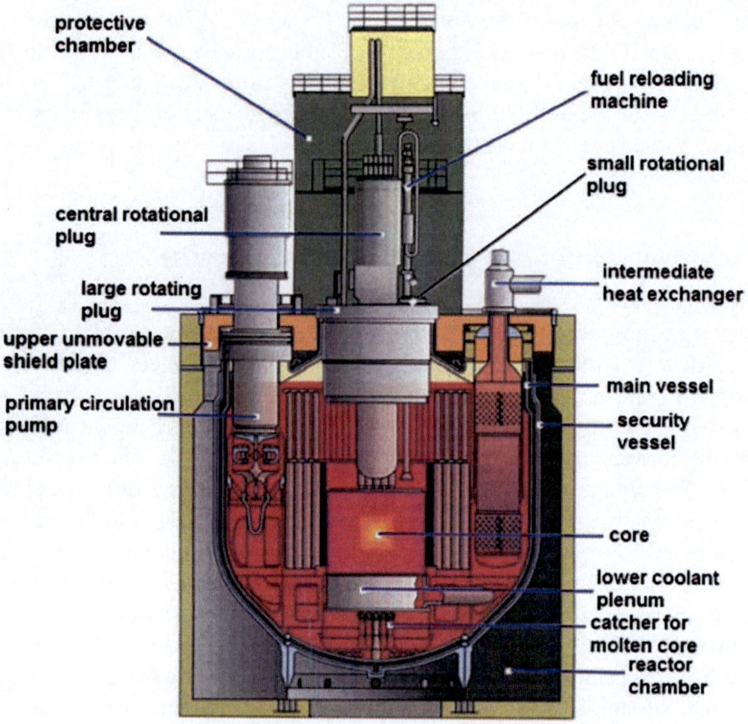

Fig. 6.8 BN 800: Russian sodium cooled fast reactor [3, 10]

Underneath the core support structure and sodium inlet plenum a tank internal core catcher tray structure is located. These core catcher structures are supposed to cool molten core fuel and avoid recriticalities. The whole pool tank is contained in a cavity of reinforced concrete lined with a leaktight steel liner.

6.5.3 Commercial Size SFBR Designs

Numerous studies on commercial size FR designs with a power output of 1,200–1,500 MW(e) were performed in Europe, Russia and Japan since about 1990 [20–25]. The objective of such studies was to investigate the technical and economical feasibility of such large FRs and how they could be introduced into the already existing market of nuclear power plants. One of the more recent design proposals representing a sodium cooled loop type fast reactor (JSFR—Japanese Sodium cooled Fast Reactor) will be described in this section. It incorporates a number of new design ideas for SFR which can become important for future commercial size SFRs.

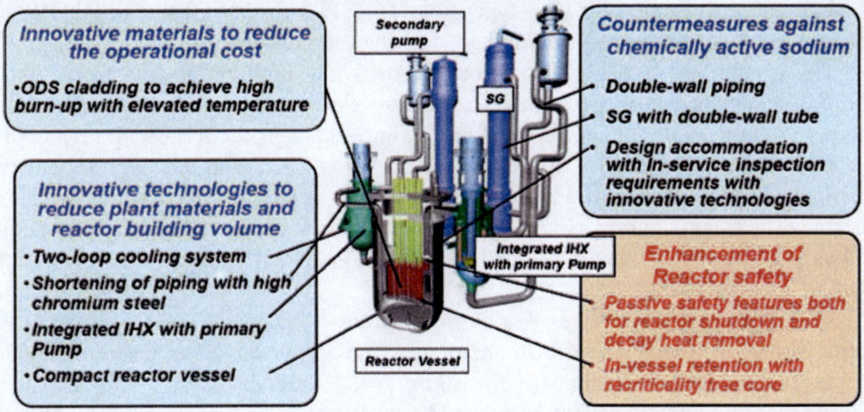

Fig. 6.9 JSFR two loop design with intermediate heat exchangers (IHX) and steam generators (SG) [21]

JSFR shall generate a thermal power of 3,570 MW(th) and an electrical power output of 1,500 MW(e). The reactor core rests on steel support structures and is housed in a sodium filled reactor tank.

PuO_2/UO_2 mixed oxide is used as fuel in the core. The reactor core has two radial zones with different enrichment for radial power flattening. The inner core zone has an enrichment of 18.3% plutonium. The outer zone enrichment is 20.9% plutonium. This leads to a fissile plutonium core inventory of 8.5 t. The breeding ratio is 1.10. The cladding is made of oxide dispersion strengthened steel which must withstand a neutron fluence of 5×10^{23} n/cm^2. It allows a burnup of the core fuel up to 150,000 MWd/t or an average core blanket burnup of 90,000 MWd/t and an operation cycle period of about 26 months. One fourth of the core is unloaded and replaced after these 26 months. The reactor core and radial blanket are surrounded by a core barrel which restrains the core radially to fulfill earthquake safety design requirements.

The free surface of the sodium is covered by argon gas at a pressure of slightly above 0.1 MPa. The reactor tank is covered by a thick shield cover plate with a single rotating plug. The guide structures of the control and absorber rods penetrate this shield plug from above. The fuel element loading and transfer machine rests on one of the excentrically rotating plugs.

The primary radioactive sodium coolant enters the reactor tank with a temperature of 395°C and flows from the lower entrance plenum upward through the core. It is heated up in the core to an outlet temperature of 550°C and flows to the intermediate heat exchangers.

JSFR has only two loop cooling circuits. The primary pumps are integrated into the intermediate heat exchangers. In the secondary non-radioactive loop sodium is pumped to two steam generators where steam at 19.2 MPa and 497°C is produced. The thermal efficiency of the JSFR plant is 42% (Table 6.4).

In comparison to earlier loop type prototype SFRs (MONJU) the 1,500 MW(e) JSFR has much shorter sodium pipings outside of the reactor vessel (Fig. 6.9).

This is achieved by using high chromium steel and simplified geometric configurations with inverse L-shaped pipes. The reactor vessel and all primary and secondary sodium piping are double walled to avoid sodium leackage out of the sodium boundaries. The space in between double walled piping is filled with nitrogen gas which can be heated. Electrical trace heating on sodium piping can be avoided by this design. The steam generators are equipped with especially developed double walled tubes to avoid sodium-water interactions and mitigate their consequences in case of failing steam generator tubes.

JSFR has two independent diverse shutdown systems, one of which is designed with flexible joint absorber parts. This allows absorber insertion under robust restraint conditions in case of earthquakes. A third shutdown system is based on the thermomagnetic properties of ferromagnetic alloy in the control rod holding structures. The shutdown function is initiated passively when the sodium outlet temperature exceeds the Curie point of the holding magnets. This third shutdown system prevents sodium boiling and cladding failure assuming anticipated transients without scram. A re-criticality free core concept is pursued by utilizing fuel subassemblies with an inner duct structure to discharge molten fuel.

Multilayered molten core debris tray structures are arranged underneath the reactor core support structures. This core catcher shall retain molten core fuel, avoid so-called recriticalities and cool the molten fuel (Chap. 12).

Decay heat removal under normal operating conditions can be accomplished by natural convection of the sodium in the primary and secondary coolant circuits. Under accident conditions additional emergency decay heat removal systems start passively. They act on the basis of natural convection of the sodium with sodium-air coolers and dampers. No pumps, no pony motors and no air blowers are needed in such cases.

The inner reactor containment (Fig. 6.10) is a concrete containment clad with inside and outside steel plates which can resist all mechanical and thermal loads in case of accidental conditions. The surrounding outer containment must be designed against external loads, e.g. earthquakes, flooding etc.. As modern pressurized water reactors, e.g. EPR or boiling water reactors, e.g. SWR-1,000 also future SFR containment must have extremely low leakage conditions, and filter systems. Thus, evacuation or relocation of the population outside of the plant can be avoided even in case of severe accidents.

6.6 Lead-Bismuth Cooled FBRs

Based on experience with submarine reactors lead-bismuth cooled FRs were proposed first in Russia and later investigated also in Japan and in Europe [26–38]. Lead-bismuth as coolant had been ruled out in the USA, Europe and Japan because

Table 6.4 Main design characteristics for the Japanese 1,500 MW(e) JSFR (JAEA) [21]

Thermal power (MW(th))	3,530
Electrical power (MW(e))	1,500
Plant efficiency (%)	42
Coolant system	Loop type, sodium
Primary	
Sodium inlet (°C)	395
Sodium outlet (°C)	550
Secondary	
Sodium inlet (°C)	335
Sodium outlet (°C)	520
Tertiary	
Steam pressure (MPa)	19.2
Steam temperature (°C)	497
Core fuel	PuO_2/UO_2[a]
Pu-enrichment	
Inner core zone (%)	18.3
Outer core zone (%)	20.9
Pu fissile inventory (t)	8.5
Breeding ratio	1.10
Maximum discharge burnup (GWd(th)/t)	147
Core fuel elements	
Operation cycle length (months)	26
Maximum neutron fluence (n/cm^2)	5×10^{23}
neutron energy >0.1 MeV	
Burnup reactivity (%)	2.3
Sodium void reactivity (core) ($)	5.3

[a]Low decontaminated TRU-MOX

of corrosion problems. But these problems could be overcome by intensive research and developments [26, 27].

6.6.1 Lead-Bismuth Coolant Properties

Lead-bismuth (44.5% lead and 55.5% bismuth eutectical alloy (LBE)) has a melting point at 125°C and a boiling point at 1670°C [26, 27, 31]. Its density at 400°C is 10.24 g/cm^3. Its thermal conductivity at 400°C is 13.7 W/(m°C) and its heat capacity is 0.146 kJ/(kg·K). Due to its low neutron moderation capabilities the mean neutron energy in an LBE cooled FR is in the range of 200 keV. The excellent thermal properties allow a relatively high power density in the core similar to the case of SFRs. The corrosion properties of LBE require oxygen control and special cladding surface treatment and protection layers [28–30]. LBE does hardly react with oxygen or water and, therefore, simplifies the design of lead-bismuth FRs.

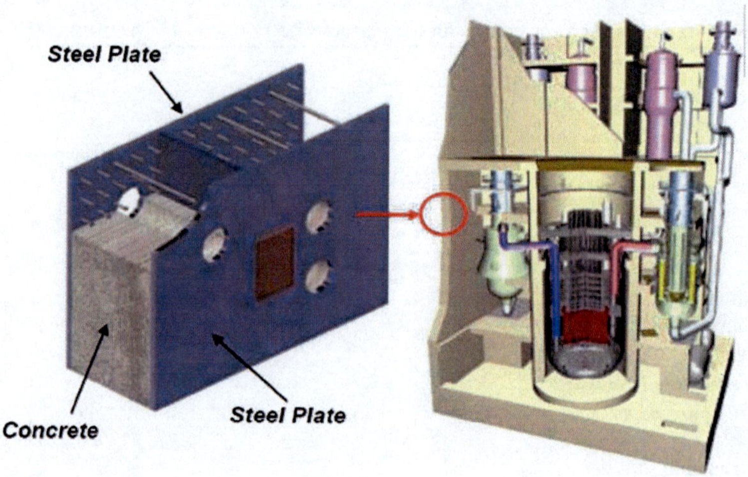

Fig. 6.10 Inner containment structures of JSFR [21]

6.6.2 Design Proposals for Lead-Bismuth FRs

The reactor core of LBE cooled FRs is similar to the core design of SFRs [31–38]. Due to the corrosion concerns the LBE coolant velocity is somewhat lower than 2 m/s.

LBE has a smaller heat transfer coefficient than sodium due to its lower thermal conductivity of 11 W/(m°C). This leads to lower power densities in the core (150–180 kW(th)/l) for LBE instead of 300–400 kW(th)/l for sodium. In order to attain reasonable pumping power, the pressure drop over the fuel element length should be small to counterbalance somewhat the higher density of LBE. Both characteristics: smaller pressure drop and smaller power density, lead to wider spacing in the fuel rod lattice of the subassembly, which is acceptable from the reactor physics point of view because of the low moderation power of heavy liquid metal coolants.

The coolant circuits can be drastically simplified, due to the very low chemical affinity between lead-bismuth and oxygen as well as water.

The steam generators can be directly integrated into a pool type tank (Fig. 6.11). Boiling of LBE does not need to be considered in a safety analysis, because of its high boiling point of 1,670°C. But above temperatures above 1,400°C molten steel could move upwards out of the core due to its lower density and may induce reactivity effects similar to those caused by sodium voiding in sodium cooled FBRs.

Neutron interaction with bismuth results in the formation of the alpha emitter Polonium-210 with a half-life of 138 days. Experience with operation of LBE cooled submarine reactors showed that this aerosol hazard can be reduced at lower temperatures and by solidification at cooler surfaces. Repair of leakages and removal of LBE can be done without exessive radioactive exposures to personnel [26, 27].

Table 6.5 Core design parameters of LBE cooled FRs [26–38]

	SVBR-75/100	BREST-300	LBE-JAEA forced convection
Thermal power (MW(th))	280	700	1,980
Electrical power (MW(e))	106	300	750
Thermal efficiency (%)	38	38	38
Steam pressure (MPa)	9.5	24.5	6
Steam temperature (°C)	306	520	400
Core conversion/			
Breeding ratio	0.87	1.05	1.1
Coolant inlet			
Temperature (°C)	345	420	285
Coolant outlet			
Temperature (°C)	495	540	445
Core diameter (cm)	165	230	445
Core height (cm)	90	110	70
Number of fuel elements	61	145	534
Number of shim and control/shut-down elements	37	28+45[a]	19
Maximum coolant velocity (m/s)	1.8	1.8	2
Average power density (kW(th)/l)	160	150	180
Core fuel	UO_2	PuN/UN	PuN/UN
Maximum fuel burnup (MWd/t)	106,000	92,000	150,000
Operation cycle (years)	8	1	1.5

[a]Hydraulically suspended rods

LBE cooled FRs were proposed for power sizes of 100, 300 and 1,200 MW(e) in Russia [26, 34–40] as well as 750 MW(e) in Japan [32, 33]. However, small size modular lead-bismuth cooled FRs of 100 MW(e) will be constructed first in Russia. Table 6.5 gives the main design characteristics of the Russian small modular type SVBR 75/100 LBE-FR design, of the Russian BREST-300 design and of the 750 MW(e) Japanese design. Figures 6.11 and 6.12 show the main design details of the Russian small modular type SVBR-75/100. The core of the SVBR-75/100 has 61 hexagonal fuel elements which have UO_2 fuel rods with 16.1% U-235 enrichment. PuO_2/UO_2 oxide fuel and PuN/UN nitride fuel can be loaded as well. PuN/UN nitride fuel needs the isotope N-15 for fabrication of this fuel in order to avoid the production of carbon-14. This would request N-15 enrichment plants to be built on a larger scale than available at present.

In case of enriched UO_2 a conversion ratio of $CR = 0.87$ is attained. In case of MOX or Pu/U nitride fuel a breeding ratio of $BR \sim 1$ or slightly above can be achieved. The core diameter is 1.65 m, the core height is 0.9 m. The void effect of the reactor core is negative. The maximum coolant loss reactivity is −2.9 $. Also, the power coefficient is strongly negative (-3.1×10^{-5}/MW(th)).

The hexagonal fuel assemblies have a central position for 22 compensating absorber rods, 12 control and compensating absorber rods, 12 emergency protection rods and 2 automatic control/shutdown rods as well as 6 emergency shut down rods

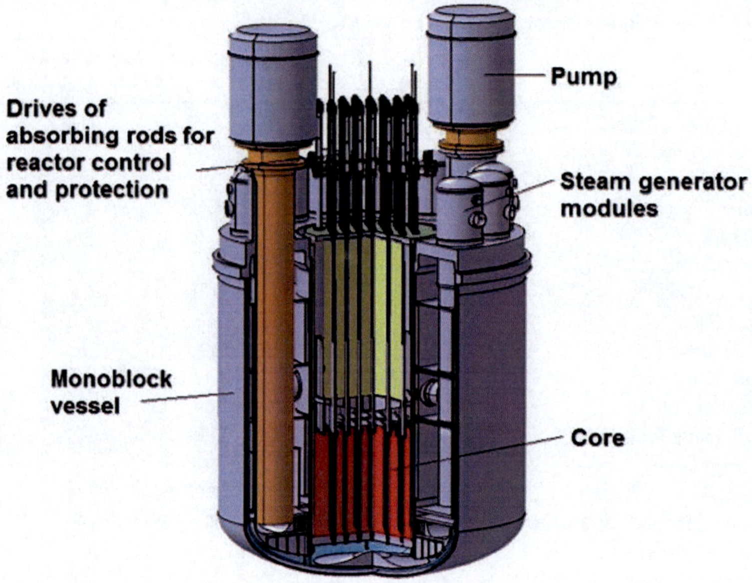

Fig. 6.11 Design of LBE cooled SVBR-75/100 (IPPE) [38]

(Fig. 6.12). Seven assemblies are without absorber rods. The absorber rods contain boron with 50% enriched B-10. In case of overheating of the lead-bismuth coolant (>700°C) fuses connecting the absorber rods and the drive shafts allow a separation of the absorber rods which drop into the core by gravity.

The coolant volume fraction in the fuel assemblies is as low as 25–30% and the coolant velocity is about 2 m/s which is less than that of a typical SFR. The core remains on power for 8 years and is then replaced as a monoblock (no partial fuel reloadings).

Two primary pumps and 12 steam generators are integrated into the LBE filled reactor tank (Fig. 6.11). Intermediate heat exchangers as in the case of SFRs are not necessary. In case of failing primary pumps the heat can be dissipated by natural convection within the reactor tank and to the steam generators. In addition the after-heat can be conducted through the walls of the reactor tank to an outside water basin. Figure 6.13 shows the overall design arrangements of the SVBR-75/100 housed in inner concrete containment structures. An outer concrete containment surrounding the reactor hall etc. protects the reactor system from external events as in case of LWRs, CANDUs, HTRs and as described in Chap. 5. Filter systems will restrict radioactive particle emissions from penetrating to the environment. Several small SVBR-75/100 LBE-FBR plants can also be assembled in a cluster of SVBR-75/100 plants to a large 800 MW(e) cluster or even to a 1,600 MW(e) plant [38].

Design proposals for the Japanese 750 MW(e) lead-bismuth cooled FBR show a similar design approach. Three primary pumps and six steam generators together

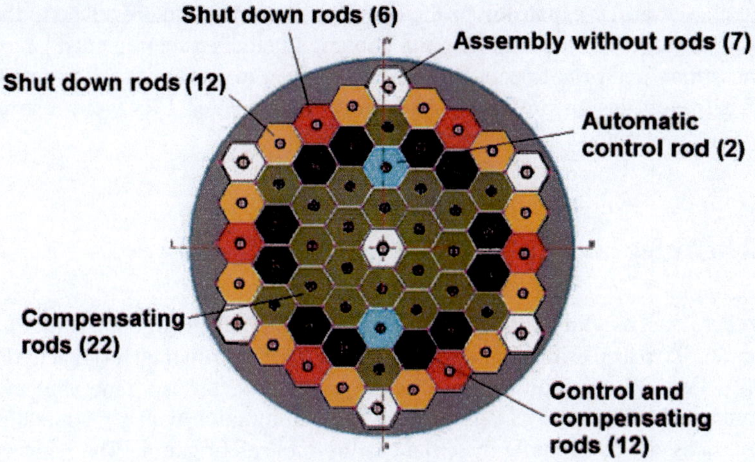

Fig. 6.12 Core, fuel elements and absorber control and shut down rods [35, 36]

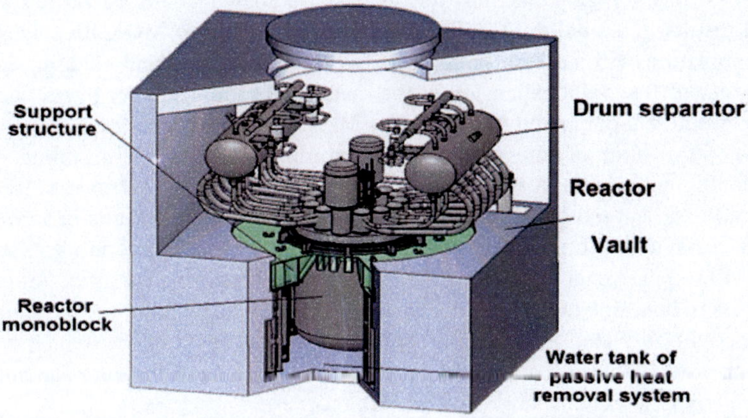

Fig. 6.13 Arrangement of SVBR-75/100 LBE cooled reactor with containment [35, 36]

with the core were also integrated in the LBE filled 14.5 m diameter reactor tank. This LBE-FBR design was proposed as a 550 MW(e) design applying natural convection cooling and as a 750 MW(e) design applying forced convection of the LBE for cooling of the core. The main design parameters of the forced convection design are given in Table 6.5. This design also uses similar systems for shutdown and emergency cooling as described for the commercial size JSFR in Sect. 6.5.3.

The reactor is equipped with three small emergency cooling systems relying entirely on natural convection as described in case of the SFR. The reactor is shut-down by independent and diverse shut down systems. A third shutdown system relies on extended thermal expansion of the control rod guide structures. Due to the relatively high density of the lead-bismuth coolant all steel structures must have larger thickness than in SFRs. Special care must be given to earthquake resistant design. For that reason design proposals for lead-bismuth cooled FRs were restricted to 750 MW(e) in Japan [32, 33].

6.7 The Integral Fast Reactor (IFR)

The integral fast reactor (IFR) concept is being developed by Argonne Natioinal Laboratory (ANL) based on the long experience with the experimental fast reactor EBR-II [41, 42]. The IFR is a sodium cooled pool type fast reactor using metallic uranium-plutonium-zirconium alloy (U-Pu-Zr)-fuel in combination with pyro-metallurgical fuel reprocessing and remote injection casting fuel refabrication. The reactor plant, the pyro-processing plant and the metallic fuel refabrication plant are collocated at one site. The pyro-metallurgical fuel reprocessing and the injection casting fuel refabrication concepts are described in more detail in Sect. 9.6.1.

The use of U-Pu-Zr metal alloy fuel leads to a somewhat harder neutron spectrum which results in a smaller Doppler coefficient compared to MOX fuel. This leads in combination with sodium bonding between the fuel rod and cladding and the possibility of free axial expansion of the metallic fuel to a smaller overall negative power coefficient compared to UO_2/PuO_2 MOX fuel cores [43, 44].

The optimization of other reactivity temperature coefficients, e.g. sodium expansion coefficient, structural expansion coefficients including control rod drive line expansion and natural convection flow of the coolant sodium results in an inherent control behavior of the reactor without reliance on control rod scram systems [44, 45]. Off normal events with very low probability of occurence, e.g. loss of coolant flow, loss of heat sink or run-out of a control rod followed by failure of the shut down systems will only lead to a sodium coolant rise up to about 600°C which is about 200°C below the boiling point of sodium. The decay heat of the core can be safely removed by natural convection flow also in this case.

The low melting point of the U-Pu-Zr metallic alloy does not lead to violent reactions between molten or dispersed fuel and the coolant. The post-dispersal debris from molten fuel will remain in the vessel and will be coolable in molten core retention structures.

IFR core designs were reported for 340, 600 and 1,350 MW(e). Table 6.6 and Fig. 6.14 give an impression of the core design and main design characteristics of a 900 MW(th) or 340 MW(e) IFR design. The IFR was designed as an LMFBR [45–47] with core internal blanket fuel elements or as an FR burner reactor (Advanced burner reactor for the incineration of the transuranium elements (plutonium, neptunium, americium, curium).

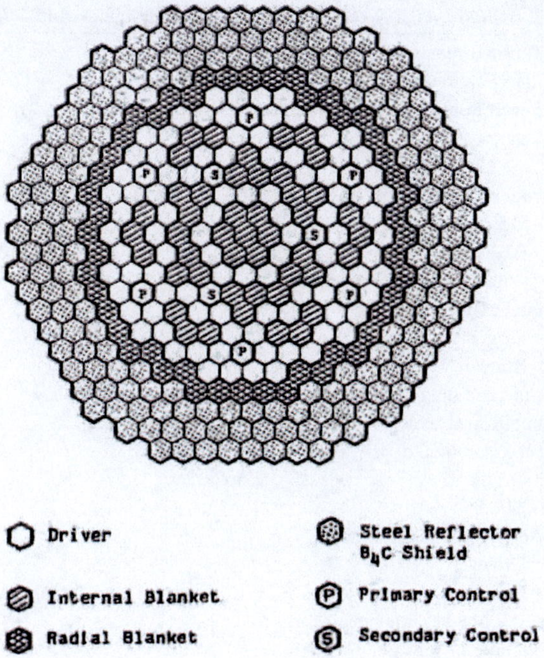

Fig. 6.14 The IFR core design with internal and external radial blanket elements [45–47]

6.8 Structural Materials for LMFBRs

The successful operation of LMFBRs is dependent to a large extent on the performance of structural materials which are used for components of the coolant circuits and of the reactor core. These materials have to withstand sodium corrosion and the various rigours of operation to provide basic integrity [39, 48]. Dimensional stability and integrity must be maintained under conditions of 350 and about 700°C. In case of the fuel elements the high fast neutron flux and very long (2–3 years) in-reactor residence times (resulting in neutron fluences in excess of 10^{23} n/cm^2), elevated temperatures (typically between 350 and 700°C), various mechanical loads (fission gas pressure, interaction stresses between components in contact), and potentially corrosive environments (liquid metal coolant, aggressive fission products). All of these phenomena have to be taken into account in the selection of structural materials (Sect. 6.8.2).

6.8.1 Steels for Coolant Pipes, Pumps, Intermediate Heat Exchangers and Steam Generators

LMFBR primary coolant circuit components (pipes, pumps, intermediate heat exchangers) are characterized by low pressure loads under normal operation, but

Table 6.6 900 MW(th) IFR Core Design Parameters [45–47]

Thermal reactor power MW(th)	900
Electrical reactor power MW(e)	340
Reactor outlet temperature (°C)	510
Reactor ΔT (°C)	135
Core Concept	Heterogeneous
Fuel residence time (cycles)	
Driver	4
Blanket[a]	4
Cycle length (full-power days)	292
Fuel material	
Driver	U-Pu-10% Zr
Blanket	U-10% Zr
Clad and duct material	HT-9
Fuel smear density (% theoretical density)	
Driver	75
Blanket	75
Active fuel height (cm)	
Driver	91
Blanket	112
Axial blanket thickness (cm)	–
Number of pins per assembly	
Driver	271
Blanket	169
Fuel pin diameter (cm)	
Driver	0.72
Blanket	1.0
Pin pitch/diameter ratio	
Driver	1.18
Blanket	1.09
Cladding thickness (cm)	0.056
Duct wall thickness (cm)	0.36
Interassembly sodium gap (cm)	0.98
Assembly lattice pitch (cm)	15.4

[a]Refers to internal and radial blanket

relatively high temperature stresses due to large temperature differences. For pipes, pumps and intermediate heat exchangers, the austenitic steels equivalent AISI 304 and AISI 316 are applied. For compact designs of intermediate coolant circuits of the Japanese JSFR, also high chromium martensitic steel is used. Under normal operation high pressure stresses will only occur in steam generators (19 MPa pressure in steam generator tubes). Short-term accidental pressure loads may occur in primary, secondary and tertiary coolant circuit components as a consequence of sodium-water reactions in steam generators or of dynamic pressure developing in the reactor vessel in the course of a core melt.

The stresses and strains have to be evaluated for all circuit components in the light of material characteristics. At relatively low temperatures comprehensive background information is offered, e.g. by the ASME code, Section III [39], and by the Nuclear Piping Code. At material temperatures in excess of some 430°C (austenitic steels), creep effects have to be taken into account. This is necessary, since at such elevated temperatures creep effects may lead to intolerably high deformations or even to rupture. Moreover, time dependent failure modes such as creep fatigue failure interaction must be considered. For this purpose, in addition to the number and size of anticipated load changes, also the cycle durations and hold times of individual load cycles must be taken into account.

6.8.2 Steels for Claddings and Subassembly Ducts

At an early stage in the development of materials for cladding and subsassembly duct applications, it was recognized that substantial damage to the crystalline lattice of these materials occurs as a result of prolonged exposure to high fluxes of fast neutrons. A convenient measure of this damage is the number of times every atom is displaced from its lattice site during the in-core lifetime of a component: 1 dpa (displacement per atom corresponds to 2×10^{21} n·v·t for E >0.1 MeV). For commercial LMFBRs of the future a value in the 200–250 dpa range is required. Large commercial size FR cores have lower neutron leakage and, therefore, lower average plutonium enrichment. To attain a similar linear rod power of 400–500 W/cm the average neutron flux must be higher than in small size FRs. This higher average neutron flux leads to the higher neutron fluence of 100–250 dpa.

The principal cause of ductility loss in stainless steels irradiated at high temperatures (above 550°C) is helium embrittlement mainly caused by the formation of tiny bubbles of helium. The helium arises from (n, α) reactions and migrates to grain boundaries where it can form bubbles under the influence of an applied stress. The increased helium concentration in the grain boundaries reduces the strength and ductility simultaneously and also reduces the stress-rupture life [49]. In addition, irradiation can also lead to significant changes in the creep strength. The important factor for fuel pin design is that despite grain boundary weakness, ductility saturates at low values for normal operating conditions [48, 49].

Until 1967, the major problem of damage to cladding materials was embrittlement. However, in 1967, evidence was published of considerable swelling taking place in austenitic stainless steels irradiated to high fluences in DFR. This phenomenon, void swelling, has since then tended to dominate the attention of groups involved in the development of cladding and duct alloys and to guide the principal lines of development. Swelling is caused by the production of small cavities as a result of vacancy condensation; it is a direct consequence of the large number of times atoms are displaced from their lattice sites during a component's in-reactor lifetime [48, 49].

In alloys, such as austenitic stainless steels, void swelling occurs only above a fluence threshold of about 10^{22} n/cm^2 (corresponding to a stage at which every atom will have experienced about 5 displacements from its lattice site).

6.8.3 The Different Time Phases of Steel Development for LMFBR Cores

6.8.3.1 First Phase of Austenitic Steel Development

The first generation of steels for claddings and subassembly ducts were austenitic steels. In Western Europe the cold worked austenitic steels [15-15 NiMoTiB (France) or DIN 4970 (Germany)] reached good irradiation experience in about 10,000 fuel rods of Phenix. Peak cladding temperatures of 650°C and neutron fluences up to a maximum of 140 dpa were attained without endurance failures. The mechanical and corrosion resistance properties of these titanium stabilized austenitic steels were extensively investigated [49].

In the USA the cold worked titanium stabilized steel D9 was tested in the FFTF reactor. About 3,000 fuel rods in fuel subassemblies reached a maximum fluence of 140 dpa and peak cladding temperatures, of 675°C. However, at higher neutron fluences the D9 steel exhibited swelling and volume increases of 37%. This type of cladding material was also prone to embrittlement at higher fluences [50].

In Japan and Russia equivalent austenitic steels were developed and applied.

6.8.3.2 Second Phase of Ferritic Martensitic Steel Development

In most countries developing FRs attention was, therefore, focused on ferritic martensitic steels (France, Russia and the USA). These ferritic martensitic steels have considerable smaller void swelling at higher neutron fluences. This is important for large commercial type FBRs with fuel burnups of about 150,000 MWd/t.

In the USA, the Sandvik HT-9 alloy was tested at FFTF and reached record level neutron fluences of 200 dpa without any cladding failure (peak cladding temperatures were 600°C). The HT-9 is considered the reference steel cladding and subassembly duct material for future FBRs [50]. Figure 6.15 shows of the low swelling behavior of HT-9 in comparison to D9 and titanium stabilized austenitic steel CW 316.

6.8.3.3 Third Phase of Oxide Dispersed Steel Development

Ferritic martensitic alloys have been optimized improving their high temperature strength by adding TiO_2 or YO_2. This lead to the oxide dispersion strengthed (ODS) steels. These ODS steels have been particularly improved by JAEA (Japan) and

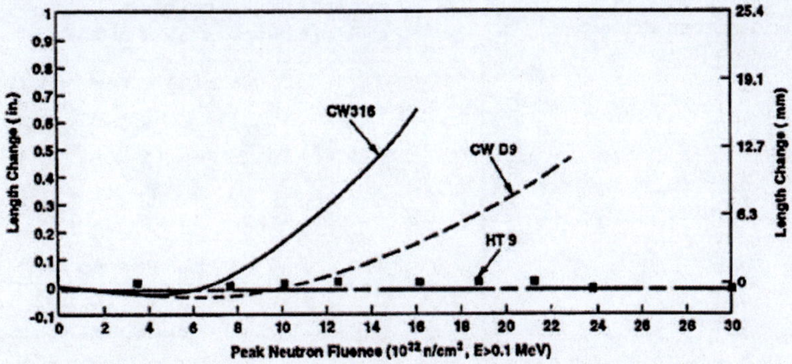

Fig. 6.15 Experience of FFTF subassembly length change with various types of duct materials [50]

in Russia and France. Oxide dispersion strengthening is a mechanism which will enable cladding temperatures well above 650°C. They are considered as the most promising candidate for cladding and subassembly duct material for future large scale commercial SFRs in Japan and Russia [51, 52]. Figure 6.16 shows a comparison of austenitic steels as well as ferritic martensitic steels and ODS for displacement doses up to 160 dpa obtained in Phenix [52].

6.8.4 Conclusions on Cladding and Subassembly Duct Material Development

World-wide research and development work on various cladding materials have reached a very high level of understanding of the basic phenomena involved as well as of the operational requirements to be met by a well-designed fuel element. Austenitic alloys have proved their ability to reach exposure rates as high as 150 dpa and their advanced versions are very promising candidates for target doses up to 170 dpa. Ferritic/Martensic alloys for subassembly ducts are able to fulfill the highest objectives of target doses up to 200 dpa. For more ambitious targets (over 200 dpa), a large amount of R&D is still required for qualification of the most promising candidates, namely the ODS steels.

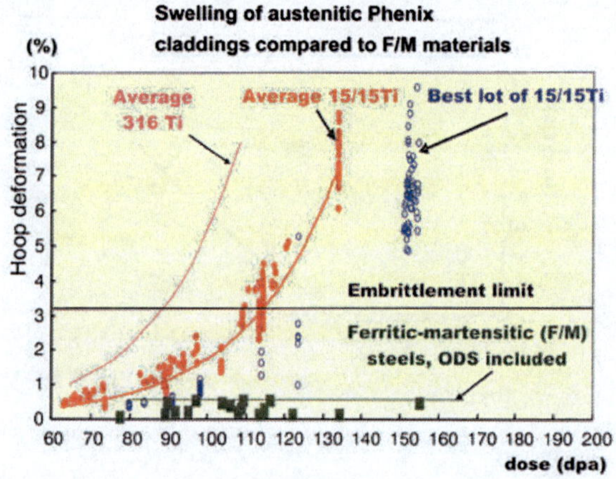

Fig. 6.16 Comparison of austenitic, ferritic-martensitic and ODS steels [52]

6.9 LMFBR Cores With Advanced Oxide, Carbide, Nitride and Metallic Fuels

6.9.1 Mixed Oxide PuO_2/UO_2 Fuel

A large amount of experience on the irradiation of mixed oxide PuO_2/UO_2 fuel pins has been gained during 30 years of FBR operation [3, 11].

In Europe, more than 7,000 fuel rods have reached a burnup of 15 at.% or 150,000 MWd/t. In addition some experimental fuel rods with solid or annular pellets have attained a burnup of 16.9 at.% (Phenix) and 23.5 at.% (PFR) [49].

In the USA, more than 63,500 fuel rods were irradiated during 16 years operation of FFTF and more than 3,000 fuel rods reached a maximum burnup of 24.5 at.% [50, 53].

In Russia, a large amount of experience was gained with vibrocompacted MOX fuel [54]. A record burnup of 33 at.% was reached with one fuel element at BOR 60.

This large statistical experience demonstrates that about 20 at.% burnup (170–190 dpa) can be safely reached with annular pellets (95% of theoretical density and smear density 85%), 520 W/cm max. linear power and 650°C max. cladding temperature.

Nevertheless, some confirmatory evidence is necessary with respect to cladding material with higher than 170 dpa and fuel cladding interaction.

The further development of mixed oxide fuel is, therefore, mainly aimed at achieving higher burnup, slightly higher power ratings of the fuel rods, and slightly smaller wall thicknesses of the cladding tubes and fuel element cans.

6.9.2 Mixed Carbide PuC/UC Fuel

Mixed carbide PuC/UC fuel development has been underway for many decades within the major LMFBR development projects. Compared to mixed oxide fuel, the main advantages of carbide fuel are its higher density of about 12.4 g/cm^3, smaller fraction of moderating material and higher thermal conductivity, resulting in a rated power of 800–1,100 W/cm of fuel rod length which is a factor of two higher than attainable in mixed oxide fuel [55, 56].

219 (U, Pu)C fuel rods were tested during 12 years operation of the FFTF fast reactor in the USA. In the BOR 60 reactor (Russia) PuC/UC carbide fuel was tested up to 10.4 at.% burnup at 700 W/cm linear rod power and a max. cladding temperature of 680°C.

This leads to smaller cores with lower fissile inventories, lower fuel temperatures and higher breeding ratios, than in advanced mixed oxide fuel. However, mixed carbide PuC/UC fuel is pyrophoric (problems during fabrication). In addition the claddings tend to be enriched with carbon during irradiation.

6.9.3 Mixed Nitride PuN/UN Fuel

PuN/UN mononitride fuel has even somewhat higher density of about 13.1 g/cm^3 and almost the same high thermal conductivity if compared to PuC/UC carbide fuel. This is the reason why it is considered for LBE-FBR projects. In contrast to mixed carbide fuel it is not pyrophoric. However, nitride fuel results in the formation of C-14. Carbon-14 can lead to considerable environmental hazards. Therefore, this problem must be overcome by using nitrogen enriched in N-15. This would require the buildup of considerable industrial enrichment capacity in the future [55, 56].

54 nitride (Pu,U) N fuel pins were successfully tested in the FFTF test reactor in the USA. Also some nitride fuel rods were irradiated in JOYO and Phenix (France). In the Russian BR-10 about 66 fuel pins were successfully irradiated up to 9 at.%.

6.9.4 Metallic U/Pu Fuel

Metallic fuel in the form of U-10%Zr alloy has been used as fuel in early small FBRs, e.g. EBR-II and DFR. This metallic fuel has a density of about 14 g/cm^3 and a thermal conductivity which is roughly a factor of 2 higher than for the mono carbide or mono nitride mixed U/Pu fuel. This leads to lower fuel temperatures, smaller core inventories and higher breeding ratio. Metallic fuel was proposed as U-Pu-10%Zr metallic fuel for the integral fast reactor project (IFR) by Argonne National Laboratory (ANL) in the USA [55, 57–59].

In FFTF (USA) 1,000 fuel rods with Pu-U-Zr and the ferritic martensitic steel HT-9 as cladding and duct materials were irradiated. The highest burnup reached in EBR-II with Pu-U-Zr-fuel was 20 at.%. In the FFTF test reacor (USA), the metallic fuel rods attained a burnup of 15 at.% and 110 dpa at a lower max. fuel pin power of 548 W/cm.

References

1. Häfele W et al (1970) Fast breeder reactors. Ann Rev Nucl Sci 20:393–434
2. Kessler G (1983) Nuclear fission reactors. Springer, Wien
3. Status of liquid metal cooled fast breeder reactors (1985) Technical report series No. 246, IAEA, Vienna
4. Nuclear Engineering International (1971) PHENIX—a special feature on the french prototype fast breeder reactor. Nucl Eng Int 16:557–580
5. Kazachkovskij OD et al (1977) The present status of the fast reactor programme in the USSR. In: Nuclear power and its fuel cycle, Proceedings of the International Conference, Salzburg, IAEA, Vienna, 2–13 May 1977
6. Nuclear Engineering International (1971) PFR—a special feature on the British prototype fast breeder reactor. Nucl Eng Int 16:629–650
7. Nuclear Engineering International (1972) Fast Flux Test Facility (FFTF)—A special survey. Nucl Eng Int 17:613–628
8. Leipunskii AI et al (1968) A nuclear power station with the BN-600 reactor. Sov At Energy (Atomnaya Energiya) 25:1216–1221
9. Carle R (1975) SUPERPHENIX: first commercial plant of the fast breeder line. J Br Nucl Energy Soc 14:183–190
10. IAEA Fast reactor data base, IAEA-TECDOC-1531 (2006)
11. Adamov EO, Fuji-ie Y (2011) Developments and tendencies in fission reactor concepts. In: Vértes et al. (eds) Handbook of nuclear chemistry, vol 6. Springer, Heidelberg, p 2663–2730
12. Waltar AE, Reynolds AB (1981) Fast breeder reactors. Pergamon Press, New York
13. Stacey W (2001) Nuclear reactor physics. Wiley, New York
14. Hummel HH, Okrent D (1979) Reactivity coefficients in large fast power reactors. American Nuclear Society, Hinsdale
15. Nicholson RB, Fischer EA (1968) The doppler effect in fast reactors. Adv Nucl Sci Technol 4:109–195
16. Häfele W (1963) Prompt überkritische Leistungsexkursionen in schnellen Reaktoren. Nukleonik 5:201–208
17. Judd AM (1981) Fast breeder reactors—an engineering introduction. Pergamon Press, Oxford
18. Yevick JG, Amorosi A (1966) Fast reactor technology. The MIT Press, Cambridge
19. Tang YS et al (1978) Thermal analysis of liquid metal fast breeder reactors. American Nuclear Society, LaGrange Park
20. Ichikawa K et al (2009) Conceptual design study of JSFR-reactor cooling system. In: Proceedings of the international conference on fast reactors and related fuel cycles: challenges and opportunities (FR09), Kyoto, Japan, 7–11 Dec 2009
21. Kotake S et al (2005) Feasibility study on commercialized fast reactor cycle systems, current status of FR systems design. In: Proceedings of GLOBAL 2005, Tsukuba, Japan
22. Ogura M et al (2009) Conceptual design study of JSFR: overview and core concept. In: Proceedings of the international conference on fast reactors and related fuel cycles: challenges and opportunities (FR09), Kyoto, Japan, 7–11 Dec 2009
23. Etoh M et al (2009) Conceptual design study of JSFR—reactor system. In: Proceedings of the international conference on fast reactors and related fuel cycles: challenges and opportunities (FR09), Kyoto, Japan, 7–11 Dec 2009

24. Ohyama K et al (2009) Decay heat removal by natural circulation for JSFR. In: Proceedings of the international conference on fast reactors and related fuel cycles: challenges and opportunities (FR09), Kyoto, Japan, 7–11 Dec 2009
25. Kurome K et al (2009) Steam generator with straight double-walled tube. In: Proceedings of the international conference on fast reactors and related fuel cycles: challenges and opportunities (FR09), Kyoto, Japan, 7–11 Dec 2009
26. Subbotin V et al (1994) Lead-bismuth cooled fast reactors in nuclear power of the future. In: Proceedings of the international topical meeting on advanced reactor safety, ARS'94, Pittsburgh, pp 524–529
27. Tupper R et al (1991) Polonium hazards associated with lead-bismuth used as reactor coolant. In: Proceedings of the international conference on fast reactors and related fuel cycles, Kyoto
28. Weisenburger A et al (2009) Corrosion, corrosion barrier development and mechanical properties of steels foreseen as structural materials in liquid lead cooled nuclear systems. In: Proceedings of 17th international conference on nuclear engineering (ICONE 17), Brussels
29. Müller G et al (2003) Control of oxygen concentration in liquid lead and lead-bismuth. J Nucl Mater 321:256–262
30. Weisenburger A et al (2006) Behaviour of chromium steels in liquid Pb-55.5 Bi with changing oxygen content and temperatures. J Nucl Mater 358:69–76
31. Li N (2008) Lead alloy coolant technology and materials-technology readiness level evaluation. Prog Nucl Energy 50:140–151
32. Enuma Y et al (2003) Conceptual design of a medium scale lead-bismuth cooled fast reactor, GENES4/ANP2003, Kyoto, Japan
33. Hayafune H et al (2005) Conceptual design study of Pb-Bi cooled fast reactor plant system in the feasibility study in Japan. In: Proceedings of GLOBAL 2005, Tsukuba, Japan, 9–13 Oct 2005
34. Dollezhal N (2010) A naturally safe, cost effective reactor BREST with electric power of 300 and 1200 MW(e). http://www.nikiet.ru/eng/structure/mr-innovative/brest.html
35. Todreas NE et al (2004) Medium-power lead-alloy reactors: missions for this reactor technology. Nucl Technol 147:305–320
36. Zrodnikov A et al (2004) SVBR-75/100 multipurpose modular low power fast reactor with lead-bismuth coolant. At Energy 97(2):528–533
37. Gromov B et al (1997) Use of lead-bismuth coolant in nuclear reactors and accelerator driven systems. Nucl Eng Des 173:207–217
38. Zrodnikov A et al (2008) Innovative nuclear technology based on modular multi-purpose lead-bismuth cooled fast reactors. Prog Nucl Energy 50:170–178
39. ASME Boiler and Pressure Code (2007) Nuclear plant components. Section III, rules for the construction of nuclear power plant components
40. Power reactors and subcritical blanket systems with lead and lead-bismuth as coolant and/or target material. IAEA-TECDOC-1348 (2003)
41. Wade D et al (1997) The decision rationale of the IFR. Prog Nucl Energy 31(1/2):13–42
42. Wade D et al (1988) The integral fast reactor concept: physics of operation and safety. Nucl Sci Eng 100:507–524
43. Wade D et al (1997) The safety of the IFR. Prog Nucl Energy 31(1/2):63–82
44. Wigeland R et al (2009) Mitigation of severe accident consequences using inherent safety principles. In: Proceedings of the international conference on fast reactors and related fuel cycles: challenges and opportunities (FR09), Kyoto, Japan, 7–11 Dec 2009
45. Wade D et al (1988) Trends versus reactor size of passive reactivity shutdown and control performance. In: Proceedings of the international conference on reactor physics, Jackson Hole, USA, 18–21 Sept 1988
46. Kim TK et al (2009) Core design studies for a 1000 MW(th) advanced burner reactor. Ann Nucl Energy 36:331–336
47. Dunn FE et al (2006) Preliminary safety evaluation of the advanced burner test reactor, ANL-AFCI-172

48. IAEA (1985) Status of liquid metal cooled fast reactors. Technical report series No. 246, IAEA, Vienna
49. Brown C et al (1993) Status of LMR fuel development in Europe. J Nucl Mater 204:33–38
50. Leggett R et al (1993) Status of LMR fuel development in the United States of America. J Nucl Mater 204:23–32
51. Yoshitake T et al (2004) Ring-tensile properties of irradiated oxide dispersion strengthened ferritic/martensitic steel claddings. J Nucl Mater 329–333(Part 1):342–346
52. Yvon P et al (2009) Structural materials challenges for advanced reactor systems. J Nucl Mater 385:217–222
53. Baker R et al (1993) Status of fuel, blanket, and absorber testing in the fast flux test facility. J Nucl Mater 204:09–118
54. Herbig R et al (1993) Vibrocompacted fuel for the liquid metal reactor BOR-60. J Nucl Mater 204:93–101
55. Pahl RG et al (1992) Irradiation behavior of metallic fast reactor fuels. J Nucl Mater 188:3–9
56. International Nuclear Fuel Cycle Evaluation (1980) Fast breeders. Report of INFCE Working Group 5, International Atomic Energy Agency, Vienna
57. Lahm C et al (1993) Experience with advanced driver fuels in EBR-II. J Nucl Mater 204: 119–123
58. Pitner A et al (1993) Metal fuel test program in the FFTF. J Nucl Mater 204:124–130
59. Bridges AE et al (1993) A liquid-metal core demonstration experiment using HT-9. Nucl Technol 102:353–366

Chapter 7
Technical Aspects of Nuclear Fuel Cycles

Abstract After discharge from the reactor core, the fuel elements are stored in a fuel element storage pool onsite for several years to allow radioactivity decay and after heat decrease. Spent fuel elements are shipped then in special fuel transport casks to either intermediate storage facilities or to the storage pool of a reprocessing plant. After a total cooling period of about 7 years LWR spent fuel elements can be chemically reprocessed. The spent fuel elements are moved from the storage pool into the disassembly cell, where they are cut up by large bundle shears into small pieces. These pieces fall into a dissolver basket filled with boiling nitric acid. The PUREX process is used to chemically separate the dissolved spent fuel into uranium, plutonium and higher actinides with fission products. The final products are uranylnitrate, plutonium nitrate and high level waste. The total capacity of commercial reprocessing facilities is currently about 4,500 t_{HM}/year in France, UK, Russia, Japan and India. The uranium and plutonium products can be converted into oxides and fabricated into Uranium/Plutonium mixed oxide fuel elements. The latter can be loaded into light water reactor or fast breeder reactor cores. Thorium/uranium fuel can be reprocessed using the THOREX process. The thorium/uranium-233 fuel also can be fabricated into mixed oxide fuel elements and loaded into light water reactors or fast breeder reactors. The remaining wastes are classified into high level waste, medium level waste and low level waste. The high level waste after concentration is vitrified by giving it first into a calcinator and then mixing it with borosilicate glass frits and melting this mixture to a glass. The result is a vitrified high level glass in a steel container. The fuel rod hulls and end pieces of fuel elements as well as insoluble residues are compacted by a 250 MPa press into a cylindrical container. Low level organic waste is sent to a medium temperature pyrolysis system and then to a calcination system. The end product is mixed with pastes, grouts or concrete and filled into low level waste containers. The medium and low level waste packages are sent to medium/low level waste repositories which are already in operation in France, Japan, Spain, Sweden, Finland and the USA since the early 1990s. High level waste packages are foreseen to be disposed into deep geological repositories. For the direct disposal concept of spent fuel elements either the fuel elements or only the

G. Kessler, *Sustainable and Safe Nuclear Fission Energy*, Power Systems,
DOI: 10.1007/978-3-642-11990-3_7, © Springer-Verlag Berlin Heidelberg 2012

fuel rods are loaded in high level waste containers and foreseen to be disposed in a deep geological repository. Up to now no deep geological repository is in operation, but test sites are explored and under investigation.

7.1 Discharge and Storage of Spent Fuel Elements

After discharge from the reactor core, the fuel elements are stored in the fuel element storage pool on site for a period of up to about three years or somewhat more to allow for radioactivity decay and cooling. The reactor plant usually has a fuel storage capacity of at least three years discharge volume in addition to a full standby core inventory. The use of compact storage racks with neutron absorbers allows this storage capacity even to be extended to nine years discharge volume. The fuel element storage pool must be located within the outer containment (see Chaps. 5 and 11) which protects also against external events, e.g. earthquakes, floods, tsunamis etc. Spent fuel elements are then transported in spent fuel transport casks either to intermediate storage facilities or to storage pools at reprocessing plants. The intermediate storage facilities can be also located at the site of the reactors. Table 7.1 shows the fuel characteristics of discharged fuel elements for different converter reactors and LMFBRs.

7.1.1 Shipping Spent Fuel Elements

Spent fuel elements are shipped in special fuel transport casks, which weigh between 60 and some 120 t and have load capacities for about 12 t of spent fuel (Fig. 7.1; Table 7.2). Fuel casks are usually transported by special trucks on the road or on special rail cars. Also barge shipments on both inland waterways and oceans are made. The spent fuel elements are cooled within the casks either by air (dry casks) or water (wet casks) [1–5].

The transport casks contain the necessary shielding with steel, lead and water or borated water. They are cooled by natural airflow over fins on the outer surface or by forced air circulation. Spent fuel transport casks are designed to withstand severe accident conditions during shipment. Releases of radioactivity under such conditions must be rendered impossible. Therefore, the transport casks must be able to withstand such impacts as thermal tests (fire), drop tests under gravity, penetration tests, and water immersion tests before being cleared for actual shipment. Special international shipping regulations have been elaborated [6, 7].

Table 7.1 Fuel element characteristics for different nuclear reactors after full burnup and discharge from the reactor core (INFCE) [1]

Characteristics of fuel elements	LWR				HWR CANDU	Gas cooled reactor		LMFBR
	PWR		BWR					
	KWU-1,300	EPR	BWR-1,280	SWR-1,000		AGR	HTR 200	Superphenix
Power (MW_{el})	1,300	1,600	1,280	1,250	680	620	200	1,250
Total length (mm)	3.9	4.2	3.74	3.0	0.495	1.05		5.4
Diameter (mm)	–	–	–	–			60 (sphere)	
Cladding material	Zircaloy 4 or M5		Zircaloy 2		Zircaloy 4	Stainless steel	Graphite	Stainless steel
Heavy metal weight per element(kg)	648	635	173	205	13.4–19.8	42.7		155
Fuel	UO_2	UO_2	UO_2	UO_2	UO_2	UO_2	UO_2	UO_2/PuO_2
Design burnup ($MWd_{(th)}$/kg)	55	70	56	65	6.5–8.1	10–25	100	70–100
Total activity after unloading (10^{10} Bq/kg)								
150 days	1.7×10^4		1.4×10^4			4.4×10^3	5.2×10^4	2.4×10^4
1 year	8.5×10^3		7×10^3		2.9×10^3	2.2×10^3	2.0×10^4	1.4×10^4
10 years	1.2×10^3		1.1×10^3		3.1×10^2	3.8×10^2	4.1×10^3	2.6×10^3
Decay heat after (W/kg):								
150 days	24.3		18.7			4.9	60	27
1 year	10.4		8.2		3.15	2.4	26	14
10 years	2.3		2.2		0.22	0.3	4.2	1.3
Annual discharge:								
Number of fuel elements	40		130–150		4,560	820	1.2×10^5	154

Table 7.2 Design characteristics of transport casks (GNS) [2–4]

Country	Type	Number of fuel subassemblies (FSA)	Total weight (t)	Height/ diameter (m)
Germany	CASTOR V/19	19 PWR-FSA	121	5.86/2.44
	CASTOR V/52	52 BWR-FSA	123	5.45/2.44
	CASTOR IIa	9 PWR-FSA	116	6.01/2.48
	CASTOR 440/89	84 WWER–440 FSA	116	4.08/2.66
France	TN 13/2	12 PWR-FSA	105	5.60/2.5
Great Britain	Excellox 4	7–15 PWR-FSA	91	5.6/2.2

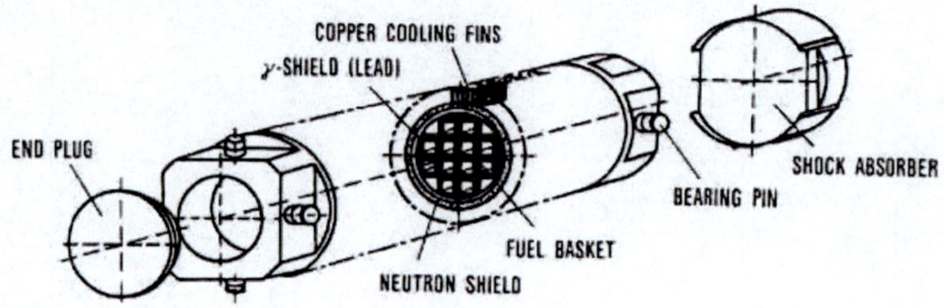

Fig. 7.1 Spent fuel shipping cask for LWR fuel elements [8]

7.1.2 Interim Storage of Spent Fuel Elements

Spent fuel elements can be stored for interim periods in water pools (wet storage), air cooled vaults (dry storage) or in special containers [1, 4, 5, 7]. For wet storage in intermediate storage pools or storage pools of reprocessing plants, the spent fuel elements are arranged in racks or baskets kept in water pools. The water serves as a heat transfer medium for the heat generated in the fuel elements and provides the necessary shielding of the fuel elements. It is maintained at a sufficiently high level to provide shielding during all fuel handling operations. The walls and floors of storage pools are made of reinforced concrete lined with stainless steel. The outer reinforced concrete walls protect the water pools against external events, e.g. earthquakes, floods etc. (Chap. 11). Water pools with a capacity of up to several 1,000 t of spent fuel are technically feasible. An intermediate storage facility may be equipped with several such water pools.

Figure 7.2 shows an intermediate storage facility with water pools for LWR spent fuel elements. It has a storage capacity of 1,500 t of uranium or 5,400 LWR fuel elements and consists of four water pools for intermediate storage, two pools for receiving and two pools for discharging spent fuel elements. All water pools are equipped with heat exchange systems to keep the temperature at about 40°C. The water is purified by ion exchange to ensure the specified water quality and good visibility. Fission and corrosion products are eliminated by the purification system.

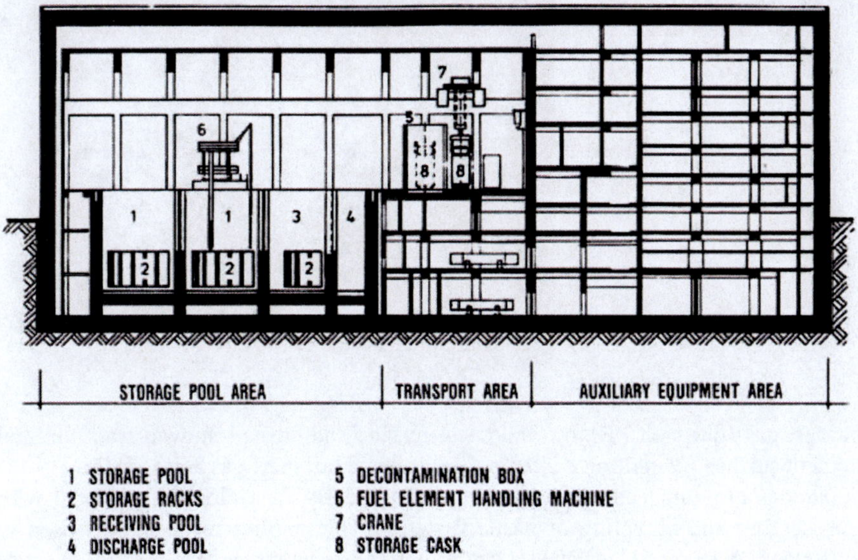

Fig. 7.2 Spent fuel intermediate storage in water pools [5]

7.2 The Uranium-238/Plutonium Fuel Cycle

The intermediate storage facilities are equipped with lifting, handling and transfer devices to handle the spent fuel elements and heavy fuel transport casks. Criticality aspects have to be taken into consideration in the design and construction of fuel element storage racks or baskets for enriched fuel elements in storage pools. LWR spent fuel elements can be stored, if needed, in water pools for many decades [9]. During this time period, the fuel elements will not experience remarkable water corrosion on their outer surfaces.

Dry storage of LWR spent fuel elements is also feasible in air cooled storage casks made of cast iron. These cast iron spent fuel casks take up to 19 PWR or 52 BWR fuel elements. They are equipped with outside cooling fins and can be stored in large intermediate storage buildings over many decades (Fig. 7.3) [7].

Dry storage is also used for HWR spent fuel elements and HTGR graphite fuel elements. Spherical graphite fuel elements of HTRs can be stored under dry conditions in gastight cans.

After years of intermediate storage [9], spent fuel elements may be conditioned for permanent storage without reprocessing (see Sect. 7.6.2) or they are sent to a reprocessing plant for chemical reprocessing (see next section).

LMFBR fuel elements are kept first in sodium cooled storage tanks within the reactor containment (Chaps. 6 and 11). For intermediate storage they are filled in cans, cooled either by sodium and then stored under water or only cooled by air or

Fig. 7.3 Intermediate storage facility (dry storage) with CASTOR casks [7]

an inert gas (nitrogen). Before reprocessing, the sodium is removed from the fuel element surface by melting or steam cleaning in a hot inert gas atmosphere.

Natural uranium can be utilized more efficiently in a closed fuel cycle with reprocessing and recycling of fissile material. This applies to spent fuel used in LWRs or HWRs. For HTGR/HTR spent fuel, reprocessing is very difficult because the fuel particles are coated by pyrolytic carbon and silicon carbide. These layers must be crushed. Prior to these procedures the graphite must be removed (Sect. 7.3.1). For near-breeders and FBRs, the closed fuel cycle is imperative. Technical aspects of reprocessing and recycling (refabrication) in the uranium/plutonium fuel cycle will be described in the following sections.

7.2.1 Reprocessing of Spent UO_2 Fuel Elements

Spent fuel elements with UO_2 fuel and stainless steel or zircaloy claddings are transported to the reprocessing plant and stored there prior to chemical reprocessing. The steps of disassembly of such fuel elements, dissolution of the fuel as well as chemical separation are the same in principle for all fuel elements of converter reactors operated on UO_2 fuel (PWR, BWR, HWR, AGR). It is therefore sufficient to describe, as representative example, the technical steps of fuel disassembly and dissolution in the so-called head end of the reprocessing plant for spent UO_2 fuel from an LWR.

7.2.1.1 Radioactive Inventories of Spent Fuel from LWRs

Figure 7.4 shows the inventories of uranium, plutonium, fission products for 1 t of spent LWR fuel after a burnup of 60,000 MWd/t. These data are based on 1 t of heavy metal (HM) fuel. In that case, roughly 1.14 t of UO_2 or UO_2/PuO_2 correspond to 1 t_{HM}. When loaded into the core, 1 t_{HM} of fresh LWR fuel in an equilibrium cycle with 5% U-235 enrichment contains 50 kg of U-235 and 950 kg of U238. When

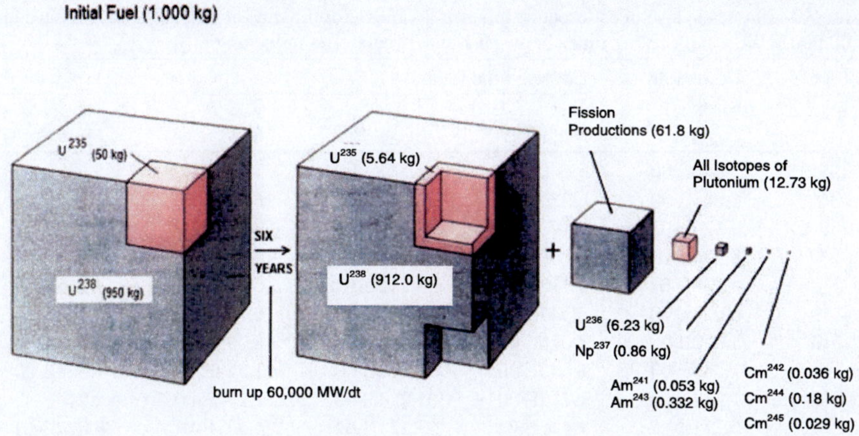

Fig. 7.4 Mass content in 1 ton of fresh LWR and spent LWR fuel after a burnup of 60,000 MWd/t (prior to unloading from the core) [10]

unloaded from the LWR core after a burnup of 60,000 MWd/t, one ton of spent fuel still contains about 5.64 kg of U235 and 912.0 kg of U-238, but 6.23 kg of U-236, some 12.73 kg of different plutonium isotopes, about 61.8 kg of fission products. 0.86 kg of Np-237 and 0.053 kg Am-241 and 0.332 Am-243 as well as 0.036 kg Cm-242, 0.18 kg Cm-244 and 0.029 Cm-245.

Figure 7.4 shows the masses of the actinides (isotopes of neptunium, uranium, plutonium, americium and curium) as well as of the fission products in one ton of spent fuel after a burnup of 60,000 MWd/t. After unloading for transfer to the spent fuel storage pool and later transport, the actinides continue to decay. This is shown by Table 7.3 where the masses are listed for cooling times of 1–7 years. The shortest half-lives are valid for the isotopes Cm-242 (half-life 163 days), Pu-241 (half-life 14.6 years), Cm-244 (half-life 18.1 years), Cm-243 (half-life 32 years), Pu-238 (half-life 88.9 years), Am-241 (half-life 433 years). The masses of these isotopes will decrease (mainly Pu-238, Pu-241, Cm-242, Cm-244) whereas others increase, e.g. Am-241.

The decay of Pu-241 and the build-up of Am-241 must be accounted for when the spent fuel is reprocessed 5–7 years after unloading and the plutonium is utilized for refabrication of MOX fuel elements.

7.2.1.2 LWR Fuel Element Disassembly and Spent Fuel Dissolution

In a reprocessing plant (Fig. 7.5 shows the head end of such a plant), the storage pools are arranged close to the fuel element disassembly cells [11–14]. The fuel elements are moved by means of a crane from the storage pool into the disassembly cell above it. In this cell, LWR fuel elements are cut up by large bundle shears. After the end caps have been removed from the fuel elements, the fuel rod bundles are

Table 7.3 Masses in kg/t of different isotopes of actinides for the time of unloading after a burnup of 60,000 MWd/t and cooling times between 1 and 7 years after unloading [10]

Isotope	Unloading Mass (kg/t)	Cooling time (years) 1 Mass (kg/t)	3 Mass (kg/t)	5 Mass (kg/t)	7 Mass (kg/t)
U-235	5.640E 00	5.640E 00	5.641E 00	5.641E 00	5.641E 00
U-236	6.233E 00	6.233E 00	6.234E 00	6.234E 00	6.235E 00
U-238	9.120E 02	9.120E 02	9.120E 02	9.120E 02	9.129E 02
Np-237	8.596E-01	8.767E-01	8.775E-01	8.770E-01	8.804E-01
Pu-238	4.839E-01	5.118E-01	5.110E-01	5.033E-01	4.955E-01
Pu-239	6.028E 00	6.145E 00	6.145E 00	6.145E 00	6.145E 00
Pu-240	3.070E 00	3.076E 00	3.088E 00	3.099E 00	3.109E 00
Pu-241	1.936E 00	1.845E 00	1.676E 00	1.522E 00	1.378E 00
Pu-242	1.214E 00	1.214E 00	1.214E 00	1.214E 00	1.214E 00
Am-241	5.271E-02	1.435E-01	3.121E-01	4.647E-01	6.027E-01
Am-243	3.324E-01	3.328E-01	3.327E-01	3.326E-01	3.326E-01
Cm-242	3.645E-02	7.768E-03	3.493E-04	1.572E-05	7.250E-07
Cm-243	1.175E-03	1.147E-03	1.092E-03	1.040E-03	9.910E-04
Cm-244	1.801E-01	1.737E-01	1.609E-01	1.491E-01	1.381E-01
Cm-245	2.938E-02	2.937E-02	2.937E-02	2.936E-02	2.936E-02

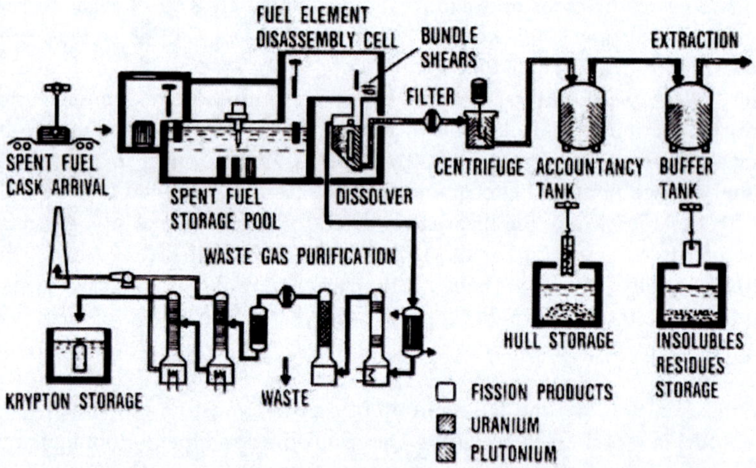

Fig. 7.5 Head end and waste gas purification system of a chemical reprocessing plant [8]

chopped into pieces approximately 5 cm long. The bundle shear is operated remotely and is designed so that it can also be repaired by remotely operated tools. The fuel element and fuel rod sections drop through a chute directly into a dissolver basket located in the dissolver cell underneath. The basket is filled with boiling nitric acid, which leaches the fuel out of the chopped fuel rod hulls. After leaching of the fuel, the remaining hulls and fuel element sections are dumped from the basket into a container, and the container is moved into the hull storage facility.

The fuel solution still contains small solid parts, such as zirconium or steel chips from chopping. Moreover, it includes undissolved particles of fission and corrosion products, such as ruthenium, palladium, rhodium, molybdenum, technetium and zirconium. The undissolved fraction of plutonium (compounds of plutonium with rhodium and palladium), which may be contained in these undissolved particles, is about 1%. The undissolved solid particles are removed through coarse filters or by centrifuges. Solid particles separated in the dissolution of LWR fuel elements may contain up to 0.3 kg/t fuel of zirconium chips and some 3.3 kg/t fuel of undissolved fission and corrosion products.

7.2.1.3 Gas Cleaning and Retention of Gaseous Fission Products

During the processes of chopping and dissolution of the fuel, gaseous and volatile fission products are released. They must be removed together with water vapor, nitrous gases (NO, NO_2, N_2O) and the nitrogen which may have been applied as a scavenging gas in fuel element chopping. This mixture of volatile fission products, vapors and gases must be treated in the waste gas cleaning system. Gaseous and volatile fission products are made up of the following components:

- Tritium is produced by ternary fission and by (n,t)-reactions in light atomic nuclei. Some 40% of the tritium generated remains in the metal structure of the zircaloy cladding tubes. The other 60% are released as tritiated water, HTO, during dissolution and may enter the gaseous effluent section together with water vapor. Less than 1% of the tritium is found there as gaseous tritiated hydrogen, HT.
- Carbon, C-14. is produced by an (n,α)-reaction from O-17 and by the (n,p)-reaction of N-14. Krypton is generated as a gaseous fission product. Some 7% of the krypton fission products produced consist of radioactive Kr-85 isotopes.
- Xenon is another gaseous fission product. Most of its isotopes are stable or fairly short-lived, e.g. Xe-135 with a half-life of 9 h, so that only traces of the Xe-133 isotope must be considered, because it has also a relatively short half-life of 5.27 days.
- All the other fission product noble gases generated are either stable or have very short half-lives.
- I-129 and traces of I-131 are partially volatile isotopes initially found in dissolved fuel. They may be carried into the gas stream through boiling and as a result of passing an inert gas through the dissolved fuel solution, which entraps the iodine in the gas stream.
- Ru-106 may volatilize as ruthenium tetroxide evaporating from strong nitric acid solutions, but only some 10^{-4} fractions of Ru-106 enter into the gaseous effluent stream. In a similar way, small traces of such β-emitters as strontium or α-emitters as uranium and plutonium can penetrate into the gaseous effluent as aerosols.

However, only some 10^{-4}–10^{-6} fractions of the fuel inventory are carried into the gas stream as aerosols.

The gaseous effluents are first passed through a condenser. Afterwards, the nitrogen oxides are oxidized and washed out. This already removes 99% of the aerosols. The remaining aerosol fractions only amount to 10^{-6}–10^{-8} times the inventory. Scrubbers and high-efficiency particulate aerosol (HEPA) filters are used next to remove the aerosols. Iodine is retained very efficiently in silver impregnated ($AgNO_3$) filter materials. Tritium present as HT hydrogen is converted into HTO water. Tritium as HTO contained in water vapor and $^{14}CO_2$ are retained in molecular sieves as described in Sect. 7.5. The removal of Kr-85 can be achieved by means of low temperature rectification. In the same process, the xenon noble gas can also be removed. The separated krypton can be stored in gas cylinders under high pressure. Alternatives may be the entrapment in zeolites (crystallized silicates) and ion implantation in metals.

7.2.1.4 Chemical Separation of Uranium and Plutonium

Although a number of chemical separation techniques have been proposed and developed in the past few decades, the most efficient process to date has remained the PUREX process (Plutonium and Uranium Recovery by EXtraction) [15]. The PUREX process uses tri-n-butyl phosphate (TBP), which may be diluted, for instance, by kerosene or n-paraffin (hydrocarbon) solvents as organic solvents to extract uranium and plutonium. TBP is stable in nitric acid and can selectively extract tetravalent and hexavalent uranium and plutonium nitrate complexes. However, this selective extraction capability of TBP does not apply to trivalent plutonium nitrate complexes.

For extraction, the fuel solution acidified with nitric acid and containing uranium, plutonium, higher actinides and fission products is moved from the middle of column I (Fig. 7.6) in a liquid–liquid countercurrent extraction flow past the specifically lighter organic solvent (TBP in kerosene) rising from the bottom. In that process, the organic solvent extracts uranium and plutonium, while the fission products and actinides remain in the aqueous solution. The solution with nitric acid leaves the column at the bottom as high level aqueous waste (HLW). It contains the fission products and higher actinides. The aqueous waste is evaporated to recover the nitric acid. The remaining concentrate is further treated as high level waste concentrate (HLWC).

The rising organic solvent contains uranium and plutonium and small traces of fission products, which are removed by a nitric acid solution injected at the top of the column. The organic solvent leaves the column at the top and is introduced into column II, where the tetravalent and hexavalent plutonium is reduced to trivalent plutonium by means of a reducing agent stream, e.g. U (IV) nitrate with hydrazine nitrate, hydroxylamine nitrate or, formerly, Fe (II) sulfamate. (The most elegant method developed recently, uses electrolytic reduction within the extraction apparatus). This trivalent plutonium is sparingly soluble in organic TBP-kerosene and, as a consequence, is re-extracted into the aqueous phase, while hexavalent uranium

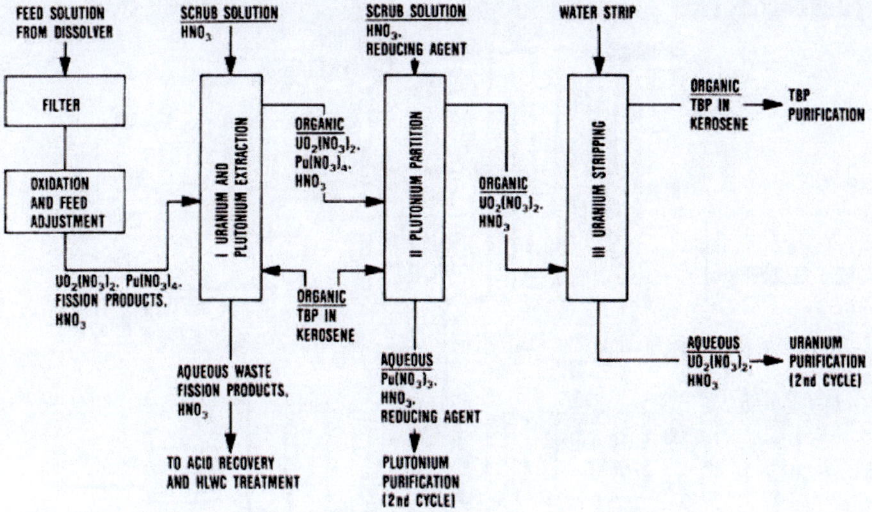

Fig. 7.6 Simplified PUREX process flowsheet [12]

remains in the organic TBP-kerosene phase. Small amounts of re-extracted uranium are extracted again by organic TBP-kerosene introduced at the bottom of the second column. The aqueous plutonium product stream leaves the second column at the bottom, while the organic uranium product stream leaves at the top and enters the third column, where it is met by a countercurrent stream of diluted nitric acid as an aqueous re-extraction solution flowing from the top. The uranium product stream with nitric acid then leaves column III at the bottom, while the organic solvent leaves at the top. After removal of organic decomposition products and fission products by washing, the organic solvent can be recycled into the system.

For sufficient decontamination of uranium and plutonium, the uranium and plutonium product streams are required to pass through two further decontamination cycles as shown in Fig. 7.7. The final products, after concentration and purification, are plutonium nitrate, $Pu(NO_3)_4$, and uranyl nitrate, $UO_2(NO_3)_2$. The resulting waste streams must be treated separately, as will be described in Sect. 7.5.

The extraction apparatus shown in Fig. 7.7 can be used in three different technical designs, i.e., as pulsed perforated plate columns, mixer-settlers, or centrifugal contactors. Centrifugal contactors have very short contact times for the aqueous and the organic phases, largely protecting the organic solvent from degradation by radiation. This makes them particularly suitable for fuel with short cooling time and high burnup, such as LMFBR fuel. Pulsed columns also have relatively short contact times of the organic solvent, permitting the installation of heterogeneous lattices of neutron-absorbing materials for criticality control. Mixer settlers are very reliable, flexible and simple systems with longer contact times. They have been used most successfully for reprocessing fuel elements with low burnups and in the second and

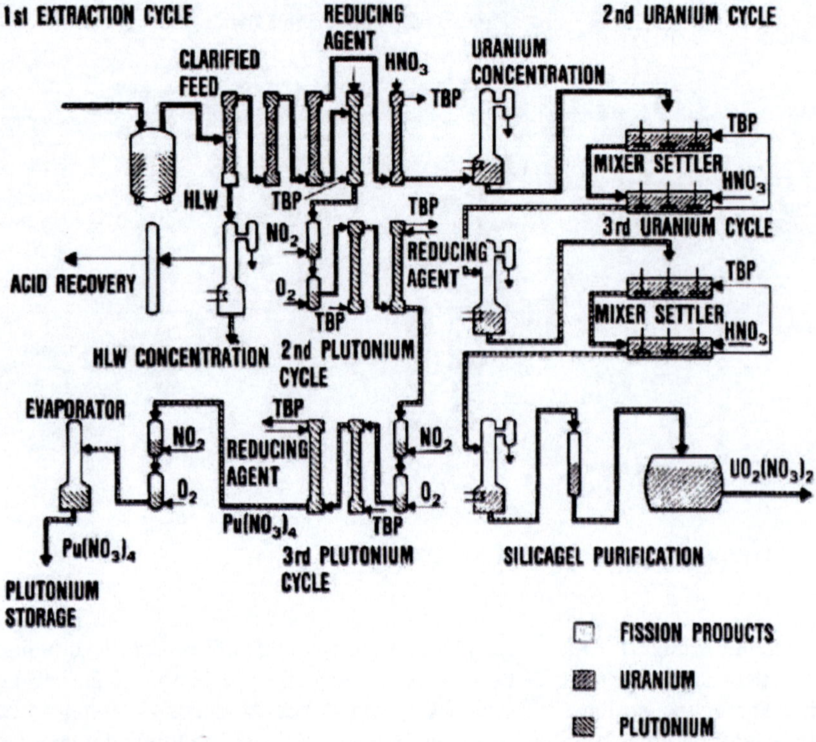

Fig. 7.7 Simplified flowsheet of the uranium/plutonium extraction cycles of a reprocessing plant [8]

third uranium purification cycles. However, in the first decontamination cycles and in the plutonium purification cycles, preferably pulsed columns are used.

7.2.1.5 Mass Flows of Radioactive Material in a Model LWR Fuel Reprocessing Plant

After this short and simplified description of the chemistry of reprocessing LWR fuel elements, the individual mass flows of fuel and nuclear waste will now be considered in a commercial scale model reprocessing plant with a throughput of about 4 t/day or 1,000–1,200 t/a of LWR fuel, as described in Fig. 7.8. A reprocessing plant of this capacity can handle spent fuel discharged from some 40–50 GW(e) LWRs. The model plant is designed for a maximum burnup of LWR fuel of approx. 40,000 MWd/t fuel with an initial fissile material enrichment of some 4% [14].

The fuel is assumed to have been kept in intermediate storage for an average of three years before being reprocessed (Fig. 7.8). During spent fuel dissolution,

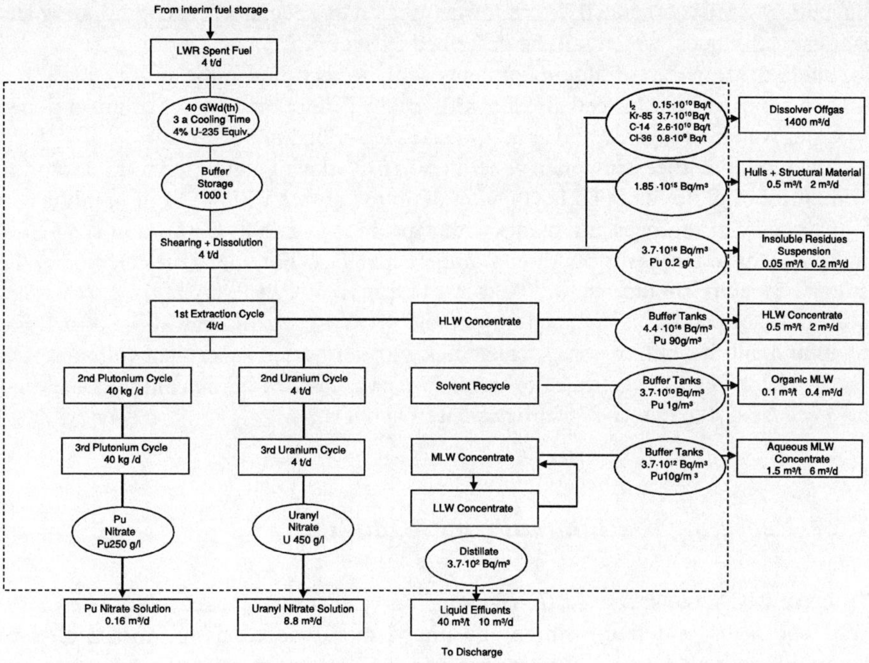

Fig. 7.8 Mass flow in a model LWR fuel reprocessing plant with a capacity of 4 t/d of spent UO_2 fuel [8]

a model plant of this type will produce some 1,400 m^3/day of gaseous effluent, some 2 m^3/day of hulls and structural materials containing 0.8×10^9 Bq/t of Cl-36, and roughly 0.2 m^3/day of sludge of insoluble fuel residues. In the gaseous effluent, Kr-85 has the highest radioactivity with 3.7×10^{10} Bq/t of fuel, while I-129 and C-14 with 0.15×10^{10} Bq/t and 2.6×10^{10} Bq/t of fuel, respectively, only make minor contributions. The hulls and structural components of the fuel elements are radioactive after prolonged exposure to neutrons in the reactor core. In addition, the hulls contain small amounts of undissolved uranium and plutonium. The radioactivity of the hulls and structural material amounts to 1.8×10^{15} Bq/m^3. The insoluble residues also contain small amounts of plutonium and have a bulk radioactivity of roughly 3.7×10^{16} Bq/m^3.

In the first extraction cycle, the fission products, higher actinides and small quantities of unextracted U/Pu amount to 2 m^3/day of HLWC with a radioactivity of 4.4×10^{16} Bq/m^3. In the second and third uranium and plutonium decontamination cycles, some 0.4 m^3/day of organic solvent is produced as organic medium level waste (MLW), which contains small traces of U/Pu. It has a radioactivity of approximately 3.7×10^{10} Bq/m^3. Moreover, about 6 m^3/day of aqueous medium level waste of 3.7×10^{12} Bq/m^3 are produced, which also contains small traces of U/Pu. All these quantities of HLWC and MLW are further treated by special waste conditioning and storage techniques, which will be described in Sect. 7.5.

Some 160 m^3/day of liquid effluents with a very low radioactivity of 3.7×10^2 Bq/m^3 can be discharged directly without any further treatment. Tritium enriched in water is being recycled and conditioned as described in Sect. 7.5.

The model reprocessing plant generates 0.16 m^3/day of plutonium nitrate solution with 40 kg of plutonium and 8.8 m^3/day of uranyl nitrate with 3.96 t of uranium.

Commercial reprocessing plants with capacities of 1,000–1,200 t/a of LWR fuel must also have the corresponding storage capacities. For spent fuel elements, the storage capacity should be 1,000–3,000 t U, for HLWC, 1,000–2000 m^3, for aqueous MLW, 1,500–10,000 m^3, and for organic MLW, approximately 200–500 m^3. For plutonium nitrate, buffer storage capacities with very small subcritical volumes each are provided. For uranyl nitrate the buffer storage capacities can be volumes of several m^3 because of the low U-235 enrichment of about 0.7%.

7.2.2 Recycling of Plutonium and Uranium

To make MOX fuels, e.g. for LWR cores operated in the Pu-recycle mode or for FBR cores, the plutonium nitrate and uranyl nitrate solutions are mixed already within the reprocessing plant to form a so-called master-mix. This master-mix can have already the plutonium enrichment which is needed for the MOX fuel. This plutonium-uranium nitrate master-mix (Fig. 7.8) must be converted chemically into uranium and plutonium oxides. This is done at the end of the reprocessing process.

7.2.2.1 Mixed Oxide Fuel Fabrication

Plutonium recycling in thermal reactors, e.g. LWRs, requires the fabrication of MOX pellets [16–20]. The mixed oxide powder is precompacted and granulated into a freely flowing powder. This is turned into cylindrical pellets, which are first sintered at temperatures of 1,000 and 1,700°C and then ground to the required dimensions. The pellets are heated to remove their gas contents. Finally, they are loaded into zircaloy or steel tubes, to which end caps are welded, and assembled into fuel assemblies [18, 20].

During longer storage the Pu-241 isotope decays to Am-241, which emits 60 keV γ-radiation. This requires either chemical separation of Am-241 prior to fabrication or appropriate shielding against γ-radiation during the fabrication process, if the PuO_2/UO_2 master mix powder had been stored for more than two years.

Also, a certain fraction of the mixed oxide pellets will be imperfectly fabricated and are rejected during control and inspection procedures. Such material and grinder fines, which are designated clean rejected oxide, can be recycled directly into the manufacturing process. However, a small fraction of pellets and powder are contaminated by corrosion products, etc. These constitute the so-called dirty scrap requiring chemical purification. Dirty mixed oxide scrap is dissolved in a nitric acid/fluoride solution and, along with filtrates from wash-leach processes, is treated chemically as in a PUREX reprocessing step to recover the uranium and plutonium. Other process residues include such contents as metal scrap, plastics, rubber, cellulose, cleaning materials and organic substances. These wastes are treated as low level waste (see Sect. 7.5).

As MOX fuel must be reprocessed after having attained its design burnup, the fuel fabrication process must guarantee high solubility (>99%) of the irradiated MOX fuel in nitric acid. Such high solubility is required to minimize the plutonium losses in the residues during chemical reprocessing.

7.2.2.2 Additional MOX Fabrication Processes

Three other refabrication processes can be applied. The sol-gel process, described in more detail in Sect. 7.3.3, allows the direct fabrication of spherical MOX particles, which can be pressed and sintered into fuel pellets. The AUPuC (ammonium (U,Pu) carbonate) refabrication process also allows the fabrication of relatively coarse grain MOX powder. This crystal powder is fabricated essentially free of Am-241 and then pressed and sintered into MOX fuel pellets. Both the sol-gel and the AUPuC fabrication processes avoid the generation of plutonium dust within the glove boxes of the fuel fabrication line and therefore lead to relatively lower γ-radiation doses to the staff working in MOX fuel refabrication plants.

A third MOX fabrication process is based on vibro compaction. The MOX fuel is broken into small fuel particles of different size. These are filled into the cladding tube and compacted by vibro compaction [19].

A MOX fuel fabrication plant with an annual fabrication capacity of 300 t_{HM}/a roughly corresponds to the plutonium mass flow produced by the 4 t/d or 1,000–1,200 t_{HM}/a model reprocessing plant described in Sect. 7.2.1.5.

7.2.3 Status of Uranium Fuel Reprocessing Technology

The basic technology of the PUREX process, of plutonium storage, handling and transport is well established [14, 21].

A large experience is available in reprocessing spent metallic fuel of gas graphite reactors. A reprocessing plant with a capacity of 1,500 t_{HM}/year is operating at Windscale, UK. Similar reprocessing plants were operated at Marcoule and LaHague, France.

Small scale pilot reprocessing facilities for LWR-UO_2 fuel had been operating, e.g. in Belgium, Germany and Japan. Table 7.4 indicates the annual capacities of present large scale commercial reprocessing plants in operation in the world.

The development of spent fuel reprocessing as a commercial industry was stopped in the USA, during the late 1970s as a consequence of the non-proliferation policy.

Table 7.4 Reprocessing plant capacity for spent fuel in the World [14]

Country	Plant	Fuel type	Reprocessing capacity (t_{HM}/year)
France	Cap de la Hague	LWR	1,700
United Kingdom	Sellafield	LWR	1,200
	Windscale	AGR	1,500
Japan	Tokai-mura	LWR	90
	Rokkasho-mura	LWR	800
Russia	Mayak	LWR	500
India	Tarapur	CANDU	100
	Kalpakkam	FBR	100
China	Lanzhou	LWR	50

Large scale LWR spent fuel reprocessing plants are operating at Cap de la Hague (France), at Sellafield (United Kingdom) at Rokkasho-mura (Japan) and Mayak (Russia). Smaller scale reprocessing plants are operating at Tokai (Japan) and Tarapur as well as Kalpakkam (India). China operates a small reprocessing plant at Lanzhou. This adds up to a total reprocessing capacity of somewhat more than 4,000 t_{HM}/a, for spent LWR fuel in the world.

7.2.4 Status of Experience in MOX Fuel Fabrication and Reprocessing

A considerable amount of MOX fuel has been fabricated already in small scale refabrication plants in countries including the USA, UK, France, Belgium, Japan and the Federal Republic of Germany [14, 21]. MOX fuel elements have been irradiated successfully in LWRs and other types of reactors up to burnups of 55,000 MWd/t. Some of these MOX fuel elements have also been successfully reprocessed.

Spent MOX fuel differs from low enriched spent UO_2 fuel in its slightly different content of fission products and in its content of higher actinides. Plutonium fuel builds up a higher percentage of americium and curium as a consequence of neutron capture and decay processes.

The higher plutonium content only requires a slight adjustment of the flowsheet of the reprocessing plant described in Sect. 7.2.1. If fabrication specifications for the MOX fuel are observed carefully (high solubility and homogeneity), there is no great difference between the dissolution capability of spent MOX fuel and that of spent low enriched UO_2-LWR fuel. The increased fractions of americium and curium isotopes become part of the HLW, where they will cause a somewhat higher decay heat generation. This must be taken into account in waste conditioning.

Table 7.5 indicates the MOX fuel fabrication plant capacities in the world by 2010.

Table 7.5 MOX fuel fabrication plant capacities by 2010 [14]

Country	Location	Type of fuel	Plant capacity t_{HM}/year
France	MELOX, Marcoule	LWR-MOX	195
United Kingdom	Sellafield	LWR-MOX	120
Japan	Rokkasho-mura	LWR-MOX	130
	Tokai	FBR-MOX	20
USA	Savannah River	LWR-MOX	70[a]
Russia	Zheleznogorsk	FBR-MOX	60
	Tomsk	LWR-MOX	70[a]
India	Tarapur	FBR-MOX	50

[a] Weapons plutonium

7.2.5 Safety of Reprocessing and MOX Re-Fabrication Plants

7.2.5.1 Safety Design Measures in Reprocessing Plants

Unlike nuclear reactors, reprocessing plants are characterized by the following differences in hazard potential:

- The nuclear fuel is dissolved in nitric acid and is not arranged in a neutronically critical geometry ($k_{eff} < 1$). The fuel is not used for power generation and, correspondingly, is only at a low temperature and low pressure.
- The radioactivity of the spent fuel decays by a factor of 65 within one year after unloading from the reactor core, as a result of the decay of fairly short-lived fission products and actinides.

Accordingly, the radioactive inventory per t_{HM} of LWR fuel in a reprocessing plant is much smaller than in a nuclear reactor. These are the reasons why large reprocessing plants can have capacities which allow the spent fuel from e.g. 30–75 GW(e) of power reactors to be reprocessed.

In reprocessing plants, as in nuclear reactors, the safety design principles of diversity and redundancy are applied in supplying electricity and cooling water to ensure the reliability of heat removal. Appropriate reserves for cooling water are taken into account in plant design. The multiple barrier principle between radioactive substances and the environment is observed. The radioactive materials are enclosed in leaktight systems of stainless steel pipes, vessels and other equipment which, in turn, are enclosed in leaktight cells with high density concrete walls up to 2 m thick. In case of a leak in a pipe or vessel, stainless steel catch pans prevent radioactive liquid from penetrating the floor of the containment cells. The inner containment is surrounded by an outer protective shield. The inner and the outer containments are operated at pressures lower than atmospheric. Rooms with the highest radioactivity levels are kept at the lowest pressures. Contaminated air is filtered by at least two redundant filter systems and treated to reduce any radioactivity to acceptable levels.

After filtering, the air is released into the environment through a stack (for details, see Chap. 10 and 11). Releases to the atmosphere or to rivers, lakes or the sea are continuously monitored.

As in nuclear reactor plants, the containment of a reprocessing plant must be designed to withstand earthquakes, floods, tornados, airplane crash impacts, shockwaves caused by explosions, fires and sabotage. Engineered safety features also include measures to prevent criticality in dissolvers, extraction columns and buffer tanks. This can be achieved by limiting the geometries of the equipment, adding such neutron poisons as boron, hafnium and gadolinium, and by strictly limiting the fissile enrichment of spent fuel elements to be reprocessed, respectively. Moreover, the fissile enrichment of solutions is continuously monitored by measurements at critical points of a facility.

Accidents to be considered are the explosion of an evaporator for a high level waste concentration tank and a criticality accident in one of the components carrying fissile material. Kerosene, TBP and nitric acid have a potential for exothermic reactions only if organic products were able to reach the evaporator and only if temperatures above 140°C were attained. Such conditions are avoided during operation by keeping the temperature of the process steam for evaporation at 130°C.

Despite such safety design measures, assumed accidents are analyzed and the design of the inner cells must limit the consequences of such accidents. If the evaporator or another container were destroyed by an explosion, the waste solution would leak into the catch pan on the cell floor and could be pumped back into another container hold in reserve. Radioactive aerosols would be retained in the filter system. The radioactive impact upon the environment would be limited. Similar design measures are taken to limit the consequences of a criticality accident. For details of risk analyses of reprocessing plants, see Chap. 11.

7.2.5.2 Safety Considerations for Mixed Oxide Fuel Fabrication Plants

The multiple barrier system is also applied to limit the release of plutonium aerosols from MOX fabrication plants. Primary confinement is provided by shielded glove boxes or hot cells containing the plutonium pellet fabrication equipment. This primary confinement is surrounded by an operating and maintenance building accessible only through locks. It also constitutes a firewall and a shielding protection. An outer shell protects these two inner confinements against natural and external events (earthquakes, airplane crashes, etc.). Pressure differentials are maintained within the different confinements such that the lowest pressure applies to the rooms with the highest plutonium concentrations. Air is exhausted only through a number of HEPA filters connected in series. The safety design basis of a mixed oxide fuel fabrication plant is determined by the potential of criticality accidents, fires and explosions in the manufacturing equipment.

7.3 The Thorium/Uranium-233 Fuel Cycle

Thorium may be used as a fertile material in converter or breeder reactors, as described in Chap. 6. In that case either U-235 must be added initially or U-233 must be recovered from spent fuel elements by chemical reprocessing. After storage on the reactor site or in interim storage facilities, spent fuel elements with U233/Th fuel are transported to the reprocessing plant.

7.3.1 Fuel Element Disassembly

The design of U-233/Th fuel elements to be used in LWRs or HWRs roughly corresponds to that of today's standard fuel elements with low enriched uranium. Consequently, the head end of the U-233/Th reprocessing plant required for chopping the fuel element into short pieces and subsequent dissolution may be the same as described in Sect. 7.2.1.

The fuel of HTGRs or HTRs consists of small fissile particles coated with pyrolytic carbon and silicon carbide; fertile particles made of ThO_2 are coated with carbon. These particles are imbedded in a graphite matrix of the fuel element (prismatic block or sphere). Prior to dissolution of the fuel, the graphite must be separated from the fuel. This is done by crushing the blocks or spheres and burning the graphite in a fluidized bed. Fissile particles coated with SiC must be crushed and the residual inner pyrolytic carbon layer must be burnt. C-14 produced by the neutron activation of carbon and by (n,p)-reactions with residual nitrogen must be specially treated in the waste gas cleaning system as ^{14}CO and $^{14}CO_2$. The ash produced is made up of the fissile particles containing U-235/U-238, plutonium and fission products and of the fertile particles containing $^{233}UO_2/ThO_2$ and fission products. For further treatment, the fertile and fissile particles are separated.

The $^{233}UO_2/ThO_2$ fuel must be treated by the THOREX (Thorium Oxide Recovery by EXtraction) reprocessing technology. The fissile fuel, which contains U235/U238, plutonium and fission products, may be reprocessed by the PUREX technology. If medium enriched uranium were mixed with thorium, this fuel would contain thorium, uranium isotopes, plutonium, fission products and higher actinides after irradiation in the reactor core. In this case, a combined PUREX-THOREX-reprocessing technology would have to be applied.

7.3.2 THOREX Process

The U-233/Th fuel is dissolved in very highly concentrated 13 M nitric acid, 0.05 M hydrofluoric acid and 0.1 M aluminum nitrate held at boiling temperature [22–24]. The residual solids are removed from the solution by centrifuging. The solution with

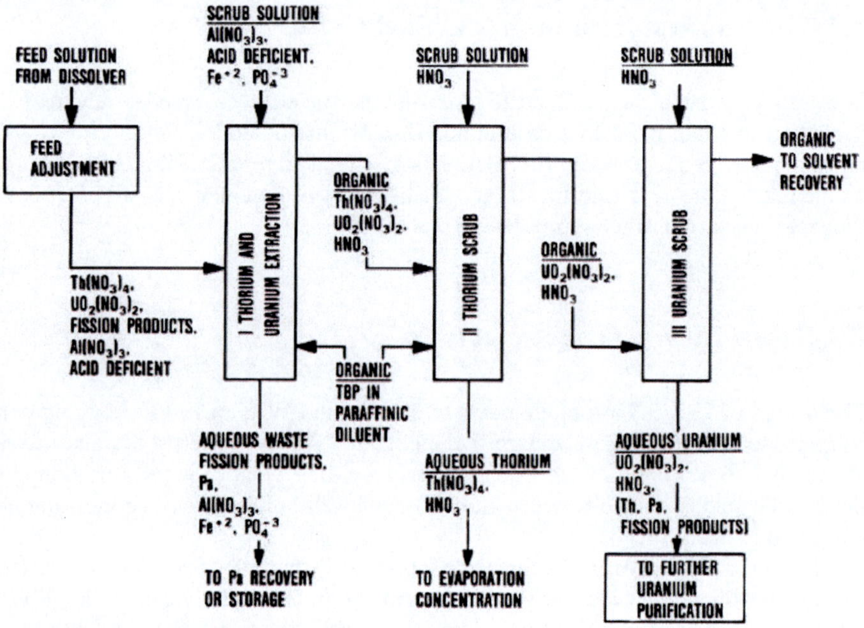

Fig. 7.9 Simplified THOREX process flowsheet [12]

$Th(NO_3)_4$ and $UO_2(NO_3)_2$ then enters the first extraction column (Fig. 7.9) and is moved in a countercurrent flow against TBP dissolved in a hydrocarbon solvent. TBP selectively dissolves thorium nitrate and uranyl nitrate while moving upward in the column. The fission products, protactinium and aluminum nitrate leave the column at the bottom together with the scrub solution, which is added at the top of the column. Careful adjustment of these chemical processes is necessary to separate fission products, especially Zr-95, from thorium. In the tetra-valent state, thorium is chemically very similar to zirconium. Pa-233 with a half-life of 27.0 days is the precursor of U-233. The fertile fuel can be cooled until Pa-233 has decayed into U-233. In the second column, $Th(NO_3)_4$ is recovered from the TBP by being moved in a countercurrent flow against diluted nitric acid. $Th(NO_3)_4$ and nitric acid leave the second column at the bottom. The organic solution together with $UO_2(NO_3)_2$ flows into the third column, where uranium is re-extracted. Uranium is then purified further in additional solvent extraction separation steps. Small traces of plutonium and neptunium may be separated by additional extraction chromatography. In case more plutonium is built up, e.g. in medium enriched U-235/U-238/Th fuel (see Chap. 6 for MEU-Th reactor cores), the separation from plutonium and neptunium by extraction chromatography is not sufficient. In these cases the plutonium must be co-extracted with uranium and thorium. This may also be achieved by solvent extraction in contact with TBP. However, this PUREX/THOREX process is more complicated than the THOREX process.

7.3.2.1 The THOREX Process: Status and Experience

The technological basis of the THOREX process is well understood, but much less experience has been accumulated in its implementation than with the PUREX process. Small THOREX pilot facilities were operated successfully at Oak Ridge, Hanford and Savannah River in the USA. Within the US LWBR and HTGR programs, approximately 870 t of thorium (mostly ThO_2) was reprocessed. In Germany, Italy, Japan and India, some experience with small pilot facilities was also obtained [25]. The availability of commercial scale reprocessing plants for thorium fuel would certainly need future development efforts.

7.3.3 Uranium-233/Thorium Fuel Fabrication

$^{233}UO_2/ThO_2$ pellet type fuels for LWRs or CANDUs can be fabricated by remote fabrication lines in shielded concrete cells [26, 27]. The fabrication process follows similar steps of powder mixing, pressing, sintering and grinding of the pellets, as described for the MOX fuel fabrication in Sect. 7.2.2.

Wet chemical processes are applied to produce thorium-uranium fuel particles for HTGR and THTR fuel. These techniques are used because, in reprocessing, thorium and uranium occur as thorium nitrate and uranyl nitrate, respectively. If thorium nitrate is brought into contact with ammonium hydroxide, thorium hydroxide and ammonium nitrate will be produced. Thorium hydroxide precipitates as an amorphous, gel type structure. Similarly, uranyl nitrate reacts to form ammonium diuranate, but special additives are necessary to modify the precipitate into a gel. The first step in fabrication is the preparation of a solution of the right viscosity (sol). The sol is then passed through a vibrating jet and dispersed into appropriately shaped spherical droplets. The sol droplets fall into a gelation bath where the gelation process into microspheres is completed (gel). The gel spheres are then washed and dried. Then the kernels are sintered at approximately 1,400°C. In this way, either ThO_2, kernels or U-233/Th mixed oxide kernels are produced. The preparation of uranium kernels requires one additional calcination step in a fluidized bed to decompose organic compounds of the broth. Carbides are produced by carbothermic reduction at temperatures up to 2,500°C.

The outer layer of the kernels of pyrolytic carbon or silicon carbide is applied in an electrically heated fluidized bed. The kernels are then assembled into graphite fuel elements. Assembly is performed remotely in shielded cells. This remote fabrication is necessary because of the U-232 contamination of uranium and the Th-228 contamination of thorium. U-232 is formed by an (n,2n)-reaction or (γ,n)-reaction with Th-232 and a subsequent (n,γ)-reaction with Pa-231 (half-life 3.3×10^4 years) as well as by an (n,2n)-reaction with U-233. Pa-231 is also formed by the decay chains initiated by the α-decay of U-235 and the (n,γ)-reaction with Th-230. After decay of Th-231 the subsequent (n,γ) reaction of Pa-231 leads to Pa-232 and subsequent β^- decay to U-232. U-232 decays to Th-228, which decays to Bi-212 and Tl-208 through a chain of short-lived isotopes. Both end products are emitters of very high-energy γ-radiation. Recycled uranium in the Th/U-233 fuel cycle will contain several 100 ppm of U-232. Its high-energy γ-radiation together with neutrons produced from (α,n)-reactions (α-decay of uranium and thorium isotopes) with light elements, such as oxygen or carbon, require shielding during refabrication.

Thorium separated from reprocessing is not recycled directly. Th-228 appearing in the separated thorium results in appreciable radioactivity. Therefore it has to be stored for some ten years before it may be recycled.

Most of the experience in U-233 UO_2/ThO_2 pellet fuel fabrication has been gained in the US LWBR project. Actual experience, however, is limited to low U232 (about 10 ppm) feed material. Experience with fresh U-235 UO_2/ThO_2 fuel for HTGRs and HTRs is available in pilot plants, but not yet for fuel refabrication with several 100 ppm U-232 content.

7.4 The Uranium/Plutonium Fuel Cycle of Fast Breeder Reactors

Like all recycling converter reactors and near-breeder reactors, LMFBRs must work in a closed fuel cycle. Their systems inventory, consisting of the core fuel inventory and the out-of-core fuel inventory passed through reprocessing and refabrication, should be as small as possible for economic reasons. On the one hand, this leads to the requirement of a small in-core fuel inventory and high burnup (long utilization of the fuel for energy generation). On the other hand, it means short ex-core times of the fuel inventory in the fuel cycle.

7.4.1 Ex-Core Time Periods of LMFBR Spent Fuel

At present, it is generally assumed that an ex-core time of two years is feasible for the LMFBR fuel cycle [28, 29]. Figure 7.10 shows the LMFBR fuel cycle, indicating the different time spans for the fuel outside the reactor plant. A model fuel cycle is assumed for this diagram, which corresponds to a fuel reprocessing capacity for roughly 10 GW(e) FBRs.

After unloading from the reactor core, the core elements and the radial blanket elements are first stored on the reactor site for some 180 days. Then they are transported to the reprocessing plant in shipping casks, which can contain six to twelve fuel elements each. Shipping the fuel elements takes about thirty days. Another thirty days are assumed for intermediate storage and pretreatment of the fuel elements prior to cutting and dissolution. Assuming a reprocessing plant with annual reprocessing capacities of about 257.7 t_{HM}/year (in more detail 170 t_{HM}/year of core and axial blanket fuel mixed with 87.7 t_{HM}/year of radial blanket fuel), the total time required

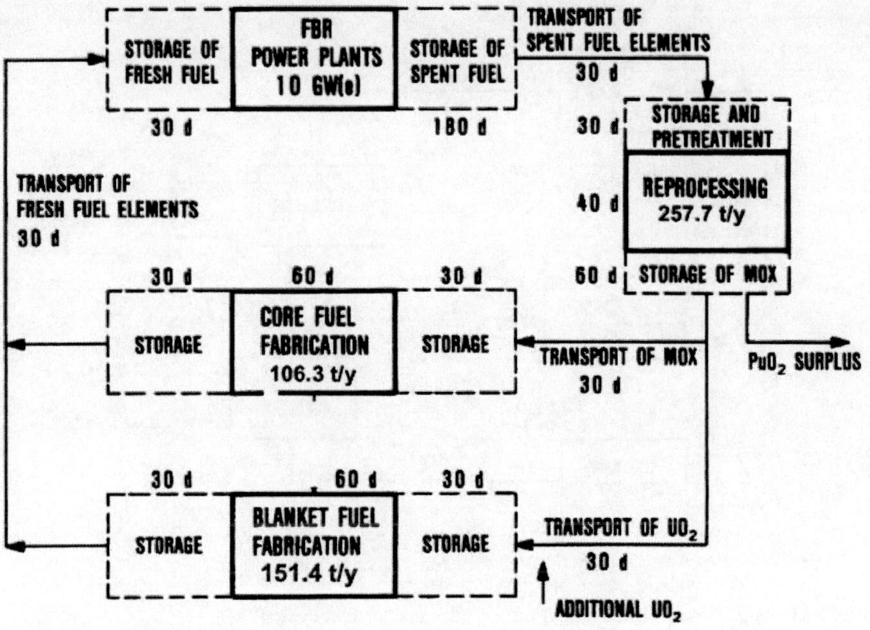

Fig. 7.10 Ex-core time of a fast breeder reactor spent fuel cycle [28]

for all steps, from chopping the fuel pins to conversion to PuO_2 and UO_2 powder, is estimated to be forty days. Sixty days are assumed for intermediate storage of the oxide powder, another thirty days for transfer to the fuel refabrication plant. The reprocessing plant and the fuel refabrication plant are assumed to be colocated at one site.

The associated PuO_2/UO_2 fuel refabrication plant will have an annual capacity of about 110 t_{HM} of mixed PuO_2/UO_2 fuel for the core and an annual capacity of about 150 t_{HM} for UO_2 blanket fuel (about 63.7 t_{HM} for the axial blankets and about 87.7 t_{HM} for the radial blanket). The UO_2/PuO_2 powder will be stored for about thirty days and then transferred in batches to the fabrication lines. The fabrication process takes about sixty days, and another thirty days are required for fuel element storage prior to shipment to the FBR power plant. Shipment requires some thirty days; another thirty days are assumed for storage on the reactor site before the fuel is loaded into the core for power generation.

These time periods add up to 550 days. Assuming another 180 days for unforeseen delays, which may arise from imperfect synchronization between the various fuel cycle operations, the total ex-core or fuel cycle time adds up to 730 days or two years. However, it is obvious that co-location of reprocessing and refabrication plants, good synchronization of the different fuel cycle activities, and reductions in storage time can help to shorten these ex-core times.

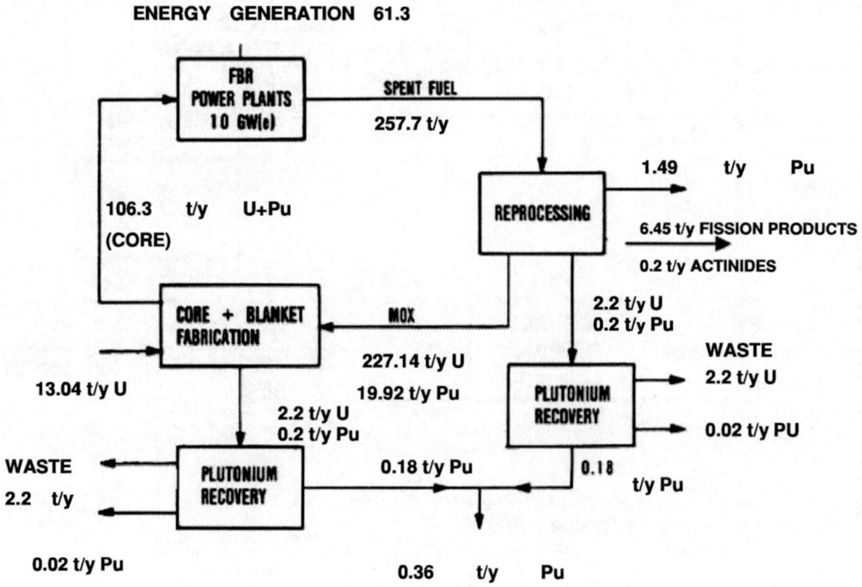

Fig. 7.11 Mass flow within a model fast breeder reactor fuel cycle [28]

7.4.2 Mass Flow in a Model LMFBR Fuel Cycle

A model fuel cycle for reprocessing the PuO_2/UO_2 fuel discharged from 10 GW(e) LMFBRs roughly corresponds to a capacity of slightly more than 1 t_{HM}/day or, at 250 full-load days, an annual capacity of 257.7 t_{HM} [28, 29]. Such a fuel cycle includes reprocessing and refabrication plants on an industrial scale. Figure 7.11 indicates the mass flows of the most important materials in this model LMFBR fuel cycle.

From the 10 GW(e) LMFBRs, at a load factor of 0.7, an annual 170 t_{HM} of uranium and plutonium in core fuel elements and 87.7 t_{HM} of uranium and plutonium in radial blanket elements are discharged and shipped to the reprocessing plant. These spent fuel and blanket elements contain 6.45 t of fission products. Of these fission products, some 5.8 t are contained in the core fuel elements and in the axial blankets, and some 0.65 t in the radial blanket elements. The distribution into different fission product isotopes differs slightly from that encountered in LWR spent fuel because of the fast neutron spectrum and the plutonium as fissile material compared to U-235 in LWRs. In addition to these quantities of fuel and fission products, there are approximately 200 kg of higher actinides (Np-237, Am-241, Am-242m, Am-243, Cm-242 and Cm-244).

In the reprocessing plant, the fission products and the actinides are separated and go into the HLWC. Some 227 t_{HM} of uranium and 21.7 t_{HM} of plutonium are recovered, of which 1.49 t of plutonium can be diverted as a breeding gain to start new LMFBRs or to feed converter reactors. Roughly 1%, i.e., 200 kg of plutonium and some 2,000 kg of uranium, initially remain as high level solid or liquid wastes accumulating in the reprocessing and refabrication plant. However, in the waste treatment step, most of the plutonium is recovered so that ultimately only some 20 kg/y of plutonium will be lost to the HLW and MLW.

In refabrication, the spent U-238 must be replaced by depleted uranium from an enrichment plant. The quantity of uranium to be replaced is equivalent to the fractions destroyed by fission, conversion of U-238 to Pu-239 and losses during reprocessing and refabrication. In this case, it amounts to approximately 13 t_{HM}, of depleted uranium per annum.

7.4.3 LMFBR Fuel Reprocessing

Also for LMFBR spent fuel, the PUREX process is used as described in Sect. 7.2.1. However, technical modifications are required to take into account the specific characteristics of LMFBR fuel elements enriched in plutonium [30–35].

Unlike LWR fuel assemblies, the hexagonal LMFBR fuel elements with outer wrapper are first dismantled. The present state of the art does not permit direct chopping of the fuel element assemblies. The end pieces of the fuel elements are cut off, and the fuel element wrapper is removed mechanically. The fuel rods are then separately cut into pieces 2.5 cm long by means of a shear. In this step, core fuel and blanket fuel are mixed. The fuel rod pieces fall into the dissolver, where the fuel is dissolved in hot nitric acid. The dissolver geometry must be carefully adapted to the higher plutonium enrichment of LMFBR fuel to avoid criticality.

The dissolution capability of the fuel is influenced by the method of fuel fabrication and by the irradiation history of the spent fuel. A small fraction of insoluble particles will remain, which is made up of ruthenium, rhodium, tellurium, molybdenum and palladium and of undissolved fuel. The noble metal fission products, ruthenium, rhodium and palladium, tend to form alloys with part of the plutonium at high burnup of the fuel. If this insoluble particle fraction contains more than 0.5% of the total plutonium, it must be dissolved in a separate step by adding hydrofluoric acid.

The fuel solution coming from the dissolver is first cleared by centrifuging or filtration, as in LWR fuel reprocessing. Then the PUREX countercurrent solvent extraction process is applied. However, unlike LWR reprocessing, the contact times of the solvent and the nitric acid solution must be shorter to limit radiolysis of the solvent. For this purpose, pulsed columns or centrifugal contactors are used.

When decontaminating plutonium and uranium, the higher plutonium concentration must be taken into account to ensure that the plutonium fraction in the waste is kept as small as possible. The process for conversion of plutonium nitrate and uranyl nitrate into PuO_2 and UO_2 is the same as in an LWR reprocessing plant. Also, plutonium nitrate and uranyl nitrate can be directly mixed to become a master-mix which is co-precipitated into mixed PuO_2/UO_2.

The modifications of the PUREX process described above, the introduction of special technical components, and the smaller dimensions of all tanks (higher plutonium enrichment, criticality) ultimately require the construction of special LMFBR reprocessing plants.

7.4.4 LMFBR Fuel Fabrication

In general, the same fabrication process is applied for LMFBR fuel as for LWR recycle (MOX) fuel (Sect. 7.2.2) [20, 28, 29, 36]. The reprocessing plant and the LMFBR fuel refabrication plant are co-located at one site. After storage in a buffer store, the MOX fabrication process begins with mechanical blending of UO_2 and master mix PuO_2/UO_2 powders to establish the desired enrichment. Afterwards, the fabrication process proceeds with pressing, sintering, grinding and drying of sintered pellets. This is followed by assembling core pellets into stacks, adding axial blanket pellets, inserting the pellets into cladding tubes and introducing an inert atmosphere. Finally, the fuel pins are welded and assembled into fuel elements.

The PuO_2/UO_2 fuel must be highly soluble in nitric acid. Even after a burnup of about 100,000 MWd/t, the LMFBR fuel must have a solubility in nitric acid of >99% in order to minimize the volume of insoluble residues and fuel losses during reprocessing. This requires special attention to be paid in the milling and sintering processes.

LMFBR fuel pellets usually have somewhat smaller diameters than LWR pellets. The pellet fabrication and pin loading operations are carried out in glove boxes. To protect the workers against γ-radiation and neutrons originating from various plutonium isotopes and their radioactive daughters (spontaneous fission and (α,n)-reactions with oxygen), shielding must be provided at the fabrication lines. Future plants are expected to be operated remotely to a large extent. Besides the present reference LMFBR fuel fabrication technology described above, also the sol-gel precipitation technique, the vibro-compaction technique, and the AUPuC process are being employed.

The sol-gel process allows the co-conversion of Pu nitrate and uranyl nitrate followed by the fabrication of spherical PuO_2/UO_2, particles, which can be pressed and sintered into fuel pellets. The AUPuC refabrication process following after co-conversion allows the fabrication of a coarse grain PuO_2/UO_2 and Am-241 free crystal powder, which can be pressed and sintered into pellets. Both refabrication processes avoid the formation of Pu dust (see Sect. 7.2.2.2). In the vibro compaction refabrication process the MOX fuel is broken into small particles which are filled into the cladding tube and vibro compacted. These refabrication processes are favored for application in future advanced refabrication plants of FBR fuel cycle centers.

In the fabrication process, different types of contaminated waste are produced. In a refabrication plant of the reference fuel cycle shown in Fig. 7.11, it must be taken into account that roughly 200 kg of Pu (1%) annually would go into the waste. Consequently, special efforts must be made to recover this plutonium. The resultant waste should not contain more plutonium then is allowed according to safeguards requirements.

7.4.5 Status of LMFBR Fuel Reprocessing and Refabrication

Several small test and pilot plants for LMFBR fuel reprocessing and MOX fuel fabrication in the UK and France, had been operated for almost twenty years [21, 28, 29, 37]. The UK operated a small pilot reprocessing plant of 5 t_{HM}/year throughput at Dounreay for reprocessing fuel from PFR. France operated a small experimental test facility at La Hague with a capacity of 1 kg_{HM}/day. It reprocessed about 1 t_{HM} of fuel from RAPSODIE with fuel burnups between 40,000 and 130,000 MWd/t. The Service de l'Atelier Pilote (SAP) reprocessing pilot plant at Marcoule, France, reprocessed some 6 t_{HM} of fuel from RAPSODIE and PHENIX. The Traitement d'Oxydes Rapides (TOR) facility at Marcoule, France, had a higher yearly capacity of up to 5 t_{HM} and could reprocess spent fuel from PHENIX and some fuel from SUPERPHENIX. Japan has small scale reprocessing facilities in operation at Tokai-mura for the fuel of JOYO and MONJU. Japan has a MOX fuel fabrication plant with a capacity of 5–10 t_{HM}/year in operation at Tokai-mura. France operated a MOX fuel fabrication plant of 20 t_{HM}/year throughput at Cadarache. Russia will start a 60 t_{HM}/year plant at Zheleznogorsk to provide the PuO_2/UO_2 fuel for BN 800.

7.4.6 Radioactive Inventories of Spent LWR Fuel

Table 3.5 in Sect. 3.7 indicated the radioactivity in Ci/t_{HM} ($1\ \text{Ci} \triangleq 3.7 \times 10^{10}\ \text{Bq}$) of the most important radioactive isotopes in spent LWR fuel with a burnup of 55,000 MWd/t_{HM} as a function of time after discharge from the core. The bulk of radioactivity is constituted by the fission products. In the reprocessing step, the fission products and all actinides, such as neptunium, americium and curium, and less than 1% Pu- and U-losses are separated and go into the high level waste, while more than 99% of the plutonium and uranium is recovered. The bulk of the fission products decays rapidly within the first about 500 years. After somewhat more than 100 years, the radioactivity of the actinides plays the dominant role in high level waste until fission products like Tc-99 and others take over for a long time. This is shown by Fig. 7.12 displaying the radioactivity in Bq/t_{HM} of the different isotopes as function of time for spent fuel in the burnup of 45 GWd/t [38].

7.4.7 Decay Heat of Spent Fuel

The decay of fission products and actinides in the spent fuel produces heat which must be accounted for waste treatment and waste storage. Figure 7.13 shows the decay heat in W/t_{HM} for the fission products and actinides. The total decay heat is about 1,000 W/t_{HM} at about 50 years after discharge from the reactor core and about 100 W/t_{HM} at about 800 years after discharge. The bulk of the decay heat is produced by the fission products during the initial 70 years. Afterwards the actinides—mainly

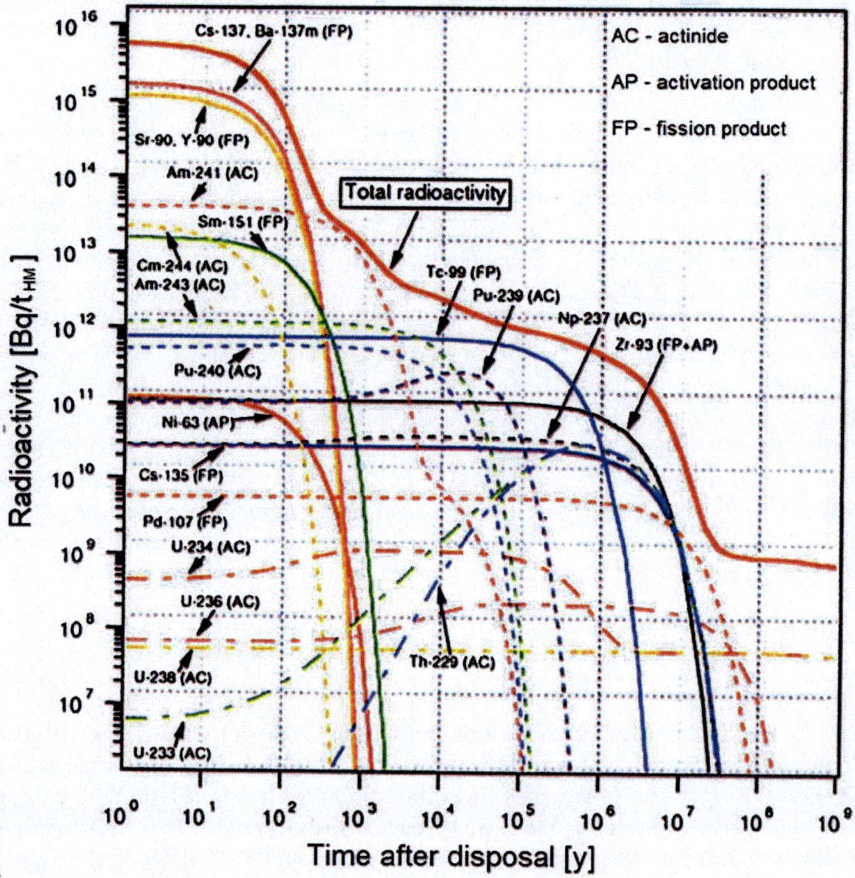

Fig. 7.12 Radioactivity [Bq/t_{HM}] as a function of time for spent LWR fuel with a burnup of 45 GWd/t (adapted from [38])

Am-241 and some of the plutonium isotopes are dominating the decay heat. The decay heat of the fission products decreases fairly rapidly and is less than 10 W/t_{HM} after about 300 years.

7.5 Conditioning of Waste from Spent LWR Fuel Reprocessing

7.5.1 Classification of Radioactive Waste

Radioactive wastes are classified internationally according to their specific radioactivity in Bq/m^3. Figure 7.14 shows the classification into:

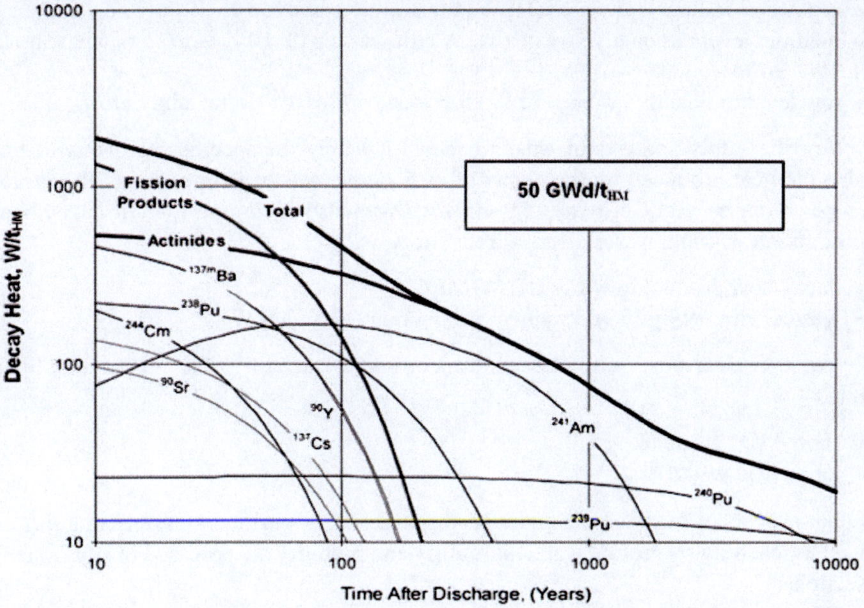

Fig. 7.13 Decay heat production of fission products and actinides after discharge of the spent PWR fuel [39]

Fig. 7.14 Classification of radioactive waste according to its specific radio-activity [7, 40]

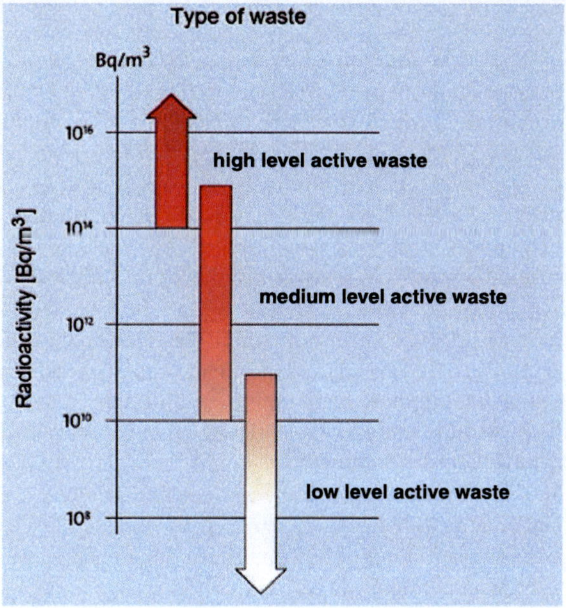

- high level radioactive waste (HLW) for all waste above 10^{14} Bq/m^3
- medium level radioactive waste (MLW) for waste with 10^{10} up to somewhat above 10^{14} Bq/m^3
- low level radioactive waste (LLW) for waste with 10^{11} Bq/m^3 and below.

For the safety analysis of waste disposal not only the specific radioactivies but also the heat produced by the decay of radioactive wastes is important. Therefore, some countries, e.g. Germany, classify the waste also according to their decay heat production within a waste disposal repository as:

- decay heat producing waste (HLW/MLW)
- wastes with negligible decay heat production (LLW/MLW).

Another classification can be made according to the half-life of the radionuclides into

- short-lived waste or
- long-lived waste.

Other aspects of the waste, e.g. material properties of vitrified glass or ceramics as well as leaching by ground water, solubility and mobility are adressed in subsequent sections.

7.5.2 Solidification and Storage of Liquid High Level Waste

The HLW solution generated in the extraction of uranium and plutonium during reprocessing of spent LWR fuel is concentrated by approximately a factor of 10 by evaporation, which gives rise to some 0.5 m^3 of liquid high level waste concentrate (HLWC) solution per tonne of spent fuel (see Fig. 7.7). This HLWC solution is first pumped into tanks of acid resistant steel (Fig. 7.15). The activity of the HLWC can reach 3.7×10^{13} Bq/l, the decay heat production can attain 7 W/l. The tanks stand in stainless steel pans which collect any leakages. Because of the strong radiation emitted by the HLWC solutions, these tanks are installed in hot cells lined with steel. The HLWC solution is recirculated in the tanks and cooled by water circulating in tube coils (<65°C) for removal of the decay heat generated by fission products and actinides. Standby tanks of sufficient capacity must be available. The HLWC solution can be kept in these tanks for at least 30 years.

For solidification of the liquid HLWC solution the first process step is calcination, i.e., the expulsion of liquid from the HLW solution at temperatures up to 400°C. This decomposes the nitrides and produces oxides. Calcinates are not resistant enough to leaching. Their thermal conductivity is low and their mechanical stability is insufficient. They are intermediate products, which must be converted into vitreous materials. Glass frit with additives of silicon, boron, aluminum, etc. can be used for conversion into glass. After melting at high temperatures (1,000–1,200°C), this results in borosilicate glass.

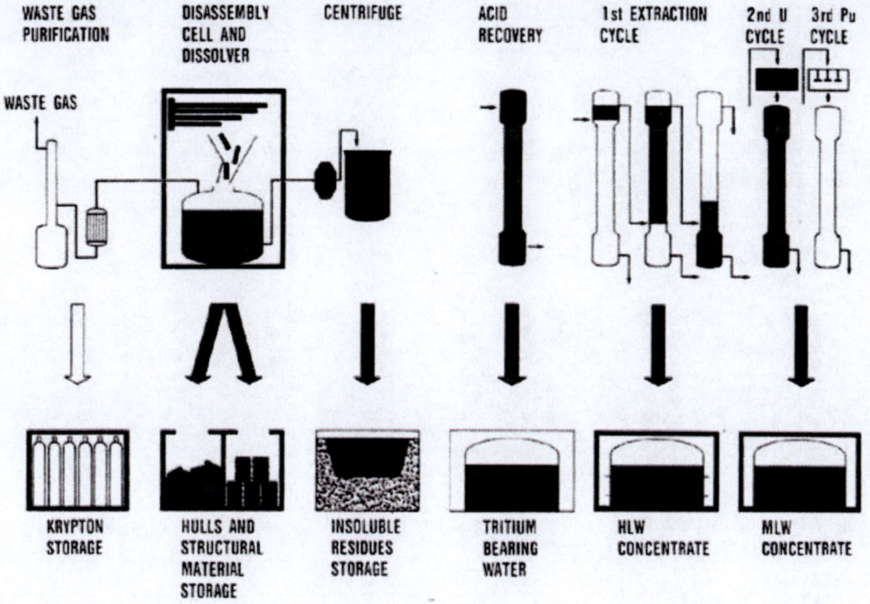

Fig. 7.15 Different waste forms from reprocessing [8]

This type of glass has a high thermal stability, a high chemical durability, low leaching rates and high resistance to self radiation. In addition, it has a high solubility at the glass melting temperatures for a broad range of different fission products, actinides, and corrosion products (Fe, Ni, Na, P).

Figure 7.16 shows the relative abundances of fission products for spent UO_2 fuel and spent MOX fuel at a burnup of 60,000 MWd/t_{HM}. As already explained in Sect. 3.7 the abundances of fission products are slightly different for spent uranium and plutonium/uranium fuel.

Borosilicate glass consists of SiO_2, B_2O_3, Al_2O_3, Na_2O and CaO. Table 7.6 shows the average composition of the R7T7 borosilicate glass fabricated at the LaHague, France, reprocessing and waste conditioning plant [41].

The incorporation rate of this type of glass for fission products is limited to 18.5% and about 0.6% for actinides. The so-called platinoids like RuO_2 crystals and the metallic phases of Pd, Rh, T have only a limited solubility in borosilicate glass. Due to their high density they tend to sediment to the bottom of the molten glass. Stirring systems in the glass melter allow incorporation of platinoids up to 3%.

7.5.2.1 HLW Vitrification Process

Figure 7.17 explains the different technical processes during the vitrification of HLWC [38, 41]. The HLWC is given into a calcinator where the nitrates of most

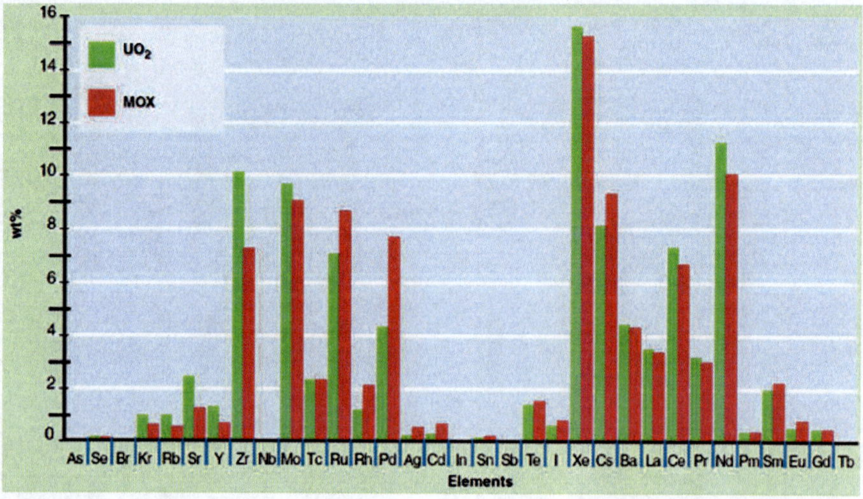

Fig. 7.16 Relative abundances of fission products in spent UO_2 and MOX-fuel with a burnup of 60,000 $MWd/t_{(HM)}$ [41]

Table 7.6 Chemical composition of the R7T7 borosilicate glass [41]

Type of oxides	Average composition (wt%)
SiO_2	45.6
B_2O_3	14.1
Al_2O_3	4.7
Na_2O	9.9
CaO	4.1
Fe_2O_3	1.1
NiO	0.1
Co_2O_3	0.1
P_2O_5	0.2
LiO_2	2.0
ZnO	2.5
Fission products and actinides	15.6

elements are decomposed to oxides at temperatures up to 400°C. Then the borosilicate glass is added in form of glass frits. The glass is molten in an induction heated metal pot at a temperature of about 1,100°C. It is then poured into stainless steel containers (Fig. 7.18). After lid welding of the container and external decontamination, the glass container can be transferred to a storage site. The waste gas from the glass melter is purified in a waste gas cleaning section by wet scrubbers and high efficiency particle filters.

Other vitrification processes avoid the initial calcination process and feed the HLWC together with glass frits directly into the glass melter. The glass is molten in this process by electrical Joule heating.

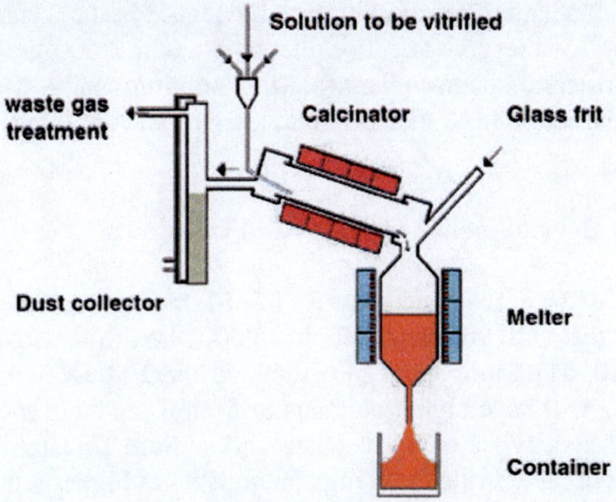

Fig. 7.17 HLW vitrification process [41]

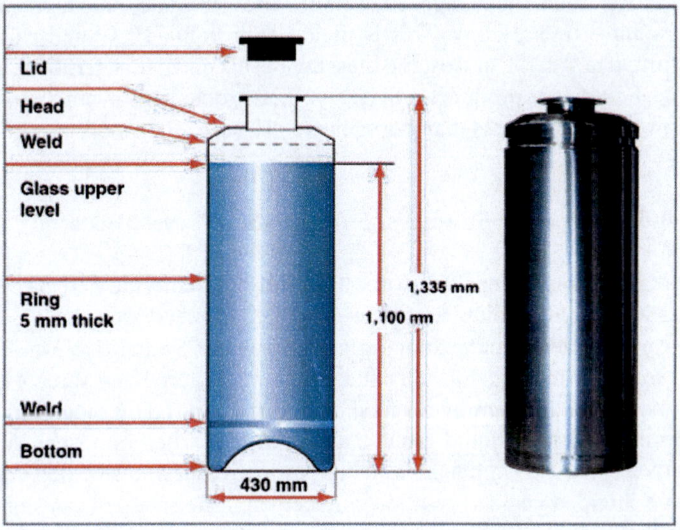

Fig. 7.18 Vitrified HLW glass container [41]

More recent developments led to the cold crucible concept. It creates a thin glass layer at the wall of the glass melter and prevents corrosion of its inner walls.

One glass container (Fig. 7.18) has an inner glass volume of $0.159\,m^3$ and an overall volume of $0.180\,m^3$ and contains 400 kg of glass. At the time of vitrification it produces about 2.5 kW heat power which decreases to 1 kW after 10 years and to 0.4 kW after 50 years.

Commercial vitrification plants operate at LaHague (France) and Sellafield (UK). The total capacity of these vitrification plants is about 2,500 glass containers per year. Smaller vitrification plants operate at Mol (Belgium) and Karlsruhe (Germany) or are under construction in Rokkasho-mura (Japan) and in China.

7.5.2.2 Further Development of Glasses and Ceramics

The composition of the borosilicate glass R7T7 of Table 7.5 was developed for HLWC from PWR spent fuel with a burnup of about 45,000 MWd/t_{HM} with actinide contents of about 0.4% [40, 41]. Since spent UO_2 fuel and spent MOX fuel with a burnup of 60,000 MWd/t_{HM} will have higher contents of fission products and higher contents of actinides, different types of borosilicate and sodium silicate glasses are under investigation which have sufficiently high solubilities at higher temperatures for the higher contents of fission products and actinides.

Ceramics contain the radioactive fission products and actinides in their crystal structure. This is different to borosilicate glass which is amorphous. Synroc—developed in Australia—is composed of hollandite, zirconite, perovskite, rutile and a small amount of metal alloy. This is mixed with liquid HLW and calcinated at 750°C to form a powder. Hot isostatic pressing of this powder at 1,150–1,200°C in a stainless steel container produces a black synthetic rock. This synthetic radioactive rock has lower leaching rates than borosilicate glass.

7.5.2.3 Solidification and Storage of Solid Medium Level Waste

Solid MLW is mainly composed of the fuel rod hulls and the end pieces of fuel elements as well as the insoluble residues generated in fuel dissolution [7, 41]. They are initially put into temporary storage silos under water. Solid MLW waste volumes amount to some 0.6 m^3/t_{HM} for hulls and structural material and some 4 kg/t_{HM} of insoluble residues. Conditioning for final storage is achieved by separating the solid waste and mixing it with liquid grout. This mixture is filled into steel canisters of 1.06 m diameter and 1.69 m height which, in turn, are put into shielded vessels. All spaces in the shielded vessels are filled with cement. The shielded canister is closed with a shielding lid.

Since the year 2010 the MLW metallic structures are compacted at the LaHague reprocessing and waste conditioning plant. The cladding hulls guide tube and nozzles are introduced into a strong metallic cylinder and compacted with a 250 MPa press [35]. This allows a volume reduction by a factor of five. The resulting package is called CSD-C container which contains between 5 and 10 of these cyclindrical compacted pancakes (Fig. 7.19). These CSD-C containers have the same outer geometry (0.43 m diameter and 1.3 m height) as the glass containers.

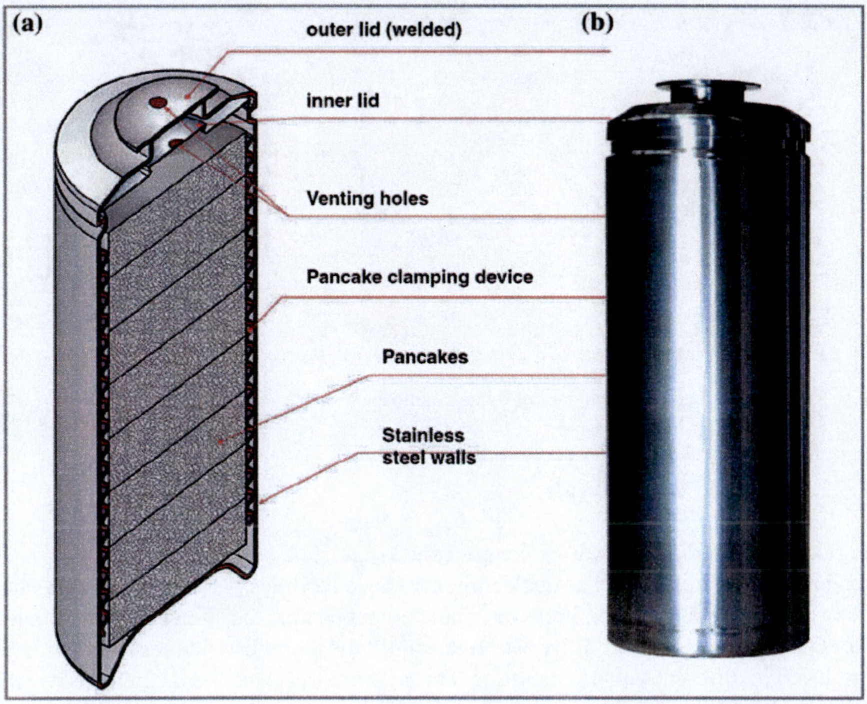

Fig. 7.19 CSD-C container with compacted HLW pancakes of cladding hulls, guide hulls etc. [41]

7.5.2.4 Treatment of Aqueous Organic Medium Level Waste

Aqueous MLW solutions are concentrated by evaporation and treated in a denitrator. This amounts to waste volumes for further conditioning of about 0.6 m^3/t_{HM} of spent fuel. Afterwards, these concentrates can be mixed with cement and filled in drums as described in Sect. 7.5.2.3.

Another technique is mixing the concentrate solution with hot bitumen. This causes any water residues to evaporate. The product is again filled in drums, which are closed with lids after cooling. Bituminization has the advantage of producing smaller waste volumes than cementing.

Organic MLW solutions of the reprocessing plant are treated by the phosphoric acid adduct method, which allows the kerosene to be purified and recycled. The remaining organic MLW is mixed with plastic granulate. A homogeneous solution is produced, which is filled in drums and sealed.

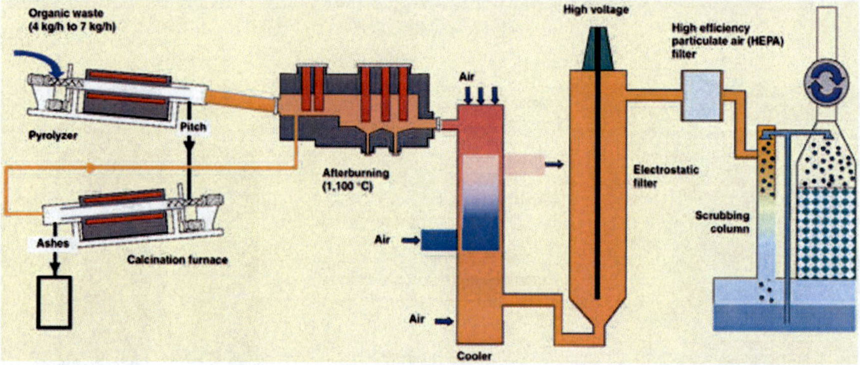

Fig. 7.20 IRIS process for treatment of organic radioactive waste [41]

7.5.2.5 Organic Waste Treatment

Various types of organic wastes are generated in nuclear reactors, reprocessing and refabrication plants as well as nuclear maintenance facilities [7, 41]. On the one side this is solid waste like α-emitter contaminated papers, plastics, ion exchange resins etc. On the other hand liquid waste like scintillating liquids, sludges, oils etc. are produced in different nuclear facilities. This organic waste can be incinerated which reduces its weight strictly to its mineral load. The gases are purified by high voltage electrostatic filters, high efficiency particulate air (HEPA) filters and scrubbing columns.

Figure 7.20 shows the IRIS (according to the name of the prototype unit Incineration Research Installation for Solid waste) process for organic waste treatment developed in France. The organic waste is first sent to a medium temperature pyrolysis system (500°C) for removal of the most corrosive compounds and then to an oxygen fed calcination system (900°C) which completes the combustion. The resulting ashes do not contain carbon any more, but concentrate about 99% of the initial radioactivity. The off-gas treatment system consists of an afterburning chamber followed by filter systems. The gases are finally released through a chimney. They contain only radioactivity below the allowable limits.

7.5.2.6 Waste from Light Water Reactors

As shown in Fig. 7.21 LWRs generate medium and low active wastes. The ion exchange resins, evaporator concentrated filters, sludges and oils can be treated like organic wastes as described in the previous Sect. 7.5.2.5 [7, 42–44]. The metallic parts are compacted, e.g. in Germany, to metallic pancakes by a 200 MPa press as described in Sect. 7.5.2.3 [7]. Their volume can be reduced by a factor of five.

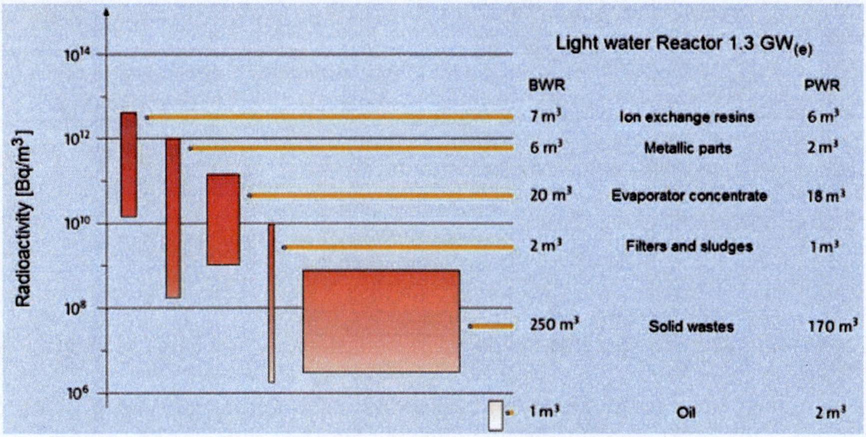

Fig. 7.21 Medium and low active waste volumes from 1.3 GW(e) PWRs and BWRs [7]

7.5.2.7 Treatment of Low Active Solid and Liquid Wastes from Reprocessing Plants and Nuclear Reactor Plants

Solid low level waste (LLW) is concentrated, e.g. in Germany, by pressing with 200 MPa and burning, whereas liquid LLW is concentrated chiefly by evaporation. The concentrated waste is solidified by cementing and bituminization, filled in drums and sealed [7, 43].

7.5.2.8 Cements as Confining Materials

Cement based materials are widely used for radioactive waste conditioning and disposal [41, 42, 44]. Several categories can be distinguished

- pure pastes of cement and water
- grouts and mortars with different sand contents
- concretes which may be reinforced by metal fibres.

The radioactive waste is mixed with cement pastes, grouts or concrete and filled in containers of different size.

7.5.2.9 Treatment of Kr-85 and Tritium

The separated Kr-85 can be forced into pressurized steel cylinders of 50 l volume [42]. The krypton cylinders are dumped into storage shafts in a krypton storage facility. The shafts can accommodate, e.g. four krypton cylinders each stacked on

Table 7.7 Waste volumes from reprocessing spent UO_2 or MOX LWR fuel normalized to 1 GW(e)·y [41]

	Type of waste	Normalized to 1 GW(e)·y of energy produced
Vitrified high level waste (m^3)	HLW	2.5
Hulls, spacers, insolubles organic waste (m^3)	MLW	5
Low level waste (m^3)	LLW	12
Noble gases number of gas cylinders	LLW	17

top of each other. The krypton cylinders are cooled by air. The half-life of Kr-85 is 10.8 a.

Tritium can be concentrated first by proper mass flow optimization in the PUREX process. For final storage, the water containing high concentrations of tritium may be stored in tanks. This process allows tritium to be removed from the biosphere for periods long enough compared to its half-life of 12.4 a.

7.5.2.10 Waste Volumes to be Stored from Reprocessing of Spent LWR UO_2 or MOX Fuel

Table 7.7 lists data for conditioned waste volumes arising from a spent LWR fuel reprocessing plant as described in Sect. 7.2.1 [41, 42]. The data are normalized to 1 GW(e)·y of energy produced. The following assumptions hold for Table 7.7:

Vitrified liquid HLW is stored in steel containers as shown in Fig. 7.18 of Sect. 7.5.2.1. Fuel claddings are compacted and filled into CSD-C containers as described by Fig. 7.19 of Sect. 7.5.2.3.

Krypton is filled into steel cylinders of 50 l volume and 5 MPa pressure.

The conditioned MLW and LLW is filled into shielded and unshielded drums.

7.5.3 Radioactive Waste from Uranium-233/Thorium Fuel Reprocessing

Solid HLW in the form of cladding hulls is compacted and will have to be conditioned as described in Sect. 7.5.2.3 [42]. Liquid HLW consists mainly of the aqueous waste from the first extraction column, which contains the fission products, small amounts of actinides as well as significant amounts of dissolved aluminum and fluorides. They will have to be concentrated, calcinated and converted into mixtures of stable oxides. Medium and low level wastes consist of aqueous waste, filter wastes, etc. and will have to be treated as described in Sect. 7.5.2.5.

Data on waste volumes arising from U/Th fuel reprocessing were collected during INFCE [1].

7.5.4 Radioactive Waste from Reprocessing Plutonium/Uranium Fuel of LMFBRs

The bulk waste quantity in the LMFBR fuel cycle is produced in the reprocessing plant [29, 42]. Liquid HLW contains the fission products and actinides and has activity levels similar to those of liquid HLW of LWR fuel. Correspondingly, also waste treatment consists of the liquid HLW storage and vitrification steps described in Sect. 7.5.2. Conditioning of solid HLW as well as MLW and LLW is also identical.

A comparison of waste volumes arising from LMFBRs and their Pu/U fuel cycle was given in INFCE [1]. For the front end (uranium refining, conversion and enrichment) the waste volume is very small for the FBR fuel cycle (no enrichment etc. needed). All other waste volumes are similar to those of the LWR UO_2 or Pu/U MOX fuel case described in Sect. 7.5.2.

7.5.5 Wastes Arising in Other Parts of the Fuel Cycle

In addition to wastes arising in fuel reprocessing and refabrication plants, waste from other parts of the fuel cycle [40, 42], i.e., uranium or thorium mining, fuel conversion, fuel enrichment, fuel fabrication and the nuclear power plant must be considered (INFCE) [1].

7.5.5.1 Wastes from Uranium Ore Processing

Mill tailings are a slurry of ore residues with some of the process chemicals. Although they only contain natural uranium or thorium, they do require careful treatment. As a result of chemical treatment in ore processing, radioactive isotopes of the U-238 and Th-232 decay chains may be released to the biosphere (see Chap. 10). Of these, Ra-226 (half-life 1,600 years) is most important, because it can both be ingested by way of the aqueous pathways and inhaled through its gaseous daughter, Rn-222 (half-life 3.8 days). Also Th-230 is relevant on a long term basis, as a precursor of Ra-226.

By comparison, the management of thorium mill tailings is a minor problem. The decay products of Th-232, such as Ra-228 (half-life 5.9 years) and Th-228 (half-life 1.9 years), have decay periods so short as to be of little relevance for the biosphere.

7.5.5.2 Wastes from Uranium Refining, Conversion and Enrichment

Both in refining and in the conversion of U_3O_8, to UF_6 for the enrichment process, small volumes of waste are produced, which contain natural uranium. In the enrichment step, depleted UF_6 is produced which, in turn, can be converted into depleted

UO_2. This depleted UO_2 can be used as a fertile material in FBRs. In a pure once-through fuel cycle, this depleted UO_2 has to be treated as waste.

7.5.5.3 Wastes from Fuel Element Fabrication and Nuclear Power Plants

Similar to the fabrication of MOX fuel, also plants fabricating low, medium or highly enriched uranium oxide fuels produce waste, which must be treated and stored.

Reactor power plants generate waste in the form of filters and ion exchanger resins etc. as shown in Fig. 7.21. Repair and maintenance of radioactive components and the replacement of absorber rods produces additional waste volumes. These wastes are reduced as described in Sects. 7.5.2.3–7.5.2.5.

Estimates of the total amount of waste produced by different reactors operating in different fuel cycle options were also estimated during INFCE [1]. This waste has to be packaged in shielded and unshielded drums, gas flasks, and canisters as described in Sect. 7.5.2.5.

7.6 Nuclear Waste Repositories

The HLW radioactive glass containers and the MLW compacted hulls produced yearly by one 1.3 GW(e) PWR can be transported in two CASTOR transport casks to an intermediate storage facility [37, 38, 45, 46]. They will have to be stored there for about 50 years for further decrease of the decay heat prior to disposal in a repository.

It is commonly accepted that HLW should be stored in suitable deep geological formations. Such geological formations should be free from circulating groundwater, have high impermeability and good heat conductivity. Thick rock salt formations meet these requirements in an almost ideal way (Table 7.8). In addition, they have a high plasticity so that fissures in the salt around waste canisters or drums are closed (self-sealed) and isostatic pressure conditions are maintained.

Granite, gneiss and basalt formations at a depth of several hundred meters as well as argillaceous formations, like clays etc., are also attractive for nuclear waste disposal. Even tectonically very stable and extremely thick sediment packs under the Atlantic and Pacific Oceans have been proposed as disposal sites.

7.6.1 Disposal of Short Lived MLW/LLW

Practical experience exists in Canada, France, the UK, USA and Russia with burying α-emitter-free MLW and LLW in specially selected and prepared burial grounds close to the surface (shallow ground burial) [42, 45].

Table 7.8 Geological formations for HLW/MLW under investigation in different countries [42, 45]

Country	Geological formation
Belgium	Clay
Canada	Crystalline igneous rock (salt, limestone, shale)
Finland	Granite, gneiss
Germany	Dome salt
France	Clay
Japan	Granite
Netherlands	Dome salt
Sweden	Granite, gneiss
Switzerland	Granite, clay
United Kingdom	Granite, clay, (salt)
USA	Tuff, bedded salt, granite

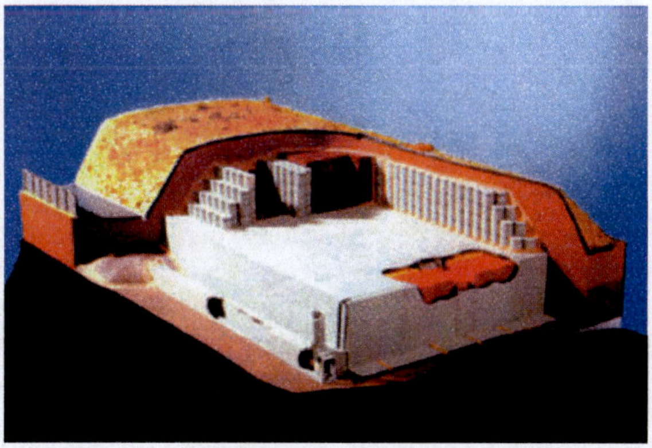

Fig. 7.22 Concrete structure for short lived α-emitter-free MLW and LLW [41, 45]

7.6.1.1 Shallow Ground Burial of Shortlived MLW/LLW

Sophisticated technical measures are currently employed to emplace containers with shortlived waste-free from α-emitters in a concrete structure close to the surface of the ground (Repository Type 1, Fig. 7.22). When several layers of containers have been put in place, a concrete ceiling is cast. Finally, this ceiling is sealed and covered with layers of clay impervious to water and with soil. Below the concrete floor of this repository there should also be layers of clay impervious to water. Accessible galleries are to allow monitoring for leakages and radioactivity. After approximately 300 years, government surveillance of this repository can be given up as the level of radioactivity will have decreased far enough by that time. A repository of this type exists, e.g. in France (Centre de l'Aube). Table 7.9 refers to it as a Type-1 Repository.

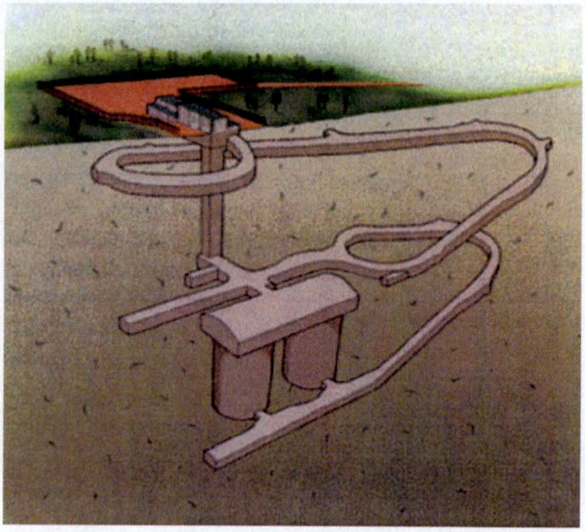

Fig. 7.23 Repository for MLW and LLW from nuclear reactor plants (Sweden, Finland) [42, 45, 47]

7.6.1.2 Final Storage of MLW/LLW in Granite Beds at a Depth of Approximately 50–100 m

A different type of repository for MLW/HLW, especially for waste originating from nuclear power plants, is developed mainly in Sweden and Finland.

This repository is located in granite rock at a depth of roughly 50–100 m (Fig. 7.23). It is accessible through a vertical shaft and ramplike galleries. Material is transported by vehicles over the ramps into the repository area. The waste containers are emplaced in concrete silos. For protection against the entry of water from cracks in the granite rock these concrete silos are sealed by layers of bentonite. This type of repository is referred to as a Type-2 Repository in Table 7.9.

7.6.1.3 Final Storage of HLW and Longlived MLW in Deep Geological Formations

HLW containers and containers with longlived heat producing MLW are to be stored at a depth of 500–1,000 m (Repository Type 3). This is achieved by building repositories in mines with vertical shafts. The HLW containers are moved into the final storage (Fig. 7.24) area through shafts and emplaced in boreholes or horizontal galleries.

A multibarrier system is installed to add to the safety of the repository. The Japanese multibarrier system for emplacement in granite formations [38] consists first of an overpack around the canister with borosilicate glass. This overpack must withstand the pressure exerted by the rock and may consist of a steel enclosure 19 cm

Table 7.9 Repositories for LLW/MLW in different countries [42, 45]

Country	Location	Type of repository	Type of waste	Begin operation
France	Aube	1	LLW/MLW	1992
Japan	Rokkasho	1	LLW/MLW	1992
Spain	El Caboil	1	LLW/MLW	1992
Sweden	Forsmark	2	LLW/MLW	1988
Finland	Olkiluoto	2	LLW/MLW	1992
	Lovisa	2	LLW/MLW	1998
USA	Clive	1	LLW/MLW	1998

Fig. 7.24 Deep HLW repository in granite rock [48]

thick. This steel enclosure is surrounded by a 6 cm cladding of titanium or copper withstanding the corrosive attack of brine or water for about 1,000 years (Fig. 7.25).

This overpack is enclosed in an 0.8 m thick layer of bentonite/sand designed to prevent the access of water or brine to the radioactive glass canister. Moreover, this bentonite layer is to act as a filter preventing the transport of any colloidal radioactive nuclides forming.

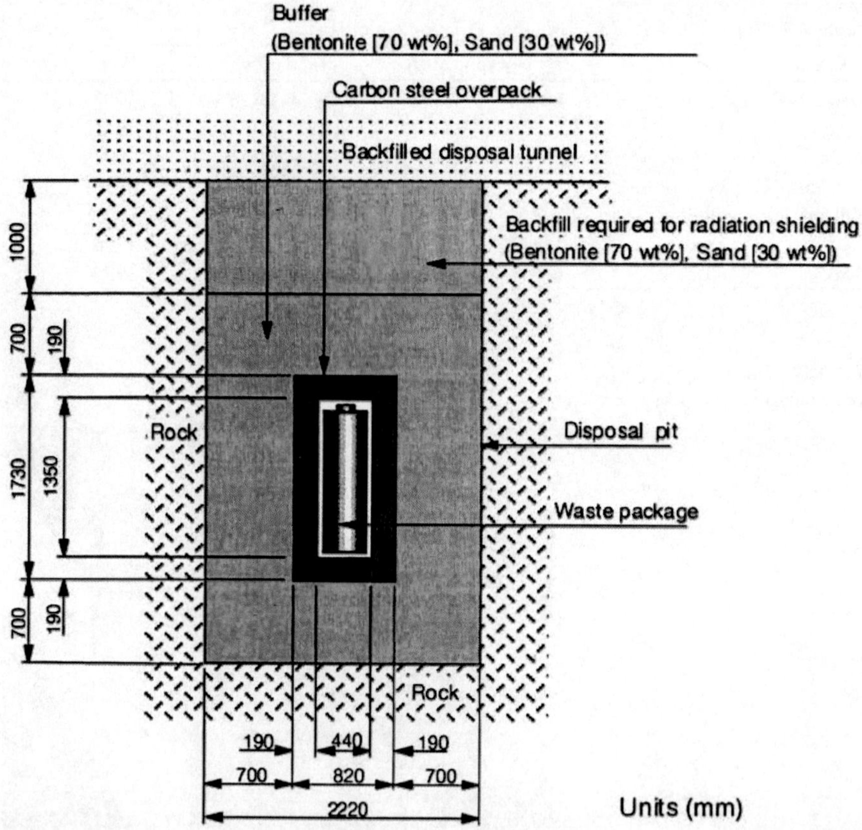

Fig. 7.25 Japanese concept for final disposal of vitrified HLW in granite rock geological formations [38]

In the German concept of final disposal of the radioactive glass cylinders, in a salt dome underground (see Fig. 7.26), the necessary overpack of the vitrified glass cylinder is to be decided yet.

At present, there is no HLW repository in deep geological formations in operation. However, there are definite plans in a number of countries to construct national nuclear waste repositories for placement of all kinds of high level radioactive waste in deep geological formations. Experience with salt mines for the disposal of MLW and LLW and test results for HLW are available only in the USA.

The first repository of type 3 for long-lived radioactive waste with low decay heat was taken in operation in the USA in 1999. This so-called Waste Isolation Pilot Plant (WIPP) is a 650 m deep underground salt repository in New Mexico (USA). It takes primarily transuranium waste from military programs.

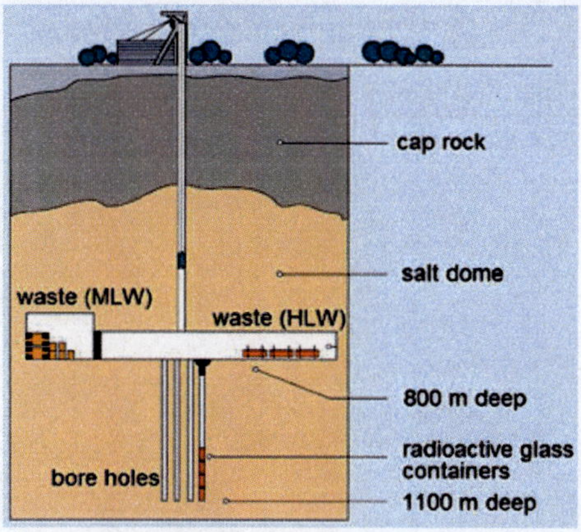

Fig. 7.26 Deep waste repository for HLW/MLW in salt formation [40, 45–47]

Figure 7.26 gives an impression of an underground salt repository. It consists of shafts, access corridors and disposal rooms excavated some 600 to 1,000 m under ground, within the salt formation. Following the excavation of the rooms, storage holes for canistered HLW are drilled in the floors of the rooms. For HLW canisters, the necessary spacing of storage holes is about several 10 m, dependent upon the storage concept applied.

When the HLW canisters will have been placed in prepared holes or horizontal galleries by remote handling techniques, the remaining void at the top of the hole is backfilled with either excavated salt or other material. Canisters with negligible heat generation rates are placed several per hole. Drums with long-lived medium level wastes are, e.g. placed in storage rooms at different locations in a repository. After an excavated room has been filled with the desired amount of waste, it will be backfilled with excavated salt or other material.

7.6.1.4 Limiting Thermal Requirement for Geological Formations

The reason for spreading HLW canisters over a certain area is their heat generation. Since this heat is only transferred by thermal conduction, the temperature at the surface of the canister and also of the rock, clay, tuff or of salt in the direct vicinity of the canister must be compatible with the allowable limits and design basis of the repository. Table 7.10 shows the allowable temperature limits in different geological formations.

The HLW glass containers or final disposal casks for spent fuel elements are therefore stored in the deep repository after about 50 years of intermediate storage. As explained already by Fig. 7.13 in Sect. 7.4.7 the heat generation by fission products decreases by several orders of magnitude over a time period of about 300 years. After a time period of about 700 years, the actinides will be mainly responsible for heat generation.

Table 7.10 Temperature limits for surface temperatures of waste canisters in different geological formations [49, 50]

Geological formation	Temperature limit °C
Canister surface temperature limits	
Salt	150–200
Clay	100
Granite	100
Tuff	200

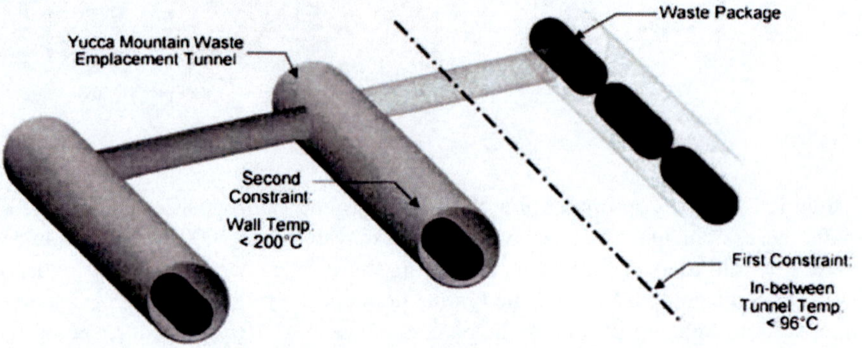

Fig. 7.27 Waste emplacement tunnels and temperature limits for Yucca Mountain geological repository [51]

7.6.1.5 Thermal Requirements for a Tuff Repository

US regulatory authorities did require the following condition for the tuff repository in Yucca Mountains (see Fig. 7.27).

The temperature in between the waste emplacement tunnels must remain below 96°C (local water boiling point) to permit water to percolate in between the tunnels, avoiding any accumulation of water above the repository. The temperature of the surroundings rock must remain below 200°C. This avoids inducing changes to the mechanical properties of the rock. The temperature in between the tunnels is caused primarily by the heat producing isotopes of plutonium and americium [50–52].

7.6.2 *Direct Disposal of Spent Fuel Elements*

In the once-through cycle concept with direct disposal, the spent fuel elements, following interim storage for a couple of years, are conditioned in such a way that they can be stored in a repository over secular periods of time [7, 40, 45]. This concept has been proposed in Canada, the USA and Sweden. Studies of the concept

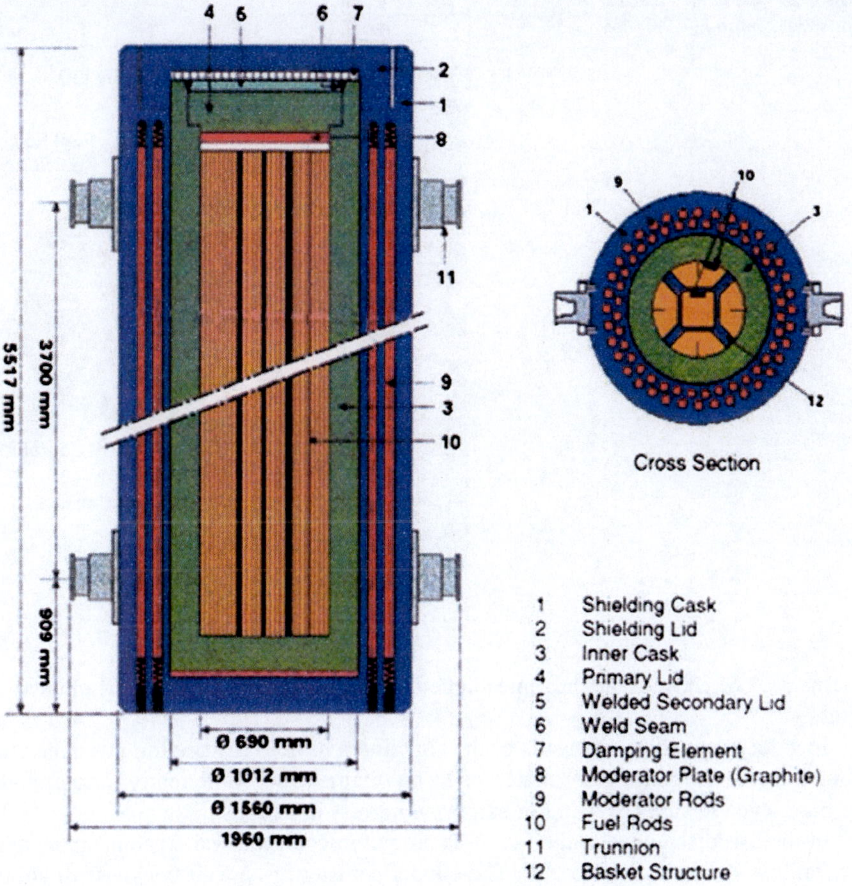

Fig. 7.28 Final disposal cask POLLUX for spent fuel elements [7, 40, 45]

were carried out also in Germany. The uranium (U-235 and U-238) contained in the spent fuel elements and the plutonium in that case are no longer available for further utilization.

In a technique proposed in Sweden or Germany, the spent fuel elements can be disassembled and, after removal of the carriers and head pieces of the fuel elements, the fuel rods can be stored directly in containers or they can be shortened.

In the German concept the long fuel rods can be encapsulated in POLLUX containers. One POLLUX container (Fig. 7.28) can take the spent fuel rods of 10 PWR fuel elements. Therefore, four POLLUX containers will be necessary for direct spent fuel rod disposal of the spent fuel rods of a 1.3 GWe LWR plant per year.

The shortened fuel rods can also be encapsulated in a so-called BSK 3 container (Fig. 7.29). Such BSK-3 containers can take the spent fuel rods of three PWR fuel

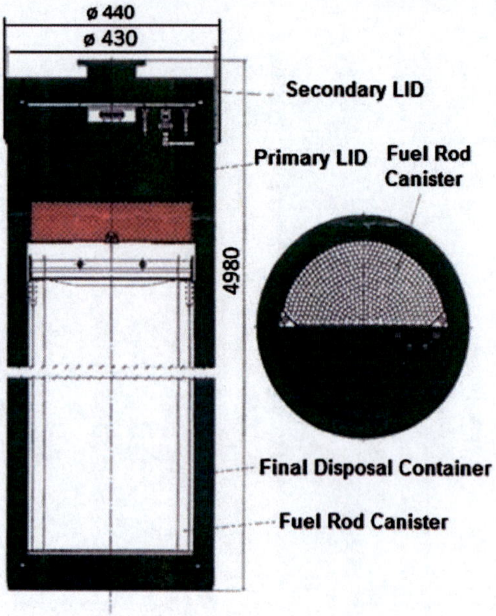

Fig. 7.29 Container BSK 3 for spent fuel rod [7, 40, 45]

elements. They have about the same diameter as the containers for vitrified radioactive glass.

In a deep repository the walls of the containers have to assume long term barrier functions for the radioactive materials to be retained over sufficiently long periods of time, even in cases of water or salt brine ingress in a repository.

In the Swedish concept, the fuel rods are put into a thick-walled copper canister, the hollows being cast with lead. The copper canister has a wall thickness of about 20 cm. After filling, the canister is welded tight by attaching three lids. The thick-walled copper sleeve is sufficiently stable against the pressures and movements of geological layers in the repository and also against corrosion due to groundwater. The copper container is to be inserted into geological strata (repository) in boreholes spaced a few m apart. The boreholes are surrounded by compacted bentonite rings, filled up then with a mixture of bentonite and sand and closed. The bentonite layer has a very low permeability to ground water.

7.6.3 Health and Safety Impacts of Radioactive Waste Disposal

Any discussion of the health and safety impacts on the environment arising from nuclear waste repositories involves the problem of very long time periods and the long term risk of geologic incidents or other events [38, 45, 48, 53–55]. Hazards to the environment can only occur if there is a release of radionuclides from the

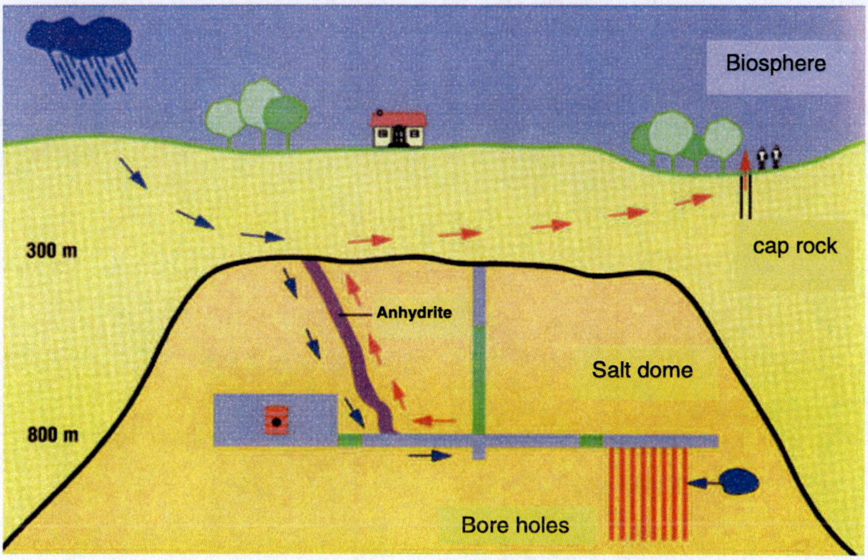

Fig. 7.30 Possibilities of water ingress for safety analyses of deep salt dome repository [45]

repository after a failure of all barriers around the nuclear waste. This can be initiated after a failure of the geologic confinement. A detailed assessment of potential modes of breaking the geological confinement as a consequence of tectonic and igneous activity, erosion, meteorite impact, and release of radioactive material from a sealed repository by sabotage or inadvertent drilling into the HLW repository has been summarized by the American Physical Society Study Group on Nuclear Fuel Cycles and Waste Management [53] as well as other research groups. The result is that such potential modes would either not lead to a failure of the geological confinement within the time period of concern or that the risk can be reduced to very low levels by the selection of the site of the repository. The movement of groundwater almost universally present in the underground is the only important medium for the transport of radionuclides from the waste repository to the biosphere.

Breaking by faulting or diapirism in salt or rock formations or undiscovered channels of anhydrite (Fig. 7.30) may create a path for groundwater towards the waste. If such water reaches the biosphere, an uptake of radioactive substances is possible by ingestion. It must be emphasized, however, that e.g. channels of anhydrite can be discovered by good geotechnical mining and vibroseismic exploration before such a deep salt dome waste repository will be taken into operation.

Similarly groundwater can penetrate into the HLW area of a granite rock deep repository (Fig. 7.31). Such groundwater penetration is more probable in a granite type repository along water conducting faults and fractures, than in a deep salt dome repository.

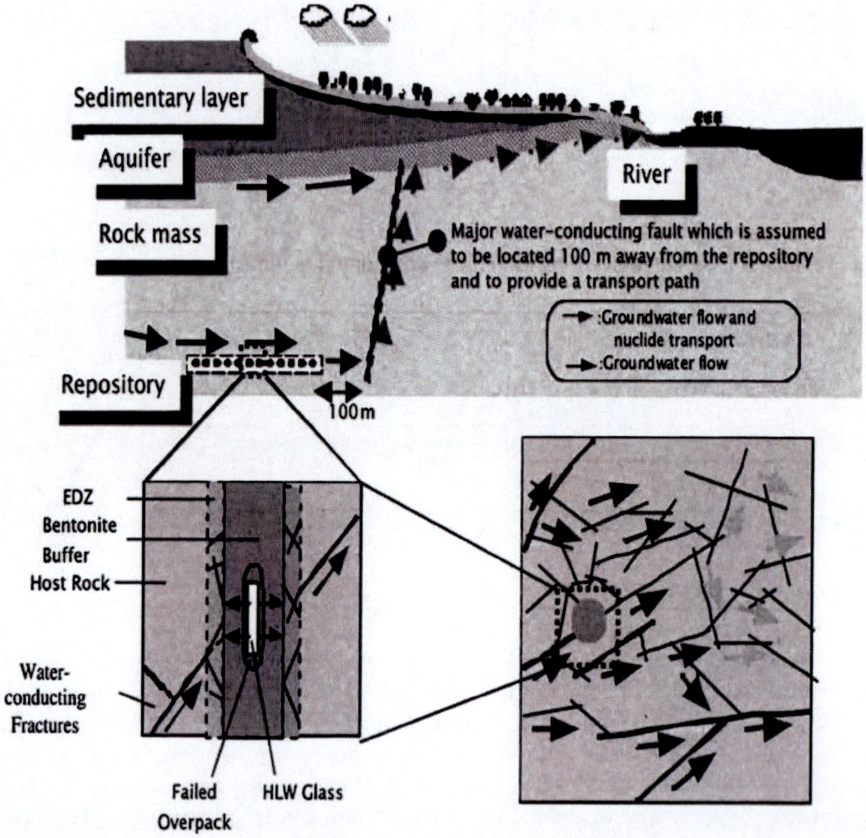

Fig. 7.31 Possibilities for water ingress in HLW area of granite rock deep repository [47]

Studies using theoretical models and experimental parameters have been performed for the hypothetical cases of water ingress in granite or in a salt repository with subsequent chemical reactions of the water or brine with the waste and transport of radionuclides by groundwater to the environment. The first barrier against the water or brine would be the overpack around the glass block or the POLLUX or BSK-3 container (in case of direct spent fuel disposal). If multi-layer (steel-lead-titanium) canisters are used, the lifetime against corrosion would be about 1,000 or several 1,000 years. During this period, strontium-90 and cesium-137 would have decayed for a long time. But very longlived fission products still have to be considered [46, 48, 56].

A near-field consideration (container and overpack) and subsequent far-field transport analysis considering transport of nuclides through granite structures or overlying sediments, lakes or rivers followed by a radioactive exposure calculation must be performed for the safety analysis.

7.6.3.1 Near Field Consideration

For a granite type repository it is assumed that a ground water level begins to establish in the area of the HLW. The overpack with bentonite (Fig. 7.25) becomes saturated with water and is expected to swell and seal gaps so that a homogeneous block is formed. Colloids produced in the failed container and radionuclides sorbed on colloids will not be transported away from the bentonite block due to filtration by the microporous structure of the bentonite. Similarly, colloid formation from waste forms in rock salts near-field conditions is not expected to be relevant [57].

Models, assumptions and experimental data on the mobilization and solubility of the different radionuclides as well as on the diffusion and sorption in the bentonite block lead to a near-field release rate which is taken over for modelling the radionuclide transport in the geosphere (far-field). The far-field model contains the description of the effects of advection, dispersion, sorption, diffusion, dilution and radioactive decay.

7.6.3.2 Nuclides of Significance for Release from the Waste Package

Studies show that the radiation dose in the biosphere from drinking water taken from wells or surface water during the time period of 10^3–10^5 years, after groundwater contact with the waste, is determined by C-14, Cl-36 and I-129. The reason is their high solubility in groundwater [54, 58–60]. During the time period of 10^5–10^6 years after closure of the repository the fission products Cs-135, Se-79, Tc-99 and others make important contributions. All the actinides (uranium, plutonium, etc.) are almost insoluble in groundwater and will make smaller contributions after about 10^6 years.

C-14

C-14 with a half-life of $T_{1/2} = 5{,}730$ years is produced by (n,p) reaction with N-14 impurities in the fuel elements and by (n,α) reaction with O-17 in the oxide fuel. It is released in the reprocessing plant as $^{14}CO_2$ during nitric acid dissolution.

Typical values of nitrogen impurities are 25 ppm which results in about 0.4×10^{10} Bq C-14 per t_{HM} fuel from LWRs and LMFBRs. In addition about 0.4×10^{10} Bq of C-14 are present in cladding waste. The C-14 can be trapped either in $Ba(OH)_2 \cdot 8\,H_2O$ or $Ca(OH)_2$ and mixed with cement for waste disposal [53].

Cl-36

Chlorine-36 has a half-life of 3×10^5 years and is produced from stable Cl-35 impurities in fuel or cladding material by neutron capture. Typical assumptions for studies are about 5 g/t_{HM} of oxide fuel and 1–5 g per tonne of Zircaloy cladding. This leads to about 8×10^8 Bq per tonne spent fuel. The solubility and transport behavior of Cl-36 are similar to those of I-129.

I-129

Iodine is a semi-volatile fission product where compounds exhibit quite complex chemical properties. It has high biological significance. For reactor operation I-131 (half-life 8.02 days) is of importance. For long term waste management and release analysis from a geological repository it is I-129 (half-life 1.57×10^7 years) which is of main significance.

About 99.9% of iodine can be removed by volatilization in the head-end of the reprocessing plant and subsequent trapping by absorption in aqueous nitric acid solutions or sorption in zeolites.

Among the immobilization methods considered are about 10% $Ba(IO_3)_2$ mixed with concrete. For radioactivity release studies a radioactivity inventory of 1.5×10^9 Bq per tonne of spent fuel is assumed.

Sr-90

Strontium-90 (half-life 29 years) together with Cs-137 (half-life 30 years) produces most of the heat generated in the high level waste.

Cs-135

Cesium-135 (half-life 2.3×10^6 years) is a very long-lived fission product which is an important contributor to the very long term radiological impact of the deep geological repository.

Nb-94

Niobium-94 (half-life 2.03×10^4 years) is an activation product of niobium present in structural materials of the fuel element.

Actinides

Uranium, plutonium and the other actinides are almost insoluble in underground water and migrate extremely slowly. They begin to contribute fairly late at about 10^6 years after closure of the repository. Their radiation dose levels are lower than that of I-129.

For the direct fuel disposal concept the groundwater contact with the waste will lead first to the release and transport of C-14, Cl-36 and I-129. Other fission products are less soluble and migrate more slowly.

For the reprocessing case studies it is often assumed that most of the C-14, Cl-36, and I-129 are already released to the environment during reprocessing. This is not an adequate solution of the high level waste problems and methods for immobilisation, e.g. in concrete as indicated above should be applied.

The RED IMPACT [60] study compares different waste disposal concepts, e.g. direct spent fuel disposal, plutonium recycling and actinide transmutation. As the production of fission products is very similar quantitatively for all fuel cycle concepts only the amounts and the composition of actinides are different. The latter play only a minor role with regard to their environmental impact after about one million years from the begin of HLW emplacement into a deep repository (Sect. 7.6.3.3). However, the volume of the waste repositories needed is very much different for different fuel cycle options [52].

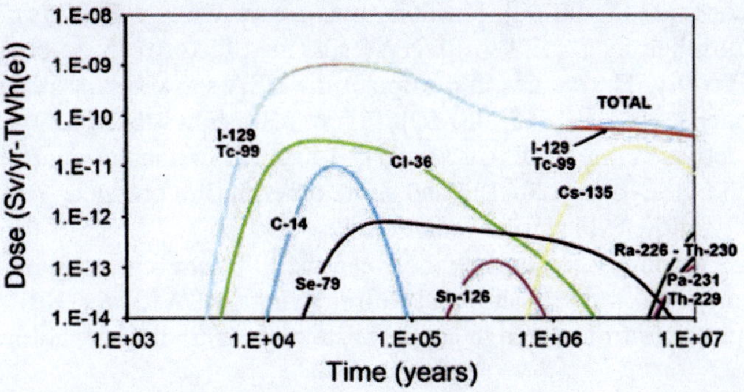

Fig. 7.32 Dose rates corresponding to releases from a model granite deep repository to the biosphere in case of direct spent fuel disposal [60]

7.6.3.3 Far Field Considerations

The transport of nuclides with the water is mainly determined by convection and dispersion effects [38, 40, 45, 48, 60]. Convection means radionuclide transport at groundwater velocity. Dispersion describes the diffusion and dilution of the contaminant in the groundwater. In addition, processes such as ion exchange, colloid filtration, reversible precipitation and irreversible mineralization must be accounted for. Some of the radionuclides, e.g. all the actinides are adsorbed very strongly in the rock or soil formations; others, like iodine and technetium, are not adsorbed at all. This holds also for Cl-36 (from activation of impurities in reactor materials) or for fission products which are water-soluble and moved through ground water pathways, e.g. I-129, Cs-135, Se-79 and Tc-99 [48, 53, 54].

For calculations of the radioactive exposure of human beings or animals in the biosphere for times after 10^4–10^6 years the following exposure pathways must be considered: drinking water or ingestion of fish, meat, milk, vegetables etc. The present radiation protection rules in the USA require that the effective dose rate does not exceed 3×10^{-4} Sv/year for all pathways except for groundwater for human beings. For groundwater in a distance of 11 miles from the repository the US Environmental Protection Agency (EPA) requires that a limit of 4×10^{-5} Sv/year should not be exceeded. This means that a person living in the vicinity of a deep repository and drinking untreated water 10,000 years from now cannot have a higher radiation exposure than 4×10^{-5} Sv/year [55].

Model studies for granite and salt repositories show that the mean arrival times of radionuclides of interest are on the order of 10^3–10^6 years. During this period of time, most of the nuclides will have decayed. The calculated effective dose rate for a human being is well below the limits defined by USEPA and European regulatory authorities [48, 60].

Figure 7.32 shows the results of the possible radiation doses of a study for a model granite repository containing waste from spent fuel elements according to the direct spent fuel concept [45, 48, 60]. The dose rates are shown in Sv/TWh(e). These units must be multiplied by 7.45 TW(e)·h/GW(e)·y (load factor 0.85) to obtain the dose rates per GW(e)·y. The studies in Europe or the USA show similar results for these disposal concepts [38, 40, 45, 48, 60]. Figure 7.32 shows that I-129 and Tc-99 as well as the activated impurities Cl-36 and C-14 appear first in the environment. They are followed by Se-79 and Sn-126 and some other fission products. After 10^6 years the actinides follow with similar radiation dose levels.

In Chap. 9 a comparison is given between the different fuel cycle options (direct spent fuel disposal, reprocessing and Pu-recycling in LWRs and FBRs) and their expected does rates from the high active waste disposal in a deep geological repository.

Uncertainties still inherent in such model analyses and parameter studies should certainly not be neglected. Therefore, active research is still going on. But uncertainties also have to be weighed against the occurrence probability of such hypothetical water intrusion events, which are assumed to occur perhaps once on a time scale of 10^6 years [53].

References

1. International Nuclear Fuel Cycle Evaluation (1980) Spent fuel management. Report of INFCE working group 6. International Atomic Energy Agency, Vienna
2. IAEA Bulletin (1979) The safe transport of radioactive materials. Special Issue. IAEA Bull 21(6):2–75
3. US Department of Energy (1977) Shipments of nuclear fuel and waste: are they really safe? US Department of Energy, Washington, DOE/EV-0004
4. Brennelementbehälter (2010) http://de.wikipedia.org/wiki/Brennelementbeh%C3%A4lter
5. Brennelement-Zwischenlager Ahaus (1979) Deutsche Gesellschaft für Wiederaufarbeitung von Kernbrennstoffen (DWK), Hannover
6. US Department of Energy (1980) Spent fuel storage factbook: facts booklet. US Department of Energy, Washington, DOE/NE-005
7. Entsorgung von Kernkraftwerken (2004) Arbeitskreis Abfallmanagement des VGB Power Tech. http://www.vgb.org/abfallmanagement-dfid-1235.html
8. Kessler G (1983) Nuclear fission reactors. Springer, Wien
9. Sindelar R (2011) Extended spent fuel storage. Nucl News 54(12):46–48
10. Broeders CHM (2011) Private communication, Karlsruhe Institute of Technology
11. Baumgärtner F (1978) Chemie der nuklearen Entsorgung. Karl Thiemig, München
12. Benedict M et al (1981) Nuclear chemical engineering. McGraw-Hill, New York
13. Bericht über das in der Bundesrepublik Deutschland geplante Entsorgungszentrum für ausgediente Brennelemente aus Kernkraftwerken (1977) Deutsche Gesellschaft für Wiederaufarbeitung von Kernbrennstoffen (DWK), Hannover
14. International Nuclear Fuel Cycle Evaluation (1980) Reprocessing, plutonium handling, recycle. Report of INFCE working group 4. International Atomic Energy Agency, Vienna
15. Cohen B (1977) High-level radioactive waste from light water reactors. Rev Mod Phys 49:1–20
16. US Nuclear Regulatory Commission (1976) Final generic environmental statement on the use of recycle plutonium in mixed oxide fuel in light water cooled reactors (GESMO). US Nuclear Regulatory Commission, Washington, NUREG-002
17. McLeod HM et al (1993) Development of mixed oxide fuel manufacture in the United Kingdom and the influence of fuel characteristics on irradiation performance. Nucl Technol 102:3–17

18. Leblanc JM, Vanden Bernden E (1978) Chemical aspects of mixed oxide fuel production. Radiochimica Acta 25:149–152
19. Herbig R et al (1993) Vibrocompated fuel for the liquid metal reactor BOR-60. J Nucl Mater 204:93–101
20. Krellmann J (1993) Plutonium processing at the Siemens Hanau fuel fabrication plant. Nucl Technol 102:18–28
21. Nuclear power in Russia (2010) http://www.world-nuclear.org./info/inf45.html
22. Merz E (1978) Wiederaufarbeitung thoriumhaltiger Kernbrennstoffe im Lichte proliferationssicherer Brennstoffkreisläufe. Naturwissenschaften 65:424–431
23. Orth DA (1979) Savannah River Plant thorium reprocessing experience. Nucl Technol 43: 63–74
24. Hebel LC et al (1978) Report to the American Physical Society by the study group on nuclear fuel cycles and waste management Part II. Rev Mod Phys 50 (1) (Pines D (ed))
25. International Nuclear Fuel Cycle Evaluation (1980) Advanced fuel cycle and reactor concepts. Report of INFCE working group 8. International Atomic Energy Agency, Vienna
26. Feraday MA (1979) Remote fabrication of (U-233/Th)O2 pellet-type fuels for CANDU reactors. Trans Am Nucl Soc 32:233–234
27. Zimmer E et al (1978) Aqueous chemical processes for the preparation of high temperature reactor fuel kernels. Radiochimica Acta 25:161–169
28. International Nuclear Fuel Cycle Evaluation (1980) Fast breeder reactors. Report of INFCE working group 5. International Atomic Energy Agency, Vienna
29. Status of Liquid Metal Cooled Fast Breeder Reactors (1985) Technical report series 246, IAEA
30. Allardice RH et al (1977) Fast reactor fuel reprocessing in the United Kingdom. In: Nuclear power and its fuel cycle, proceedings of the international conference, Salzburg, vol 3. International Atomic Energy Agency, Vienna, 2–13 May 1977, pp 615–630
31. Auchapt P et al (1980) The French R & D programme for fast reactor fuel reprocessing. In: Fast reactor fuel reprocessing, proceedings of a symposium, Dounreay, Society of Chemical Industry, London, 15–18 May 1979, pp 51–59
32. Barret TR (1980) The reconstruction of the fast reactor reprocessing plant. In: Fast reactor fuel reprocessing, proceedings of a symposium, Dounreay, Society of Chemical Industry, London, 15–18 May 1979, pp 17–35
33. Baumgärtner F, Ochsenfeld W (1976) Development and status of LMFBR fuel reprocessing in the Federal Republic of Germany. Kernforschungszentrum Karlsruhe, KfK-2301
34. Bishop JF et al (1977) Fast reactor fuel design and development. In: Nuclear power and its fuel cycle, proceedings of the international conference, Salzburg, vol 3. International Atomic Energy Agency, Vienna, 2–13 May 1977, pp 377–391
35. Sauteron J et al (1977) Technologie du retraitement des combustibles des réacteurs rapides. In: Nuclear power and its fuel cycle, Proceedings of the international conference, Salzburg, vol 3. (IAEA-CN-36/567), International Atomic Energy Agency, Vienna, 2–13 May 1977, pp 633–645
36. Funke P et al (1980) Weiterentwicklung des oxidischen Brennstoffes zum Schnellbrütereinsatz. Atomkernenergie/Kerntechnik 36:253–258
37. National Research Council, Nuclear Wastes (1996) Technologies for separations and transmutation. National Academy Press, Washington
38. H12: Project to establish the scientific and technical basis for HLW disposal in Japan (2000) Report JNC TN 1410, 2000–001. JNC, Tokai-mura
39. Hill R (2009) Fuel cycle subcommittee, overview and status, fusion-fission hybrid workshop. Gaithersburg, 30 Sept 2009
40. Radioactive waste management (2010) http://www.world-nuclear.org/info/info04.html
41. Nuclear Waste Conditioning, Nuclear Energy Division Monograph (2009) Commissariat à l'énergie atomique. Gif-sur-Yvette CEDEX, France
42. International Nuclear Fuel Cycle Evaluation (1980) Waste management and disposal. Report of INFCE working group 7. International Atomic Energy Agency, Vienna

43. US Department of Energy (1979) Technology of commercial radioactive waste management. US Department of Energy, Washington, DOE/ET-0028
44. Bröskamp H et al (2004) Endlagerung radioaktiver Abfälle in Deutschland—Abfallaufkommen und Endlagerverfügbarkeit aus EVU-Sicht. Atomwirtschaft 49:248–256
45. Closs KD (2001) Internationaler Stand der Entsorgung radioaktiver Abfälle, Radioaktivität und Kernenergie, ISBN 3-923704-26-7
46. Storck R (1992) Long time safety aspects of ultimate storage of transuranium elements. Institut für Tieflagerung, GSF-Forschungszentrum für Umwelt und Gesundheit, Neuherberg GmbH, Braunschweig
47. Physics and Safety of transmutation systems (2006) A status report. OECD-NEA
48. Luehrmann G et al (2000) Spent fuel performance assessment for a hypothetical repository in crystalline formations in Germany, GRS-154
49. Gompper K (2010) Private communication, Karlsruhe Institute of Technology, Institut für Nukleare Entsorgung
50. Morris E et al (2002) Impact of actinide removal on waste disposal in a geologic repository. In: Fifth topical meeting DOE spent nuclear fuel and fissile material management, Charleston
51. Davidson E et al (2006) Benefits of an integrated fuel cycle on repository effective capacity. In: Waste management conference, Tucson
52. Wigeland R (2006) Criteria derived for geologic disposal concepts. In: OECD/NEA 9th information exchange meeting of actinide and fission product partitioning and transmutation. Nîmes
53. Pigford TH et al (1983) A study of the isolation system for geological disposal of radioactive waste. National Academy Press, Washington
54. Pigford TH et al (1991) Effects of actinide burning on risk from high level waste. Trans Am Nucl Soc 63:80–83
55. Nuclear News (2001) EPA releases "final" radiation protection limits. Nucl News 44(8):52
56. Müller W et al (2009) Abschätzung der Standzeit von Endlagerbetten in einem zukünftigen HAW-Endlager im Salzgestein unter dem Einfluß der Korrosion. Atomwirtschaft 54:303–306
57. Geckeis H et al (1998) Formation and stability of colloids under simulated near field conditions. Radiochimica Acta 82:123–128
58. Sheppard SC et al (1996) Chlorine-36 in nuclear waste disposal. 1. Assessment results for used fuel with comparison to ^{129}I and ^{14}C. Waste manag 6(7):607–614
59. Beasly TM et al (1992) Chlorine-36 releases from the Savannah river site nuclear fuel reprocessing facilities. Groundwater 30(4):539–548
60. Impact RED (2007) Impact of partitioning, transmutation and waste reduction technologies on final nuclear waste disposal. Synthesis report, Forschungszentrum Jülich, Germany

Chapter 8
Nuclear Fuel Cycle Options

Abstract In the once-through (OT)-cycle the valuable uranium and plutonium as well as some of the minor actinides are not re-used. The spent fuel—after unloading from the reactor and being stored for about 50 years in intermediate storage facilities—is conditioned for permanent storage and then intended to be disposed in a deep geological repository. In a closed fuel cycle the spent fuel, after its radioactivity has decayed to a certain level in intermediate storage facilities, is shipped to a reprocessing plant. After chemical or pyrochemical reprocessing the fissile plutonium as well as the residual uranium are re-used to fabricate new fuel elements. The plutonium and uranium are recycled in converter and breeder reactors. Recycling is possible both in the U-238/Pu and the Th/U-233 fuel cycles. It is shown that plutonium can be recycled at least three times or even more in light water burner reactors over a time period of about 75–90 years, thereby burning about 50% of the self generated plutonium. After this time period the plutonium can be loaded into the cores of fast breeder reactors for further recycling. Different thermal converter reactor types operating in either the U-238/Pu or Th/U-233 fuel cycle are compared with regard to their natural uranium consumption and separative work requirement for fuel enrichment. For fast breeder reactors operating in U-238/Pu or Th/U-233 fuel cycle also the net fissile material gain (Plutonium or U-233) and the consumption in U-238 and thorium are given.

8.1 Fuel Cycle Options for Reactors with Thermal Neutron Spectrum

8.1.1 The Once-Through Fuel Cycle

After unloading from the reactor core, spent fuel contains a mixture of U-238 and unused U-235, newly generated plutonium, higher actinides and fission products. U-235 and plutonium are valuable materials, which can be used again for energy

G. Kessler, *Sustainable and Safe Nuclear Fission Energy*, Power Systems,
DOI: 10.1007/978-3-642-11990-3_8, © Springer-Verlag Berlin Heidelberg 2012

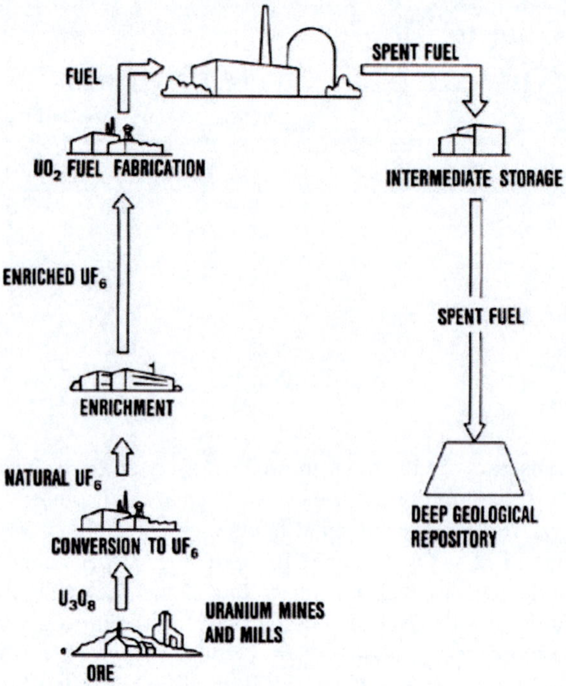

Fig. 8.1 Once-through nuclear fuel cycle for converter reactors with direct spent fuel disposal (Kessler [1])

generation, U-238 might be used as fertile material in FRs, whereas fission products are waste materials.

In the once-through (OT)-cycle (Fig. 8.1), sometimes also called OTTO cycle (once through then out), valuable uranium and plutonium are not re-used. After removal from the reactor, the fuel is kept in intermediate storage facilities for some time (a couple of years up to several decades). This intermediate storage is only an interim solution. In the long run, spent fuel elements must be conditioned such that a permanent storage in deep geological repositories is possible (see Sect. 7.6.2). In this so-called "direct spent fuel disposal" solution, plutonium, uranium and the minor actinides remain in the fuel elements and their radioactivity decays rather slowly due to their relatively long half-lives, while the radioactivity of the fission products decays gradually and practically almost disappears after a few centuries. Uranium, plutonium and the minor actinides are no longer available for later power generation in reactors.

Nuclear reactors presently in operation are operated mainly in the OT-cycle. This OT-cycle concept implies a high natural uranium consumption. Commercial reprocessing of the fuel elements is only available in Europe, Russia and Japan (Sect. 7.2.3).

Table 8.1 shows the main fuel cycle data for nuclear reactors in the OT-cycle. PWRs are operated with low enriched uranium (LEU). HWRs can be operated both

Table 8.1 Fuel cycle design data for thermal spectrum reactors operating in the once-through fuel cycle (30 years operating time, load factor 0.85) [2, 3]

		PWR	HWR	HWR	HTGR	HTR (pebble bed) (modular type)
Total thermal power	MW(th)	3,765	3,425	3,425	3,360	3,000
Net electric power	MW(e)	1,229	1,000	1,000	1,332	1,240
U-235 enrichment	%	3.8 (LEU)	Natural uranium	1.2 (LEU)	10 (LEU)	8.5 (LEU)
Fraction of core replaced per year		0.25	Online	Online	0.33	Online
Fuel residence time	d	1,020	276	790	766	1,120
Average discharge burnup	GWd(th)/t	44	7.3	20.9	111	100
Natural uranium requirement	t/GW(e)					
• Initial core		367	131	256	186	178
• Annual eq. reload		155	147	99	143	121
• 30 years cumulative		4,687	4,512	3,299	4,184	3,759
Separative work requirement	t SWU/GW(e)					
• Initial core		257		91	180	189
• Annual eq. reload		135	–	35	156	129
• 30 years cumulative		3,990		1,147	4,590	3,983
Fissile material in spent fuel: enrichment						
• U-235	wt%	0.80	0.23	0.1	1.3	1.6
• Pu-fissile		0.76	0.28	0.34	1.0	1.1
Annual eq. discharge	kg/GW(e)					
• U-235		135	276	40.7	58	84
• Pu-fissile		121	331	140	45	57

with natural and low enriched uranium. HTGRs and HTR pebble bed reactors, are designed for about 8–10% U-235 enrichment. They have a higher U-235 enrichment of 1.3–1.6% U-235 in their spent fuel compared to PWRs with 0.8% U-235 with exceeds the U-235 content in natural uranium.

Among the light water reactors, PWRs have a natural uranium consumption of 4,687 t over an operating period of thirty years (load factor 0.85). BWRs consume almost the same quantity of natural uranium.

Because of their better neutron economy HWRs with natural uranium fuel consume slightly less natural uranium (4,512 t over 30 years (load factor 0.85)). If the fuel is slightly enriched, their natural uranium consumption can be cut to 3,299 t. HTGRs and HTR pebble bed modular reactors have natural uranium consumption levels between 4,184 and 3,759 t.

Spent fuel of nuclear reactors run on U-235/U-238 fuel always contains plutonium. The largest amount of plutonium per GW(e) and year is generated in HWRs fueled with natural uranium.

8.1.2 Closed Nuclear Fuel Cycles

In a closed fuel cycle the spent fuel, after its radioactivity has decayed to a certain level in intermediate storage facilities, is shipped to a reprocessing plant for chemical reprocessing. After chemical reprocessing, the fissile plutonium as well as the residual uranium can be re-used to fabricate new fuel elements and be recycled in nuclear reactors (Fig. 8.2). A small fraction of the fissile material (about 1% losses) goes into the radioactive waste during chemical reprocessing and refabrication of the fuel, where it is eventually lost. Recycling improves the utilization of fuel (see Sect. 8.1.2.3) and, consequently, decreases the consumption of natural uranium.

Recycling is possible both in the U-238/Pu and the Th/U-233 fuel cycles. Fissile plutonium or U-233 from several nuclear reactors may be collected and used exclusively in special recycle nuclear converter reactors. Also, every reactor can recycle its own plutonium or U-233 generated in preceding operation cycles. For both cases this is then called self-generated recycling (SGR) [4–6].

8.1.2.1 Plutonium Recycling in the SGR Mode

At the beginning of the first recycle phase of a PWR core, i.e., after unloading spent LEU fuel with a burnup of 50 GWd/t, the plutonium fuel has an isotopic composition as indicated in Table 8.2 (see also Fig. 3.10a, b). Further recycling of the same plutonium would lower the fraction of fissile Pu-239 and Pu-241 isotopes [7–12]. Therefore, the SGR mode is usually applied. In this SGR mode, continuous mixing of recycled plutonium with "fresh" plutonium from spent LEU fuel elements leads to a higher percentage of the fissile plutonium isotopes Pu-239 and Pu-241.

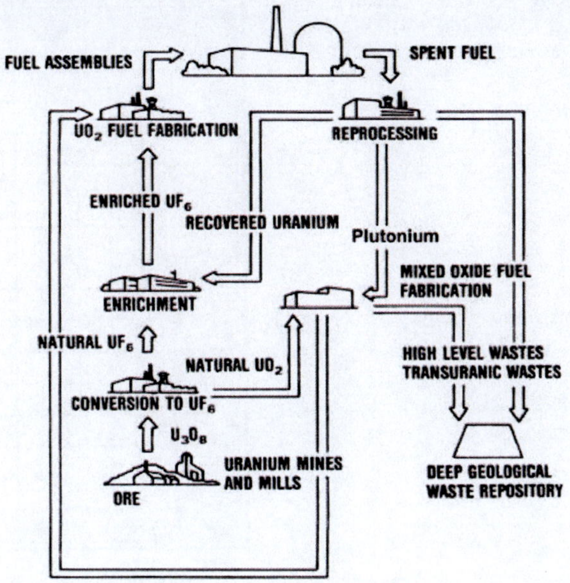

Fig. 8.2 Closed nuclear fuel cycle (recycling of converter reactors) (Kessler [1])

Plutonium bearing fuel elements have the same structural design as LEU UOX fuel elements. Each fuel rod of such a fuel element contains PuO_2/UO_2 mixed oxide (MOX) fuel pellets. The average fissile plutonium enrichment is chosen such that the same discharge burnup as in the LEU UOX fuel elements can be achieved. Radial enrichment zoning or so-called water rods within the fuel element are necessary to avoid unacceptably high local power peaks. Enrichment zoning is achieved by giving the fuel rods at the periphery of the fuel subassemblies a lower Pu enrichment while all remaining fuel rods of the inner part of the fuel element have a somewhat higher enrichment.

Figure 8.3 shows a cross section of a 17 × 17 rods MOX fuel assembly design with 264 MOX fuel rods and different plutonium fissile enrichments. The corner fuel rods have a plutonium fissile enrichment of 3.44%, whereas the remaining MOX fuel rods along the outer sides of the fuel assembly have 6.44% enrichment. All other MOX fuel rods inside of the MOX fuel subassembly have a plutonium fissile enrichment of 7.44% corresponding to 12% plutonium total. Neutron poison is added to the fuel pellets in form of gadolinium (Gd_2O_3) [7, 11, 12].

The MOX fuel subassembly also contains 25 positions for guide tubes. These guide tubes are used as locations for rod cluster control assemblies (RCCAs) and for incore instrumentation [7].

Another design of MOX fuel assemblies, originally proposed by Barbrault [10], is shown in Fig. 8.4. It contains 36 water rods for neutron moderation to compensate for hardening of the neutron spectrum induced by the plutonium isotopes. The water rods are evenly distributed over the cross section of the MOX fuel subassembly. This leads to an increased moderator to fuel volume ratio of 2.5. This allows the same

Fig. 8.3 Cross section of a 17 × 17 PWR MOX fuel assembly [7]

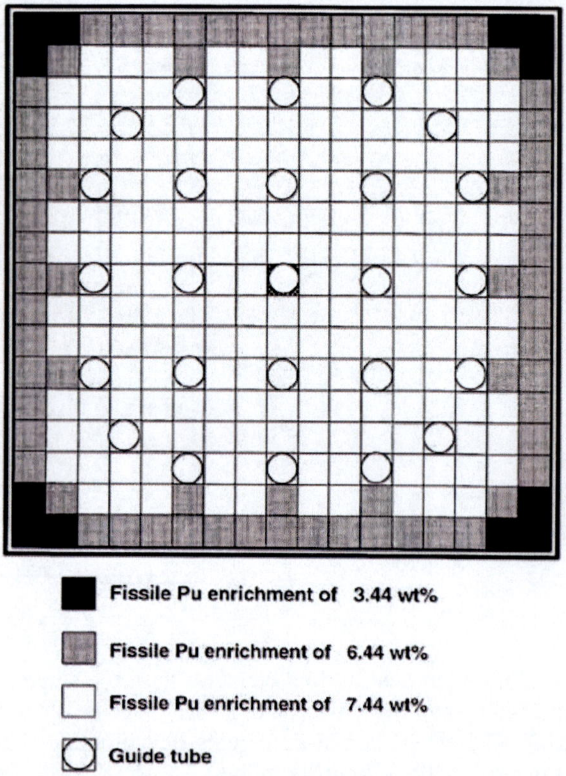

plutonium fissile enrichment for all fuel rods. The design also contains 25 positions for guide tubes with RCCAs or incore instrumentation. The MOX fuel pellets also contain gadolinium as neutron poison [13].

8.1.2.2 Reactor Physics of MOX-PWR Cores

Compared with the uranium isotopes in UOX PWRs the plutonium isotopes in MOX PWR cores have large resonance cross sections for neutrons in the low epithermal eV energy range[13–18]. These large absorption cross sections of the plutonium isotopes lead to hardening of the neutron flux spectrum (less neutrons in the thermal energy range). This can be seen from Fig. 8.5 which compares the neutron flux spectra as neutron flux $\phi(u) = E \cdot \phi(E)$ per unit lethargy u. The lethargy u is defined as ln (E_0/E), where E_0 is the upper limit of the energy scale. This logarithmic energy is suggested by the fact that the average logarithmic energy loss per elastic collision of a neutron with a nucleus is an energy independent constant. Figure 8.5 compares three cases [17].

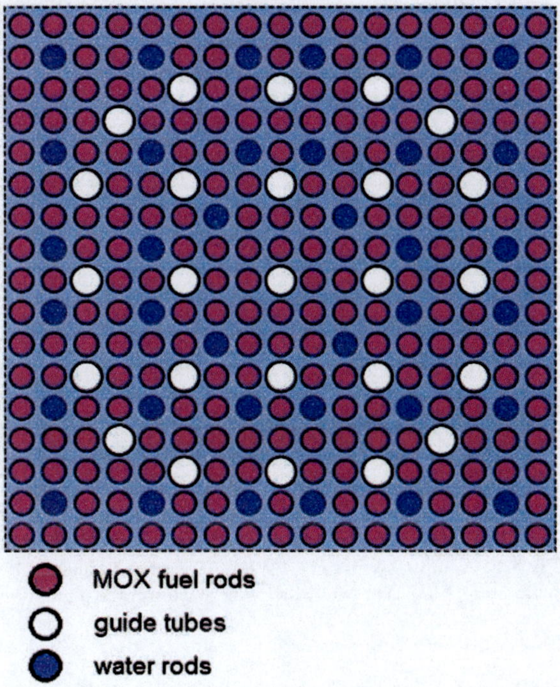

Fig. 8.4 Cross section of a 17 × 17 PWR MOX fuel assembly with 25 water rods [13]

- A UO_2 fueled PWR core.
- A PWR core fueled one third with weapon-grade plutonium (WGPu) and two thirds with UO_2 fuel.
- A full PuO_2/UO_2 MOX PWR core with reactor-grade plutonium (RGPu).
- Due to the spectrum hardening, the control materials (control rods, soluble boron acid, burnable poisons, e.g. gadolinium) become less efficient and need enrichment. Also AgInCd control shut down rods may have to be replaced by B_4C rods with higher B-10 enrichment.
- The Doppler coefficient becomes larger negative and requires additional shut down reactivity margins.
- The coolant/moderator temperature coefficient can become less negative or even positive at higher plutonium contents in the MOX fuel. However, this can be counterbalanced to a certain extent by wider fuel rod spacings (higher moderator/fuel volume ratio) or additional water rods (higher moderator/fuel volume ratio) [14, 17, 19].
- The delayed neutron fraction is decreased from above $\beta = 0.65\%$ for UO_2 PWR cores to about $\beta = 0.4\%$ for full MOX PWR cores.
- The radioactivity of unloaded spent MOX-PWR fuel elements is higher than for spent UOX-PWR fuel elements. Whereas UO_2-PWR spent fuel elements decay

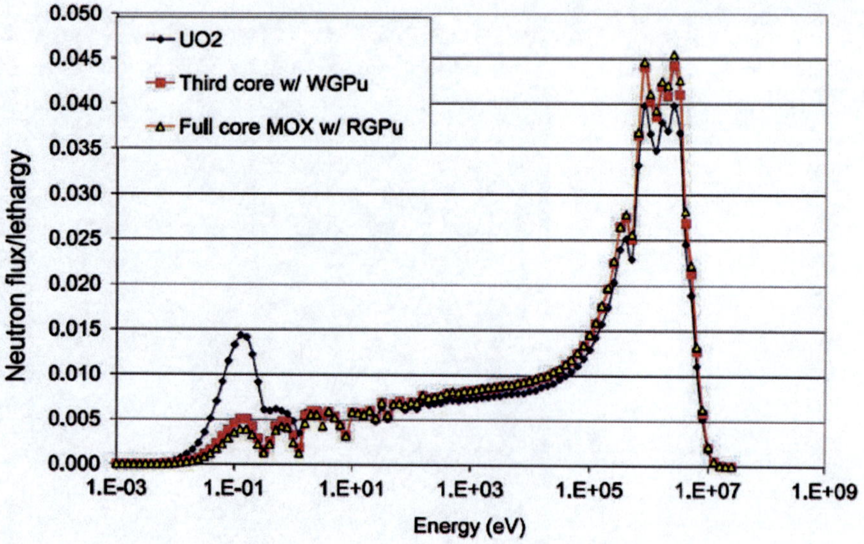

Fig. 8.5 Neutron spectra in a PWR core with UO_2 fuel or UO_2 fuel with one third weapon-grade plutonium (WGPu) or MOX fuel with reactor-grade plutonium (RGPu) [17]

to 5.5×10^5 Ci within a cooling time of three years the spent MOX PWR fuel elements need about seven years to attain this radioactivity level.

- The heat power of spent MOX PWR fuel elements is by about 10% higher than for UOX-PWR spent fuel elements.

Compared with the OT-fuel cycle, a Pu-SGR type MOX PWR has a natural uranium consumption of 3,315 t/GW(e) over 30 years operating time and a load factor of 0.85 as the uranium and plutonium are recycled. This corresponds to savings on the order of 35%. Accordingly, only 3,023 t SWU/GW(e) over 30 years are needed, which saves 25% of separative work requirements. Data similar to those found for PWRs also result for BWRs. The Pu-SGR mode can be adopted also for reactor lines other than LWRs, e.g., HWRs [2, 4].

A plutonium recycle LWR fully loaded with PuO_2/UO_2 fuel is called a plutonium burner. In such a plutonium burner, all plutonium bearing fuel assemblies have the same plutonium enrichment. Since LWR plutonium burners obtain the plutonium for their initial core as well as for the Pu_{fiss}, make-up from other UO_2 LWRs, their natural uranium requirement and their separative work requirement is nil for the first Pu-recycling step and relatively low for the 2nd and 3rd Pu-recycling steps (addition of low enriched uranium needed). They need only depleted uranium or reprocessed uranium in the first Pu recycling step.

The whole strategy encompassing LWRs with LEU fuel and MOX LWR plutonium burners has about the same requirements of natural uranium and separative work as one LWR with partial MOX fuel loading operating in the Pu-SGR mode.

LWRs with partial MOX fuel loading are operating already in several countries mainly in Europe and in Japan.

8.1.2.3 Plutonium Incineration in PWRs During Several Recycling Steps

Recycling of plutonium over more than one recycling step is sometimes questioned. However, recycling of plutonium over several recycling steps is possible if the spent MOX fuel from PWRs is always mixed with the plutonium from UOX PWRs during reprocessing. In this way plutonium recycling over several full recycling steps becomes feasible [14, 15, 18].

On the basis of a pin cell model Broeders [15] analyzed recycling of plutonium in PWRs applying the SGR mode with LEU-UOX PWRs and MOX PWR burners. In his scenario the plutonium from reprocessing of spent fuel elements of several LEU MOX PWRs is collected until a first full MOX PWR can be started. This is explained in Fig. 8.6. The assumptions of Broeders [15] for this SGR recycling strategy were:

- 4.5% U-235 enrichment of the fuel element of the UOX PWRs,
- 6% max. Pu_{fiss} enrichment for MOX fuel (low enriched U-235 is added to fulfil criticality conditions),
- the initial plutonium isotopic composition is shown by Table 8.2,
- Am-241 originates from the decay of Pu-241 after reprocessing,
- 6 burnup cycles within 10 years, i.e. 20 months per burnup cycle,
- 7 years cooling time of the spent fuel after unloading from the core and time for reprocessing,
- 3 years time for refabrication of the MOX fuel including time for transport.

For a maximum burnup of 50 GWd/t and the above time periods for reactor operation, reprocessing, refabrication and transport it is appropriate to consider a cluster of M = 8 PWRs. Initially these PWRs are operated with LEU UOX fuel having 4.5% U-235 enrichment. Their spent fuel is reprocessed and MOX fuel elements with 6% Pu_{fiss} enrichment are fabricated and transported to the MOX PWR. This needs a time period of 10 years. These MOX fuel elements with an isotopic mixture M1 given in Table 8.2 are loaded into a first MOX PWR (Fig. 8.6). As one UOX reactor with a power of 1 GW(e) generates about 0.24 t of plutonium per GW(e) year, the cluster of M = 8 UOX reactors needs about 5 years to produce the first core inventory of 8.9 t of plutonium for the first MOX PWR of one GW(e) power. Accounting for the time of intermediate storage of spent fuel, reprocessing, refabrication and transport of the fuel this MOX PWR can start in the 16th year from the begin of the scenario. The plutonium from the spent fuel elements of this full MOX-PWR burner is again mixed with plutonium coming from 7 operating UOX PWRs. This mixing procedure is then continued until after 26 years a new plutonium isotopic mixture M2 can be loaded to the MOX PWR burner (Fig. 8.7).

After 36 years a second full MOX PWR burner can be added (Fig. 8.6). These two full MOX PWRs operate with plutonium isotopic mixture M3 between 36 and 45

Table 8.2 Isotopic composition of PWR spent fuel after 50 Gwd/t burnup and 10 years after unloading of the spent fuel

Isotopic composition	Plutonium in wt% 10 years after unloading (50 GWd/t)
Pu-238	2.8
Pu-239	55.1
Pu-240	23.3
Pu-241	9.7
Pu-242	7.6
Am-241	1.5

This plutonium composition differs somewhat from the plutonium in Table 12.2 due to different cross section sets used

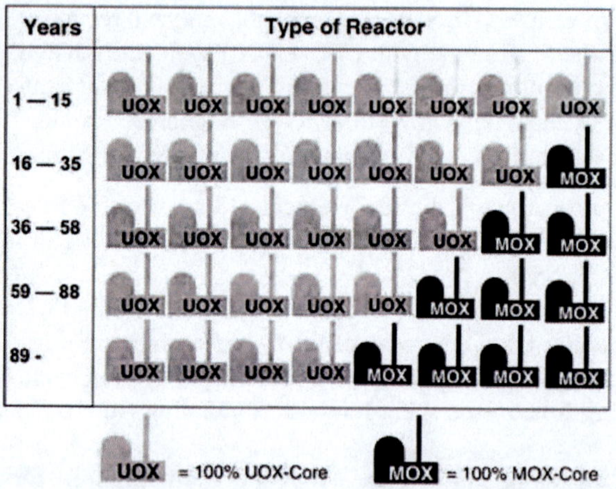

Fig. 8.6 Scenario for plutonium multi-recycling in the SGR mode for a cluster of M = 8 PWRs [15]

Table 8.3 Isotopic compositions M1–M6 for SGR recycling strategy [15]

Plutonium compositions (wt%)	Pu-238	Pu-239	Pu-240	Pu-241	Pu-242	Am-241
M1	2.8	55.1	23.3	9.7	7.6	1.5
M2	3.5	49.4	26.2	10.0	9.4	1.5
M3	3.9	46.8	27.9	9.2	10.8	1.4
M4	4.3	43.1	28.9	9.9	12.3	1.5
M5	4.6	41.5	29.3	9.5	13.6	1.5
M6	4.8	40.4	29.6	9.1	14.7	1.4

This table is cut off after composition M6 in comparison to Fig. 8.6

years, with plutonium isotopic mixture M4 between 46 and 55 years, with plutonium isotopic mixture M5 between 56 and 65 years and with plutonium isotopic mixture M6 between 65 and 75 years. Table 8.3 shows the plutonium isotopic composition for the MOX fuels M1–M6 (Fig. 8.7) (This Fig. 8.7 is cut after 75 years in comparison to Fig. 8.6).

Time year	Cycle	Reactors							
		1	2	3	4	5	6	7	8
1	1	U	U	U	U	U	U	U	U
	2	U	U	U	U	U	U	U	U
	-	U	U	U	U	U	U	U	U
15	6	U	U	U	U	U	U	U	U
16	7	U	U	U	U	U	U	U	M1
	-	U	U	U	U	U	U	U	M1
25	12	U	U	U	U	U	U	U	M1
26	13	U	U	U	U	U	U	U	M2
	-	U	U	U	U	U	U	U	M2
35	18	U	U	U	U	U	U	U	M2
36	19	U	U	U	U	U	U	M3	M3
	-	U	U	U	U	U	U	M3	M3
45	24	U	U	U	U	U	U	M3	M3
46	25	U	U	U	U	U	U	M4	M4
	-	U	U	U	U	U	U	M4	M4
55	30	U	U	U	U	U	U	M4	M4
56	31	U	U	U	U	U	U	M5	M5
	-	U	U	U	U	U	U	M5	M5
65	36	U	U	U	U	U	U	M5	M5
66	37	U	U	U	U	U	U	M6	M6
	-	U	U	U	U	U	U	M6	M6
75	36	U	U	U	U	U	U	M6	M6

U: PWR with UOX fuel; Mi: PWR with MOX fuel of generation i = 1....6

Fig. 8.7 Scenario for multi recycling of plutonium in full MOX PWR cores in a pool of 8 PWRs is adequate for a target burnup 50 GWd/t_{HM} [15] (This figure is cut off after 75 years if compared to Fig. 8.6)

This SGR mode plutonium recycling procedure can be terminated after a certain time and the plutonium can be loaded into the cores of FRs having better neutronic properties for the incineration of plutonium. Broeders [15] noticed that with a fuel to moderator volume ratio of 2 as in present PWRs (Fig. 8.4; Sect. 9.8) already plutonium isotopic compositions like M3 or M4 would result in untolerable coolant temperature coefficients. Such untolerable coolant temperature coefficients can be avoided by choosing a wider spacing of the fuel rods [14].

As can be seen from Table 8.3 the percentage of Pu-238 increases steadily from 2.8% (M1) to 4.8% (M6) whereas Pu-239 decreases from 55.1% (M1) to 40.4% (M6). Also, the Pu-240 increases from 23.3% (M1) to 29.6% (M6), the Pu-241 from 7.6% (M1) to 9.1% (M6) and Pu-242 even from 7.6% (M1) to 14.7% (M6). These changes in plutonium compositions occur over 60–70 years of plutonium recycling in operating MOX PWRs.

Table 8.4 Low enriched uranium to be added to the plutonium mixtures M1–M6 [15]

Uranium added to plutonium	Enrichment wt%
M1	0.7
M2	1.5
M3	2.0
M4	2.5
M5	3.0
M6	3.3

This table is cut off after composition M6 in comparison to Fig. 8.6

In order to fulfil the requirements that the fissile part of the plutonium does not exceed the 6% and that a certain initial criticality value k_{eff} must be attained also low enriched uranium must be added to the MOX fuel (Table 8.4).

8.1.2.4 Balance of Plutonium Inventory and Incineration of Plutonium

The balance of the plutonium inventories in the cluster of M = 8 PWRs reveals that a considerable part of the plutonium which is generated by the UOX PWRs is incinerated by the full MOX PWRs. Figure 8.8 shows the plutonium inventory of the cluster of M = 8 PWRs for two cases:

- The straight full line represents the OT cycle with direct spent fuel storage strategy (no reprocessing). The plutonium is accumulating steadily and as part of the spent fuel would have to be disposed in a deep geological repository.
- The SGR plutonium recycle strategy is represented by a set of straight dotted lines which are bending more and more, according to the incineration of the plutonium in the full MOX PWRs with plutonium MOX fuel isotopic mixture M1–M6. The difference between the full line and the dotted lines represents the plutonium which is incinerated by the MOX PWRs normalized to 1 GW(e). The essential result is that over 60–80 years of SGR mode operation about 50% of the plutonium generated by the UOX PWRs could be incinerated by the MOX PWRs, i.e. employed for energy production. This is due to the fact that one full MOX PWR can incinerate about 430 kg Pu/GW(e) y [15].

A similar Pu-recycle scenario can be analyzed by replacing the full MOX PWRs by FR burners (Pu incinerators). FR-burners can achieve a plutonium incineration rate of about 570 kg Pu/GW(e) y (Sect. 9.8.3).

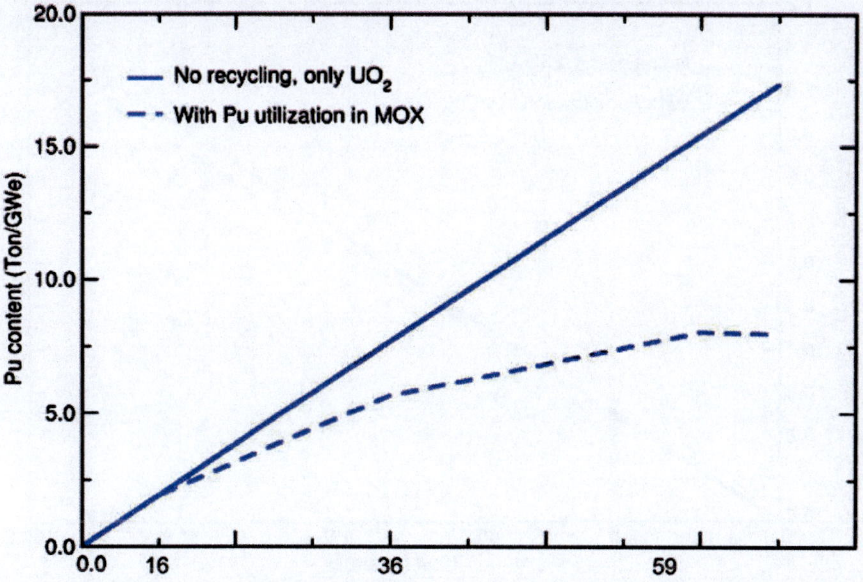

Fig. 8.8 Balance of plutonium collected from a cluster of M = 8 UOX PWRs with 10 GW(e) total power operating in the OT cycle with direct spent fuel disposal (full line) and inventory of plutonium in the cluster of M = 8 UOX operating in symbiosis with MOX PWRs with 10 GW(e) power (Pu inventory normalized to 1 GW(e)) [15]

8.1.2.5 Neptunium and Americium Generation in the SGR Plutonium Recycle Scenario

Figure 8.9 shows that the neptunium generation in the SGR plutonium recycle scenario is only slightly different between the UOX-PWR OT cycle with direct spent fuel disposal and the UOX and MOX PWRs operating in the SGR mode.

The Am-243 production via neutron capture in Pu-242 and beta-decay of Pu-243 is shown for the SGR plutonium recycling case and for the UOX direct spent fuel disposal case in Fig. 8.10. In the SGR scenario considerably more (factor 2.5 after 50 years and a factor of 3.5 after 70 years) Am-243 is produced.

8.1.2.6 The Thorium/Uranium Fuel Cycle

The generation of U-233 must be started with U-235/U-238 or plutonium as fissile and Th-232 is fertile fuel [2, 20–24]. Since in the first case this fuel also contains fertile U-238, plutonium will be produced besides U-233. The production of plutonium can be restricted by limiting the amount of U-238 in the fuel. There is, however, the condition set by IAEA safeguards (INFCE/153) [25] that the enrichment of U-235 in uranium must be $\leq$20% and the enrichment of U-233 must be $\leq$12% [26]. After

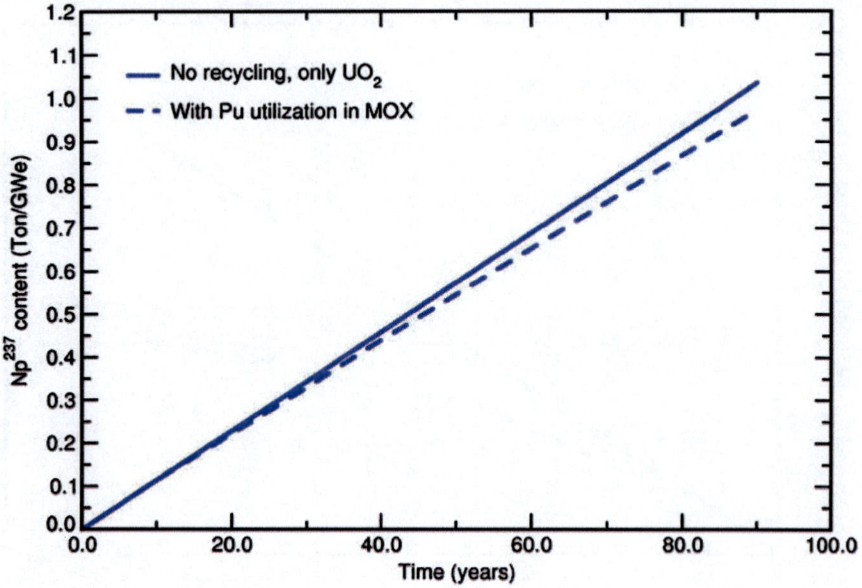

Fig. 8.9 Amount of neptunium generated by a cluster of M = 8 UOX PWRs (*full line*) and amount of neptunium generated by a cluster of M = 8 UOX and MOX PWRs operating in the SGR-mode (*dotted curve*). The amounts are normalized to 1 GW(e) [15]

unloading, cooling and chemical separation the fissile fuel U-233 and plutonium can be recycled.

PWRs Operating in the Once-Through Cycle with Uranium/Thorium and Plutonium/Thorium Fuel

Such PWRs with uranium-thorium fuel need either enriched uranium or plutonium as fissile fuel. Table 8.6 [2] shows a PWR design with 20% enriched U-235/U-238 fuel and thorium. The annual reload of U-235/U-233 fissile fuel is 677 kg.

If instead of enriched uranium as shown in Table 8.6 only plutonium separated from spent fuel of UOX PWRs is combined with thorium, an initial plutonium enrichment is required as shown by Table 8.5 [22, 23, 27, 28]. Such a Pu/Th fueled PWR can incinerate about 40% of the initially loaded plutonium over a burnup period of 60 GWd/t.

These MOX-Pu/Th fuelled PWRs can incinerate about 15% more plutonium as similar MOX-Pu/U fuelled PWRs [27, 28].

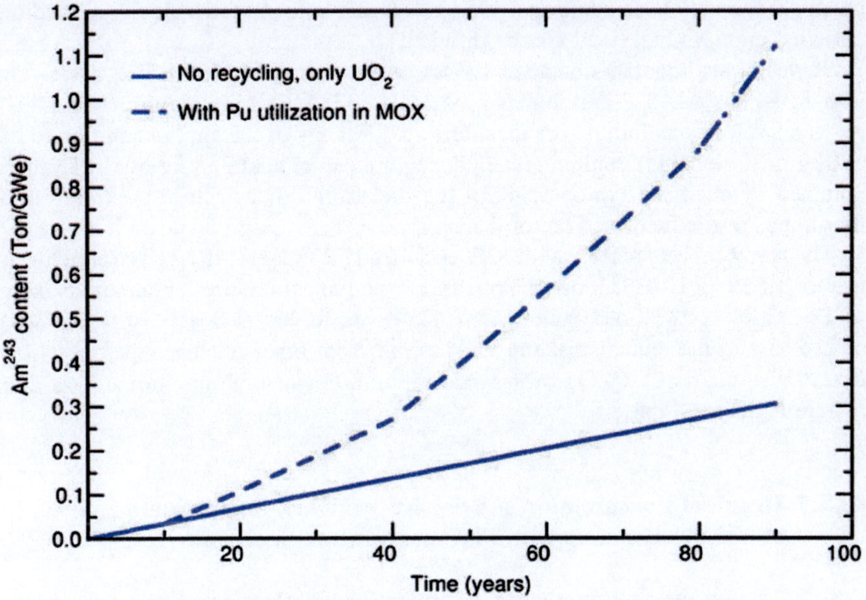

Fig. 8.10 Amount of americium-243 generated by a cluster of M = 8 UOX PWRs operating in the OT cycle (*full line*) and amount of americium-243 generated by a cluster of M = 8 UOX and MOX PWRs (*dotted curve*) operating in the SGR mode (Amounts are normalized to 1 GW(e) [15]

Table 8.5 Fuel cycle data for PWRs with plutonium/thorium fuel

Maximum burnup [GWd/t]	60
Initial Pu-239 and Pu-241 enrichment (wt%)	8.7
Initial total plutonium (wt%)	15.1
Total plutonium fraction at max. burnup (wt%)	8.94
Amount of plutonium incinerated (%)	40

The initial uranium, plutonium and americium isotopic composition is 0.088% U-234, 0.004% U-235, 0.008% U-236, 3.72% Pu-238, 47.26% Pu-239, 27.36% Pu-240, 9.4% Pu-241, 10.66% Pu-242 and 1.5% Am-241 [28]

PWRs and HWRs Operating in the Uranium/Thorium Recycle Mode

Table 8.6 shows the fuel cycle design data of PWRs operating in the U-235/U-238/Th-232 recycle mode or in the U-233/Th-232 recycle mode. In addition fuel cycle design data for HWRs operating in the U-233/Th-232 recycle mode are given [2, 22, 23].

The PWR reactor, with 20% U-235/Th-232 fuel (MEU) , attains a natural uranium consumption of 2,685 t of natural uranium over thirty years of operation (load factor 0.85). The corresponding separative work requirement is 3,326 t SWU. Although the fissile U-233 produced plus the unused U-235 is recycled in the reactor core after chemical separation, an additional 455 kg of U-235 must be fed annually as medium enriched uranium fuel (load factor 0.85) [2].

Table 8.6 also contains data for LWRs operated in a U-233/Th-232 cycle. The initial fuel has a U-233 enrichment of 12% U-233 (MEU) in uranium. This U-233 recycle LWR has an initial core inventory of 1,904 kg U-233 and, in addition to the 580 kg of fissionable uranium recycled, requires an annual 358 kg of U-233 to be obtained by chemical reprocessing of the fuel from other nuclear reactors run on thorium as fertile fuel (load factor 0.85) [2].

The HWR recycle reactor with 12% enriched U-233 fuel (MEU) has a core inventory of 1,859 kg U-233. It needs no natural uranium. It requires an annual makeup of 129 kg of U-233 (load factor 0.85). However, its initial inventory of 1,859 kg of U-233 and the annual makeup must come from other nuclear reactors which, in turn, use the fissile U-235 of natural uranium. For natural uranium and thorium requirements see Table 8.6.

8.1.2.7 Uranium Consumption and Separative Work Requirement of Various Reactors with Thermal Neutron Spectrum

Table 8.7 shows the natural uranium consumption and the separative work requirements for different types of nuclear reactors capable of working in the once-through or recycle modes. Improved neutron economy and higher conversion ratios as well as recycling of the fissile material generated reduce the natural uranium consumption. All nuclear reactors operating in the recycle mode require technically mature reprocessing in the U-238/Pu fuel cycle or in the Th/U-233 cycle (see Chap. 7).

8.2 Fuel Cycle Options for Liquid Metal Cooled Fast Breeder Reactors

All breeder reactors must work in a closed fuel cycle [20, 25]. Liquid Metal Cooled Fast Breeder Reactors (LMFBRs) as described in Chap. 6 are presently designed to operate in the U-238/Pu fuel cycle, since the highest breeding ratios are attained with a fast neutron spectrum (see Sect. 3.6). However, in principle it is also possible to design FBRs with U-233/Th fuel, which still attain breeding ratios above 1.

Initially, thermal and fast breeder reactors can be started with plutonium or U-233 available from chemical reprocessing of spent fuel of thermal neutron spectrum reactors. Later, when a fission breeder reactor economy will have developed, sufficient plutonium or U-233 can be generated also by the FBRs themselves to start additional FBR plants.

Table 8.6 Fuel cycle design data for reactors (PWR or HWR) operating with U-235/U-238/Th-232 or U-233/Th-232 (30 years operating time, load factor 0.85) [2]

Reactor type		PWR recycle	PWR-recycle	HWR-recycle
Total thermal power	MW(th)	3,765	3,800	4,029
Net Electric Power	MW(e)	1,229	1,270	1,260
Fuel cycle		MEU-235/Th	MEU-233/Th	MEU-233/Th
Equilibrium recycle enrichment—recycle feed	%	20(U-235)	12(U-233)	12(U-233)
Fraction of core replaced per year		0.33	0.33	0.5
Average burn-up	MWd(th)/kg	33.5	33.4	14
Fissile requirements	kg/GW(e)	2,544	1,904	1,859
Annual recycled eq. fissile	kg/GW(e)	677	580	1,039
Annual makeup eq. fissile	kg/GW(e)	455	358	129
Natural uranium requirements	t/GW(e)			
• Initial core		493	Not adequate	Not adequate
• 30 years cumulative		2,685	Not adequate	Not adequate
Thorium requirements	t/GW(e)			
• Initial core		63	57.9	105.4
• 30 years cumulative		827	655	1,030
Separative work requirements for MEU[a]	t/SWU/GW(e)			
• Initial core		715	Not adequate	Not adequate
• 30 years cumulative		3,326	Not adequate	Not adequate
Fissile material in spent fuel: enrichment	%	9.3	7.9	11.2
Annual discharge	kg/GW(e)			
• U_{fiss} eq.		530	478	848
• Pu_{fiss}		68	65	22.8

[a]MEU means 20% U-235 in U-238 or Th-232 and 12% U-233 in U-238 or Th-232 (IAEA)

Table 8.7 Comparison of 30 years' cumulative requirements of natural uranium and separative work for different reactors with thermal neutron spectrum (load factor 0.85) [2]

	Natural uranium (t/GW (e))	Separative work (t SWU/GW(e))
Once-through cycle		
LWR-LEU (3.8% U-235)	4,687	3,990
HWR-nat. uranium	4,512	0
HWR-LEU (1.2% U-235)	3,299	1,147
HTGR-LEU (10% U-235)	4,184	4,590
HTR-LEU (10% U-235)	3,759	3,983
Closed fuel cycle		
LWR-U-238/Pu recycle	3,315	3,023
LWR-MEU-235/Th[a]	2,685	3,326
HTGR-MEU-235/Th[a]	2,643	3,219

[a]Thorium requirements—see Table 8.6

8.2.1 The Uranium-Plutonium Fuel Cycle for LMFBRs

In Tables 6.1 and 6.3 are listed the core design and fuel cycle data of SUPERPHENIX and BN 800 as examples of commercial size LMFBR plants. Table 8.8 indicates the core design and fuel cycle data of advanced mixed oxide fuel LMFBR cores. Such advanced LMFBR cores would have a higher core power density. Higher power densities are expected to be applied in later commercial size LMFBR plants [29]. Compared to SUPERPHENIX, with a fissile core inventory of 4.8 t Pu_{fiss}/GW(e), these advanced mixed oxide cores have fissile core inventories of only 3.1 t Pu_{fiss}/GW(e). At the same time, the breeding ratios could increase from BR = 1.18 (Superphenix) to 1.32.

The ex-core time of the fuel amounts to some two years, as described in Sect. 7.4.1. It is expected that the ex-core time can be shortened in the future. This will lead to relatively low total inventories of the FBR fuel cycle, which will be important for an expanding nuclear energy economy based on FBRs.

8.2.2 The Thorium/Uranium-233 Fuel Cycle

Besides the abundant reserves of fertile U-238 as depleted uranium available from U-235 enrichment plants at 0.2% U-235, the reserves of thorium represent an even higher energy potential when used as a fertile fuel in breeder reactors. Thorium and U-233 could be used in FBRs as mixed oxide fuels.

Starting from Pu/U-238 fueled LMFBRs, thorium may first be used in the radial blanket instead of U-238. This would breed U-233 from Th-232 following neutron capture. In addition, the present Pu/U-238 fuel could then be replaced in the FBR core by U-233/Th fuel. Table 8.8 shows characteristic data of an advanced oxide fuel

Table 8.8 Core characteristics of LMFBRs with alternative fuel cycles (Reference plant with 1 GW(e) power output, mixed oxide core and 0.75 load factor) [20, 29]

Core fuel		Pu/U-238	Pu/U-238	U-233/Th
Ax. blanket fuel		U	U	Th
Rad. blanket fuel		U	Th	Th
Total breeding ratio		1.32	1.31	1.10
Core fissile inventory				
U-233	kg	–	–	3,304
Fissile Pu	kg	3,158	3,184	
Annual net fissile gain				
U-233	kg/y	–	133	43
Fissile Pu	kg/y	245	81	
Annual consumption				
U-238	kg/y	1,420	1,359	–
Th-232	kg/y	–	324	1,374

LMFBR core. All variants indicated in Table 8.8 have the same design parameters for the fuel elements as the reference cores.

From Table 8.8 it can be noted that the breeding ratio is not affected very much when thorium is utilized in the radial blanket instead of U-238. When U-233 is substituted for plutonium in the LMFBR core, the smaller η-value of U-233 (See Sect. 3.5) is mainly responsible for the reduction in breeding performance. The breeding ratio of a U-233/U-238/Th fueled core would be about halfway between the Pu/U-238 reference core and the U-233/Th fueled core. This can be explained by the higher fast fission contribution of U-238 and the improved η-value arising from a buildup of Pu-239 (conversion of U-238) during reactor operation. As a conclusion, it can be stated that FBRs operating in the Th/U-233 fuel cycle would have breeding ratios approximately 20% lower than the reference FBR core with mixed oxide PuO_2/UO_2 fuel operating in the U-238/Pu fuel cycle [20, 29].

In the initial phase of designing LMFBRs the smaller breeding ratio was considered a disadvantage with respect to fissile fuel doubling time assuming a rapidly expanding breeder installation for nuclear power production.

8.3 Natural Uranium Consumption in Various Reactor Scenarios

Developing the nuclear energy potential of the U-238 and Th-232 fertile materials requires first that the plutonium or U-233 fissile materials are generated by nuclear reactors using fuel containing U-235. This plutonium or U-233 may then be used to start FBRs. When a sufficient number of FBRs are built and operated it is also possible to produce in their blankets plutonium from U-238 or U-233 from Th-

232. This development strategy is followed because many thermal neutron spectrum reactors, e.g. LWRs producing fissile plutonium are operating already. If this were not the case, FBRs operating in the U-238/Pu fuel cycle could also be started with U-235 enriched UO_2 fuel and breed their own plutonium. In any case, this implies a relatively high natural uranium consumption until a condition will have been reached in which FBRs in symbiosis with thermal neutron spectrum reactors can use the U-238 or Th-232 fertile materials. FBRs which produce a surplus of plutonium or U-233 can make this fuel available for the construction of new FBRs or they can be operated in a symbiosis with thermal neutron spectrum reactors, which also contain plutonium (MOX fuel) or U-233 as fissile materials.

Thermal neutron spectrum reactors operating in symbiosis with FBRs are able to limit the uranium consumption to a certain asymptotic value. In this way all U-238 and Th-232 can be used following the breeding process. Nuclear energy can then be provided for many thousand years.

References

1. Kessler G (1983) Nuclear fission reactors. Springer, Vienna
2. International Nuclear Fuel Cycle Evaluation. Advanced Fuel Cycle and Reactor Concepts (1980) Report of INFCE working group 8. International Atomic Energy Agency, Vienna
3. Reutler H (1988) Plant design and safety concept of the HTR module reactor. Nucl Eng Des 109:335–340
4. International Nuclear Fuel Cycle Evaluation, Reprocessing, Plutonium Handling, Recycle (1980) Report of INFCE working group 4. International Atomic Energy Agency, Vienna
5. Final Generic Environmental Statement on the Use of Recycle Plutonium in Mixed Oxide Fuel in Light Water Cooled Reactors (GESMO) (1976) NUREG-002. US Nuclear Regulatory Commission, Washington, DC
6. Report to the American Physical Society by the Study Group on Nuclear Fuel Cycles and Waste Management (Pines D (ed)) (1978) Rev Modern Phys 50 (1):Part 11
7. Sengler G et al (1999) EPR core design. Nucl Eng Des 187:79–119
8. Aniel-Buchheit S et al (1999) Plutonium recycling in a full-MOX 900-MW(electric) PWR: physical analysis of accident behaviors. Nucl Technol 128:245–256
9. Tommasi J et al (1995) Long-lived waste transmutation in reactors. Nucl Technol 111:133–148
10. Barbrault P (1996) A plutonium-fueled high-moderated pressurized water reactor for the next century. Nucl Sci Eng 122:240–246
11. Burtak F et al (1996) Advanced mixed oxide fuel assemblies with higher plutonium contents for pressurized water reactors. Nucl Eng Des 162:159–165
12. Taiwo TA et al (2006) Assessment of a heterogeneous PWR assembly for plutonium and minor actinide recycle. Nucl Technol 155:34–54
13. Leppänen J (2005) Preliminary calculations on actinide management using advanced PWR MOX technology, Report Pro 1, P 1007/05, VTT Processes, Finland. http://www.virtual.vtt.fi/virtual/proj4/kyt/vr_anttila2.pdf
14. Kloosterman JL et al (2000) Plutonium recycling in pressurized water reactors: influence of the moderator-to-fuel ratio. Nucl Technol 130:227–241
15. Broeders CHM (1996) Investigations related to the buildup of transurania in pressurized water reactors, FZKA 5784, Forschungszentrum Karlsruhe
16. Languille A et al (1995) CAPRA core studies, the oxide reference option. In: Proceedings of international conference on sustainable nuclear energy systems for future generation (GLOBAL 1995). European Nuclear Society, Versailles, 11–14 Sept 1995

17. Trellue HR (2006) Safety and neutronics: a comparison of MOX vs UO_2 fuel. Prog Nucl Energy 48:135–145
18. Youinou G et al (2005) Plutonium recycling in standard PWRs loaded with evolutionary fuels. Nucl Sci Eng 151:25–45
19. Märkl H (1976) Core engineering and performance of KWU pressurized water reactors. Kraftwerk Union, Erlangen
20. Chang YI et al (1977) Alternative fuel cycle options, ANL-77-79, Argonne National Laboratory
21. Data Base for a CANDU PHWR Operating on the Thorium Cycle (1979) AECL-6595. Atomic Energy of Canada, Chalk river
22. International Nuclear Fuel Cycle Evaluation, Fuel and Heavy Water Availability (1980) Report of INFCE working group 1. International Atomic Energy Agency, Vienna
23. Shapiro NL et al (1977) Assessment of thorium fuel cycles in pressurized water reactors. EPRI-NP-359, Electric Power Research Institute, Palo Alto
24. Weaver KD et al (2003) Performance of thorium-based mixed oxide fuels for the consumption of plutonium current and advanced reactors. Nucl Technol 143:22–36
25. INFCIRC/153 Corrected (1972) INFCIRC/153 corrected, the structure and content of agreements between the agency and states required in connection with the treaty on the non-proliferation of nuclear weapons. International Atomic Energy Agency, Vienna
26. Forsberg CW et al (1998) Definition of weapons-useable U-233, ORNL/TM-13517
27. Gruppelaar H et al (ed) (2000) Thorium as a waste management option, EUR 19142 EN
28. Neuhaus I et al (1999) Comparison of uranium- and thorium-based plutonium-recycling with pressurized water reactors, Bericht JUEL 3640, Forschungszentrum Jülich
29. International Nuclear Fuel Cycle Evaluation, Fast Breeder Reactors (1980) Report of INFCE working group 5. International Atomic Energy Agency, Vienna

Chapter 9
Minor Actinides: Partitioning, Transmutation and Incineration

Abstract Plutonium isotopes, but also the isotopes of minor actinides: mainly neptunium, americium and curium can be fissioned by neutrons in the core of nuclear reactors. They also can be transformed as non-fissile isotopes by neutron capture into fissile nuclides (transmutation). Incineration of 99% of the plutonium, neptunium, americium and curium would decrease the long term radiotoxicity of the high active waste (HLW) such that the radiotoxicity level of natural uranium would be underrun already after about 3×10^4 years. This requires chemical separation of plutonium, neptunium, americium and curium and the fabrication of fuel elements with these actinides. These chemical separation methods and the fuel refabrication methods were already developed by research and development programs and demonstrated in pilot plants or at laboratory scale. The possible incineration rates for the different actinides in different reactor types (light water reactors, liquid metal cooled fast breeder reactors and accelerator driven systems) have been thoroughly investigated. Reactor strategies with light water reactors operating in symbiosis with liquid metal cooled fast breeders or accelerator driven systems are feasible. The different reactor and fuel cycle strategies have different radioactivity loads and different radiotoxicity levels within the different parts of their fuel cycle. Whereas the radiotoxicity can be drastically decreased in the back end of the fuel cycle, the masses of plutonium and minor actinides and their radioactivity and radiotoxicity can be higher during reprocessing and refabrication. Transmutation and destruction of long-lived fission products is only feasible with reasonable efficiency for Iodine-129 and Technetium-99.

9.1 Introduction

Safety and health impact analyses of the emplacement in deep geological repositories of HLW described in Sect. 7.6.3 have shown the C-14, Cl-36, and Tc-99 as well as I-129 nuclides to be first in reaching groundwater close to the surface because of their solubility in water. The time when these nuclides will turn up in lakes,

G. Kessler, *Sustainable and Safe Nuclear Fission Energy*, Power Systems,
DOI: 10.1007/978-3-642-11990-3_9,

rivers or drinking water wells is a function of the technical barriers enclosing the HLW container. All safety studies for deep geological repositories so far have indicated that point in time to be on the order of a few thousand years (Sect. 7.6.3). The radioactive burden possible at that time is below the strict regulations of USEPA and of European licensing authorities. After that period of time, Cs-135, Se-79, Nb-93m, and Zr-93 could reach the aquifers close to the surface in 10^4–10^6 years without, however, causing a radioactive burden higher than that of C-14, Cl-36, Tc-99 and I-129. Later, the neptunium, uranium, plutonium and other actinide nuclides would reach the biological environment. That point in time depends on whether these longlived nuclides will form so-called colloidal species after an intrusion of water, and on the efficiency with which colloidal species of the longlived actinides can be retained by an engineered barrier (near-field) during their transport through the cap rock of a repository [1–4].

Plutonium, but also neptunium, americium, and curium, can be split directly by neutrons in the core of a nuclear reactor or transformed into other fissile nuclides (transmutation). The use of plutonium as a fissile material has been described in Chap. 8 (plutonium recycling). Besides generating energy, plutonium recycling entails the advantage that plutonium will no longer exist in the HLW (except for minor losses in chemical reprocessing and in refabrication of the MOX fuel). This applies similarly to the minor actinides (neptunium, americium, curium) when recycled like plutonium.

For this reason, studies have been performed internationally for a number of years to find out what would be the advantages of separating the longlived actinides and of their transmutation and incineration in nuclear reactors. These would be the advantages gained:

- The risk of human intrusion into a repository and the misuse of plutonium and neptunium (nuclear proliferation problem) would be non-existent.
- A safeguards concept extending over very long periods of time to monitor the HLW, especially the spent fuel elements in direct disposal, would not be necessary.
- The radiotoxic inventory of longlived HLW would be reduced drastically.
- The thermal load acting on the repository structures due to decay of actinides, e.g. plutonium and americium, would be diminished.
- The actindes could be transmuted and split, especially in reactor systems with fast neutron spectra, which would allow the generation of additional energy.

Especially breeder reactors with a fast neutron spectrum could exploit the vast potential of U-238 and Th-232 by way of the breeding process and fission of plutonium and U-233, thus ensuring energy generation for the world over many thousands of years.
However, some drawbacks are referred to as well [5, 6]:

- Recycling plutonium and neptunium causes problems in IAEA safeguards and problems of potential proliferation of plutonium and neptunium (plutonium economy).

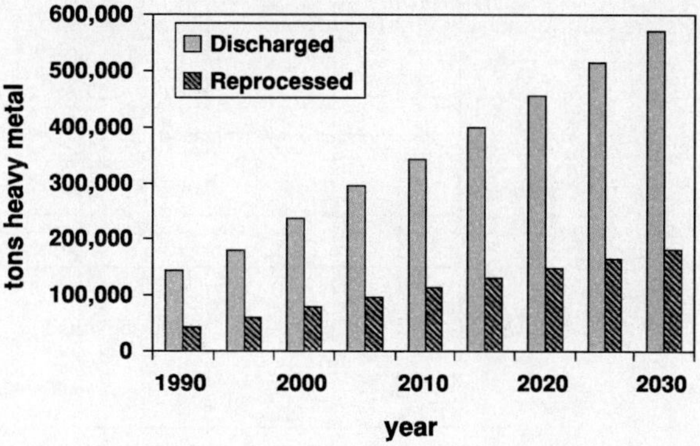

Fig. 9.1 Spent fuel discharged and spent fuel already reprocessed from nuclear reactors operating world wide [7]

- Larger quantities of radioactivity need to be handled in the nuclear fuel cycle than is the case in direct disposal of spent fuel elements.
- In a deep geological repository, the risk of radionuclide release after water intrusion is dominated by the C-14, Cl-36, and fission products with high solubility like Tc-99 and I-129 released first. On the other hand, the release of actindes and their contribution to a potential radioactive burden on the environment is not going to manifest itself until after about 10^6 years, and must be considered a lower risk of additional radiation exposure.
- Chemical methods of separating the actinides (plutonium, neptunium, americium, curium) need to be developed on a technical scale.
- Methods of fabricating new fuels must be developed which contain, in particular, plutonium, neptunium, americium, curium besides uranium and thorium.
- This adds to the cost of the fuel cycle and in some cases to the radiation exposure of workers.

9.2 Worldwide Inventories in Spent Fuel Elements of Uranium, Plutonium, Neptunium, Americium

The civil use of nuclear power has given rise to approximately 340,000 t of spent fuel elements with 2,300 t of plutonium by 2010. Roughly one third of these spent fuel elements were chemically reprocessed with the reprocessing capacity available worldwide (Fig. 9.1). By 2010, approximately half of this reactor-grade plutonium was processed into plutonium/uranium MOX fuel elements and is being recycled in nuclear reactors.

Figure 9.2 shows an IAEA estimate for the nuclear power installed world wide and of the quantities of reactor-grade plutonium the spent fuel elements contain. Moreover, the quantities of separated plutonium and MOX fuel elements are shown [7].

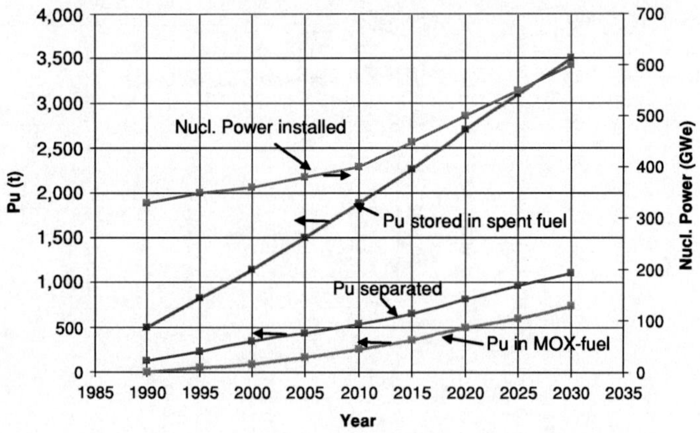

Fig. 9.2 Nuclear power installed as well as plutonium stored in spent fuel elements, plutonium separated and plutonium in MOX fuel until 2010. Projections are also made until 2030 [7]

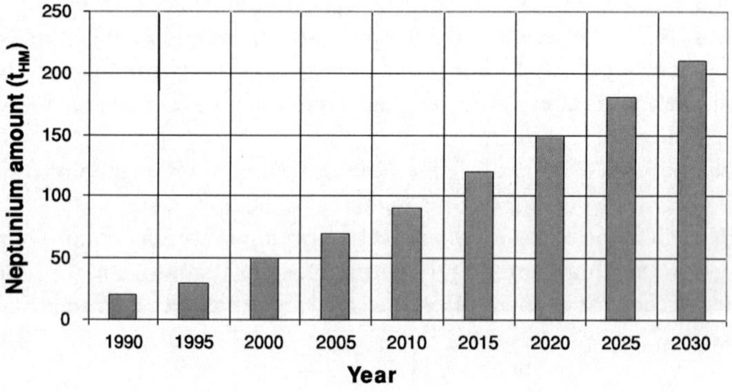

Fig. 9.3 World wide neptunium stored in spent fuel elements or high active waste until 2030 [7]

Figures 9.3 and 9.4 show IAEA estimates of the quantities of neptunium and americium generated in spent fuel elements and to be generated by 2030, respectively [7]. With one third of the fuel elements already reprocessed chemically, these quantities of neptunium and americium are to be found in liquefied or vitrified HLW. The quantities of curium generated or to be generated in the future in the case of spent LWR fuel elements are approximately a factor of 11 lower than those of americium.

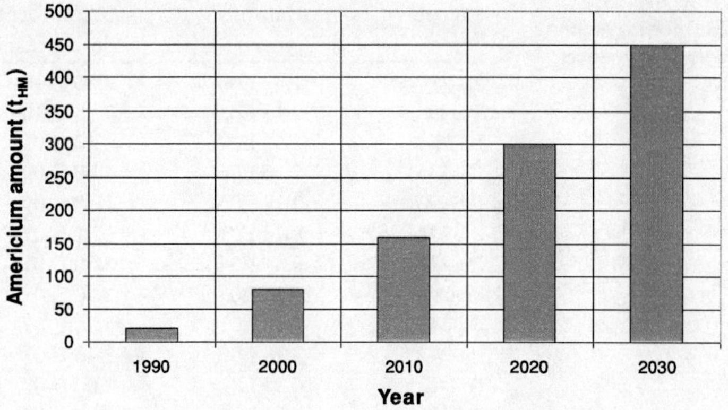

Fig. 9.4 World wide americium stored in spent fuel elements or in high active waste until 2030 [7]

9.3 Radiotoxicity of HLW

Radiotoxicity is a measure of the health hazard posed by a radionuclide. It depends on the type and energy of the radiation emitted by the radionuclide and, in addition, on the resorptivity in the organism and the residence time of the radio-nuclide in the body. The radiotoxicity of a radionuclide is expressed by the effective dose in Sievert per Becquerel (Sv/Bq). For airborne radionuclides arising, for instance, in reactor accidents, the effective dose for incorporation and external exposure is important. For HLW in the repository and associated safety analyses, however, only the effective dose for ingestion is important. Table 9.1 shows the effective dose values for ingestion for some selected fission products and actinides [8].

Table 9.1 can be used to determine, from the known radionuclide quantities and their half-lives, the radiotoxicity of radionuclides in 1 tonne of spent LWR fuel as a function of time.

This is shown for the radiotoxicity for ingestion in Fig. 9.5 for 1 tonne of spent fuel of a PWR fuel element with a burnup of 40,000 MWd/t. The fresh fuel had an enrichment level of about 4% U-235. The contributions to radiotoxicity stemming from the fission products, neptunium, plutonium, americium, and curium can be seen. Moreover, the radiotoxicity of the radioisotopes of the decay chains of plutonium, americium, and curium is shown. The radiotoxicity of 1 tonne of natural uranium is represented by a line parallel to the abscissa at 2×10^4 Sv/t_{HM}.

The radiotoxicity due to direct disposal of spent fuel elements is represented by the total sum of all contributions. Curium falls below the radiotoxicty of natural uranium after only approximately 200 years, while the fission products underrun this line of natural uranium not before 600–700 years. The americium isotopes reach this state at approximately 3×10^4 years, plutonium isotopes, afer 3×10^5 years [8].

Table 9.1 Effective dose values (Sv/Bq) for ingestion for selected fission products and actinides

Isotope	Half-life (years)	Factor for effective dose ingestion (Sv/Bq)
Zr-93	1.53 E06	2.80 E-10
Tc-99	2.13 E05	7.79 E-10
I-129	1.57 E07	1.10 E-07
Cs-135	2.30 E06	2.00 E-09
Np-237	2.14 E06	1.10 E-07
Pu-238	8.78 E01	2.31 E-07
Pu-239	2.41 E04	2.50 E-07
Pu-240	6.57 E03	2.50 E-07
Pu-241	1.44 E01	4.70 E-09
Pu-242	3.74 E05	2.40 E-07
Am-241	4.33 E02	2.00 E-07
Am-243	7.37 E03	2.00 E-07
Cm-244	1.81 E01	1.20 E-07
Cm-245	8.51 E03	2.10 E-07

9.4 Various Strategies of Partitioning and Transmutation with Incineration of Actinides

The influences of various possible strategies including partitioning of plutonium (plutonium recycling) and additional partitioning of americium and curium are shown in Fig. 9.6 [8]. These are the scenarios compared:

- Direct disposal of spent fuel elements. This strategy results in the highest radiotoxicity over very long periods of time.
- Partitioning and incineration of plutonium/uranium. This strategy of Pu/U recycling, with an incineration efficiency of 99 or 99.9%, leads to the radiotoxicity level of natural uranium being reached after approximately 6×10^4 years.
- Partitioning and incineration of U, Pu, Am, Cm with an efficiency of 99% results in underrunning the radiotoxicity level of natural uranium after some 3×10^4 years. Only partitioning and incineration of U, Pu, Am, Cm with an efficiency of 99.9% would cause the radiotoxicity level of natural uranium to be underrun after some 800 years.
- Incineration of curium could also be replaced by interim storage of this minor actinide for a period of about 200 years awaiting the α-decay of curium isotopes into plutonium isotopes and subsequent incineration in nuclear reactors (Sects. 9.7.6 and 9.8.1). However, this is an objective very difficult to achieve on a technical scale.

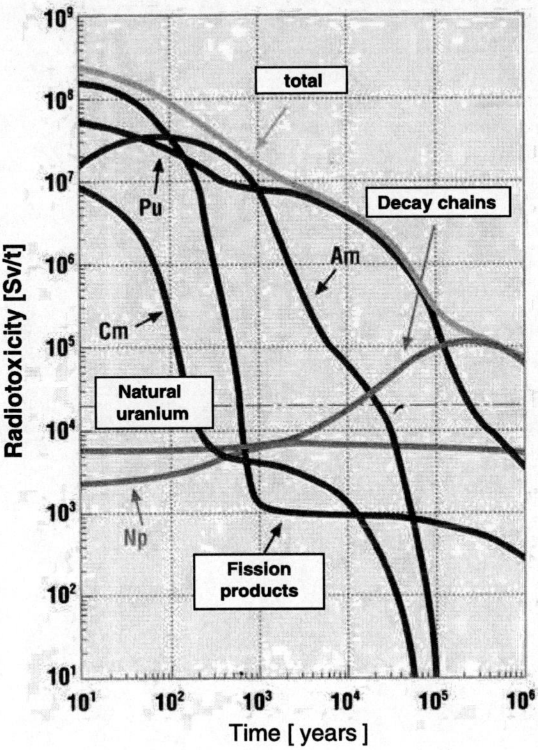

Fig. 9.5 Radiotoxicity for ingestion of 1 tonne of PWR spent fuel with a burnup of 40,000 MWd/t (enrichment of fresh fuel 4% U-235) [8]

9.5 Chemical Separation of Actinides

9.5.1 Joint Chemical Separation of Plutonium and Neptunium from Spent Fuel

The chemical separation of uranium and plutonium from the fission products and minor actinides (reprocessing, PUREX process) was described in Sect. 7.2.1 above.

The chemical separation of uranium and plutonium is achieved in present-day large reprocessing plants, such as LaHague, France, or Sellafield, United Kingdom, by means of tributyl-phosphate (TBP) with an efficiency of approximately 99.8–99.9% [8–11].

In the PUREX process, neptunium together with uranium is put through the first separation stages and separated from uranium only in the uranium purificatioin step. The PUREX process thus can be modified in such a way that neptunium is separated together with plutonium. For this purpose, neptunium must be oxidized chemically from the pentavalent to the hexavalent state (Table 9.2) and can then be co-extracted with uranium (VI) and plutonium (IV) in the first separation cycle by means of

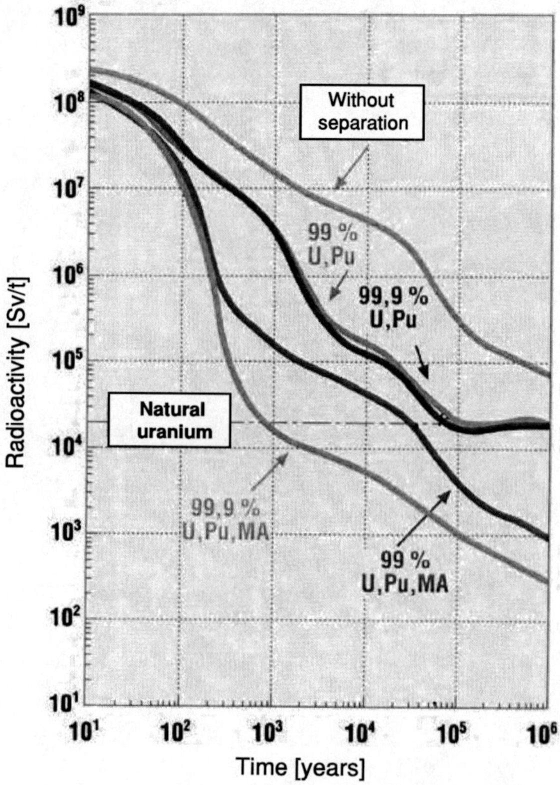

Fig. 9.6 Radiotoxicity for ingestion of 1 tonne of PWR spent fuel with a burnup of 40,000 MWd/t for different partitioning, transmutation and incineration strategies [8]

tributylphosphate. This is achieved with nitric acid and by adding special agents, such as vanadium (V) compounds [8]. Purification of neptunium after co-extraction with uranium and plutonium can be achieved in the second uranium cycle [8, 9, 12].

This modified PUREX process was demonstrated by the Japan Nuclear Cycle Development Institute (JNC) and the reprocessing plant LaHague, France, with a separation efficiency of 99% [9, 12].

However, it is also possible to separate neptunium subsequently from the liquid HLW of present reprocessing plants [8].

9.5.2 Separation of Americium and Curium Together with the Lanthanides

Americium and curium as well as the lanthanides (Lns) are found in the trivalent state in the liquid HLW (Table 9.2) [9]. Therefore, they must be separated together. This can be achieved by the following chemical processes:

Table 9.2 Valence or oxidation state of different elements [9]

Element	Valence			
	III	IV	V	VI
Uranium				•
Plutonium		•		
Neptunium			•	•
Americium	•			
Curium	•			
Lanthanides	•			

9.5.2.1 DIDPA Process

The DIDPA process was developed by the Japan Atomic Energy Research Institute (JAERI) [8, 13]. It is based on the organophosphorus agent, diisodecyl phosphoric acid (DIDPA). For this purpose, the liquid HLW must be reduced in nitric acid concentration from 2–3 to 0.5 mol/l. This can be done, e.g., by denitrating the HLW with formic acid. In this process, nitric acid is decomposed into gaseous products. A mix of DIDPA and TBP is then used to extract the actinides, including neptunium, and the lanthanides, and separate them from the remaining fission products. Americium, curium, and the lanthanides are then re-extracted together with 4 mol/l nitric acid. In this step, they are separated from the organic phase containing neptunium and residues of plutonium and uranium.

The nitric acid concentration is then reduced again to 0.5 mol/l [13, 14] and americium and curium are separated as described under Sect. 9.5.4.

9.5.2.2 The TRUEX Process

The TRUEX process was developed at the Argonne National Laboratory (ANL) in the United States [15]. It is based on CMPO, an organophosphorus extraction solvent. No pretreatment of the HLW is necessary, and the actinides and lanthanides can be extracted from HLW acidified with 0.7–5 mol/l nitric acid. However, as the efficiency of actinide and lanthanide separation is not sufficient, the Japan Nuclear Cycle Development Institute proposed a combination of the TRUEX and DTPA processes.

9.5.2.3 TRPO Process

The TRPO process for actinide and lanthanide separation was developed at the Tsinghua University, Beijing [16]. TRPO (trialkylphosphinoxide) dissolved in kerosene is used. Separation factors of 99.9–99.99% (laboratory scale) were achieved at a nitric acid concentration of 1 mol/l. However, as in the TRUEX and DIPA processes, an additional treatment step is required. As re-extraction from TRPO is carried out at a higher nitric acid concentration, another neutralization step is necessary for the separation of actinides and lanthanides.

9.5.2.4 DIAMEX Process

The DIAMEX process was developed by the French Commissariat à l'Energie Atomique (CEA) [17]. This chemical separation process employs dimethyl-dibutyl-tetradecyl malonamide (DMDBTDMA) dissolved in aliphatic hydrocarbon. The trivalent minor actinides and lanthanides are separated at a nitric acid concentration of 3–5 mol/l, while re-extraction is conducted at 0.1 mol/l. The DIAMEX process was successfully tested by the CEA in the ATALANTE facility at Marcoule on HLW with separation factors of 99.9% [12, 18].

9.5.3 Chemical Separation of Actinides from the Lanthanides

The chemical separation processes referred to above allow the trivalent lanthanides and americium and curium, respectively, to be separated only together as trivalent species. Special processes had to be developed to separate the lanthanides from americium and curium.

9.5.3.1 DIDPA–DTPA Process

In the second part of the DIDPA process selective stripping with diethylenetriamine-pentaacetic acid (DTPA) separates americium together with curium from the lanthanides (fission products).

The DIDPA process was tested with HLW. All actinides (neptunium, plutonium, americium, curium) were recovered 99.9% [13, 14].

9.5.3.2 Chemical Separation with Dithiophosphinic Acids

In connection with the TRPO process of the Tsinghua University of Beijing, the CYANEX 301 solvent was used together with TBP. This achieved separation factors in excess of 99.9% for separation of the lanthanides from americium and curium. The process was developed further by the Jülich Research Center in Germany by modifying the CYANEX 301 solvent [19–21].

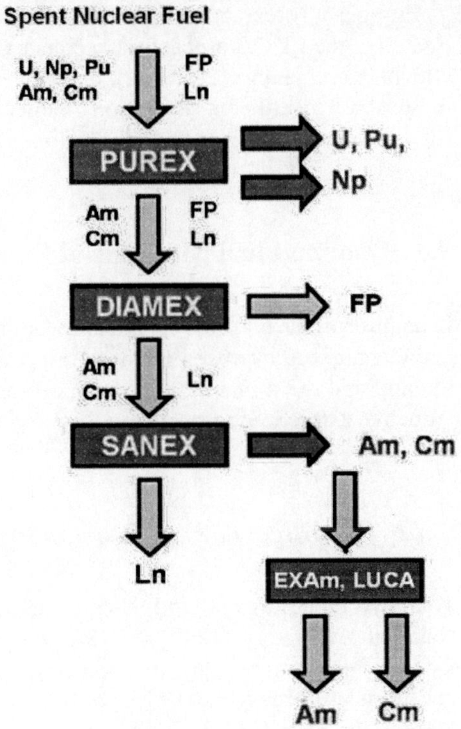

Fig. 9.7 Chemical separation processes developed in Europe for partitioning U, Pu, Np, Am and Cm

9.5.3.3 SANEX process

In connection with the DIAMEX process, extraction solvents based on bistriacylpiridine (BTP) were developed at the Karlsruhe Research Center and successfully tested by the Institute of Transuranium Elements in Karlsruhe, Germany, and the CEA in the ATALANTE facility in Marcoule. They resulted in the development of the SANEX separation process (Fig. 9.7) by the CEA, by means of which americium and curium, on the one hand, can now be separated from the lanthanides, on the other hand, with 99.9% efficiency [12, 22, 23].

9.5.4 Chemical Separation of Americium from Curium

Studies of the separation of americium from curium seem to be very successful. The TRPO process of the Tsinghua University, Beijing, is able to separate americium from curium with a separation factor of 99.9% [19].

With the TODGA and CyMe4BTBP agents separation of americium was achieved with an efficiency of 99% [24–26].

European research institutes developed the simplified GANEX process and EXAm or the LUCA processes to achieve this separation of americium from curium with high efficiency [12, 26–32]. Figure 9.7 shows the different processes developed in Europe. The efficient separation of americium and curium is absolutely necessary for a number of actinide fuel fabrication processes (Sect. 9.7).

9.6 Pyrochemical Methods of Separating Minor Actinides

In addition to the aqueous chemical separation methods described above, pyrochemical techniques have been under development for decades by which uranium, plutonium, and the minor actinides can be separated by electrolytic fractionation and reductive extraction, respectively.

9.6.1 The Integral Fast Reactor Pyroprocessing Process

This pyrometallurgical and electrochemical process is developed at Argonne National Laboratory (ANL) in USA in combination with the Integral Fast Reactor (IFR) programme [33]. It is an evolution of the pyroprocessing methods which had been utilized in the 1960s for the metallic fuel of the experimental breeder programme EBR-II [34]. These pyroprocessing technologies were improved by the development of electrorefining methods [35] for the separation of actinides from the fission products. The current improvements of the ANL pyropressing methods aim at separation efficiencies of >99.9%. However, the plutonium and minor actinides cannot be partitioned, since the plutonium and all other minor actinides (neptunium, americium, curium) in combination with about 30% uranium always remain together. From the nonproliferation point of view this is considered to be an advantage [33, 36].

The pyroprocessing method is a batch mode process, whereas the aqueous partitioning processes (Sect. 9.5) operate in a continuous mode.

After dismantling of the fuel assemblies and chopping of the fuel rods including claddings, the short fuel rod segments are loaded into perforated steel baskets and placed in the electrorefiner vessel (Fig. 9.8). The electrorefiner vessel is covered at the bottom by a thick layer of liquid cadmium (melting point 321°C). This cadmium layer is again covered by a thick layer of an eutectic mixture of lithium chloride, LiCl, and potassium chloride, KCl, (melting point 350°C) acting as electrolyte salt. The electrorefiner is operated at a temperature of about 500°C.

The perforated steel baskets with fuel segments are lowered into the electrolyte salt and act as the anode. The actinides from the spent fuel are transported from the anode basket to two kinds of cathodes [a solid cathode and a liquid cadmium cathode (Fig. 9.8)] by means of an applied electrical current. Pure uranium is collected at the solid cathode and a mixture of uranium, plutonium, americium and curium is collected at the liquid cadmium cathode suspended in the electrolyte salt. The fission

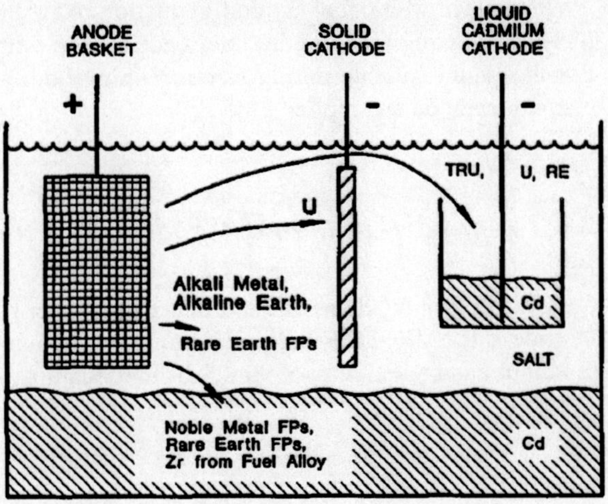

Fig. 9.8 IFR pyroprocessing scheme [36]

products remain in the electrolyte salt and collect in the liquid cadmium layer at the bottom.

After the desired amount of actinide materials has been collected, these deposites at the cathode are recovered in a high temperature vacuum furnace (cathode processor) by melting. Any volatile products are removed by vaporization. These include electrolyte salts and cadmium. They are collected in a condenser and recycled.

The metal ingots resulting from the cathode processing operation are rather free of impurities (traces of solid fission products remain) and sent to the injection casting station for fabrication of new metallic fuel rods (Sect. 9.7.5) [36, 37].

Due to the higher radiation resistance of the electrolysis in molten salts and remotely controlled technology, pyroprocessing of short-cooled spent fuel is possible. Cooling times as short as several months of the spent fuel seem possible compared to the present two years for the LMFR fuel cycle or seven years needed for aqueous reprocessing of LWR spent fuel.

9.6.2 *Electro-Reduction and Refining of Spent UOX and MOX Fuel to Metallic Fuel*

In addition to the pyroprocessing of spent metallic fuel also methods for electrolytic reduction and electrorefining of spent UOX and MOX fuel to metallic fuel were developed. Efficiencies for the electrolytic reduction of spent oxide fuel to metallic fuel of 99.7% were demonstrated [36, 37].

Similar research and demonstration experiments as at ANL were reported by CRIEPI (Japan) for electroreduction and refining as well as pyroprocessing of metallic plutonium/uranium fuel [38–40]. The pyroprocess can start with either spent

LWR UOX or MOX fuel (lithium reduction) or with liquid HLW from spent LWR fuel reprocessing applying an intermediate step with denitration and chlorination. For the liquid cathode during electrorefining both a liquid cadmium and a liquid bismuth cathode are applied [38].

9.6.3 Actinide/Lanthanide Separation Using Aluminum

Conocar et al. [41] demonstrated that a one stage reduction process using molten fluoride salt (AlF_3-LiF) and an aluminum solid anode resulted in a separation of plutonium and americium of 99.3% from the lanthanides.

9.6.4 Pyro-Processing of Fast Reactors PuO_2/UO_2 Fuel in Russia

A similar pyroprocess approach is developed for fast reactors in Russia. Here, the PuO_2/UO_2 is the product instead of U-Pu-Zr Pu-metal. In the Russian DOVITA-process [42, 43] the mixed oxide fuel is converted into chlorides and separated by electrolysis in a melt of NaCl–KCl at 650°C. The transuranium elements are precipitated sequentially as oxo-chlorides or oxides out of the NaCl–KCl melt by gassing with Cl_2/O_2 and adding Na_2CO_3.

9.7 Fuel Fabrication for Transmutation and Incineration of Actinides in Nuclear Reactors

The fabrication procedures of fuels for actinide transmutation and incineration are presently based on existing fabrication technologies for mixed oxide, e.g. $(U/Pu)O_2$ fuel or mixed nitride (U/Pu)N fuel. In principle the minor actinides can be mixed into the MOX fuel. However, due to the increased gamma and neutron radiation caused by americium or curium the fuel fabrication facilities will need heavy shielding.

Therefore, dust free aqueous processes like SOL-GEL techniques or other liquid–solid conversion processes for the fabrication of granulates or microspheres are applied. In addition infiltration methods are under development. In this case americium or curium nitrates are infiltrated in a porous medium, e.g. magnesium aluminate spinel pellets etc. [44, 45].

Metallic fuel has a better thermal conductivity than oxide fuel. Metallic U-Pu-Zr fuel with sodium bonding between the cladding and fuel is being developed by Argonne National Laboratory in the USA as part of the IFR program (Sect. 9.6.1).

Uranium free fuels (inert matrix fuel) on the basis of oxides like ZrO_2, Y_2O_3, MgO etc. are also developed in Europe and Russia [44, 46].

Table 9.3 Particle size distribution for a smear density of 85% theoretical density

Particle size (μm)	vol.%
1200	60
300	15
30	25

The ALFA fuel manufacturing facility for actinide bearing fuel elements is under construction at Marcoule, France [47].

9.7.1 Pellet Fabrication with SOL-GEL Microspherical Particles

In the SOL-GEL droplet-to-particle conversion process droplets are generated by passing a feed solution of nitrides of actinides over the edge of a cylindrical cone rotating at high speed. The droplets are collected in an ammonia bath where gelation occurs. After washing, drying and calcination, spherical particles of different size (20–150 μm) are obtained. The spherical particles can be pressed and sintered to pellets [48, 49]. As an example the following actinide fuels can be produced with different enrichments

$$(U,Np)O_2 \text{ or } (U,Pu,Am)O_2 \text{ or } (U,Pu,Np)O_2$$

9.7.2 Fuel Fabrication by Vibrocompaction

Microspheres produced by the SOL-GEL process or small granulates produced by crashing can be rinsed into tubes of fuel rod claddings and be compacted by vibration. Smear densities of 85% of the theoretical density can be achieved by carefully selecting the grain sizes. Table 9.3 shows three particle sizes and their volume distribution applied to achieve a smear density of 85% theoretical density [45, 48–53].

The remaining open porosity provides space for helium from alpha-particle decay and for gaseous fission products. This process was originally applied and is still used in Russia (VIPAC process) [49, 50] for MOX fuel fabrication of LMFBRs.

More recent developments in Europe use SOL-GEL microspheres (SPHE-REPAC process [48, 51, 52].

The granulates or microspheres must be produced from powders or actinide solutions having already the enrichment used in the fresh fuel.

9.7.3 Inert-Matrix Fuel

Inert matrix fuel is free of uranium or thorium. The actinides are distributed as a separate phase in a so-called inert matrix. Oxides, nitrides as well as metals can be considered as the inert matrix [44, 53, 54].

Oxides like ZrO_2, Y_2O_3, MgO, $MgAl_2O_4$ or $Y_3Al_5O_{12}$ were proposed as inert matrix [44, 53]. Different fabrication processes are considered:

- coprecipitation is based on the dissolution of the starting materials in nitric acid and the precipitation of all these materials. The resulting powder, after washing, drying and calcination is directly used for the pellet production.
- the mixing of particles and powder is based on the fabrication of microspheres or particles containing the actinides by the SOL-GEL technique followed by mixing these particles with the powder of the inert matrix.

Swelling of inert matrix materials during irradiation and induced α-particle production resulting from radioactive decay of, e.g. curium, may need particular attention [55, 56].

9.7.4 Infiltration Method

This process requires a porous medium in which the actinides can be infiltrated. The porous medium can be, e.g. a magnesium aluminate pellet formed by powder and subsequent calcination. Instead of porous pellets also porous beads produced by the SOL-GEL method can be used. In this case a higher loading of the beads with actinides can be achieved [44, 45].

9.7.5 Metallic Fuel

The most common fabrication process for metallic fuel is so-called remotely controlled injection casting of the U-Pu-An-Zr alloy [36]. In case of pyro-chemical reprocessing (Sect. 9.6.1) the product from the liquid cathode is already an alloy of uranium, plutonium and minor actinides. Uranium from the solid cathode is added to achieve the required fuel composition. This fuel batch is induction heated under vacuum and homogenized. Then the system is pressurized and the fuel alloy is injected into closed end molds which are rapidly cooled. The molds are removed, the fuel slugs cut to length and inserted into claddings with a small amount of sodium for bonding between the fuel and cladding [35, 36, 57, 58].

9.7.6 Intermediate Storage of Curium

Curium is a mixture of the isotopes Cm-242 (half-life 163 days), Curium-243 (half-life 29 years), Curium-244 (half-life 18 years) and Curium-245 (half-life 8,500 years) and minor amounts of higher Cm isotopes. Curium 242 decays already almost completely to Pu-238 during reactor operation and subsequent cooling [59, 60].

Because of the high radiation and thermal loads during fabrication of curium containing fuel elements, it was proposed that curium should be stored until curium-243 and curium-244 have decayed to Pu-239 and Pu-240, respectively.

Curium solutions can be infiltrated into porous beads. These are calcined and sintered and then poured into vessels especially designed for interim storage over about 100–200 years. After this time period the fuel can be reprocessed and the separated Pu-239 and Pu-240 can be incinerated in FRs (see Sect. 9.4).

Curium transmutation in actinide fuel and subsequent irradiation in nuclear reactor cores would lead to extreme difficulties if aqueous reprocessing and subsequent refabrication were applied. The high neutron radiation and high alpha-particle heat production of Cm-244 are mainly responsible for these extreme difficulties.

This is different for pyrochemistry where subsequent metallic fuel refabrication to metallic fuel element is possible as described above. This process is performed entirely under remote handling (Sect. 9.7.5).

9.7.7 Irradiation Experience with Fuel Containing High Plutonium Contents or Neptunium and Americium

Experience is already available with fuel based on high plutonium contents for incineration of plutonium in FR burners [61]. Also, irradiation experience with neptunium and americium was obtained from experiments in JOYO and Phenix [62–67]. Whereas neptunium containing fuel behaves very similar to plutonium/uranium mixed oxide fuel, americium needs more care for its fuel design. Neutron capture in americium leads to the build up of curium isotopes. The helium production from the decay of Curium-242 to Pu-238 and Cm-244 to Pu-240 needs special design provisions in order to avoid too high gas pressures in the fuel rod. In addition to the large experience with the irradiation of metallic fuel in EBR-II and FFTF in USA, experience was obtained with U/Pu/MA fuel in the French fast reactor Phenix [64].

Inert matrix fuel containing americium was tested in irradiation experiments in the Phenix reactor [64] and to very high burnup of 19 at.% in the Russian BOR 60 [46].

Table 9.4 Generation of plutonium, neptunium, americium and curium of an LWR having a fuel burnup of 51 GWd/t

Actinide	kg/GW(e)· y
Plutonium	238
Neptunium	13.2
Americium	19.5
Curium	1.38

9.8 Incineration of Minor Actinides in Nuclear Reactors

9.8.1 Introduction

An LWR core containing fresh UOX fuel with an enrichment of 4.3% U-235 generates the following amounts of plutonium, neptunium, americium and curium (Table 9.4) after a fuel burnup of 51 GWd/t and 10 years cooling time of the spent fuel [68]:

For the incineration of plutonium, neptunium, americium or curium, fuel elements containing minor actinides can be loaded into the cores of PWRs, FRs, ADSs or other nuclear reactors. Most of such investigations have been performed for PWRs, FRs and ADSs so far [69, 70].

Neutron absorption reactions lead to transmutation or fission of minor actinides. Fuel assemblies containing minor actinides can be distributed homogeneously over the core or be arranged heterogeneously at the periphery of the core. The presence of minor actinides influences the initial fissile enrichment, the safety parameters, e.g. Doppler coefficient and void coefficient and changes the decay heat and radiation characteristics of spent fuel elements [69, 70].

The amount of minor actinides to be loaded into the core of a nuclear reactor requires a detailed analysis of safety parameters. This leads to recommendations for upper limits. For homogeneous loading of minor actinides into the core of PWRs an upper limit of 1% for each of the minor actinides neptunium and americium was proposed. For large cores of LMFRs an upper limit of 2.5% of neptunium or americium was recommended [69].

The high radiation caused by curium makes recycling of this minor actinide very difficult except for the case of pyrochemistry combined with metallic fuel fabrication of the IFR (Sect. 9.7.5).

9.8.1.1 Curium Recycling in PWRs

In addition to the high radiation and heat loads (Sect. 9.7.6) during fabrication of curium containing fuel elements, recycling of curium in PWRs leads to the build up of Cf-252 [68, 69, 71]. This causes an increase of the decay heat and gamma radiation of spent fuel by a factor of 3 and an increase of the neutron radiation by a factor of 8,000 if compared to a spent MOX-PWR fuel assembly [72]. Therefore, such spent fuel containing Cf-252 (half-life about 2 years) would have to be stored intermediately for about 2 decades until Cf-252 will have decayed to sufficiently low concentrations to Cm-248.

Therefore, curium should be stored for about 100–200 years until the most important curium isotopes will have decayed to plutonium isotopes (Sect. 9.7.6) and the build up of Cf-252 can be minimized.

9.8.2 Transmutation and Incineration of Neptunium and Americium

Neptunium can be loaded homogeneously to the core fuel of PWRs and FRs. Neutron capture in Np-237 results in build up of Pu-238 with high spontaneous neutron radiation and high alpha decay heat power. There is presently a limit of about 5% Pu-238 in plutonium set by radiolysis during aqueous reprocessing and neutron radiation exposure during MOX fuel refabrication. Neutron capture in americium results in the build up of large amounts of curium (Cm-242, Cm-243 and Cm-244) which are strong neutron and alpha-particle emitters. In order to avoid deterioration of the safety parameters, heterogeneous loading of a certain number of americium containing fuel assemblies at the core periphery is preferred [69].

9.8.3 Neutronic Analysis for Potential Destruction Rates of Neptunium and Americium in PWRs and FRs (One Cycle Irradiation)

Neutronic analysis for destruction rates of plutonium, neptunium and americium for one cycle irradiation were reported by [69, 70, 73]. If neptunium and americium are admixed to the fuel together with plutonium, build up of Pu-238 via neutron capture in Np-237 or the decay of Cm-242 to Pu-238 (after neutron capture in Am-241) or the decay of Cm-244 to Pu-240 (after neutron capture of Am-243) decrease the destruction rate of plutonium compared to those reported in Sect. 8.1.2.3 (fissile fraction of plutonium 6%) for the case of recycling plutonium only.

Table 9.5 shows the destruction rates for a 1.3 GW(e) PWR and a 1.5 GW(e) FBR (burner) for the cases of

- plutonium only (fissile plutonium fraction 7.3% in PWR and 17.7% in FBR)
- plutonium and 1% neptunium (PWR) or 2.5% neptunium (LMFBR, homogeneously) (fissile plutonium fraction 10.2% in PWR and 17% in FBR)
- plutonium and americium 1% (PWR) or 2.5% (LMFBR) distributed heterogeneously at the core periphery.

These data were determined [69] with an accompanying analysis of all safety parameters for the fresh core only. This differs somewhat from the data given in Sects. 8.1.2.3

and 9.8.4 which are based on a determination of all safety parameters over the full burn up cycle. However, the results of Sect. 8.1.2.3 were based on a moderator to fuel ratio of 2.

The destruction rates for the case "plutonium only" are about 420 kg/GW(e)·y in a MOX-PWR and about 570 kg/GW(e)·y in an LMFBR. If 1% neptunium (PWR) or 2.5% neptunium (LMFBR) are admixed homogeneously to the MOX fuel the plutonium destruction rate decreases somewhat to 359 kg/GW(e)·y in case of the PWR and 525 kg/GW(e)·y in case of the LMFBR. The neptunium destruction rate is 85 kg/GW(e)·y in a MOX-PWR and 78 kg/GW(e)·y in an LMFBR.

If americium is loaded in special fuel assemblies at the periphery of the core of PWRs or LMFBRs the destruction rate is 39 kg/GW(e)·y in case of the MOX-PWR and 110 kg/GW(e)·y for the LMFBR case.

Slightly different results are also reported [72–75] for cases with higher than 1% loading of neptunium or overmoderated fuel assemblies in PWR cores. Also matrix fuel loaded with neptunium or americium can achieve somewhat higher destruction rates [73, 74].

9.8.4 Multi-Recycling of Plutonium, Neptunium and Americium in PWRs

Multi-recycling in PWRs of only plutonium or of plutonium with neptunium as well as plutonium with neptunium/americium was investigated in [68, 75]. As already described in Sect. 8.1.2.3 for multirecycling of plutonium in full MOX-PWRs the fraction of plutonium in the MOXfuel must be restricted because of the tendency to develop a positive moderator temperature coefficient. The required k_{eff} is achieved by adding low enriched U-235/U-238 to the MOX fuel. Therefore the MOX fuel assembly structure of Fig. 8.4 with a fuel to moderator volume ratio of 2.5 was selected. The following results were given in [68]:

- Multi-recycling of plutonium
  By restricting the fraction of total plutonium to 10% and adding low enriched U-235/U-238 slightly increasing during multi-recycling up to 3.58% U-235, the incineration rates of plutonium would vary between 532 kg/GW(e)·y (first cycle) and 420 kg/GW(e)·y (10th cycle) (load factor 0.85 assumed) [68].
- Multi-recycling of plutonium together with neptunium
  By restricting the total plutonium/neptunium fraction in the fuel to 8% and adding low enriched U-235/U-238 with slightly increasing enrichment of up to 4.45% U-235 the incineration rates for plutonium would vary between 340 kg/GW(e)·y (first cycle) and 290 kg/GW(e)·y (10th cycle). Accordingly, the neptunium incineration rates would vary between 31 kg/GW(e)·y (first cycle) and 16 kg/GW(e)·y (10th cycle) (load factor 0.85 assumed) [68].
  The Pu-238 isotopic fraction in the plutonium—would vary between 6% (first cycle) and 8.5% (10th cycle) [68].

Table 9.5 Destruction/production rates for plutonium, neptunium and americium [69]

Actinide	PWR 1.3 GW(e) Destruction/production rate [kg/GW(e)·y]			LMFBR 1.5 GW(e) Destruction/production rate [kg/GW(e)·y]		
	MOX-PWR only plutonium	MOX-PWR plutonium 1% neptunium homogeneously	MOX-PWR plutonium americium heterogeneously	MOX-LMFBR burner only plutonium	MOX LMFBR plutonium 2.5% neptunium homogeneously	MOX LMFBR plutonium americium heterogeneously
Plutonium	−420	−359	−400	−570	−525	−538
Neptunium	+3.2	−85	+0.3	+2.5	−78	+0.75
Americium	+45	+61	−39	+26	+23	−110
Curium	+20	+20	+8	+4	+4	+24

Load factor 0.85; 1 GW(e)·y = 7.45 TW(e) h

These latter incineration rates are smaller—compared to Table 9.4—since both the fraction of total plutonium with 7.58% (first cycle) and that of neptunium with 0.42% (first cycle) are smaller than the 10.2% fissile plutonium and 1% neptunium used in the previous section.

- Multi-recycling of plutonium and neptunium/americium
  By restricting the plutonium/neptunium/americium fraction to 8% and adding low enriched U-235/U-238 with slightly increasing enrichment up to 6.61% U-235, the incineration rates of plutonium would vary between 235 kg/GW(e)·y (first cycle) and 124 kg/GW(e)·y (10th cycle). Accordingly the neptunium incineration rate would vary between 24.9 kg/GW(e)·y (first cycle) and 6.8 kg/GW(e)·y (10th cycle). The americium incineration rate would vary between 5 kg/GW(e)·y (third cycle) and 10 kg/GW(e)·y (10th cycle). In the first and second cycle there would be an americium production of 39.6 kg/GW(e)·y and 8 kg/GW(e)·y respectively (0.85 load factor assumed) [68].
  The Pu-238 isotopic fraction in the plutonium would increase to 8.5% (first cycle) and 17% (10th cycle). This is mainly caused by neutron capture in Np-237 leading directly to Pu-238 or neutron capture in Am-241 leading to Pu-242 (15%) or Cm-242 which decays to Pu-238 (75%) [68].
  These incineration rates are smaller—compared to Table 9.5 and 9.7—for the same reasons as mentioned already above for the case of multi-recycling of plutonium together with neptunium.

The most important reactivity (boron worth) and temperature coefficients (fuel Doppler coefficient and moderator temperature coefficient) for multi-recycling of plutonium only can be derived from [75] where these safety coefficients were reported for a restricted total plutonium content of 8 and 12%. These data are given in Table 9.6 for the subassembly shown in Fig. 8.4 with a fuel to moderator volume ratio of 2.5 (compared to the standard PWR UOX fuel element with a fuel to moderator volume ratio of 2).

The smaller boron worth coefficients for the MOX PWR would necessitate higher B-10 enrichment. The more negative fuel Doppler temperature coefficients (FDC) and moderator temperature coefficients (MTC) of the MOX PWRs guarantee good control and safety behaviour of these reactor types.

For MOX PWRs containing also neptunium and americium a careful analysis for these reactivity and temperature safety coefficients would be needed.

9.8.4.1 The Seed and Blanket PWR Using Plutonium and Thorium

This seed and blanket design concept [76] for PWRs uses the same fuel assemblies as a PWR. These fuel elements can be either quadratic as in Western PWRs or hexagonal as in Russian PWRs. The assembly consists of an inner seed assembly containing plutonium mixed oxide fuel, whereas the surrounding blanket assembly contains thorium dioxide as fertile fuel. During operation the plutonium in the inner seed

Table 9.6 Boron worth (BW), fuel Doppler temperature coefficient (FDC) and moderator temperature coefficient (MTC) for standard UOX PWRs and MOX PWRs (multi-recycling)

	UOX PWR		MOX PWR 8% tot. Pu		MOX PWR 12% tot. Pu	
	BOC	EOC	BOC	EOC	BOC	EOC
BW per ppm boron	-6.5×10^{-5}		-3×10^{-5}		-2×10^{-5}	
FDC per °C	-2.5×10^{-5}		-3×10^{-5}		-3×10^{-5}	
MTC per °C	-10^{-4}	-5.5×10^{-4}	-3.5×10^{-4}	-6×10^{-4}	-3×10^{-4}	-5.5×10^{-5}

BOC Begin of cycle, *EOC* End of cycle

Table 9.7 Incineration rates of different plutonium burner reactors

Nuclear reactor type	MOX-PWR	FR-burner (CAPRA)	ADS
Incineration rate kg/GW(e)·y	420	570	700

elements is incinerated, whereas in the outer blanket elements U-233 is generated by neutron capture in Th-232.

Incineration rates of up to 800 kg of plutonium per GW(e)·y were claimed [76].

9.8.5 Recycling of Plutonium and Minor Actinides in ADSs

Destruction rates of plutonium and TRU (neptunium, americium, curium) in a Pb/Bi (LBE) cooled 320 GW(e) ADS and in a sodium cooled 336 GW(e) ADS loaded with metallic fuel containing uranium, plutonium, neptunium, americium and curium from pyrochemistry (Sect. 9.6) are reported in [74, 77, 78]. The destruction rates are 670 kg/GW(e)·y plutonium and 74 kg minor actinides (neptunium, americium, curium) in the LBE case and 796 kg/GW(e)·y TRU (plutonium, neptunium, americium, curium) in the sodium cooled ADS.

9.8.6 Plutonium Incineration by Multi-Recycling in MOX-PWRs, FR-Burners and ADSs

Section 8.1.2.4 described already a strategy with MOX-PWRs incinerating the plutonium generated by a certain cluster of UOX-PWRs. If instead of this MOX-PWR strategy FRs or accelerator driven systems (ADSs) are loaded with the plutonium of UOX PWRs similar results are obtained. However, the plutonium incineration rates of FR-burners and ADS are higher than those of MOX-PWRs. Table 9.7 shows the higher incineration rates of FR burners, e.g. so-called CAPRA-FRs and the plutonium incineration rates of ADSs compared to those of MOX PWRs as assumed in [79, 80].

Assuming these plutonium incineration rates, similar analyses as described in Sect. 8.1.2.4 can be performed. This leads to Fig. 9.9 which shows the plutonium inventories in the fuel cycle of a cluster of M = 8 UOX-PWRs operating in symbiosis with either MOX-PWRs or FR burners (CAPRA type) or ADSs. The inventories are normalized in tonnes per GW(e). The straight line represents the once through fuel cycle with direct spent fuel storage in a deep geological repository. The plutonium is accumulating as a function of time following an almost straight line in Fig. 9.9.

The plutonium recycle strategies for MOX-PWR FR-burner and ADS in the assumed scenario are represented by piecewise straight lines where each new line

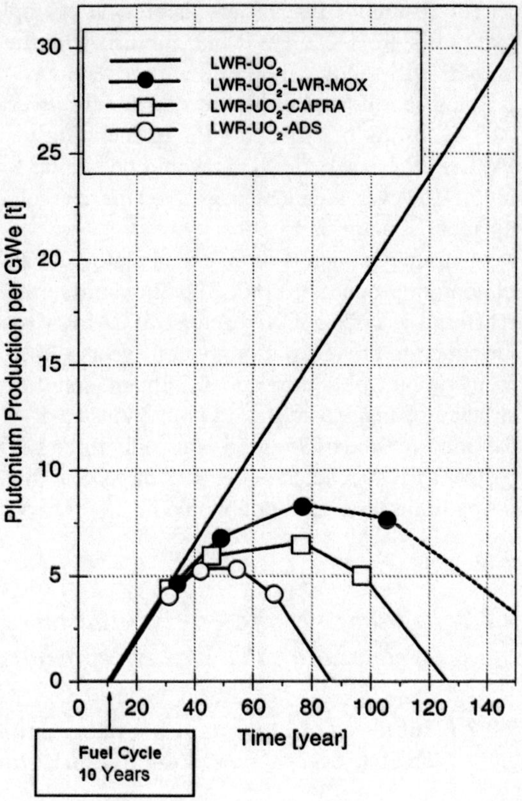

Fig. 9.9 Balance of plutonium (normalized to 1 GW(e)) from a cluster of M = 8 UOX-PWRs with 10 GW(e) total power operating in either the once through direct spent fuel disposal mode (*straight line*). The polygon type lines show the plutonium inventory (normalized to 1 GW(e)) for the cases of UOX-PWRs operating in symbiosis with either MOX-PWRs or FR-burners (CAPRA-type) or ADSs [79, 80]

represents the introduction of an additional MOX-PWR or FR-burner (CAPRA-type) or ADS. The difference in tonnes of plutonium between the full line (once through cycle) and the lines for the UOX PWR strategy with MOX-PWRs or FR-burners (CAPRA) or ADS represents the plutonium inventory which is incinerated by these recycling burner reactors. As to be expected the FR-burners and even more the ADSs incinerate the plutonium produced by the UOX-PWRs more efficiently and faster than MOX-PWRs [79, 80].

For the MOX-PWR some lines are shown as dotted lines from about 75 years on, because the coolant/moderator temperature coefficient, as an important safety related reactivity coefficient, could become intolerable (Sect. 8.1.2.3) for a fuel to moderator volume ratio of 2.0. For a fuel to moderator volume ratio of 2.5 the safety related reactivity coefficient would be acceptable (Sect. 9.8.4). Both the FR-burner and ADS strategies show a potential for incineration of all plutonium produced by the UOX-PWRs in a time frame of about 125 years (FR burners) or about 85 years (ADSs).

The results of Fig. 9.9 are theoretical examples following a certain strategy. In reality first the UOX-PWR in symbiosis with the MOX-PWR are started. Later FR-burners will follow when FRs will be deployed on large scale. This strategy might perhaps be followed by the introduction of several ADSs.

The results of strategies for incinerating neptunium and americium in MOX-PWRs, FR-burners or ADSs would be similar to those shown in Fig. 9.9 for plutonium. However, the masses involved for neptunium and americium would be smaller by about a factor of 15–20.

A strategy which is often called plutonium stabilization (no more increase of net plutonium production) [60, 81, 82] would terminate the replacement of LWR-UO_2 reactors by LWR-MOX reactors or CAPRA reactors after about 75 years and the introduction of ADSs after about 55 years (Fig. 9.9). The plutonium production and incineration would then remain contant (stabilization) for the following time. As the strategy of Fig. 9.9 is started with a cluster of M = 8 LWR-UO_2 reactors and up to the point in time of 75 or 55 years only three LWR-UO_2 reactors would be replaced by LWR MOX reactors, CAPRAs or ADSs. The fraction of Pu incinerating reactors for plutonium stabilization would be 3/8 or about 37%.

9.8.7 Influence of the Transmutation of Actinides on the Fuel Cycle and on the Waste Repository

9.8.7.1 Influence of Plutonium Recycling and Minor Actinide Recycling on Reprocessing and Fuel Refabrication

It was shown already in Sect. 7.4 for plutonium recycling in FBRs that the amounts of plutonium to be handled in the fuel cycle increases in comparison to the once-through case. Similarly, the amount of plutonium and of the minor actinides (Np, Am) increases in the fuel cycle for strategies with transmutation and incineration of these actinides. Plutonium and the minor actinides are collected after reprocessing of spent fuel in order to fabricate MOX fuel elements or other fuel subassemblies containing, e.g. oxides of neptunium and americium. This leads to a concentration of these actinides in fuel fabrication and reprocessing facilities applying the different chemical separation processes for actinide transmutation and incineration described in Sects. 9.5–9.7 [72, 81, 83, 84]. These amounts of actinides, their decay heat and their radiotoxicities are higher than those of uranium and plutonium in the once-through case.

The different nuclear characteristics for the calculation of the radioactivity (Ci/g), heat production (W/g) and radiotoxicity (using dose coefficients (Sv/Bq)) is given in Table 9.8 [72]. On the basis of these data the radioactivity, the heat production or the radiotoxicity of any composition of fresh (refabrication) or spent fuel (reprocessing) can be determined. High contents of Pu 238, Pu-241, Am-241, Am-242 or Am-242m lead to high specific activities of MOX and americium containing fuel (Table 9.8).

High contents of Pu-238 and americium (as well as Cm-244) lead to relatively high internal heat production in the fuel. High contents of Pu-238 and Pu-242 (as well as curium isotopes) are responsible for relatively high neutron radiation.

As already explained in Sect. 9.8.1 curium recycling is only feasible with pyrochemistry and related fuel refabrication methods in fast reactors.

Spent LWR MOX fuel can have a 6–7 times higher α-radioactivity than spent UO_2 fuel [60, 81, 82]. Detailed investigations for heterogeneous recycling of plutonium and of the minor actinides, neptunium and americium with the French CORAIL fuel subassembly were reported by [72, 83]. According to Table 9.8 the specific activities are mainly determined by Pu-238, Pu-241 and the americium isotopes Am-241 and Am-242m (Am-242 has a half-life of 16 h) as long as MOX fuel refabrication without curium is considered. The decay heat is determined by Pu-238 and the americium isotopes. The neutron emission is dominated by the plutonium isotopes Pu-238, Pu-240 and Pu-242 as long as MOX fuel mixed with neptunium and americium is considered.

Spent MOX and minor actinide fuel have only somewhat higher activity since the activity of the fission products is dominating that of the plutonium and minor actinide isotopes for the cooling periods of 5–10 years as long as curium is not recycled.

Present aqueous reprocessing technology is considered to be applicable up to Pu-238 contents of about 5% [83]. Present MOX fuel refabrication on the basis of glove box technology can be applied up to Pu-238 contents of about 4%.

As shown in the previous Sect. 9.8.4 recycling of plutonium, neptunium and americium can lead to Pu-238 contents in the plutonium of 8% and more. This means that not only present reprocessing technologies based on the PUREX process must be modified, but also new reprocessing facilities based on the chemical partitioning processes described in Sect. 9.5 must be developed. Also new refabrication facilities based on refabrication processes described in Sect. 9.7 must be deployed [72].

9.8.7.2 Influence of Transmutation and Incineration of Actinides on the Radioactivity, the Radiotoxicity, and on the Heat Load of Waste in a Deep Geological Repository

The different strategies for transmutation and incineration of plutonium and minor actinides have different radioactivity loads in Bq/TW(e)·h and different radiotoxicity levels in Sv/TW(e)·h (Figs. 9.10, 9.11) [72, 83]. The highest radioactivity levels and the highest radiotoxicity levels are shown for A1, the Once through strategy (direct disposal of spent PWR fuel) and for A2, the Once (mono) recycling strategy for Pu followed by direct disposal of the Pu-MOX spent fuel elements.

The next lower curve shows the multi-recycling strategy of Pu in fast reactors (FRs) and the lower curves for both radioactivity and radiotoxicity levels are represented by the Integral Fast Reactor (IFR) strategy with recycling of Pu and of the minor actinides (Np, Am, Cm) and the UOX-MOX-PWR with ADS strategy recycling also Pu and the minor actinides.

Table 9.8 Properties of transuranium nuclides [72]

Nuclide	Half-life	Energies of primary emissions (MeV)		Specific activity			Dose coefficients (Sv/Bq)
		α	β	(Ci/g)	(W/g)	(Neutrons s^{-1} g^{-1})	
Np-237	2.14×10^6 years	4.78		7.07×10^{-4}	2.07×10^{-5}	$<1.2 \times 0^{-4}$	1.1×10^{-7}
Np-238	2.10 days		0.25	2.61×10^5	1.27×10^3		
			1.24				
Np-239	2.359 days		0.332	2.32×10^5	5.86×10^2		8.0×10^{10}
			0.427				
Pu-238	87.404 years	5.49		17.2	0.570	2,583	2.3×10^{-7}
Pu-239	2.4413×10^4 years	5.15		6.13×10^{-2}	1.913×10^{-3}	0.0225	2.5×10^{-7}
Pu-240	6.580 years	5.16		0.227	7.097×10^{-3}	895	2.5×10^{-7}
Pu-241	14.98 years	4.9	0.02	99.1	4.06×10^{-3}		4.7×10^{-7}
Pu-242	3.869×10^5 years	4.90		3.82×10^{-3}	1.13×10^{-4}	1,588	2.4×10^{-7}
Am-241	432.7 years	5.48		3.43	0.1145	0.59	2.0×10^{-7}
Am-242	16.01 h		0.63–0.67	8.11×10^5	2.08×10^3		
Am-242m	144 years	5.207	I.T.	10.3	3.08×10^{-2}		1.9×10^{-7}
Am-243	7,370 years	5.27		0.200	6.42×10^{-3}		2.0×10^{-7}
Cm-242	162.7 days	6.11		3.32×10^3	122	2.02×10^7	1.3×10^{-8}
Cm-243	32 years	5.79		45.9	1.677		2.0×10^{-7}
Cm-244	18.099 years	5.81		80.94	2.832	1.14×10^7	1.6×10^{-7}
Cm-245	8,265 years	5.36		0.177	5.89×10^{-3}		3.0×10^{-7}
Cm-246	4,655 years	5.39		0.312	1.01×10^{-2}	9.30×10^6	2.9×10^{-7}
Cm-247	1.56×10^7 years	4.87		9.28×10^{-5}	2.94×10^{-6}		2.7×10^{-7}
Cm-248	3.397×10^5 years	5.05		4.24×10^{-3}	5.34×10^{-4}	4.30×10^7	1.1×10^{-6}
Cm-249	64 m		0.9	1.18×10^5	2.06×10^4		
Cm-250	1.74×10^4 years			8.20×10^{-2}	~0.1	1.08×10^{10}	2.9×10^{-7}

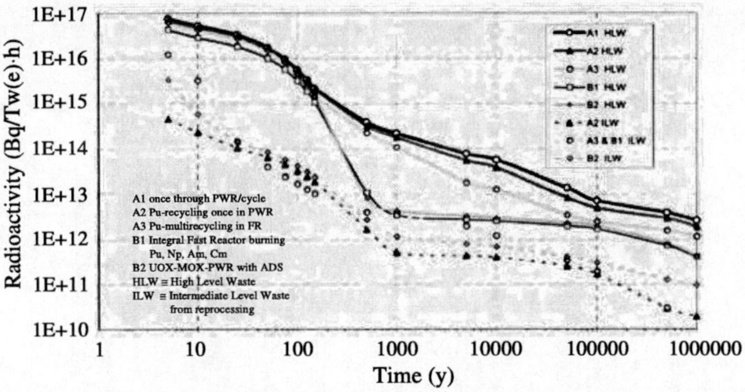

Fig. 9.10 Radioactivity (Bq/TW(e)·h) as a function of time for different fuel cycle strategies (1 Ci $\triangleq$ 3.7×10^{10} Bq) [72, 83–85]

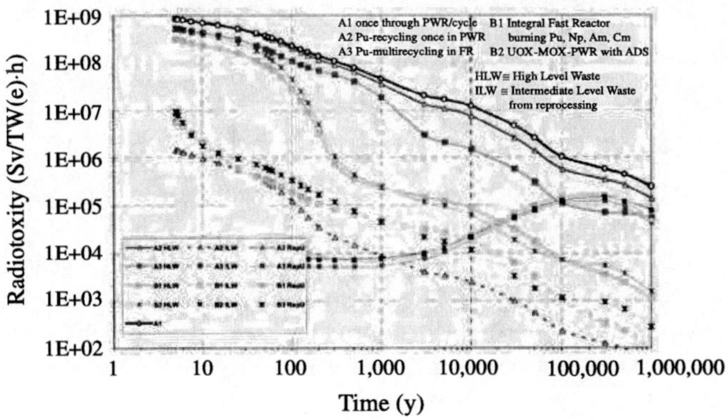

Fig. 9.11 Radiotoxicity as a function of time for different fuel cycle strategies (HLW and ILW) [72, 83–85]

As during reprocessing also intermediate level waste (ILW) is produced (Sect. 7.5) these radioactivity levels are also indicated in Fig. 9.10. In Fig. 9.11 the radiotoxicity levels for High Level Waste (HLW), Intermediate level waste (ILW) and the remaining Reprocessed Uranium (RepU) are given for the above transmutation and incinceration strategies [72, 83, 84].

The different strategies for transmutation and incineration of plutonium and the minor actinides affect strongly the heat load of HLW packages, i.e. the contribution of the minor actinides (the heat load produced by the fission products is hardly influenced). Figure 9.12 shows the expected heat loads produced by the minor actinides in the deep geological repository for different transmutation and incineration strategies (the heat loads are given in TW(e)·h [74, 83, 86, 87]. These units must be multiplied by a factor of 7.45 TW(e)·h/GW(e)y (load factor 0.85) to obtain the real heat loads per GW(e)·y) (7450 h per year if load factor is 0.85).

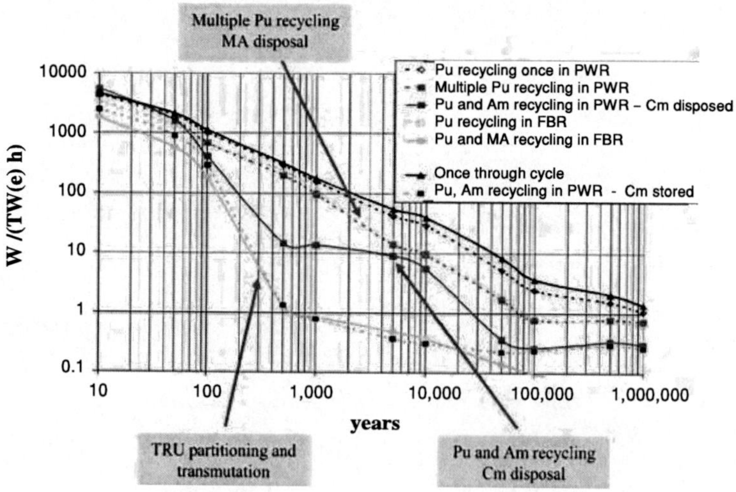

Fig. 9.12 Heat load produced by minor actinides in HLW for different recycling and incineration strategies [86]

The highest heat load is produced in HLW in case of the Once through cycle. Pu recycling once in PWRs provides only little improvement. Multiple Pu recycling in PWRs or FRs results in some improvement after about 1,000–10,000 years. However, Pu and Am recycling (with Cm disposed) results in a considerable improvement from about 100 years on. The highest improvements are obtained for Pu and MA (neptunium, americium and uranium) recycled in FRs or Pu and Am recycled in PWRs and Cm stored.

This improvement is about a factor of 100. It is, however, not only the radiotoxicity and the heat loads which are decreased by a factor of about 100 but also the masses of actinides which are decreased in the HLW packages by a factor of about 100. To achieve this goal the losses of actinides during reprocessing and refabrication must be not higher than about 0.2% [72].

These advantages achieved in the back end of the fuel cycles are, however, in contrast to higher masses of plutonium and actinides with higher radioactivity, radiotoxicity and decay heat to be handled during reprocessing and refabrication as mentioned already in Sect. 9.8.7.1.

9.8.7.3 Contributions of Cs-137 and Sr-90 to the Heat Load in a Deep Geological Repository

Cs-137 (half-life 30 years) and Sr-90 (half-life 29 years) are, besides americium, the main contributors to the heat load in the high level waste package during the first one hundred years. They determine the spacing between the high level waste packages

Table 9.9 Long-lived fission products with half-life (years), production and radiotoxicity [90]

Isotope	Half-life (years)	Production[a] ($kg/GW_{(th)}\cdot y$)	Radiotoxicity[b] (Sv/g)
^{79}Se	6.5E+4	0.066	8.259
^{90}Sr	29	6.07	1.269E+5
^{93}Zr	1.5E+6	8.04	1.045
^{94}Nb	2.0E+4	8.1E-6	1.410E+1
^{99}Tc	2.1E+5	8.54	6.056E-1
^{107}Pd	6.5E+6	2.34	1.048E-3
^{126}Sn	1.0E+5	0.30	6.306
^{129}I	1.6E+7	1.96	2.696E-1
^{135}Cs	2.3E+6	2.76	8.532E-2
^{137}Cs	30	10.65	4.190E+4
^{151}Sm	89	0.15	1.281E+2

[a] 3.2% 235U enrichment, 33 GWd/t, 20 years cooling time
[b] ICRP data (1991)

and thereby the amount of HLW to be stored in a deep geological repository [86, 87]. It is therefore considered to separate Cs-137 and Sr-90 chemically from the liquid HLW and store Cs-137 and Sr-90 for about 100–200 years separately until they will have mostly decayed to stable Ba-137 and Zr-90 [72, 88, 89].

9.8.8 Transmutation of Long-Lived Fission Products

A number of radiologically important long-lived fission products have to be taken into account in the safety assessment of a deep geological repository. The following long-lived fission products have to be assessed: Tc-99, I-129, Sr-90, Cs-135, Cs-137, Se-79, Zr-93, Nb-94, Sn-126, Sm-151 and the activation products C-14 and Cl-36.

They are listed in Table 9.9 showing their half-lives, production during reactor operation in kg/GW(th)·y and radiotoxicity in Sv/g.

The objective for transmutation of long-lived fission products is to reduce their radiotoxicity significantly, before they have to be conditioned and sent as waste to a geological repository.

The transmutation rate can be characterized by the transmutation half-life which is a measure for the time needed to destroy by neutron capture half of the initial mass. It is defined as [90]

$$T_{1/2}^{Tr} = 3.171 \times 10^{-8} \cdot \frac{\ln 2}{\sigma_c \cdot \phi}$$

σ_c microscopic neutron capture cross section in barn (1 barn = 10^{-24} cm^2);
ϕ neutron flux in $n/cm^2\cdot s$

This transmutation half-life $T_{1/2}^{Tr}$ should be considerably smaller than the natural decay half-life $T_{1/2}$.

Table 9.10 Transmutability of long-lived fission products in fast and thermal neutron fields [90]

Isotope	Capture cross section[a]		Half-life (years)	Transmutation half-life[b] (years)		Pure isotope transmutability
	Fast neutron	Thermal neutron		Fast neutron	Thermal neutron	
^{79}Se	0.002	0.33	6.5E+4	1.1E+4	666	Nontransmutable
^{90}Sr	0.01	0.08	29	2.2E+3	2.7E+3	Nontransmutable
^{93}Zr	0.09	1.03	1.5E+6	244	213	Questionable
^{94}Nb	0.22	4.22	2.0E+4	100	52	Transmutable
^{99}Tc	0.45	9.32	2.1E+5	49	24	Transmutable
^{107}Pd	0.53	2.79	6.5E+6	42	79	Transmutable
^{126}Sn	0.007	0.03	1.0E+5	3.1E+3	7.3E+3	Nontransmutable
^{129}I	0.35	3.12	1.6E+7	63	70	Transmutable
^{135}Cs	0.07	2.48	2.3E+6	314	89	Transmutable
^{137}Cs	0.01	0.03	30	2.2E+3	7.3E+3	Nontransmutable
^{151}Sm	2.09	660	89	11	0.33	Transmutable

[a]ORIGEN2 library (fast: oxide fuel liquid-metal fast breeder reactor; thermal: standard PWR)
[b]Thermal flux = 1.0×10^{14}, fast flux = 1.0×10^{15} (n/cm^2·s)

Table 9.10 presents an assessment whether or not these long-lived fission products can be transmuted in a PWR (thermal neutron spectrum), FR or ADS (fast neutron spectrum). The main parameters for this assessment are the capture cross sections of the long-lived fission products in the neutron flux of 10^{14} n/cm^2·s of a PWR core or 10^{15} n/cm^2·s of an FR or ADS core [90].

Some of these long-lived fission products, e.g. cesium exist as different isotopes like Cs-133, Cs-134, Cs-135, Cs-137 in spent fuel and would require isotope separation before transmutation.

Without application of isotope separation and considering that the production rate of Nb-94, Sn-126 and Sm-151 is small and the radiotoxicity of Pd-107 is very low, the practically remaining candidates for long-lived fission product transmutation are I-129 and Tc-99. The other fission products Se-79, Sr-90, Zr-93, Sn-126 and Cs-137 are considered nontransmutable with sufficient efficiency [90]. In addition the impact on long term radioactive dose release levels from a geological repository (Chap. 7) of the isotopes Nb-94, Pd-107 and Sm-151 are considered to be low [90].

Because of their high geochemical mobility I-129 and Tc-99 are major contributors to the biosphere release dose rates of a deep geological repository. I-129 and Tc-99 are well soluble in groundwater and are hardly adsorbed by the geological rocks (Sect. 7.6.3.2).

The extraction of Tc-99 from the HLW as TcO_4 by an advanced PUREX process is relatively easy [81]. I-129 can be extracted during reprocessing by silver impregnated filters [81, 91]. Several target materials for Tc-99 and I-129 have been studied. I-129 can be used in form of NaI, CaI_2or as silver-iodide impregnated in silica [81, 91]. Tc-99 is mostly used in metallic form [81].

In special moderated subassemblies loaded at the periphery of the core, Tc-99 is transmuted by neutron capture to stable Ru-100, whereas I-129 is transmuted to Xe-130.

The transmutation rate for Tc-99 in special subassemblies of a PWR core is too low, whereas I-129 could be transmuted such that the I-129 produced in three PWRs can be destroyed.

In FRs with their fast neutron spectrum the transmutation rates for I-129 and Tc-99 can be much higher [70, 73, 90, 92–94] than in LWRs. Special irradiation assemblies containing moderator rods (ZrH; written in short; in fact it is usually ZrH_x with $x \approx 1.7$) and rods with either BaI_2 or metallic Tc can be located at the radial boundary of the FR core. These assemblies would consist of either 37 or 127 special rods [92, 93].

9.8.8.1 I-129 Transmutation in FRs

Barium iodine, BaI_2, was chosen for its good chemical stability and manufacturing characteristics. In Table 9.11 the transmutation rates in % per year for I-129 are shown for a heterogeneous (moderator rods (ZrH) and BaI_2 rods are separated) and a homogeneous case (moderator ZrH and BaI_2 are mixed in rods) with 27 or 127 rods in the assemblies. In addition to full BaI_2 pellets in the rods also a case with hollow

Table 9.11 Transmutation rate in % per year and support ratios for I-129 and different irradiation assembly cases [92–94]

I-129 assembly type	Transmutation rate in % per year	Support ratio
Homogeneous case, 127 rods 10% BaI_2	6.7	2.6
Heterogeneous case, 27 rods	3.6	3.5
homogeneous hollow pellet, 127 rods BaI_2 10%	8.7	1.7

Table 9.12 Transmutation rate in % per year for Tc-99 and different irradiation assembly cases [92–94]

Tc-99 assembly type	Transmutation rate in % per year	Support ratio
Homogeneous case 127 pins, 10% Tc-99	6.1	3.2
Tc-99 needle case 121 pins, 0.5 mm diameter needles	5.5	2.3
Tc-99 needle case 121 pins, 0.3 mm diameter needles	7.9	1.2

pellets (smear density over the pellet of 50%) was investigated. The heterogeneous case contains 27 BaI_2 rods and 10 separate ZrH moderator rods interspersed in the assemblies. For the homogeneous case the 127 rods contain the moderator ZrH mixed with BaI_2.

The support ratio is defined as the ratio of I-129 transmuted in the special I-129 assemblies to the quantity of I-129 produced in the driver fuel of the FR [93].

The highest transmutation rates per year are attained for the homogeneous hollow pellet arrangement with 127 pins and 10% BaI_2. For another case with 67 BaI_2 and ZrH pins and 60 stainless steel even 9.5% as transmutation rate per year are reported [93]. However, these high transmutation rates belong to small initial inventories in the irradiation assemblies. Therefore the support ratios are relatively small. Small transmutation rates in % per year require several times recycling of the initial inventory. This will increase the chemical reprocessing losses to the HAW.

9.8.8.2 Tc-99 Transmutation in FRs

Technetium is difficult to mix with Zr metal. Therefore, a new irradiation pellet design was developed [93]. Between 55–124 thin Zr metal needles of about 0.3–1 mm diameter stuck in holes of an 18 mm thick ZrH pellet. This design with very thin Tc needles is chosen in order to avoid self shielding problems of the neutron flux in the absorbing material Tc-99. Table 9.11 shows the results of calculations for a homogeneous case as described above and for the thin Zr-needle cases (Table 9.12).

Again, the same observations hold for the high transmutation rates per year connected with the small support ratios (as discussed above for the destruction of I-129).

The irradiation time at the radial boundary of the FR core is limited by the radiation damage of the structural materials, e.g. austenitic steel. Dependent upon the

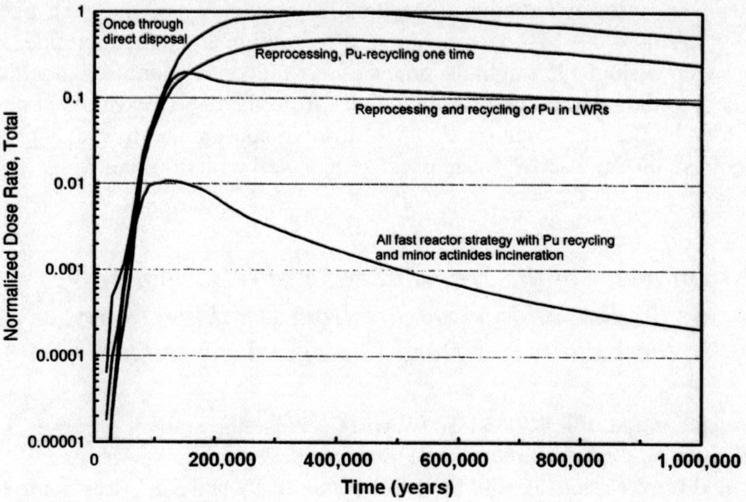

Fig. 9.13 Human exposure due to radioactivity from a well near a deep geological repository for different fuel cycle options

transmutation rate achieved and on the possible irradiation time for transmutation, several recycle steps will be required. This increases the losses of I-129 and Tc-99 going to the HAW and to the deep geological repository.

9.8.9 Comparison of Possible Radiation Exposure Rates from Drinking Water in the Vicinity of a Deep Geological Repository for Different Incineration Schemes

Similar analysis for the potential radiation exposure in the vicinity of a deep geological repository—as were reported for direct spent fuel or vitrified HAW disposal in Sect. 7.7.3.3 and Figs. 7.32 and 7.33—were also performed for different plutonium and TRU recycle and incineration schemes [86]. Figure 9.13 shows the normalized radiation dose rates for four different fuel cycle scenarios

- once through fuel cycle or direct spent fuel disposal
- reprocessing of spent fuel, but recycling plutonium only once
- reprocessing and recycling plutonium in LWRs
- all fast reactor strategy with reprocessing and MOX fuel and plutonium minor actinides recycling.

All results are normalized to the peak radiation dose rate of the once through fuel cycle case with direct spent fuel disposal of Fig. 7.32. The cases Pu-recycling in LWRs and the case all fast reactors with Pu and minor actinide recycling lead to considerably lower radiation exposures in the vicinity of the deep geological repository. This is due to the incineration of plutonium and minor actinides.

However, cesium 135 might become a dominant contributor after about one million years to the release rate of radioactivity from the deep repository if cesium is not sorbed by the surrounding geological formations (granite, tuff, clay or salt) on its way from the high active waste package to a well with drinking water [84, 85].

9.8.10 Influence of the Transmutation of I-129 and Tc-99 on the Radiation Exposure from Drinking Water in the Vicinity of a Deep Geological Repository

The transmutation of I-129 and Tc-99 will decrease the amount of these LLFPs in the HLW to be deposited into the deep geological repository. Analyses [90] were performed for the removal of 95% of I-129 and Tc-99 and 5% losses going into the HLW. In this case the total radiation exposure by drinking water from a well 20 km away from the repository can be reduced by a factor of about 3 [90] during the first 3×10^5 years.

The above results show that only two long-lived fission products (Tc-99, I-129) could be destroyed by transmutation with reasonable efficiency. Other long-lived fission products are either nontransmutable or would need additional isotope separation. The improvement obtained by transmutation and destruction of Tc-99 and I-129 with regard to the estimated radiation exposure between 10^3 and 10^6 years caused by the HLW in a deep geological repository after water ingress (Sect. 7.6.3.3) would be limited.

References

1. Kim JI et al (2001) Forschung zur Langzeitsicherheit der Endlagerung hochaktiver Abfälle. In: Radioaktivität und Kernenergie, Forschungszentrum Karlsruhe, Karlsruhe
2. Kim JI (1993) The chemical behavior of transuranium elements and barrier functions in natural aquifer systems. In: Scientific basis for nuclear waste management XVL. Interrante CG, Pabalan RT (eds), Mat. Res. Soc. Symp. Proc. 294, Pittsburg
3. Geckeis H et al (1998) Formation and stability of colloids under simulated near field conditions. Radiochimica Acta 82:123–128 (42270)
4. Kim JI et al (1999) Geochemical assessment of actinide isolation in a German salt repository environment. Eng Geol 52:221–230
5. Pigford TH (1991) Effect of actinide burning on risk from high level waste. Trans Am Nucl Soc 63:80–83
6. Sailor WC et al (1994) Comparison of accelerator-based with reactor-based nuclear waste transmutation schemes. Prog Nucl Energy 28(4):359–390
7. Fukuda K et al (2006) Prospects of inventories of uranium, plutonium and minor actinides and mass balance. Presented at 2nd consultancy meeting on Protected Plutonium Production Project, IAEA, Vienna

8. Gompper K (2001) Zur Abtrennung langlebiger Nuklide. In: Radioaktivität und Kernenergie, Forschungszentrum Karlsruhe, Karlsruhe
9. Clefs (1996) Le future du retraitement, une synergie des procédés améliorés et d'approches nouvelles, CEA-France, No. 33, p 39
10. Emin JL et al (2009) AREVA NC experience of industrial scale MOX treatment in UP2-800. In: Proceedings of GLOBAL 2009, Paris
11. Schmieder H et al (1989) IMPUREX: a concept of improved PUREX process. Radiochimica Acta 48:181–192
12. Warin D (2010) Minor actinide partitioning. 1st ACSEPT international workshop, Lisbon
13. Morita Y et al (1995) Diisodecylphosphoric acid, DIDPA, as an extractant for transuranium elements. In: International conference on evaluation of emerging nuclear fuel cycle systems (GLOBAL 95), vol 2. Versailles, France, p 1163
14. Kubota M (1999) Development of the four group partitioning process at JAERI. In: Proceedings of the 5th international information exchange meeting, Mol, Belgium, 1998, OECD, p 131
15. Schulz WW (1988) The TRUEX process and the management of liquid TRU-waste. Sep Sci Technol 23:1355–1372
16. Zhu Y (1989) The removal of actinides from high level radioactive waste by TRPO extraction. The extraction of Americium and some lanthanides. Chin J Nucl Sci Eng 9:141–150
17. Madic C et al (1998) High level liquid waste partitioning by means of completely incinerable extractants. Report EUR-18038
18. Poinssot Ch et al (2009) The CEA atalante facility: a laboratory for fuel cycle R&D. In: Proceedings of GLOBAL 2009, Paris
19. Zhu Y (1995) The separation of americium from light lanthanides by CYANEX 301 extraction. Radiochimica Acta 68:95–98
20. Hill C et al (1997) Americium(III)/trivalent lanthanides separation using organothiophosphinic acids. In: Proceedings international conference on future nuclear systems (Global 97), vol 2. Yokohama (Japan), p 1490
21. Modolo G et al (1999) Synergistic selective extraction of actinides(III) over lanthanides from nitric acid using new aromatic diorganyldithiophosphinic acids and neutral organophosphorus compounds. Solvent Extr Ion Exch 17(1):33–53
22. Vitorge P (1984) Lanthanides and trivalent actinides complexation by tripyridyl triazine. Applications to liquid-liquid-extraction, CEA-R-5270
23. Kolarik Z (1999) Selective extraction of Am(III) over Eu(III) nitrates by 2,6-ditriazolyl and 2,6-ditriazinyl piridines. Solvent Extr Ion Exch 17:23–32
24. Welden A et al (2010) 1-cycle SANEX process development studies performed at Forschungszentrum Jülich. 1st ACSEPT international workshop, Lisbon
25. Malmbeck R et al (2009) Recovery of MA using a CyMe-4-BTBP based SANEX solvent. In: Proceedings of GLOBAL 2009, Paris
26. Modolo G et al (2009) Selective separation of americium III from curium III, californium III and lanthanides III by the LUCA process. In: Proceedings of GLOBAL 2009, Paris
27. Hérés X et al (2009) Results of recent counter-current tests on AmIII/LnIII separation using TODGA extractant. In: Proceedings of GLOBAL 2009, Paris
28. Miguirditchian M et al (2009) HA demonstration in the atalante facility of the GANEX 1st cycle for the selective extraction of Uranium from HLW. In: Proceedings of GLOBAL 2009, Paris
29. Miguirditchian M et al (2009) HA demonstration in the atalante facility of the GANEX 2nd cycle for the grouped TRU extraction. In: Proceedings of GLOBAL 2009, Paris
30. Carrot M et al (2009) Extraction and stripping of actinides in TBP and TODGA/TBP systems for advanced fuel processing. In: Proceedings of GLOBAL 2009, Paris
31. Bourg St et al (2009) Toward the future demonstration of advanced fuel treatments. In: Proceedings of GLOBAL 2009, Paris
32. Warin D et al (2009) Recent progress in advanced actinide recycling processes. In: International conference on fast reactors and related fuel cycle, Kyoto

33. Chang YI (1989) The integral fast reactor. Nucl Technol 88:129–138
34. Stevenson CE (1987) The EBR-II fuel cycle story. American Nuclear Society, LaGrange Park
35. Burris LR et al (1987) The application of electrorefining for recovery and purification of fuel discharged from the integral fast reactor. AIChE symposium. Series No. 83, issue 254, pp 135–142
36. Laidler JJ et al (1997) Development of pyroprocessing technology. Prog Nucl Energy 31(1–2):131–140
37. Todd TA (2009) The US fuel cycle research and development program: separations research and development. 1st ACSEPT International workshop, Lisbon
38. Inoue T et al (1997) Recycling of actinides produced in LWR and FBR fuel cycles by applying pyrometallurgical processes. In: International conference on future nuclear systems (Global 97), vol 1. Yokohama (Japan)
39. Murakami T et al (2009) Recent achievements and remaining challenges on pyrochemical reprocessing in CRIEPI. 1st ACSEPT International workshop, Lisbon
40. Uozumi K et al (2009) Demonstration of pyropartitioning process to recover TRUs from genuine high-level liquid waste. In: Proceedings of GLOBAL 2009, Paris
41. Conocar O et al (2006) Promising pyrochemical actinide lanthanide separation process using aluminum. Nucl Sci Eng 153:253–261
42. Bychkov AV et al (1997) Pyrochemical reprocessing of irradiated FBR MOX fuel III. Experiment on high burn-up-fuel of the BOR-60-reactor. In: International conference on future nuclear systems (Global 97), Yokohama (Japan), vol 2, pp 912–917
43. Bychkov AV et al (2009) RIAR experimental base development concept: multi-purpose pyrochemical complex for experimental justification of innovative closed fuel cycle technologies. In: Proceedings of GLOBAL 2009, Paris
44. Fernandez A et al (2000) Fuels and targets for incineration and transmutation of actinides, the ITU programme. In: ATALANTE 2000 International conference on the nuclear fuel cycle on the back-end of the fuel cycle for the 21st century, Avignon
45. Nästen C et al (2007) Granulation and infiltration process for the fabrication of minor actinide fuels, targets and conditioning matrices. J Nucl Mater 362:350–355
46. Kryukov NF (2009) Results of post irradiation examination of inert matrix fuel compositions irradiated in the BOR-60 reactor up to a burnup of 19% h.a. under the Russian-French experiment BORA-BORA. In: Proceedings of GLOBAL 2009, Paris
47. Boidron M et al (2009) ALFA: atalante laboratory for actinides bearing fuel manufacturing. In: Proceedings of GLOBAL 2009, Paris
48. Pillon S et al (2003) Aspects of fabrication of uranium-based fuels and targets. J Nucl Mater 320:36–43
49. Herbig R et al (1993) Vibrocompacted fuel for the liquid metal reactor BOR-60. J Nucl Mater. 204:93–101
50. Bychkov A V (2009) Vibropac MOX-fuel for fast reactors—experience and prospects. In: International conference on fast reactors and related fuel cycles—challenges and opportunities, Kyoto
51. Pouchon MA (2010) Conversion process: internal gelation and the sphere-pac concept. 1st ACSEPT international workshop, Lisbon
52. Deptula A et al (2010) Synthesis of uranium oxides by complex SOL-GEL processes (CSGP). 1st ACSEPT International Workshop, Lisbon
53. Konings RJM et al (1999) Transmutation of actinides in inert-matrix fuels: fabrication studies and modelling of fuel behaviour. J Nucl Mater 274:84–90
54. Maschek W et al (2010) Design, safety and fuel developments for the EFIT accelerator-driven system with CERCER and CERMET cores. OECD/NEA No. 6420, p 387
55. Klaassen FC et al (2009) Spinel inert matrix fuel testing at the HFR petten. http://www.oecd-nea.org/pt/docs/iem/jeju02/session3/SessionIII-09.pdf
56. Matzke Hj et al (1999) Materials research on inert matrices: a screening study. J Nucl Mater 274:47–53

57. Walters LC et al (1999) Metallic fuels for fast reactors. In: Proceedings workshop advanced reactors with innovative fuels, Villigen, Switzerland (1998) OECD/NEA 1999
58. Kurata M et al (1997) Fabrication of U-Pu-Zr metallic fuel containing minor actinides. In: International conference on future nuclear systems (GLOBAL 97), vol 2. Yokohama, Japan, pp 1384–1389
59. Pillon S et al (2003) Aspects of fabrication of curium-based fuels and targets. J Nucl Mater 320:36–40
60. Baetslé LH et al (1998) Status and assessment report on actinide and fission product partitioning and transmutation. http://www.oecd-nea.org/pt/docs/iem/mol98/session1/Slpaper8.pdf
61. Picard E et al (2000) First in-pile experimental results of high plutonium-content oxide fuel for plutonium burning in fast reactors. Nucl Technol 129:1–12
62. Walker CT et al (1995) Transmutation of neptunium and americium in fast neutron flux: EPMA results and KORIGEN prediction for the superfact fuel. J Nucl Mater 218:129–138
63. Koyama S et al (2008) Chemical analysis of americium samples irradiated under fast neutron spectra. J Nucl Sci Technol (Suppl 6):55
64. Klaassen FC et al (2003) Post irradiation examination of irradiated americium oxide and uranium dioxide in magnesium aluminate spinel. J Nucl Mater 319:108–117
65. Fernandez A et al (1998) Fabrication of targets for the transmutation and incineration of actinides. In: Proceedings of the international conference on advances in science and technology, Florence
66. Ohta H et al (2009) Postirradiation examinations of fast reactor metal fuels containing minor actinides—fission gas release and metallography of ~2.5 at.% burnup fuels. In: Proceedings of GLOBAL 2009, Paris
67. Kasemeyer U et al (2001) The irradiation test of Inert-Matrix fuel in comparison to uranium plutonium mixed oxide fuel at the halden reactor. Prog Nucl Energy 38(3–4):309–312
68. Youinou G et al (2009) A neutronic analysis of TRU recycling in PWRs loaded with MOX-UE fuel (MOX with U-235 enriched U support). Advanced fuel cycle initiative, INE/EXT-09-16091, Idaho National Laboratory
69. Tommasi J et al (1995) Long lived waste transmutation in reactors. Nucl Technol 111:133–148
70. Messaoudi N et al (2002) Fast burner reactor devoted to minor actinde incineration. Nucl Technol 137:84–96
71. Salvatores M (2005) Nuclear fuel cycle strategies including partitioning and transmutation. Nucl Eng Des 235:805–816
72. Salvatores M et al (2011) Radioactive waste partitioning and transmutation within advanced fuel cycles. Prog Part Nucl Phys 66:144–166
73. Status and assessment report on actinide and fission product partitioning and transmutation (1999) OECD/NEA http://www.nea.fr/html/pt/pubdocs.htm
74. Physics and safety of transmutation systems, a status report (2006) OECD/NEA report No. 6090
75. Youinou G et al (2005) Plutonium recycling in standard PWRs loaded with evolutionary fuels. Nucl Sci Eng 151:25–45
76. Radkowsky A et al (1998) The nonproliferative light water thorium reactor: a new approach to light water reactor core technology. Nucl Technol 124:215–222
77. Tuček K (2004) Neutronic and burnup studies of an accelerator-driven transuranium burner in a start-up mode. Doctoral thesis, Royal Institute of Technology, Stockholm
78. Herceg JC et al (2002) Final report on the pre-conceptual design of a reference ADS for transmutation of spent nuclear fuel, ANL-AAA-056
79. Kessler G (2002) Requirements for nuclear energy in the 21st century. Prog Nucl Energy 40(3–4):309–325
80. Kessler G et al (1999) Wohin mit dem deutschen Plutonium? Atomwirtschaft 44:156–164
81. Baetslé LH (2001) Application of partitioning/transmutation of radioactive materials in radioactive waste management. Lectures given at the Workshop on Hybrid Nuclear Systems for Energy Production, Utilisation of Actinides and Transmutation of long-lived radioactive Waste, Trieste

82. Baetslé LH et al (1997) Limitation of actinide recycle and fuel cycle consequences. A global analysis, Part 2: Recycle of actinides in thermal reactors: impact of high burn up LWR-UO_2 fuel irradiation and multiple recycle of LWR-MOX fuel on the radiotoxic inventory. Nucl Eng Des 168:203–210
83. Taiwo TA et al (2005) Assessment of a heterogeneous PWR assembly for plutonium and minor actinide recycle. Nucl Technol 155:34–54
84. Nuclear Science (2011) Potential benefits and impacts of advanced nuclear fuel cycles with actinide partioning and transmutation, OECD/NEA No. 6894
85. RED-Impact (2008) Impact of partitioning, transmutation and waste reduction technologies on the final nuclear waste disposal. Synthesis report. Schriften des Forschungszentrums Jülich, Reihe Energie + Umwelt/Energy + Environment, Band/vol 15
86. Advanced nuclear fuel cycles and radioactive waste management (2006) NEA No. 5990, OECD
87. Wigeland R et al (2006) Separations and transmutation criteria to improve utilization of a geologic repository. Nucl Technol 154:95–106
88. Law JJ et al (2005) Development of technologies for the simultaneous separation of cesium and strontium from spent nuclear fuel as part of an advanced fuel cycle, INL/CON-05-00087
89. Todd TA et al (2005) Advanced technologies for the simultaneous separation of cesium and strontium from spent nuclear fuel, INEEL/CON-04-02500
90. Yang WS et al (2004) Long-lived fission product transmutation studies. Nucl Sci Eng 146:291–318
91. Modolo G et al (1997) Investigations on the partitioning of ^{129}I from silver-impregnated silica in preparation for future transmutation. Nucl Technol 117:80–86
92. Wakabayashi T et al (1998) Study of MA and FP transmutation in fast reactors. Prog Nucl Energy 32(3/4):555–562
93. Yokoyama T et al (2009) New target concepts for increase in transmutation rate of LLFPs in FBR recycle system. In: Proceedings of GLOBAL 2009, Paris
94. Wakabayashi T (2002) Transmutation characteristics of MA and LLFP in a fast reactor. Prog Nucl Energy 40(3–4):457–463

Chapter 10
Radioactive Releases from Nuclear Power Plants and Fuel Cycle Facilities During Normal Operation

Abstract During normal operation of nuclear power plants and facilities of the nuclear fuel cycle small amounts of radioactivity are released into the environment at a monitored and controlled rate. Men may be exposed to external radiation as well as radiation by inhalation and ingestion. Upper limits for the radiation exposure of individuals of the public as well as of employees during their occupational work time have been set by the International Commission on Radiation Protection as well as by state organizations. The nuclear fuel cycle begins with uranium mining and milling where the main effluents are radon and dust particles containing uranium and its decay products. Radioactive effluents are reported for both open pit and underground mining. This is followed by listing the radioactivity release and exposure rate of uranium conversion, enrichment and fuel fabrication facilities. For Pressurized Water Reactors the annual effective dose to the public is well below one micro-Sievert. For Boiling Water reactors the annual effective dose is somewhat higher. However, this is still more than a factor 100 lower than the permissible limit. Release data for radioactive nuclides from the European spent fuel reprocessing and waste treatment centers are collected by the European Commission. The radioactive exposures to the public from these facilities are well below the permissible effective radiation exposures as well. The same result is valid for the plutonium/uranium mixed oxide fuel refabrication plant MELOX in France.

10.1 Radioactive Releases and Exposure Pathways

During normal operation of nuclear power plants and other facilities of the nuclear fuel cycle, radioactivity is released into the environment at a monitored and controlled rate [1–3]. Airborne radioactivity includes the radioisotopes of the noble gases krypton, xenon, radon, of tritium, C-14, and also of fission product and fuel aerosols. Liquid effluents released into rivers, large lakes or the ocean contain tritium, fission products and other radioactive substances. Men may be exposed to ionizing radiation through various exposure pathways [1] (Fig. 10.1):

G. Kessler, *Sustainable and Safe Nuclear Fission Energy*, Power Systems,
DOI: 10.1007/978-3-642-11990-3_10,

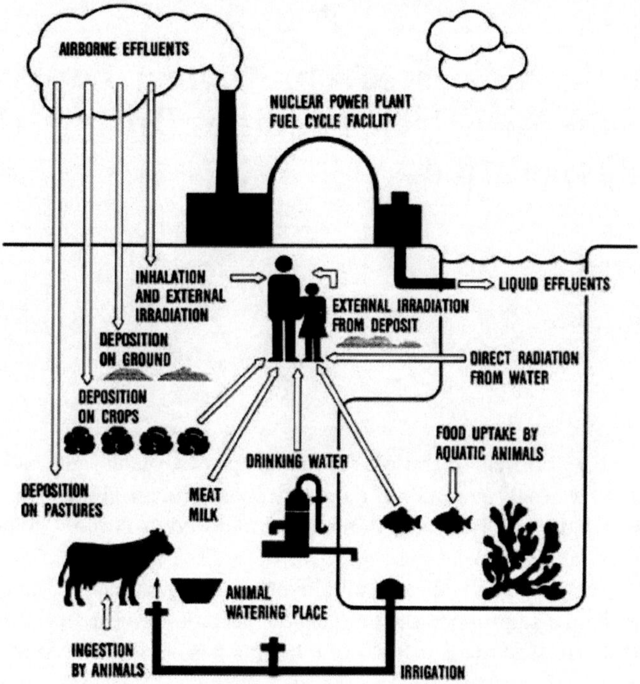

Fig. 10.1 Possible exposure pathways to men from the nuclear fuel cycle [1]

- external β- and γ-radiation of the gaseous radioactive nuclides in the atmosphere (β- and γ-submersion) or by immersion in water (swimming),
- radiation from aerosol particles deposited on the ground (soil radiation),
- internal exposure following inhalation of radioactive nuclides (inhalation),
- internal exposure as a result of the intake of contaminated food or water (ingestion).

The release rate of radioactive nuclides into the environment depends on the retention mechanisms incorporated in the engineered safety design of a fuel cycle facility. Series connection of several containment barriers with low leak rates and other technical measures allow very high retention factors or low release factors to be attained.

Gaseous nuclides or aerosols escaping from the plant or discharged in a controlled way through a stack are diluted in the ambient atmosphere. Dilution is a function of the height of the exhaust stack, the turbulence conditions of the atmosphere, and the distance from the plant [3]. Radionuclides are also deposited on the ground by dry and wet disposal. Aqueous radioactive discharges are diluted as a function of the quantitative relationships between the liquid effluents and the ambient water volume.

The individual radioactive nuclides can enter the human body on various pathways, where they again may have different radiological impacts.

10.1.1 Exposure Pathways of Significant Radionuclides

10.1.1.1 Tritium, Carbon-14 and Krypton

Tritium (half-life 12.4 years) is produced in the reactor core by ternary fission; on the average, about one in 10^4 fissions of U-235 is accompanied by the formation of tritium [1, 2]. About twice as many tritium nuclei are formed in the fission of Pu-239. In addition, tritium is generated in the coolant by neutron capture in deuterium atoms (deuterium has an abundance in natural water of 0.015 atom percent), and by the interaction of neutrons with the boron control material. It is released from nuclear reactors and reprocessing plants as HT gas or as tritiated water (HTO), either into the atmosphere or into, e.g. a river or lake or into the ocean. Gaseous tritium, HT, is very soon oxidized into HTO. Ultimately, any tritium escaping or released in a controlled manner thus will be present as tritiated water. Plants and animals may contain HTO/H_2O ratios close to those existing in the environment. Radioactive exposure of the human body then occurs as a result of the ingestion of food and drinking water. Moreover, tritiated water (HTO) can be absorbed by inhalation and through the skin of the human body. In this way, the β-radiation (maximum energy 18 keV) of tritium causes a whole body exposure.

C-14 (half-life 5,730 years) is built up in the reactor core by (n, p)-reactions with N-14, (n, α)-reactions with O-17, and (n,γ)-reactions with C-13. C-14 emits β-radiation (maximum energy 156 keV). In reprocessing plants, C-14 is oxidized to $^{14}CO_2$ as the fuel is dissolved in nitric acid. In plants and animals, $^{14}CO_2/^{12}CO_2$ ratios may be established which are very close to those in the atmosphere. Radioactive exposure of the human body then occurs mainly as a result of the ingestion of food (milk, vegetables, meat). Direct inhalation and exposure from the ambient atmosphere only play minor roles.

Kr-85 (half-life 10.7 years) is a fission product. Kr-85 emissions from a fuel cycle facility are diluted in the atmosphere. Approximately 99.6% of the Kr-85 nuclei decay by emitting ß-particles with a maximum energy of 0.67 MeV. Only 0.4% of the Kr-85 nuclei decay by emitting a β-particle (maximum energy, 0.16 MeV) and γ-radiation (0.51 MeV). There is no reduction in the airborne concentration as a result of deposition or washout. Kr-85 is only sparingly soluble in water. Its main radiological impact on the human body is due to the exposure to the skin. The inhalation of Kr-85 plays a smaller role.

10.1.1.2 Radioisotopes of Iodine

For releases from nuclear reactors the shortlived radioactive isotopes of iodine, I-131 (half-life 8 days) and I-133 (half-life 20 h), are important fission products [4]. In reprocessing plants I-129 (half-life 1.7×10^7 years) remains an important radioisotope. Radioiodines in the airborne effluents from a nuclear power plant occur partly as elemental iodine and partly as an organic compound (e.g. methyl iodide). In reprocessing plants, radioiodine is mainly released with the effluent air. Airborne iodine is deposited on the surfaces of grass or vegetables. If liquid effluents containing radioiodine are discharged into rivers or lakes etc., a possible major pathway will be their accumulation in fish or plants. The human body can take up radioiodine with the inhalation of air, ingestion of vegetables or fish, and by drinking milk. The radioiodine absorbed by the human body is concentrated mainly in the thyroid. Radioiodine emits both β- and γ-radiations.

Iodine aerosols or organic iodine compounds may, in particular under abnormal operating conditions, deposit on components and concrete walls in the reactor building. These are usually covered by suitable paintings. Nevertheless this may impair decomissionsing of these parts of the reactor plant.

10.1.1.3 Strontium and Cesium

The fission product Sr-90 (half-life 29.1 years) can be discharged into the atmosphere as aerosols from the effluent air of reprocessing plants or with liquid effluents into rivers. Through the food chain (milk, vegetables, fish, meat and drinking water), Sr-90 enters the human body. Like calcium, Sr-90 is deposited preferably in bones, representing a major burden on the blood forming organs because of the long biological residence times of 18 years and the β-radiation of 2.3 MeV maximum energy of the daughter product, Y-90 (half-life 2.7 days).

Releases of radioactive cesium by way of gaseous and liquid effluents from reprocessing plants also cause radiological exposures of the human body through the uptake of food, as in the case of strontium. Cs-134 (half-life 2.1 years) and Cs-137 (half-life 30 years) also emit γ-radiation in addition to β-radiation. Cesium can largely replace potassium in living organisms and, like the latter, is distributed throughout the body in highly soluble compounds.

10.1.1.4 Plutonium Isotopes

Plutonium isotopes may be discharged into the atmosphere with gaseous effluents as aerosols of PuO_2 or $PuNO_3$ or into rivers together with liquid effluents. The following plutonium isotopes are of main interest: Pu-238 (half-life 87.8 years): Pu-239 (half-life 24,100 years); Pu-240 (half-life 6,450 years); Pu-241 (half-life 14.4 years); Pu-242 (half-life 3.9×10^5 years). The highest burden results from inhalation, in which case plutonium is deposited in the lung. Moreover, plutonium may be ingested together with vegetables, milk, meat, fish and drinking water. Plutonium taken up by way of ingestion is deposited preferably in bone tissue.

10.1.1.5 Other Radiobiologically Significant Isotopes

In the vitrification process of high level waste, volatile oxides of the radioisotopes of ruthenium, technetium, selenium, tellurium, antimony and the higher actinides, such as americium (Am-241 (half-life 433 years); Am-243 (half-life 7,380 years) and curium Cm-242 (half-life 163 days) or Cm-244 (half-life 18.1 years)) may be discharged. The higher actinides can be absorbed in the body through similar exposure pathways as plutonium and can generate similar radiological burdens.

10.2 Radiation Dose

The measure of radiation absorbed is the energy dose. Historically the unit was called rad (radiation absorbed dose) [5, 6]. It was defined as a radiation dose that deposits 100 erg of energy per gram of absorbing material:
Absorbed dose

$$1\,\text{rad} \mathrel{\hat{=}} 100\,\text{erg/g} = 10^{-5}\,\text{Joule/g}$$

$$10\,\text{rad} = 1\,\text{Gray (Gy)} = 10^{-3}\,\text{Joule/g} = 1\,\text{Joule/kg}$$

The biological effects of the radiation dose absorbed also depend on the energy and type of radiation (γ-radiation, β-radiation, α-particles, neutrons). To take these differences into account, an equivalent dose was defined. The unit was called, historically, roentgen equivalent man (rem) and is Sievert (Sv) now
Equivalent dose

$$100\,\text{rem} \mathrel{\hat{=}} 1\,\text{Sievert (Sv)}$$

The absorbed dose and the equivalent dose are related as follows:

$$\text{equivalent dose H (Sv)} = \text{absorbed energy dose (Gy)} \times \text{radiation weighting factor, } w_R$$

The radiation weighting factor, w_R, (Table 10.1) is a measure of the relative effects of the nuclear particles in producing damage for a given energy deposition. It is defined by the International Commission on Radiation Protection (ICRP). For most of the γ- and β-radiations (photons or electrons) of fission products the radiation weighting factor can be taken as unity. For neutrons the radiation weighting factor depends on the neutron kinetic energy and varies between 2.5 and 20. (The radiation weighting factors for neutrons vary slightly between ICRP 60 (1991) [6] (step function) and ICRP 103 (2007) [7] (piecewise linear functions). For protons with an energy of >2 MeV the radiation weight factor is 5 in ICRP 60 [6] and 2 in ICRP 103 [7]. For alpha-particles entering the body and accumulating in certain tissues the radiation weighting factor is 20 according to both ICRP Publication 60 (1991) [6] and ICRP 103 (2007) [7].

Table 10.1 Radiation weighting factor, w_R, according to ICRP 60 [6] and ICRP-103 [7]

Type and energy of radiation	ICRP 60[6] radiation weighting factor w_R	ICRP 103[7] radiation weighting factor w_R
Photons (all energies) Electrons, Muons (all energies)	1	1
Neutrons	1	1
	(multiple step function)	Continuous function (see below)
<10 keV	5	
10 keV–100 eV	10	
>100 keV–2 MeV	20	
>2–20 MeV	10	
>20 MeV	5	
Protons >2 MeV	5	2
Alpha particles, fission fragments, heavy nuclei	20	20

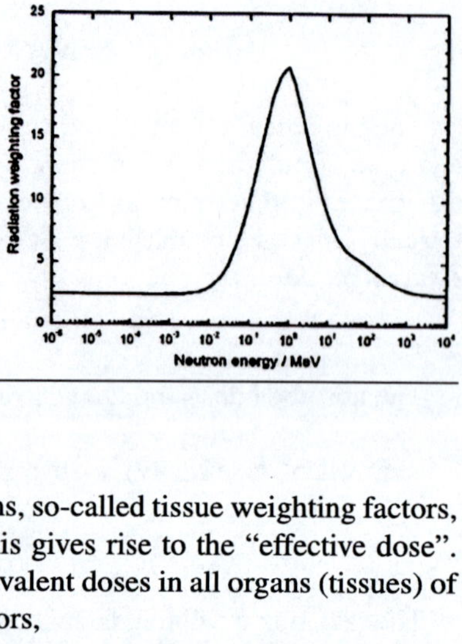

To express the different sensitivities of organs, so-called tissue weighting factors, w_T, have been introduced (see Table 10.2). This gives rise to the "effective dose". The "effective dose" is the sum total of the equivalent doses in all organs (tissues) of the body weighted by the tissue weighting factors,

$$\text{effective dose} = \sum_T w_T \cdot H_T$$

where

$H_T \triangleq$ equivalent dose in the organ (tissue),

w_T = organ weighting factor for tissue, T.

The current German Radiation Protection Ordinance (2008) [8] which will be mentioned later in Sect. 10.6 is based on ICRP 60 [6].

Table 10.2 Tissue weighting factors w_T for determination of the effective dose according to ICRP 60 (1991) [6] and ICRP 103 (2007) [7]

Issue	ICRP 60	ICRP 103
Tissue organ weighting factors, w_T		
Gonads	0.20	0.08
Breast	0.05	0.12
Red bone marrow	0.12	0.12
Lung	0.12	0.12
Thyroid	0.05	0.04
Bone surfaces	0.01	0.01
Colon	0.12	0.12
Stomach	0.12	0.12
Bladder	0.05	0.05
Oesophagus	0.05	0.05
Liver	0.05	0.05
Brain	++	0.01
Kidney	++	++
Salivary glands	++	0.01
Skin	0.01	0.01
Remainder	0.05[a]	0.12[b]

[a]see ICRP 60
[b]see ICRP 103
++included in Remainder

10.3 Natural Background Radiation

All individuals are exposed to natural background radiation, which consists of cosmic radiation, external terrestrial radiation from naturally radioactive isotopes in the soil and rocks or houses and internal radiation after inhalation and ingestion from naturally radioactive isotopes in the human body [5, 6, 9–11].

The mean natural background whole body dose on earth is 2.4 mSv/year. Accordingly, the average lifetime dose (70 years) is about 170 mSv.

The average background whole body dose in the USA is about 3.7 mSv/year (medical X-rays excluded). It varies with altitude, geographic location, etc. In Kerala, India, or in Brazil with large monazite reserves (thorium), the background radiation dose is roughly a factor of 8-100 higher than the mean background dose on earth of 2.4 mSv/year [5, 6, 10, 11].

In the Federal Republic of Germany, the corresponding average background whole body dose is 2.1 mSv/year (Table 10.3) [12, 13].

Table 10.3 Mean effective radiation dose to the population from natural background radiation in the Federal Republic of Germany during the year 2008 [12, 13]

Mean effective dose	
Radiation exposure from natural sources	mSv/year
Cosmic radiation (at sea level)	Approximately 0.3
External terrestrial radiation	Approximately 0.4
Outdoor (5 h/day)	Approximately 0.1
Indoor (19 h/day)	Approximately 0.3
Inhalation of radon and its progeny	Approximately 1.1
Outdoors (5 h/day)	Approximately 0.2
Indoors (19 h/day)	Approximately 0.9
Ingestion of natural radioactive substances	Approximately 0.3
Total natural background radiation	*Approximately 2.1 mSv/year*

10.3.1 Natural Background Exposure from Natural Sources in Germany

A major source of external radiation exposure consists of both cosmic (0.3 mSv/year) and external terrestrial radiation (0.4 mSv/year) from the natural radionuclide K-40 together with the radio-nuclides of the natural decay chains of U-238 and Th-232 [12, 13]. The internal component of radiation exposure is largely caused by the inhalation of the natural noble gas radon and its daughter nuclides (1.1 mSv/year), and partially also by the intake of natural radioactive substances with drinking water and food (0.3 mSv/year).

The annual mean value of the radon activity concentration in occupied spaces is about 50 Bq/m^3, which corresponds to a mean annual effective dose of about 0.9 mSv/year. Outdoors inhalation of radon and its progeny leads to about 0.2 mSv/year [12, 13].

10.4 Radiation Exposure from Man-Made Sources

Table 10.4 also shows data for the radioactive exposure of man-made radioactivity sources. These man-made radioactive exposures sum up to 1.9 mSv/year [12, 13].

10.4.1 Nuclear Weapons Tests

Numerous atmospheric nuclear weapons tests were carried out from 1945 to 1980, but since 1981 only underground tests have been performed [10, 12, 13]. The general level of environmental radioactivity due to former tests in the atmosphere has steadily

Table 10.4 Mean effective radiation dose to the population from man-made radiation sources in the Federal Republic of Germany during the year 2008 [12, 13]

Radiation exposure from man-made sources	mSv/year
Fallout from nuclear weapons tests	<0.01
Effects from the accident in the Chernobyl nuclear radiation accident	0.012
Nuclear installations	<0.01
Use of radioactive substances and ionising radiation in medicine	Approximately 1.9
diagnostic nuclear medicine [a]	Approximately 0.1
Use of radioactive substances and ionising radiation in research and technology	<0.01
Total of man-made radiation exposures	*Approximately 1.9 mSv/year*

[a] According to data of 2006, evaluation from 2008 [12]. The data of 2009 differ only slightly [13].

decreased since the Comprehensive Nuclear Test-Ban Treaty from 1964. At present its contribution to the total of human radiation exposure is less than 0.01 mSv/year.

10.4.2 Chernobyl Reactor Accident

In April 1986, a reactor accident occurred in the Chernobyl nuclear power plant which has had the most serious consequences of any accident in nuclear installations world-wide so far [10, 12, 13]. In the days following that accident, large amounts of radionuclides were released into the atmosphere and distributed all over Europe. In 2008, the mean effective dose was less than 0.012 mSv/year in Germany [12, 13]. It amounts to less than one percent of the natural background radiation exposure; about 90% of this radiation is caused by Cs-137 deposited on the ground. Locally also higher exposure values were found. In addition, higher concentrations of radioactive isotopes were found in mushrooms and wild animals.

10.4.3 Nuclear Installations

The emission of radioactive substances from nuclear power plants, uranium enrichment plants and fuel fabrication plants in Germany contributes only insignificantly to the radiation exposure of the population [12, 13]. The upper values for exposures to inviduals, calculated in accordance with the "General Administrative Guideline relating to §47 of the German Radiation Protection Ordinance" [8] are clearly below the required limits (Sect. 10.6.1). The annual contribution from domestic nuclear installations and other installations located close to the German borders to the mean effective dose to the population of the Federal Republic of Germany remained below 0.01 mSv in 2008 and 2009 [12, 13].

10.4.4 Medical Applications

The major part of the mean effective population dose from man-made radiation exposure is caused by medical applications of radioactive substances and ionising radiation (thyroid and sceletal scintigraphy, X-ray diagnostics, computer tomography, positron emission-tomography). It amounted to approximately 1.9 mSv/year in Germany in 2008 [12, 13].

10.4.5 The Handling of Radioactive Substances in Research and Technology

The use of ionising radiation and radioactive substances for technological and research purposes [12, 13] lead to a mean contribution to the effective population exposure in Germany of less than 0.01 mSv in 2008 [12, 13].

10.4.6 Occupational Radiation Exposure

All employees who might receive enhanced radiation doses during their occupation are subject to radiation protection monitoring. These persons are monitored through personal dosimeters. The average individual dose of 324,000 monitored employees was 0.14 mSv/year in 2008 [12, 13]. Only 57,000 monitored employees out of these 324,000 employees received an average individual dose of 0.8 mSv/year in 2008.

Aircrews received an average effective dose of at least 2.4 mSv per year from cosmic radiation during the flight during 2009 [13].

10.5 Radiobiological Effects

When ionizing radiation (α-, β-, γ-radiation, neutrons, protons, ions) hits a biological cell and penetrates it, this gives rise to ionization of the atoms in various molecules of the cell [2, 5, 8–10, 14, 15]. This may alter these molecules. Especially alterations (separations) of the DNA containing the hereditary information produce radiation consequences. The results may be

- mutation of the cell,
- death of the cell.

Each cell has a high repair potential [2, 15–17]. As a consequence, most molecular alterations will have no consequences. However, it may also happen that a mutant cell is produced which passes its modified genetic function on. A cell changed in this way may cause carcinoma or leukaemia (somatic effect). When the mutation in a gonad cell is passed on to a descendant, this is called a genetic effect.

It is commonly assumed, although not unchallenged [10], that there is no radiation dose threshold for these mutant effects. Under this assumption, the radiation dose/effect relation begins at the zero point, rises linearly [6, 7], assumes a quadratic curve shape at higher radiation doses, and then levels off again at very high radiation doses when cell death occurs.

In the lower radiation dose range, the biological effect of radiation can be measured only on a statistical basis, and is therefore refered to as a stochastic effect. When significant numbers of cells are damaged or die in the higher radiation dose range, this is called a deterministic effect of radiation.

10.5.1 Stochastic Effect

Assessing the stochastic effect quantitatively is not easy , as it is impossible, at the present state of knowledge, to recognize whether a tumor developed as a result of an exposure to ionizing radiation or for some other reason. Epidemiological studies of large populations (atomic bomb victims of Hiroshima and Nagasaki) have been used so far in attempts to determine the number of deaths from cancer exceeding the number of deaths from cancer occurring from exceeding natural background radiation. The results of the evaluation of the data from Hiroshima and Nagasaki can be related to the dose of the preceding radiation exposure, and result in the dose/effect relationship. However, the data calculated in this way do not constitute an immutable quantity. For instance, the number of deaths from cancer increases with the age of the Hiroshima–Nagasaki population under study. The International Commission on Radiological Protection, in its publication ICRP 60 [6], for the first time took into account data of the Hiroshima-Nagasaki population more than forty years after the atomic bombs were dropped, and extrapolated these data to the entire lifespan [10, 15].

For determination of the risk associated with low doses, the roughly linear dependence found at high radiation doses is now extrapolated back to the range of low radiation doses. In this way, the entire cancer risk is rather overestimated in a conservative sense.

Figure 10.2 reflects the hypotheses which can be extrapolated back from the data of Nagasaki–Hiroshima. Curve A, which is favored by most scientists, is based on a threshold level. Curve B, for which also a number of scientists argue, is based even on positive healing health effects (hormesis) of a low radiation dose [9, 10, 15].

The linear dose-effect relation is recommended in ICRP 60 and ICRP 103 as a conservative proposal.

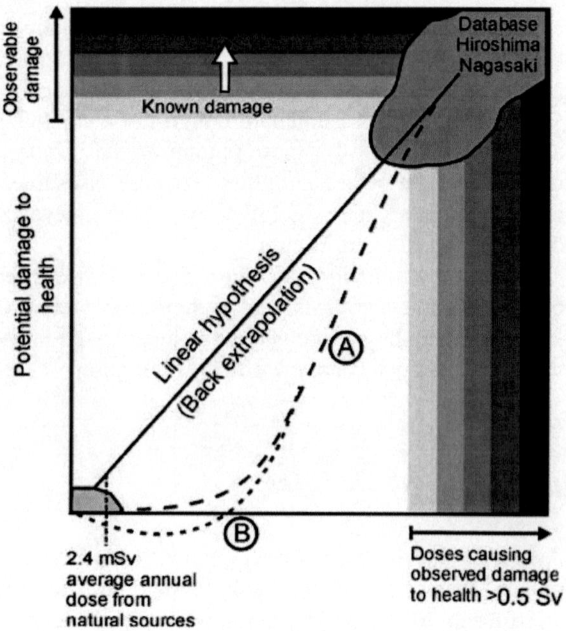

Fig. 10.2 Concept of linear dose-effect relation [9, 15]

10.5.2 Deterministic Effects of Radiation

Deterministic effects of radiation arise when a large number of cells are damaged considerably by a high radiation dose such that regeneration is not possible or the cells die. There are various threshold doses of radiation for the deterministic effect in various organs of the body.

At a radiation dose of less than 1 Gray, most tissues produce no clinical symptoms of disease (ICRP 60 [7]). However, there are exceptions:

- the male gonads:
  - >0.15 Gray causes temporary sterility,
  - permanent sterility results at >3 Gray;
- the bone marrow reacts with disorders of the blood at >0.5 Gray.

10.5.3 Acute Radiation Syndrome

Damage to the bone marrow occurs between 1 and 20 Gray [5, 10, 14, 18]. The extent of the disorder, and the therapy employed, ultimately decide whether the irradiated accident victim survives.

Gastro-intestinal disorders are produced at >2 Gray (nausea, vomiting, hyposthenia).

At more than 6–8 Gray, there are practically no chances of survival.

10.6 Permissible Exposure Limits for Radiation Exposures

The International Commission on Radiological Protection (ICRP) has published recommendations about permissible radiation exposures to man from nuclear installations (ICRP 60 [6] and ICRP 103 [7]). In those recommendations, a distinction is made between occupationally exposed persons and individual members of the public. Considerable lower levels for exposure limits of the general population were set.

10.6.1 Limits of Effective Radiation Dose from Nuclear Installations in Normal Operation

At present, the German Radiation Protection Ordinance of 2008 [8] is valid in Germany. It is based on ICRP 60 [6] and the Euratom directive of 1996 [19].

10.6.2 Radiation Exposure Limit for the Population

According to the German Radiation Protection Ordinance [8] the limit for the annual effective radiation dose due to radioactive emissions from nuclear installations to the population permissible on top of the natural background radiation is

$$1\,\text{mSv}.$$

Table 10.5 shows a comparison of ICRP 60 [6], the Euratom directive of 1996 [19], the German Radiation Protection Ordinance of 2008 [8].

10.6.3 Exposure Limits for Persons Occupationally Exposed to Radiation

The exposure limit per annum of the effective radiation dose for persons occupationally exposed to radiation is given in Table 10.6 [6–8, 12, 19].

The limit for persons occupationally exposed to radiation is 20 mSv/year. This limit can be increased to 50 mSv/year as long as the sum of exposures does not

Table 10.5 Exposure limit for the annual effective radiation dose from nuclear installations to the population permissible on top of the natural background radiation

Effective dose	ICRP 60 1991	EU 1996	German radiation Prot. Ord. 2008
Limit (population) per annum (mSv/year)	1	1	1

Table 10.6 Limits for persons occupationally exposed to radiation

	ICRP 60 1990	EU 1996	German radiation Prot. Ord. 2008
Limit for persons occupationally exposed to radiation	20 mSv/year	20 mSv[a]/year	20 mSv/year

[a] 100 mSv in five consecutive years, but not more than 50 mS/y in one year.

exceed 100 mSv over a time period of five years. This is over and above background exposure and excludes medical exposure.

The exposure doses for each organ (tissue) must be determined by special tables which are not listed here in detail [6–8, 12, 19].

10.6.4 Exposure Limits for Persons of Rescue Operation Teams During a Reactor Catastrophe

After severe reactor accidents rescue operation teams working in contaminated areas inside or outside of the reactor in high radiation fields can receive a radiation exposure up to a limit of [8, 20, 21]

100 mSv per rescue operation and year or

250 mSv once in their life

10.6.5 Life Time Occupational Exposure Limit

According to § 56 of the German Radiation Protection Ordinance [8] an occupationally exposed person can receive a maximum of

400 mSv

during its professional life. This limit can only be exceeded by 10 mSv/year if both, medical authorities and the person having reached the professional lifetime limit of 400 mSv, agree [8].

10.6.6 The ALARA Principle

Above and beyond these standards defined by ICRP and national regulations, the "as low as reasonably achievable" (ALARA) principle must be applied to all emissions of nuclear plants. This means that practically all facilities of the fuel cycle must keep below those standards [6–8, 19].

10.7 Radionuclide Effluents and Radiation Exposures from Various Parts of the Fuel Cycle

In the following sections, the radioactive effluents and the radiation exposures from important parts of the nuclear fuel cycle are discussed. The radiation exposures are analyzed on the basis of measurements and calculations for uranium mines, enrichment and fuel fabrication plants, the power reactor and the reprocessing plant with waste treatment facilities.

10.7.1 Uranium Mining and Milling

10.7.1.1 Radioactive Effluents from Mining and Milling

The main radioactive effluents from uranium mines and mills are radon and dust particles containing uranium and its decay products. All daughter products of uranium, except radon, are solid, emitting α- and β-particles, mostly in combination with γ-radiation. Radon is a noble gas trapped in the crystalline structure of rock, but distributed into the atmosphere as the rock is being opened up. Radon and the other decay products of uranium enter the lung via the inhalation process. In open pit mining, radon is directly released into the atmosphere. The radon released during underground ore mining must be removed by the ventilation system of the underground mine. After extraction of the ore, its storage above ground prior to further processing and milling constitutes another source of radon emission as well as a significant radiotoxic burden of run-off rain water.

The remaining uranium mill tailings are usually stored in a tailings impoundment located on the mill site. They and the low grade ore stockpiles can be covered with a layer of sand several meters thick. This largely prevents further releases of radon.

In NUREG and EPRI studies [22–24], a model mine-mill complex with an open pit mine of 500,000 m^2 of open area and a production capacity of 730,000 t of ore per annum was used as a basis. The associated uranium ore mill produces 930 t U_3O_8 per annum (average uranium ore compounds, 0.2%). The gaseous effluents and the release rates of radioactivity of this open pit mine-mill complex are given in Table 10.7. Studies conducted by the USEPA and the US Nuclear Regulatory

Table 10.7 Estimated radioactive effluent release rate from major sources in a model open pit mine-mill complex (EPRI) (values of Rn-222 releases may vary considerably in different mines)

Source	Effluent	Release rate (Bq)	Release rate (Bq/GW(e)·y)
Open pit mine	Rn-222	2.6E14	4.1E13
Mill stack	U-238 and daughter isotopes	1.6E08 (each)	2.5E07 (each)
Ore dust[b]	U-238 and U-234	3.3E09 (each)	5.2E04 (each)
Yellowcake	Th-230	1.6E08	2.6E07
Dust[a]	Ra-226	6.7E07	1.0E07
	Rn-222	1.4E12	2.2E11
Tailings	Rn-222	5.3E13[b] (1.6E13)	8.5E12[b] (2.4E12)
Inactive period of mine complex after 20 years of operation	Rn-222	2.2E14[b] (6.4E13)	3.5E13 (1.0E13)

[a]Dust collector efficiency more than 90%.
[b]These numbers can be reduced (values in brackets) by a factor of about 3 by 2.5 m of tailing cover.

Commission (USNRC) show radon release levels of the same order of magnitude for model open pit uranium mines of similar size. The release of radioactivity stems mainly from radon. Open pit mining and tailings are the main sources of radon emissions, while milling operations contribute only relatively little to the emissions of radioactivity.

The data in Table 10.7 apply to an open pit mine and mill complex providing U_3O_8, sufficient for about 6 GW(e)·y of PWR capacity. The underground mine also referred to in NUREG and EPRI studies [22–24] has an annual production capacity for about 2–3 GW(e)·y of PWR capacity. It is seen that underground mining, because of the higher radon emissions, represents a higher individual risk than open pit mining.

Comparisons with the radon emissions from underground mining result in radon releases by a factor of five higher than those arising in open pit mining. This factor depends on the efficiency of the ventilation of the tunnel system of underground mines.

10.7.1.2 Radioactive Exposure Pathways and Exposure Limits for Uranium Mines and Mills

The radioactive exposure pathways for people living in the vicinity of a uranium mine and mill complex are

- external radiation from radon gas and radioactive airborne dust,
- internal radiation after inhalation of radon gas and dust as well as ingestion of contaminated food products.

Since the early to mid 1990s uranium mining companies adopt the ICRP 60 recommendations [25, 26]. These recommendations are applied to both the exposure limit to the population caused by uranium mines (nuclear installations, Sect. 10.6.1) and to workers in uranium mines (exposure limits for persons occupationally exposed to radiation, Sect. 10.6.4):

- the exposure limit for the population (Sect. 10.6.2) from uranium mining activities of

  $$1\ \text{mSv/year}$$

  corresponds to a background radon level of 40 Bq/m^3 indoors and 6 Bq/m^3 outdoors assuming an indoor occupancy of 80%.
- the occupational exposure limit for workers in uranium mines is as described already in Sects. 10.6.4 and 10.6.5.

10.7.1.3 Experience with Occupational Exposures in Uranium Mines

The average annual radiation personal dose to workers at uranium mines (Canada, Australia, USA) is around 2 mSv/year ranging to a maximum of 10 mSv/year [26, 27]. This is achieved by efficient ventilation techniques and rigorously enforced procedures for hygiene, e.g. for workers handling uranium oxide concentrate [26].

These precautions with respect to radon and occupational exposure limits had not been taken in East German uranium mines between about 1950 and 1990 and underground uranium mining in the USA between 1946 and 1959. Workers died of lung cancer as a consequence of too high radiation exposures by inhalation of radon gas [27].

For radiation exposure in uranium mining also another unit is used:

The working level month (WLM) whereby 1 WLM corresponds to 10 mSv/y.

10.7.2 UF_6 Conversion, Enrichment, and Fuel Fabrication

Conversion of U_3O_8, into UF_6 and subsequent enrichment of the fuel do not entail any significant radiological burdens to the environment. In UF_6 conversion, gaseous effluents are passed through filters and wet sorbers. Discharges from the plant contain only traces of radioactive material. Solid waste, which contains small quantities of uranium or traces of thorium, is packaged and consigned to a licensed radioactive LLW burial site.

In uranium enrichment plants U-234, U-235 and U-238 radionuclides may be emitted. U-234 occurs in natural uranium with an abundance of 0.0054 atom percent. Compared with all other emissions of the fuel cycle, emissions from enrichment plants are negligibly low.

Table 10.8 Radioactive emissions from U_2O_3 conversion, enrichment and uranium fuel fabrication plants (normalized to 1 GW(e)·y of energy produced (NUREG) [23])

	Nuclide emitted	Radioactive emissions	
		Air borne (Bq/GW(e)·y)	Liquid (Bq/GW(e)·y)
U_3O_8 conversion plant	Uranium	7.8E07	2.2E10
	Ra-226		1.7E09
	Th-230		7.4E08
Enrichment plant	U-234	3.4E07	1.0E06
	U-235	1.0E06	4.1E04
	U-238	4.1E06	8.9E05
Fuel element fabrication plant	Th-234	3.7E06	8.9E08
	U-234	3.0E07	7.4E09
	U-235	8.5E05	1.9E08
	U-238	3.7E06	8.9E08

Table 10.9 Release data for alpha activities and effective doses to the German population in the vicinity of the advanced nuclear fuel fabrication plant (ANF) and the URENCO centrifuge enrichment plant [12]

	Air borne α-releases (Bq)	Liquid α-releases (Bq)	Effective dose to population (μSv)	
			Airborne releases	Liquid releases
ANF	<1.5E04	–	<0.1	<0.1
URENCO	2.5E04	2.3E03	<0.1	<0.1

In the fabrication of uranium fuel, the enriched UF_6 is first chemically converted into UO_2. Afterwards, the UO_2 powder is pelletized. Liquid and gaseous effluents from the fabrication plant contain certain concentrations of U-234, U-235, U-238 and Th-234 and must therefore be reprocessed and filtered. Estimated data for radioactive effluents from UF_6 conversion, enrichment and fuel fabrication are given in Table 10.8 [12].

The estimated data of Table 10.8 are based on earlier systems studies [23]. Data from measurements at the German Advanced Nuclear Fuel Fabrication plant (ANF) (650 t_{HM}/year) serving about 27 GW(e) LWRs and the centrifuge enrichment plant of URENCO with 1.8×10^6 kg SWU/y at Lingen serving about 11 GW(e) LWRs are reported in [12]. They are listed in Table 10.9 These real measurements are by a factor of about 10^3 lower than those estimated in the earlier NUREG studies [23].

10.7.3 Nuclear Power Plants

Most of the radioactive inventory of a nuclear power plant is made up of fission products. Gaseous fission products, such as noble gases (especially krypton and xenon) and tritium, can enter the coolant through leaks in the claddings of fuel rods. They are passed through the primary coolant purification system and the exhaust air system into carbon filter lines and into the exhaust air stack from where they are released into the environment. Emissions of shortlived isotopes, such as Kr-88 (half-life 2.8 h), can be minimized by adequate holdup of gaseous effluents in storage and decay tanks before release.

The tritium produced in the core migrates along the grain boundaries of the fuel into the fission product gas plenum of the fuel rod. The zircaloy tubes of LWR fuel rods bind some 60% of the tritium inventory. Moreover, an oxide layer building up on the outer wall of the zircaloy cladding tube acts as a diffusion barrier to the tritium. As a consequence, more than 99.9% of the tritium formed is retained in the LWR fuel rod. Only if cladding tube failures occur, will releases of tritium into the cooling water be increased. Some of the tritium is discharged with the gaseous effluents. In water cooled reactors most of it remains in the coolant as tritiated water. Some of the tritiated water is released at a controlled rate. Improved methods under study are the concentration of tritiated water by evaporation and its prolonged storage in decay tanks. With a half-life of tritium of about twelve years, some 90% will have decayed after forty years.

Besides the radioactive noble gases and tritium, also such elements as rubidium, strontium, technetium, ruthenium, silver, tellurium, antimony, iodine, cesium, barium, rubidium, lanthanum and cerium are radiologically significant. Except for iodine, cesium and rubidium, these elements have only low volatilities. They may enter the primary coolant through defective fuel rod claddings. Non-volatile fission products can enter the liquid effluent only through the primary coolant purification system.

10.7.3.1 Radioactive Effluents from PWRs and BWRs

Table 10.10 shows emission data of typical German PWRs and BWRs. This set of data was reported in the yearly report for 2008 on radioactive releases and radiation exposure from German nuclear installations [12, 13]. The data collected for 2009 differ slightly [13]. The PWR data are given for two PWR plants with a total power output of 2.2 GW(e) at the site of Neckarwestheim (Germany). The BWR data are valid for two BWR plants with a total power output of 2.7 GW(e) at the site of Gundremmingen (Germany). For comparison also the half-lives of the different emitted isotopes are given.

Although there are inherent design differences between LWRs and CANDU-PHWRs or HTGRs also these types of power reactors have radioactivity releases below the radiation limits described in Sect. 10.6. Nuclear power plants are equipped with instruments to measure continuously the amounts of gaseous and liquid radioactive effluents. These data must be reported to national environmental protection agencies or nuclear regulatory commissions. Such reports are made available to the public.

Table 10.10 Radioactive emissions (airborne and liquid) from a PWR- and a BWR-plant site with two reactors [12]

Isotope	Half-life	PWR plant site 2.2 GW(e) Neckarwestheim	BWR-plant site 2.7 GW(e) Gundremmingen
Airborne effluents Bq/y			
Tritium	12.3 years	4.5E11	5.3E11
C-14	5,730 years	1.1E11	8.1E11
Ar-41	1.8 h	5.1E11	2.6E11
Co-60	5.3 years	5.3E04	–
Kr-85m	4.48 h	–	6.8E09
Kr-85	10.8 years	2.2E11	4.0E11
Kr-88	2.8 h	1.5E08	2.1E09
I-131	8.02 days	–	1.1E07
Xe-131m	11.9 days	1.9E10	8.8E10
Xe-133m	2.2 days	1.6E08	3.1E09
Xe-133	5.2 days	4.0E09	1.4E11
Xe-135	9.1 h	2.4E10	2.9E11
Xe-137	3.8 min	4.8E07	6.1E11
Liquid effluents in Bq/y			
Fission + activation products		1.6E06	8.6E08
Tritium	12.3 years	2.6E13	3.4E12
α-emitters		Below measurement limit	Below measurement limit

10.7.3.2 Radioactive Effluents from LMFBRs

The operating experience of prototype LMFBRs (PHENIX, SUPERPHENIX etc.) and estimates of emission data of LMFBRs show radioactive effluents to be somewhat lower than those of PWRs. Tritium, after diffusion through the stainless steel cladding of fuel rods, is chemically bound by the sodium coolant. Other non-gaseous fission products from failed fuel rods are also bound by the sodium coolant and eliminated together with radioactive corrosion products in cold traps by the sodium purification system; from there they pass into the solid waste treatment system.

On the whole, radioactive effluents from LMFBRs are somewhat smaller than those of LWRs [28].

10.7.3.3 Occupational Radiation Exposure of Workers in Nuclear Power Plants

The average occupational exposure of 30,238 workers in nuclear power plants (PWRs and BWRs) was [12]

$$0.5\ \text{mSv/y}$$

in Germany in 2009. This average radiation exposure is much lower than the occupational exposure of aircraft personal (Sect. 10.4.6) [12, 13].

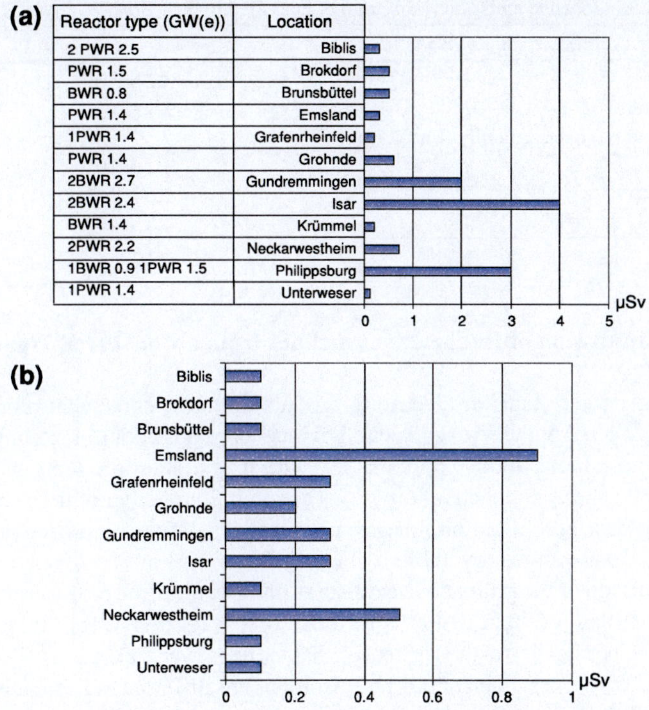

Fig. 10.3 **a** Effective dose from airborne radioactive effluents of German PWRs and BWRs in 2008 [12]. **b** Effective dose from liquid radioactive effluents of German PWRs and BWRs [12]

10.7.3.4 Radiation Exposures Caused by Radioactive Emission from Nuclear Power Plants

In calculating radiation exposures it is assumed that gaseous effluents are released into the environment from a stack of 100 m height. Moreover, liquid effluents are introduced into the cooling water of a nuclear power plant and further diluted in the main canal with a water flow of 250 m^3/s. Taking into account statistical data about the weather conditions and following the different exposure pathways, it is possible to determine the radiation exposure in the specific environment of a plant. The exposure results below were obtained on the basis of German rules and regulations. In this respect, it is assumed that a person stays in the same place throughout the year and ingests both drinking water and food from the immediate environment. Figures 10.3a, b present the exposure data of German PWRs and BWRs for gaseous and liquid radioactive effluents [12].

For PWRs the annual effective dose is well below 1 μSv. For BWRs—due to the fact that the steam produced in the reactor core goes directly to the steam turbine—the effective annual dose is somewhat higher in case of air borne radioactive effluents. However, it is still more than a factor 100 lower than the permissible limit.

Table 10.11 Radioactive emissions (airborne) of coal fired plants (Bq/GW(e)·y) [29]

U-238 natural radioactive family (Bq/GW(e)·y)	
Rn-222	3.2×10^{10}
Other daughters	2×10^{9}
Th-232 natural radioactive family (Bq/GW(e)·y)	
Rn-220	2.1×10^{10}
Other daughters	0.6×10^{9}
K-40	0.8×10^{9}

10.7.3.5 Emmission of Radioactive Nuclides from a Coal Fired Plant

A coal fired power plant for electricity production burns somewhat more than two million tons of coal per GW(e) and year. This coal contains about 1 ppm of U-238 with its daughter products, about 2 ppm Th-232 with its daughter products as well as the isotope K-40. These radioactive impurities are emitted together with the combustion gases of the coal fired plant into the environment [29]. These radioactive emissions (Bq/GW(e) ·y) are shown by Table 10.11.

A comparison of the radioactive exposures of these airborne radioactive emissions of coal fired plants (Table 10.11) with those of PWRs or BWRs (Table 10.10) on a GW(e)·y basis leads to the difficulty that both power generating systems emit different radioisotopes. Each of this radioisotopes has different radiotoxicity (Sv/Bq). In addition the whole fuel cycle for nuclear energy must be included. A comparison is, therefore, only possible on the basis of collective dose equivalents [29]. On this basis it was shown that coal fired plants on a 1 GW(e)·y basis cause about equal collective doses compared to the whole nuclear fuel cycle [29].

10.7.4 Spent Fuel Reprocessing and Waste Treatment Centers

In the reprocessing and waste treatment plant described in Chap. 7, most of the radioactive substances released during chopping and dissolution of the fuel are retained in the facility by various engineered safeguards measures. Releases of radioactivity into the environment are controlled. These are [1, 30–33]:

- During chopping and dissolution of the fuel elements, a certain fraction of tritium enters the gaseous effluents and is released through the stack. Another part of the tritium may be retained in the zircaloy cladding hulls and will go into the solid high level wastes. The residual tritium remains in the nitric acid solution as tritiated water. After removal of the nitric acid, the remaining liquid effluent is concentrated by evaporation. Part of it can be released in a controlled way. In modern large reprocessing plants, the prolonged storage of tritium in storage tanks may be applied.

- C-14 is currently released without any retention systems. In the future, retention is possible in connection with low temperature rectification. Noble gases can largely be retained by means of a low temperature rectification process. More than 99.9% of the Iodine isotopes can be held back to a large extent by silver nitrate filters [34].

Other radionuclides generated as aerosols in the dissolver offgas are:

- Mn-54, Co-60 and Ni-63, which are activated in the structural materials of the fuel elements,
- fission products, such as Sr-90, Tc-99, Ru-106, Sb-125, Cs-137, etc.
- aerosols of the higher actinides and of the fuel, such as the isotopes of plutonium, americium and curium.

The technical processes of treating radioactive waste have been described in Chap. 7. Radioactive emissions play significant roles, especially in gaseous effluent treatment of high level waste vitrification plants. Gaseous effluents from melters can have temperatures up to 1,200°C and contain aerosols, besides water vapor and nitrogen oxides. They are passed through wet scrubbers to retain ruthenium tetroxide. Then they flow through a nitrogen oxide adsorption column and high efficiency particulate air (HEPA) filters for ruthenium aerosol retention.

All gaseous effluent streams of reprocessing, refabrication and waste treatment plants are passed through HEPA filter lines for removal of particulates. In addition, the application of several containment barriers with low leak rates, storage of radioactive liquids or gases in decay tanks or pressurized gas cylinders allows high retention factors and, conversely, low release factors to be achieved. The retention factor denotes the ratio between radioactive inventories and radioactive releases. The release factor is the reciprocal value of the retention factor. By series connecting several containment barriers and filter lines it is possible to attain retention factors of more than 10^9–10^{10} and release factors less than 10^{-9}–10^{-10}, respectively, for such aerosols as strontium, ruthenium, plutonium and higher actinides [1].

Table 10.12 shows the air borne radioactive effluents and Table 10.13 gives the liquid radioactive effluents from large scale European reprocessing facilities LaHague (France) and Sellafield (UK). It should be remembered that the reprocessing plant at LaHague with a capacity of 1,700 t_{HM}/year is able to reprocess the spent fuel of about 70 GW(e) LWRs. The reprocessing plant Sellafield (UK) with a capacity of 1,200 t_{HM}/year can reprocess the spent fuel of about 50 MW(e) LWRs (on the basis of 24 t_{HM}/year spent fuel per GW(e)·year of LWRs).

Table 10.12 Airborne radioactive effluents in Bq/y from large scale European reprocessing facilities LaHague (France) and Sellafield (UK) in 2008 [30]

Isotope	Half-life	La Hague (France) (1,700 t_{HM}/year)	Sellafield (UK) (1,200 t_{HM}/year)
Airborne radioactive releases (Bq) of European reprocessing plants			
Tritium	12.3 years	4.64E13	1.41E14
C-14	5,730 years	1.35E13	6.86E11
Co-60	5.27 years	6.55E06	–
Kr-85	10.8 years	1.55E17	2.57E16
Sr-90	28.6 years	–	3.76E07
Ru-106	374 days	6.57E07[a]	1.39E09
Sb-125	2.77 years	2.17E07	3.6E09
I-129	1.57E07 years	6.76E09	5.74E09
I-131	8.02 days	2.07E08	6.26E08
Cs-134	2.06 years	4.76E06	–
Cs-137	30.2 years	4.05E06	1.31E08
Pu-238	87.7 years		
Pu-239	2.41E04 years		2.27E07 (Pu-238, Pu-239, Pu-240)
Pu-240	6,563 years		
Pu-241	14.4 years		2.62E08
Am-241	432 years	1.83E06 (Pu-238 to Cm-242)	
Cm-242	163 days		1.71E04 (Am-241, Cm-242)

[a]The data for Ru-106 were reported as Ru-106 und Rh-106

10.7.4.1 Radioactive Exposure to Population Due to Radioactive Effluents from Reprocessing Plants

The radioactive exposure to the population due to the radioactive effluents of the large scale reprocessing plants are:

LaHague (1, 700 t_{HM}/year)	0.02 mSv/year	(due to liquid and gaseous discharges) [31, 33]
Sellafield (1,200 t_{HM}/year)	0.15 mSv/year	Due to liquid discharges [32].

These exposures are well below the permissible effective radiation exposures of 1 mSv/year (Sect. 10.6.2).

10.7.4.2 Long Range Accumulation of Carbon-14 and Krypton-85

Unlike the xenon and iodine isotopes (except I-129), which have sufficiently short half-lives, C-14 and Kr-85 may accumulate in the atmosphere over prolonged periods of time. Consequently, the potential severity of radioactive exposures arising from future accumulations of these radioisotopes were studied in the light of an expanding nuclear power economy including reprocessing plants.

Table 10.13 Liquid radioactive effluents from large scale European reprocessing facilities LaHague (France) and Sellafield (UK) in 2008 [30]

Isotope	Half-life	La Hague (France) 1, 700 t_{HM}/year	Sellafield (UK) 1, 200 t_{HM}/year
Liquid radioactive releases (Bq/y)of European reprocessing plants			
Tritium	12.3 years	8.19E15	7.78E14
C-14	5,730 years	6.24E12	7.19E12
Mn-54	312 days	2.27E09	–
Co-57	272 days	7.31E07	–
Co-58	70.9 days	6.41E07	–
Co-60	5.27 years	1.18E11	7.21E10
Ni-63	100 years	6.44E10	–
Sr-90	28.6 years	1.69E11	1.70E12
Tc-99	2.1E05	7.41E10	2.37E12
Ru-106	374 days	6.74E12	1.39E12
Sb-125	2.77 years	3.80E11	3.10E12
I-129	1.57E07 years	1.04E12	1.99E11
I-131	8.02 days	1.09E10	–
I-133	20.8 h	4.90E09	–
Cs-134	2.06 years	7.50E10	1.15E11
Cs-137	30.2 years	1.07E12	5.11E12
Ce-144[a]	285 days	1.51E08	3.54E11
Eu-154	8.8 years	5.55E08	–
Eu-155	4.76 years	8.08E07	–
Np-237	2.14E06 years	4.27E08	4.30E10
Pu-238	87.7 years	5.39E09	–
Pu-239	2.41E04 years	} 1.69E09	} 1.08E11
Pu-240	6,563 years		
Pu-241	14.4 years	1.20E11	2.44E12
Am-241	432 years	2.74E09	2.97E10
Cm-242	163 days	2.14E07	–
Cm-243	29.1 years	–	} 2.92E09
Cm-244	18.1 years	1.36E09	

[a] Ce-144 includes Pr-144

The reprocessing plants of LaHague (France), Sellafield (UK) and Rokkasho-mura (Japan) emit the gaseous isotopes Tritium, Carbon-14 and Krypton (Table 10.12) into the atmosphere and the corresponding liquid effluents into the oceans. The liquid effluents are carefully surveyed by French, British and Japanese health authorities [32, 33] by taking periodical samples of fish etc. from the oceans. The radioactive impact of the gaseous releases to the atmosphere are carefully monitored as well.

The impact of the releases of Tritium, Carbon-14 and Krypton into the atmosphere was studied by USEPA already in 1979 [1, 2]. Tritium is produced in the atmosphere as a result of cosmic radiation interacting with oxygen and nitrogen atoms. According to estimates, this natural process provides for an overall inventory of some 50 million Curies of tritium on the earth at all times. Atmospheric thermonuclear weapons tests conducted until the early sixties generated another 1,700 million Curies of tritium which decayed already to a large extent. Studies of USEPA have indicated that a growing worldwide nuclear power economy of a theoretical capacity of 2,000 GW(e) in the future with the necessary fuel cycle plants, would release another 450 million Curies of tritium into the environment per annum. In similar studies of the USEPA it was further assumed that all C-14 and Kr-85 would be released into the atmosphere under the same assumptions of a growing nuclear power economy.

The tritium, C-14 and Kr-85 released was then assumed by USEPA to distribute and accumulate in the atmosphere of the northern hemisphere, where most nuclear power plants will be located. The tritium, C-14 and Kr-85 would enter the human body on the pathways described in Sect. 10.1.1. This would lead to an additional radiological exposure per person of

0.15 μSv/y	to the whole body	from tritium
9 μSv/y	to the skin	from Kr-85
0.5 μSv/y	to the lung	
0.25 μSv/y	to the whole body	
1.4 μSv/y	to the whole body	from C-14

From these very low radiation exposure data it is generally concluded that Tritium, C-14 and Kr-95 can be released into the atmosphere without serious impacts on radiation exposures.

For comparison, it should be recollected that the average whole body dose from natural background radiation in the world is 2.4 mSv/year (Sect. 10.3).

For reprocessing plants not located at an ocean the liquid tritium release could be restricted by concentrating tritiated water and storing it in tanks for a sufficiently long time. If needed also technical processes, such as cryogenic distillation, fluorocarbon absorption and carbon adsorption for the retention of C-14 and Kr-85 would be available. The Kr-85 separated can be filled as a compressed gas in steel cylinders or embedded in a metal matrix and then stored for about 150 years. If for C-14 a similar release limit as for Kr-85 would be imposed, such a limitation is technically feasible in connection with the low temperature rectification or Kr-85. C-14 dioxide would then be converted to calcium carbonate and could be stored over long periods of time [35–37].

10.7.4.3 Radiation Exposures in the Environment of the MOX-Refabrication Plant MELOX (France)

The MELOX refabrication plant for MOX fuel at Marcoule, France, has a capacity of 195 (t_{HM}/year). The environmental radiation exposure of the population was measured be to 0.00056 μSv/year in 2009 compared to a permissible exposure of 1.7 μSv/year set by French licensing organizations [38, 40]. This is very much lower than the 1 mSv/year effective radiation dose set for the population (Sect. 10.6.2).

10.7.4.4 Occupational Radiation Exposures of Workers in Reprocessing Centers

Occupational exposures were steadily decreasing in the reprocessing facilities LaHague (France) and Sellafield (UK) during the past decades.

The average occupational exposure was
0.25 mSv/year at LaHague in 1995 [33]
and 3 mSv/year at Sellafield in 1991 [39].

10.7.4.5 Occupational Radiation Exposures of Workers in the MOX Refabrication Plant MELOX

The average occupational radiation exposure for workers in the MELOX plant was 1.8 mSv/year in 2004 and 1.35 mSv/year in 2009 [38].

10.8 Summary of Radiation Exposures Caused by the Nuclear Fuel Cycle

The results reported in the previous Sect. 10.7 show that the exposure limit for the population of

$$1\ \mathrm{mSv/y}$$

is not exceeded. Uranium mining and (Radon emission, insufficient ventilation etc.) milling had been a serious problem for underground uranium mining in Eastern Germany between 1950 and 1990 and in some mining areas of the USA between 1946 and 1959 (Sect. 10.7.1.3). This changed since the uranium mining industry adopted the ICRP-60 recommendations from the mid 1990s on. The uranium mining and milling industry is able now to keep the radiation exposure for workers at an average of 2 mSv/y which is well below the limit of 20 mSv/year (Sects. 10.7.1.2 and 10.7.1.3).

The radiation exposures of uranium enrichment and uranium fuel fabrication is far below the limits both for the population and for workers (Sect. 10.7.2). This also holds for nuclear power plants (Sect. 10.7.3.2).

Also the large scale European reprocessing plants at LaHague (France) and Sellafield (UK) remain well below the radiation exposure limits for the population and for workers. This also holds for the MOX fabrication plant MELOX at Marcoule, France.

These radioactive emissions of the nuclear fuel cycle can be compared to those of coal fired plants (Sect. 10.7.3.5). Studies show that a comparison of radioactive exposures is only reasonable on the basis of collective dose equivalents [29]. They show that the collective doses of coal fired plants on a 1 GW(e)·y basis is about equal to the collective dose for the whole nuclear fuel cycle including uranium mining, enrichment, fuel fabrication, nuclear reactors, fuel reprocessing and refabrication [29].

References

1. Kessler G (1983) Nuclear fission reactors. Springer Verlag, New York
2. Glasstone S et al (1980) Nuclear power and its environmental effects. American Nuclear Society, LaGrange Park
3. Commissariat à l'énergie atomique (1999) La dispersion des pollutants dans l'atmosphère, CLEFS CEA, vol 42. Commissariat à l'énergie atomique, Paris, p 2
4. Nuclear Energy Agency, Committee on the Safety of Nuclear Installations (2007) State of the art report on iodine chemistry, NEA/CSNI/R(2007) 1
5. Krieger H (2007) Grundlagen der Strahlungsphysik und des Strahlenschutzes. Vieweg+Teubner Verlag, Berlin
6. International Commission on Radiation Protection (ICRP) (1991) Annals of the ICRP: ICRP Publication 60—1990 recommendations of the international commission on radiological protection. Pergamon Press, Oxford
7. International Commission on Radiation Protection (ICRP) (2007) Annals of the ICRP: ICRP Publication 103—the 2007 recommendations of the international commission on radiological protection. Elsevier, Amsterdam
8. General Administration Guideline relating to §47 of the Radiation Protection Ordinance (Strahlenschutzverordnung) (2008) Bundesministerium für Umwelt. Naturschutz und Reaktorsicherheit, Bonn
9. Cohen BL (1983) Before it's too late. Plenum Press, New York, p 33
10. Jaworowski Z (1999) Radiation risks and ethics. Phys Today 52:24–29
11. Sohrabi M et al (1990) High levels of natural radiation. International Atomic Energy Agency, Vienna, p 39
12. Umweltradioaktivität und Strahlenbelastung, Jahresbericht (2008) Bundesministerium für Umwelt, Naturschutz und Reaktorsicherheit (BMU), Bonn, Germany
13. Umweltradioaktivität und Strahlenbelastung, Jahresbericht (2009) Bundesministerium für Umwelt, Naturschutz und Reaktorsicherheit (BMU), Bonn, Germany
14. UNSCEAR (1998) Sources, effects and risks of ionizing radiation, UNSCEAR, report to the General Assembly, with annexes, United Nations Printing Office, Vienna, Austria
15. Commissariat à l'énergie atomique (2000) Les rayonnements, l'ADN et la cellule, Clefs CEA, vol 43. Commissariat à l'énergie atomique, Paris, p 6
16. Kesavan PL (1996) In: Wie L, Sugahara T, Tao Z (eds) High levels of natural radiation. Elsevier, Amsterdam, p 111
17. Zamboglou N et al (1981) Low dose effect of ionizing radiation on incorporation of iododeoxyuridine into bone marrow cells. Int J Radiat Biol 39:83–93
18. United Nations Scientific Committee on the effects of atomic radiation (2001) Report to the General Assembly, with scientific annex, Hereditary Effects of Radiation, New York
19. Directive 96/29 Euratom-ionizing radiation (2011). http://osha.europa.eu/en/legislation/directives/exposure-to-physical-hazards/osh-directives/73
20. Iter consult (May 2011) Independent technical evaluation and review, Fukushima Daiichi nuclear accident first considerations, preliminary report. http://www.iter-consult.it/ITER_Report_Fukushima_Accident.pdf
21. Feuerwehr-Dienstvorschrift FWDW 500, Einheiten im ABC-Einsatz, Ausgabe August 2004. http://www.bfs.de/de/bfs/recht/rsh/volltext/4_Relevante_Vorschriften/4_5_fwdv500_erl_2004.pdf

22. Jackson P et al (1979) Radon-222 emissions in ventilation air exhausted from underground mines. US Nuclear Regulatory Commission, NUREG-CR-0627, Washington
23. Nielson KK et al (1979) Prediction of net radon emission from a model open pit uranium mine. US Nuclear Regulatory Commission, NUREG-CR-0628, Washington
24. Electric Power Research Institute (1979) Status report on EPRI fuel cycle accident risk assessment. Electric Power Research Institute, EPRI-NP-1128, Palo Alto
25. Nuclear radiation and health effects (2010). http://www.world-nuclear.org/info/inf05.html
26. Occupational safety in uranium mining (2010). http://www.world-nuclear.org/inf/inf24.html
27. Diehl P (2000) Uranium mining in Eastern Germany, The WISMUT Legacy. http://www.wise-uranium.org/uwis.html
28. International Atomic Energy Agency (1985) Status of liquid metal cooled fast breeder reactors. Technical report series No. 246, International Atomic Energy Agency, Vienna
29. Halbritter G et al (1982) Beitrag zu einer vergleichenden Umweltbelastungsanalyse am Beispiel der Strahlenexposition beim Einsatz von Kohle und Kernenergie zur Stromerzeugung, Kernforschungszentrum Karlsruhe, KfK 3266
30. Van der Stricht S et al (2010) Radiation Protection 164, radioactive effluents from nuclear power stations and nuclear fuel reprocessing sites in the European Union, 2004–08. European Commission
31. Levels of radioactivity around the AREVA site in LaHague—AREVA (2010). http://www.areva.com/EN/operations-1228/levels-of-radioactivity-around-the-areva-site-in-la-hague.html
32. e-digest statistics about: radioactivity, UK defra. Department for Environment, Food and Rural Affairs (2010). http://www.defra.gov.uk/evidence/statistics/environment/radioact/radliquid.htm
33. Ledermann P et al (1995) Operational experience of industrial reprocessing facilities UP2 and UP3 at LaHague. In: Proceedings of the international conference on evaluation of emerging nuclear fuel cycle systems, GLOBAL 1995, Paris
34. Wilhelm J (1975) Spaltjodabtrennung in Kernkraftwerken und Wiederaufarbeitungsanlagen, Kernforschungszentrum Karlsruhe, KfK 2244
35. Ayers JW et al (1974) Pressure-swing sorption concentration of Krypton-85 for permanent storage. In: Proceedings of the 13th AEC air cleaning conference, San Francisco, USA
36. Laser M et al (1974) AKUT—A process for the separation of aerosols, Krypton and Tritium from burner off-gas in HTR-fuel reprocessing. In: Proceedings of the 13th AEC air cleaning conference, San Francisco, USA
37. Stephenson MJ et al (1974) Absorption process for removing Krypton from the off-gas of an LMFBR reprocessing plant. In: Proceedings of the 13th AEC air cleaning conference, San Francisco, USA
38. Rapport de sûreté nucléaire et de radioprotection (2009) Etablissement de MELOX, MELOX Service communication. www.areva.com
39. Coyle A et al (1991) Occupational exposure at the nuclear fuel reprocessing plant at Sellafield in Cumbria. Radiat Prot Dosim 36(2–4):173–175
40. Rannon A et al (2005) La radioprotection des travailleurs, bilan 2004, Rapport DRPH/2005-09 IRSN-Direction de la radioprotection de l'homme, Fontenay-aux-Roses, France

Chapter 11
Safety and Risk of Light Water Reactors and their Fuel Cycle Facilities

Abstract The safety of light water reactors is based on long term international research programs. The objective is to protect the operational personnel, the environment and the population against radioactivity releases during normal operation and in case of accidents. The safety concept is based on multiple containment structures (multi-barriers) as well as engineered safeguards components and other measures combined in a staggered-in-depth concept of four safety levels. The light water reactor plant and its protection system must be designed and built according to the design basis concept. Those design basis accidents which are part of the licensing process must be accommodated by the protection system, the inherent safety features and by the emergency cooling systems of the nuclear plant. Probabilistic safety analysis are supplements to this deterministic approach. They show that European light water reactors have a frequency of occurrence of about 10^{-5} to 10^{-6} per reactor year for core meltdown. Reactor risk studies which had been performed during the 1970s (USA) and 1980s (Europe) showed that the risk arising from light water reactors as a result of core melt down is well below the risk of other power generating or traffic systems. However, the Chernobyl accident in 1986 resulted—in addition to a not well known number of fatalities—in large scale land contamination by cesium-137 with a half-life of about 29 years. Similarly, the Fukushima accident (2011) resulted in land contamination by radioactive cesium isotopes. New research programs on severe accident consequences were initiated around 1990s. Their results lead to a revision of the results of the early risk studies of the 1980s and a new safety concept for modern light water reactors, e.g. the European Pressurized Water Reactor (EPR) and the European Boiling Water Reactor (SWR-1000). This new reactor safety concept allows to limit the severe accident consequences to the plant site itself. Also the introduction of additional severe accident management measures for existing light water reactors resulted in a considerable improvement of the prevention and mitigation of severe accident consequences. The safety concept of fuel cycle plants, e.g. spent fuel storage facilities, reprocessing facilities and waste treatment facilities is based on similar containment and engineered safeguards measures. However, the risk

G. Kessler, *Sustainable and Safe Nuclear Fission Energy*, Power Systems,
DOI: 10.1007/978-3-642-11990-3_11, © Springer-Verlag Berlin Heidelberg 2012

of these fuel cycle facilities is much smaller as the fuel is at much lower temperatures in reprocessing and refabricataion plants.

11.1 Introduction

The purpose of reactor safety is the protection of personnel in reactors and facilities of the nuclear fuel cycle as well as the protection of the environment of these plants and of the population [1, 2]. Failures leading to radioactivity releases must be excluded in the design of the nuclear plant as far as possible or, should defects occur nevertheless, their consequences must be limited reliably.

The radioactive inventory of the core of a 1 GW(e) reactor with high burnup fuel is approximately 10^{21} Bq. It arises chiefly from the fission products present in the fuel elements during reactor operation. Most of the fission products are contained in the fuel elements (fuel matrix and fuel cladding). This does not apply to some fission product gases which are accumulated in the fission gas plenum of the fuel elements.

The fuel elements can only be destroyed by overtemperatures, e.g. melting of the fuel or rupture of the fuel rod cladding due to overpressure. This first causes fission product gases to be released, e.g. tritium, carbon-14, argon, krypton, xenon, and then also highly volatile fission products, such as I-131, Cs-137, Sr-90 etc.

Overtemperatures arise from an imbalance between the heat production and the heat removal in the reactor core during reactor operation (power or cooling transients).

However, imbalances between the heat produced and after which the heat removed can arise also when the reactor is shut down, as the radioactive substances generate heat by radioactive decay. This decay heat power (afterheat) (Sect. 3.1.8) is roughly 6% of the nominal reactor power shortly after shutdown of the reactor, and 1% after approximately 6 h, 0.3% after one week, 0.1% after three months and 0.04% after 1 years and 0.006% after 3 years for a burnup of 50 MWd/kg [3]. This decay heat power is slightly dependent on burnup and increases somewhat with higher burnup for cooling time periods up to about 100 years [4].

Against the background of these considerations, the goals of protection listed below are required which should be ensured in a nuclear reactor under all conditions.

11.2 Goals of Protection for Nuclear Reactors and Fuel Cycle Facilities

In case of a disturbance during power operation, controlling, limiting and safety shut down systems intervene as foreseen in the plant design and reduce the power level or shut the nuclear power plant down [5]. However, even after shutdown of the nuclear chain reaction, the reactor core needs to be cooled because of the decay heat

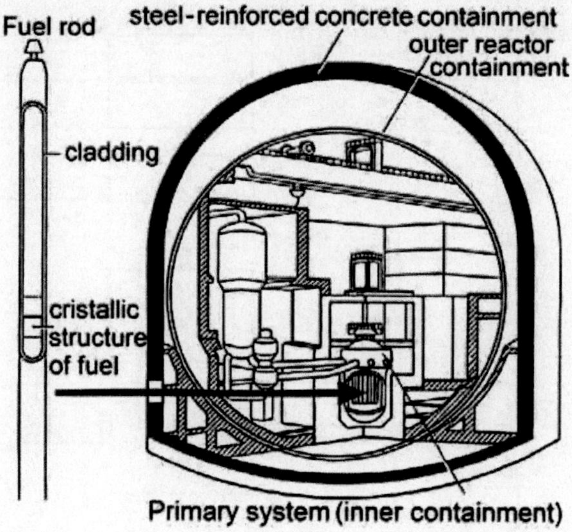

Fig. 11.1 Multiple barrier containment concept against release of radioactivity (shown for a German PWR) [5]

(afterheat) produced. Out of these reactor physics characteristics arise the following basic engineered safeguards requirements (goals of protection) which must be fulfilled at all times:

- Safe shutdown of the nuclear power plant: It must be possible to shut the reactor core down safely at any time and hold it in this shut down condition.
- Core cooling: The reactor must be cooled sufficiently at all times during operation and after shutdown.
- Safe containment structures, i.e. protection from malfunction-induced releases of radioactivity; limitation of radiation exposure to workers inside the reactor containment and of the population outside.

11.3 Safety Concept of Nuclear Reactor Plants

The safety concept of nuclear reactor plants is based on multiple containment structures around the radioactive materials in the reactor core (multi-barrier concept) as well as on engineered safeguards components and measures ensuring such containment.

11.3.1 Containment by Radioactivity Enclosures

The radioactive substances are enclosed in several radioactivity enclosures. In a German PWR shown by way of example (Fig. 11.1) these are

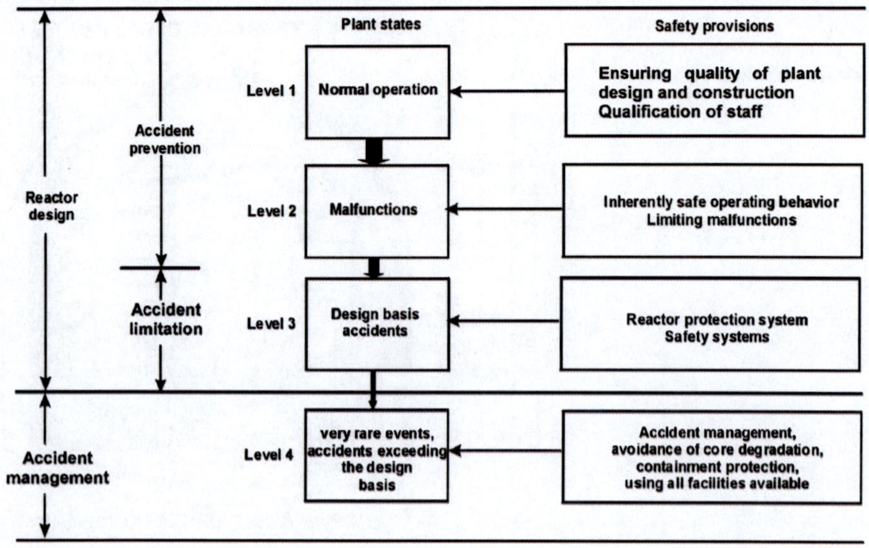

Fig. 11.2 Schematic representation of the multiple level safety principle providing safety in nuclear power plants [5] adapted

- the oxide crystal lattice of the ceramic fuel pellets (UO_2 or mixed $UO_2(PuO_2)$ oxide),
- the zircaloy cladding tubes of the fuel (welded gastight),
- the reactor pressure vessel with the closed cooling system,
- the gastight and pressure resistant steel containment enclosing the cooling systems, and the concrete structures shielding against radiation,
- the outer shell of steel-reinforced concrete. It has a limited sealing function. It also protects the plant against external impacts.

11.3.2 Multiple Level Safety Principle

In addition, the safety of a nuclear power plant is ensured by multiple levels of safety superimposed upon each other (safety concept staggered in depth) (Fig. 11.2) [5].

11.3.2.1 First Safety Level: Reactor Physics Design, Basic Safety, Quality Assurance

Nuclear reactor plants must be designed to be safe in terms of reactor physics. This includes, e.g. reactivity safety coefficients, such as the negative power coefficient, the negative Doppler coefficient, the negative coolant temperature or void coefficient, and the proper setting of the reactivity range for the shutdown system as a function of the ranges of temperature and power, as well as the proper setting of the boron or gadolinium poisoning concentrations for the range of fuel burnup envisaged.

All nuclear components of the reactor core, and the components of the cooling system are designed with high safety margins and must meet stringent requirements with respect to the choice of materials and the quality of manufacture (e.g. basic safety, leak-before-break criterion). In-service inspections and, if necessary, replacement of the components as well as great care in plant operation must ensure a high standard of technical safety quality throughout the entire operating life. This is required to make malfunctions extremely unlikely.

Basic rules in technical safety must be applied such as the failsafe principle, redundant design of cooling systems and safety systems, as well as the principle of diversity to avoid common-mode failures.

More details are given in Sect. 11.6.

11.3.2.2 Second Safety Level: Measures of Accident Prevention

Measurement and detection systems (instrumentation), control, monitoring and limiting systems (e.g. for temperatures, power, pump speeds, pressures, etc.) prevent accidents by early detection of malfunctioning. Limiting and control systems take credit of inherent safety properties to counteract disturbances in an adequate way. After correction of the malfunction, continued operation of the nuclear power plant is easily possible. Even in such cases of malfunction, the release levels for radioactive substances permitted in normal operation, must not be exceeded.

More details are discussed in Sect. 11.6.3.

11.3.2.3 Third Safety Level: Design Basis and Measures to Limit Accident Consequences

As the occurrence of an accident cannot be excluded, nuclear power plants are equipped with safety systems. These safety related systems include, e.g. the reactor protection and shutdown system, the emergency cooling system and residual heat removal system, and the containment. After having been initiated by the reactor protection system, the safety systems operate largely automatically so as to meet the goals of protection referred to above (Sect. 11.1) and limit the damage arising from an accident. The design takes into account that one redundancy level of the safety system may be under repair and another system may not be available on request (n + 2 principle).

The plant must accommodate a number of design basis accidents which must be proved by analysis during the licensing procedure of the reactor plant (Sect. 11.4).

More details are discussed in Sect. 11.6.6.

11.3.2.4 Fourth Safety Level: Measures Taken to Reduce Damage If Design Basis Is Exceeded

On the fourth safety level accident management measures take credit of existing design margins up to failure of in-plant systems and of additionally installed components which will be used in case of failure of the plant protection system.

Measures are taken to minimize the potential damage caused by external, man-made impacts, e.g. airplane crash, chemical explosion etc.

Accident management measures also take credit of existing in-plant systems which are not classified as safety systems.

11.4 Design Basis Accidents

The reactor plant and the protection systems must be designed and built on the basis of the design basis accident concept. This includes a number of design basis accidents which must be accommodated safely, even if a fault independent of the original cause of accident initiation occurs. The safety must be demonstrated by advance calculation (design basis accident analysis). This analysis must be based on conservative assumptions wherever uncertainties exist.

Selected design basis accidents require proof to be supplied, that certain limits (temperatures of the fuel elements, pressures, stresses and strains in components of the primary cooling systems) are not reached, with the provision that this requires no manual measures to be taken in the first thirty minutes. The dose levels listed in the Radiation Protection Ordinance for accidental radioactivity releases (Sect. 11.6.6) must not be exceeded.

11.4.1 Events Exceeding the Design Basis

Sequences of events exceeding the design basis and leading to severe accidents—despite measures taken by severe accident management—must have a probability of occurrence of less than 10^{-5} to 10^{-6} per year. This must be shown by probabilistic safety analysis (Sect. 11.7).

11.4.2 Probabilistic Safety Analyses

Probabilistic safety analyses (PSA) are not part of the valid licensing procedures, which use only deterministic criteria. However, they have proved their value as supplements to the deterministic approach. Probabilistic safety analyses begin with the assessment of initiating events (malfunction and defects in plant components) as well as external events like earthquakes etc. Analyzing the sequence of events requires exact knowledge of all safety systems of a nuclear power plant.

The result of a PSA are frequencies per reactor year for the failure of specific components of the safety systems and for the occurrence of specific accident sequences. In this way, weak spots in technical safety can be identified and the engineered safeguards design of a nuclear power plant can be optimized. When used in risk analyses, these values also serve for relative comparisons, e.g. of various other energy systems.

The results of PSA show that the frequency of events gradually decreases from safety level 1 to safety level 2 to safety level 3 to safety level 4. The calculated occurrence of core meltdown upon failure of the safety systems lies beyond safety level 4 in the range of a target of 10^{-5} to 10^{-6} per year to be reached. The results of probabilistic safety analyses are associated with uncertainties stemming from uncertainties of the data, assumptions, and methods applied in various ways.

11.5 Atomic Energy Act, Ordinances, Regulations

Most countries operating nuclear reactors or fuel cycle facilities issued an Atomic Energy Act. The "Act on the Peaceful Uses of Nuclear Energy and the Protection against Its Hazards" (Atomic Energy Act) establishes the legal frame for the peaceful utilization of nuclear power.

The provisions of the Atomic Energy Act are supplemented by additional ordinances (in case of Germany as an example), such as

- the Radiation Protection Ordinance,
- the Ordinance on Safety Commission and Reporting Duties under the Atomic Energy Act,
- the Ordinance on Insurance Cover under the Atomic Energy Act.

In addition, there are regulations about technical safety which serve as a basis in the licensing practice of the licensing and supervisory authorities, such as

- safety criteria and guidelines for the design of nuclear power plants,
- guidelines, e.g. about the specialized knowledge required for nuclear power plant personnel.

11.6 Detailed Design Requirements at Safety Level 1

At safety level 1 (Sect. 11.3.2.1), the key role is played by the thermodynamics, neutron physics and mechanical design of the nuclear reactor and the properties of used materials in components, such as the reactor pressure vessel, pumps and pipes. In addition, training of the operating personnel must be ensured.

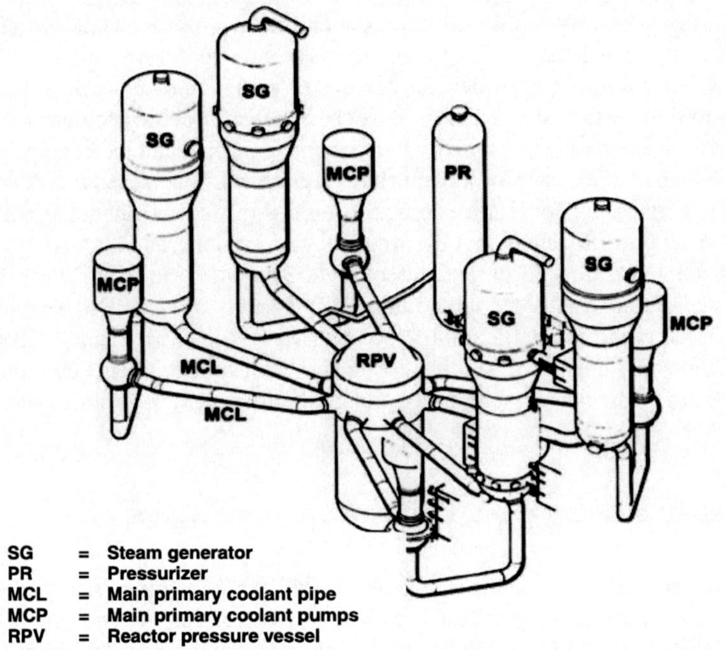

Fig. 11.3 Primary coolant circuit system of a PWR (fourfold redundant primary cooling system) [1]

11.6.1 Thermodynamic Design of LWRs

For achieving higher redundancy, the plant design is split into several identical systems for heat removal from the reactor core [1, 6–8]. Present pressurized and boiling water reactors have three or four identical cooling circuits with coolant pumps, steam generators, feedwater systems, emergency core cooling systems etc. connected to the reactor pressure vessel. Figure 11.3 shows the primary coolant circuit system of a modern PWR. This includes the pressurizer for coolant pressure control and stabilization. In the pressurizer, electric heaters increase pressure through evaporation of the pressurized water. A pressurized-water spray system in the pressurizer condenses the steam, thereby lowering the pressure. When the pressure becomes too high, relief valves above the pressurizer can release steam into an expansion vessel in the reactor containment and thus prevent overpressure failure of the primary cooling system.

The coolant pressure in the cooling circuits and in the pressure vessel of a PWR is chosen such (15.5 MPa) that nominal power of the fuel rods of the reactor core cannot give rise to local or subcooled boiling. In addition, there must be a sufficient margin relative to the critical heat flux. Because of corrosion and embrittlement problems, the temperature of the zircaloy cladding of a fuel rod should not exceed 350°C.

Critical heat flux on the surface of a fuel rod would give rise to departure from nucleate boiling (DNB). At this critical level of the heat flux a vapor film is produced on the surface of the fuel rod. This causes the temperature on the surface of the fuel rod to rise so strongly as to cause failure (break) of the zircaloy cladding. The Departure from Nucleate-Boiling Ratio (DNBR) is defined as the ratio between the critical heat flux and the actual heat flux on the surface of the fuel rod:

$$\mathrm{DNBR} = \frac{q^{11}(\text{critical})}{q^{11}(\text{actual})}$$

where q^{11} is the heat flux (W/cm^2) on the surface of the fuel rod.

In the design of the PWR core, this ratio is chosen as DBNR = 1.80. The critical heat flux is determined on the basis of empirical relations [6, 7].

As a design criterion for the maximum power of a fuel rod, the associated central fuel rod temperature must not reach the melting point of UO_2 fuel (2,865°C) under any condition. In a loss-of-coolant accident (LOCA), the maximum cladding temperature of the fuel rod should not exceed 1,200°C [9].

11.6.2 Neutron Physics Design of LWRs

Preparatory critical experiments must be run to confirm the calculated enrichment by fissile U-235 in the fuel, the number of fuel elements, and criticality, k_{eff} [5–8]. This implies the slight supercriticality necessary at the beginning, which must be compensated by burnable poisons (boron, gadolinium), boric acid and poisons in partly inserted control rods. This initial positive excess reactivity compensated by burnable poisons is consumed in reactor operation mainly by the negative reactivity arising from the buildup of fission products etc. This is followed by tests confirming the negative reactivity of control/shutdown rods.

The aggregate negative reactivity of all shutdown rods must compensate the positive reactivity arising between full load (including a positive reactivity assumed in accident analysis) and zero power. At zero power, a relatively low coolant temperature must be assumed.

These are the most important engineered safeguards design parameters of LWRs:

The negative Doppler fuel temperature coefficient and the negative coolant temperature coefficient.

For a 1.3 GWe PWR, these are a

- Doppler coefficient in the range of -2.5×10^{-5} (K^{-1})
- coolant temperature coefficient in the range of -2×10^{-4} (K^{-1})

For a 1.3 GWe BWR, they are a

- Doppler coefficient in the range of -2×10^{-5} (K^{-1})
- coolant void coefficient in the range of -1.3×10^{-3} (per % steam volume)

The effective prompt neutron life time l_{eff} is about 2.5×10^{-5} (s) for all LWR cores. The negative safety coefficients, together with the delayed neutrons (see

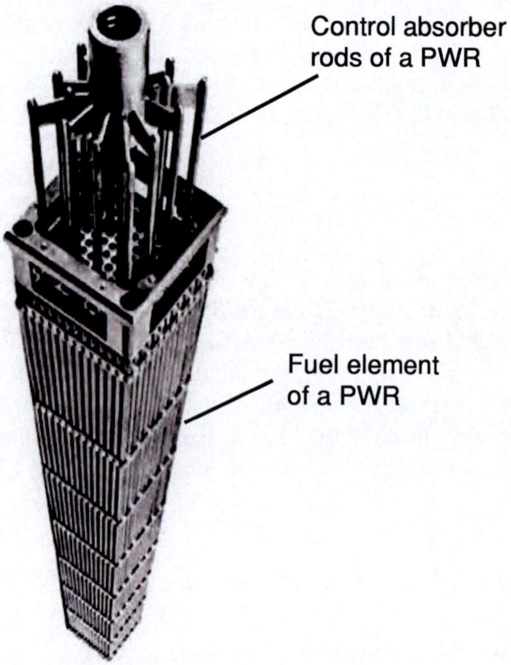

Fig. 11.4 Fuel element of a pressurized water reactor with a rod cluster control element [5, 10]

Sect. 3), guarantee a safety-oriented feedback and control behavior of LWRs. This will be demonstrated for two examples below.

11.6.2.1 Stable Time Behavior of Power when Absorber (Control) Rods are Withdrawn in a PWR

Starting from a constant reactor power level, P_0, of a PWR, which is lower than the nominal power, limited withdrawal of the absorber (control) rods by a few cm (Fig. 11.4)—as an example—shall produce a positive reactivity ramp type increase with a final positive reactivity, $\Delta\,\rho_{CR} = 0.002$ resulting in a $k_{eff} = 1.002$ within an interval of 20 s (Fig. 11.5) [1, 2]. Initially, this raises the relative power, $P(t)/P_0$, and the fuel temperature in the reactor core. After a delay of several seconds, radial heat conduction in the fuel rods of the reactor core also increases the cladding tube temperature and, as a consequence, the coolant temperature T_K as well. An increase in fuel temperature by $\Delta T_F(t)$, through the negative fuel Doppler coefficient, practically instantaneously causes a negative Doppler reactivity,

$$\Delta\,\rho_D = \Delta T_F \times (-2.5 \times 10^{-5})$$

and, after a short delay, through radial heat conduction in the fuel rods, the coolant temperature increase ΔT_K causes a negative coolant temperature reactivity,

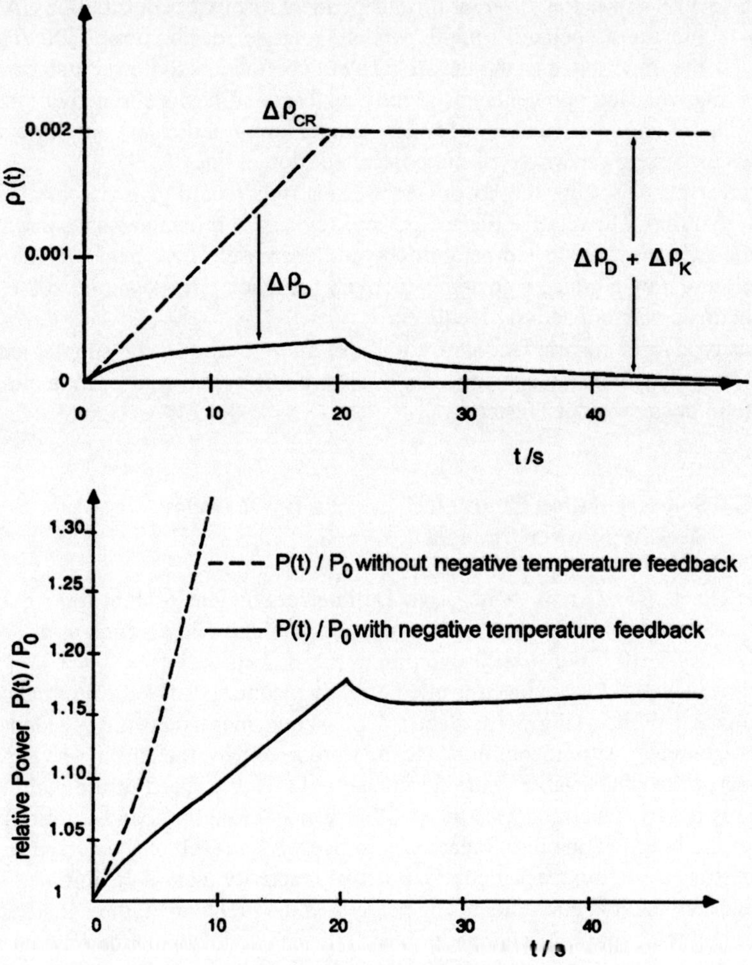

Fig. 11.5 Reactivities and relative power $P(t)/P_0$ as a function of ramp type axial movement of absorber (control) rods (without and with temperature feedbacks of reactivity)

$$\Delta \rho_K = \Delta T_K \times (-2 \times 10^{-4}).$$

Both negative feedback reactivity contributions counteract the positive initial reactivity produced by withdrawal of the absorber (control) rods until the total reactivity becomes zero. This stabilizes the reactor power at a slightly higher level of roughly $P(t)/P_0 = 1.17$ when

$$\Delta \rho_{CR} + \Delta \rho_D + \Delta \rho_K = 0$$

is reached.

Figure 11.5 shows the different time-dependent reactivity contributions, ρ_{CR}, ρ_D, and ρ_K, and the associated time-dependent relative reactor power, $P(t)/P_0$. This indicates the importance of the negative safety coefficients to power stabilization at higher relative reactor power levels. If the negative coefficients of reactivity (Doppler fuel temperature coefficient and coolant temperature coefficient) did not exist, the relative reactor power would rise uncontrolled (dotted line).

Conversely, inserting the absorber (control) rods would give rise to a negative reactivity ramp. The relative reactor power $P(t)/P_0$ will drop and, as a consequence, also the fuel and coolant temperatures would decrease. These negative changes of temperature now produce a positive reactivity feedback. The power decrease is limited by this positive reactivity feedback.

This type of automatic feedback control of a PWR core can be supplemented by minor movements of the control rods. Such support by the control rods allows the change in power to occur faster.

11.6.2.2 Self-Regulation Characteristics of a BWR Under Required Power Changes

BWRs (Sect. 11.6.2) have a negative Doppler coefficient of fuel temperature of approximately $-2 \times 10^{-5}\,K^{-1}$ and a negative coolant void coefficient of approximately -1.3×10^{-3} per % steam volume [1, 2, 8, 11].

In this way, BWRs can be controlled directly via the speed of the internal coolant recirculation pumps (Fig. 5.13; Sect. 5.1.2). When more power is needed at the turbine-generator system or more steam is required by the turbine requiring an opening of the steam valve at the turbine (Fig. 11.6), the speed of the recirculation pumps is raised. This increases the cooling water throughput, and the axial lower level of the boiling zone in the reactor core rises. As a result of the decreasing void volume fraction in the reactor core, a positive reactivity increase is initiated via the negative coolant void coefficient. The power and the fuel temperatures increase. The raising fuel temperatures in connection with the negative Doppler coefficient of fuel temperature cause a negative reactivity feedback which stabilizes the process at a higher reactor power level. In this way a BWR can be controlled in a power range above about 60% of nominal power solely by varying the speed of the axial coolant pumps located in the annulus between the reactor core and the inner wall of the reactor pressure vessel [2].

In lower power ranges control by the recirculation pumps is supported by movements of the absorber (control) rods.

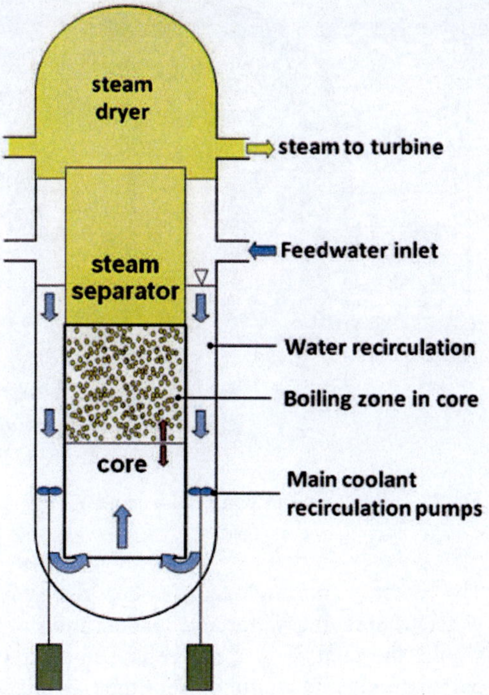

Fig. 11.6 Self-regulation characteristics of reactor power in a BWR through changes in speed of the internal main coolant pumps

11.6.3 Instrumentation, Control, Reactivity Protection System (Safety Level 2)

Instrumentation implies monitoring important measured data by

- in-core instrumentation, such as the aeroball system, miniature fission chamber detectors, continuously measuring self-powered neutron detectors [6, 12–14];
- out-of-core neutron flux measurements covering the whole range of power from startup to nominal power output; the out-of-core neutron flux instrumentation furnishes important signals to the reactor protection system; it comprises the pulse range at zero power, a medium range, and the power range [14];
- measurement of temperatures, pressures, pump speeds, water levels in the reactor pressure vessel, the pressurizer, and in the steam generators.

The thermal reactor power is determined by measurement of the inlet and outlet temperatures and the coolant flow in the four cooling circuits [1, 6, 12–16].

Disturbances and off-normal conditions, respectively, initiate countermeasures by the control systems. The control system is no safety system. Its actions counteract the course of disturbances. In case of power changes, the control system supports self-regulation of the reactor by control rod movement (PWR) or by changing the speed of the main coolant pumps and initiate control element movements (BWR).

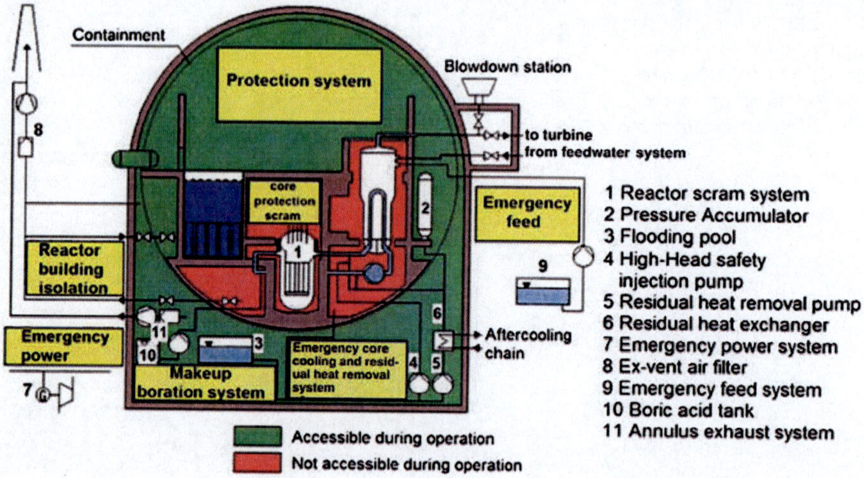

Fig. 11.7 Reactor protection system of a PWR [5]

The pressure and the level of water in the pressurizer are regulated by heating the water or spraying water for condensation of the steam in the pressurizer. On the whole, the control system keeps a number of important safety-related measured parameters stable within preset limits, in this way preventing unnecessary actuation of the reactor protection system.

When limits of normal operating ranges are exceeded, e.g. 112% of nominal power, the reactor protection system automatically intervenes to support the control system (Fig. 11.7). It can shut down the reactor by dropping the absorber rods (scram) or reduce reactor power specifically by feeding boric acid. In the case of a reactor scram, the reactor protection system at the same time automatically initiates emergency core cooling, emergency power supply, and isolation of the reactor building (containment).

The reactor protection system captures the data necessary for accident detection, e.g. reactor power too high, water levels too low, pump speed too low, etc. It has triple redundancy and operates in a 2-out-of-3 logic, i.e. when the initiation criteria are exceeded in two out of three redundant lines, the reactor is shut down.

11.6.4 Mechanical Design of a PWR Primary Cooling System

The primary cooling system of a PWR consists of the reactor pressure vessel, steam generator, pumps, a pressurizer, and the piping connected to the reactor pressure vessel (Fig. 11.3) [1, 5, 9, 17–23]. In normal operation, this system is under a coolant pressure of 15.5 MPa. The reactor pressure vessel is made of high-strength 16 MND5 steel in the EPR (Sect. 5.1.1.2) or 22NiMo37 in a German BWR (Sect. 5.1.2.1), with a stainless steel liner on the inside. Also the pipes, pump casings, pressurizer, and parts of the steam generators have stainless steel liners.

11.6.4.1 Reactor Pressure Vessel Design

German reactor pressure vessels made by KWU for the PWR-1300 design are made up of forged rings joined by circumferential welds (Fig. 11.8). The pressure vessel is designed to a pressure of 17.6 MPa, a temperature of 230°C, and a neutron fluence of 10^{19} nvt for neutrons with a kinetic energy >0.1 MeV. The wall of the pressure vessel is 25 cm thick in the cylindrical part.
The mechanical stresses in the pressure vessel wall are caused by

- loading as a result of dead weight,
- internal pressure,
- thermal stresses due to temperature gradients.

The mechanical stresses produced must be determined in accordance with the ASME Boiler and Pressure Vessel Code [17] and the RSK Guidelines [9]. A distinction must be made between primary and secondary stresses. Primary stresses are caused by the internal pressure and dead weight and cannot be relieved by plastic deformation. Secondary stresses are relieved by plastic deformation (thermal stresses). One important criterion in the ASME codes [17] and RSK Guidelines [9] is that the primary stresses in the undisturbed part, e.g. the cylindrical wall, must not exceed the value of $0.33 \cdot \sigma_B$ or $0.67 \cdot \sigma_{0.2}$ at operating temperatures (σ_B = compressive stress, $\sigma_{0.2}$ = stress at 0.2% strain). This must be demonstrated by stress and fatigue analyses for all load cases occurring.

Compliance with these criteria is able to exclude so-called failure by ductile fracture or brittle fracture (failure by crack growth with limited leakage) in a reactor pressure vessel. This conclusion is supported, inter alia, by the fact that higher internal pressures of approximately 24–26 MPa cause the seal of the top shield to leak as a result of straining of the top shield screws. The leakage area then would correspond to up to a 69 cm^2 leak.

Compliance with the provisions of the ASME code [17] must always be verified by several independent expert consultant organizations. Those provisions were laid down internationally between 1970 and 1980 after the basic principles had been confirmed in experimental programs (Heavy-Section Steel Technology (HSST) program in the United States [19–23]).

Unlike the four cooling circuits of a PWR, the reactor pressure vessel cannot be built redundant (Fig. 11.3). For this reason, the rules valid today were elaborated with particular care. A special role in this effort was played by brittle fracture behavior and the changes in brittle fracture characteristics as a result of welding processes during manufacture, or materials fatigue due to corrosion and neutron exposure.

The steel of the reactor pressure vessel or the welds could contain minute cracks or slag inclusions. When certain stresses in the material are exceeded, such minor

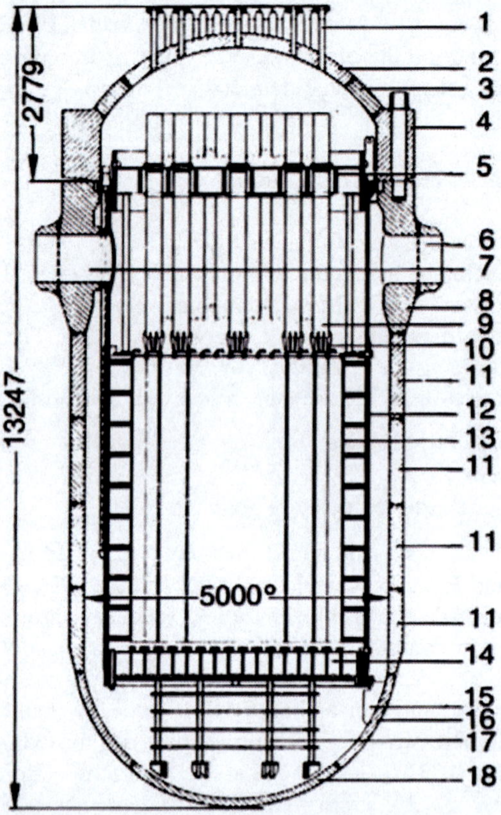

1 Control rod nozzle
2 Top hemispherical dome
3 Vessel closure head ring
4 Vessel closure head flange ring
5 Top grid plate
6 Coolant inlet nozzle
7 Coolant outlet nozzle
8 Core shroud flange ring
9 Control rod shroud tube
10 Grid plate
11 Forged ring
12 Core barrel
13 Core baffle
14 Bottom grid plate
15 Core barrel support structure
16 Bottom zone ring
17 Core support structure
18 Bottom hemispherical dome

Fig. 11.8 Reactor pressure vessel of a PWR-1300 with internals [5]

cracks could become unstable (continue to grow larger) and cause the vessel or other components of the cooling circuits to fail. These problems were clarified through the development of fracture mechanics techniques ([1, 18, 24–29]) and in many notch impact tests. The change in the so-called Nil Ductility Temperature (NDT) for assessing notch impact toughness can be verified by advance specimens in the reactor pressure vessel during reactor operation. In this way, it is also possible to

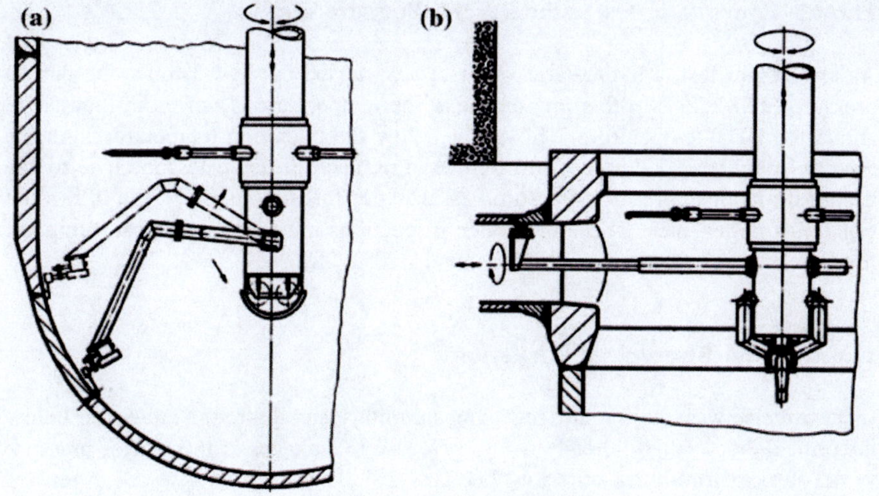

Fig. 11.9 Manipulators for ultrasonic inspection in the hemispherical bottom (**a**) and in the region of the inlet nozzle (**b**) of the reactor pressure vessel [1]

determine in advance the point when the maximum permissible neutron fluence is reached for the wall of the reactor pressure vessel.

The NDT is influenced, above all, by the existence of small fractions of copper, phosphorus, and sulphur in the steel alloy. As a consequence, these elements must be kept below preset concentration levels (Cu <0.1%, S <0.01%, P <0.01%) [1, 18, 27].

Precisely defined notch impact specimens of the base metal of the reactor pressure vessel must exhibit no less than 68 J notched bar impact work at a temperature of NDT + 33 K [1, 5, 9]. The prescribed minimum temperature for the reactor pressure vessel under nominal pressure during reactor operation or under accident conditions is 50°C [5].

11.6.4.2 Quality Assurance and In-Service Inspections (Basis Safety)

In addition to the design conditions and rules referred to above, quality assurance throughout the manufacturing process must ensure that all components have the required toughness of the base metal and the welds. For purposes of fracture mechanics the size of any small cracks must be below critical lengths [9, 27–29].

Non-destructive test methods (such as surface crack inspection and ultrasonic inspection) are used to determine crack size and crack distribution. Figure 11.9 shows the manipulators for ultrasonic inspection within the reactor pressure vessel and the manipulators for ultrasonic inspection of the coolant inlet and coolant outlet nozzles in the upper region of the reactor pressure vessel.

11.6.4.3 Hydrostatic Test of the Reactor Pressure Vessel

As an integral test, a hydrostatic water test is carried out at 1.3 times the design pressure (22.8 MPa). In this pressure test, the temperature is not more than 55°C above the NDT temperature. This is far below the operating temperature. As the fracture toughness decreases with temperature, such pressure test is close to the design limits of the reactor pressure vessel. After this pressure test, the ultrasonic tests must be repeated. These in-service inspections must be repeated during plant life time at regular intervals of eight years [1, 5, 9, 28].

11.6.4.4 Leak-Before-Break Criterion

In accordance with the fracture mechanics findings, small through cracks far below the critical crack length already would give rise to leakages of the reactor pressure vessel (leak-before-break criterion) [1, 5, 9, 28]. These can be detected in reactor operation.

11.6.4.5 Experimental Findings About Pressure Vessel Failure

Within the HSST Program of the Oak Ridge National Laboratory in the United States [1, 19–23], model pressure vessels with large artificial cracks were made to rupture at high overpressure. However, the vessel material was found to be so tough in these experiments that major plastic deformation occurred before break. Such deformations made the cracks applied less sharp-edged, thereby reducing the stress peaks arising from notch action. Ductile failure without any artificial crack faults took at least twice the design pressure level.

11.6.5 Reactor Containment

The cooling systems, which carry the high primary coolant pressure of 15.5 MPa, must be enclosed in an outer containment with the following functions and capabilities (see Figs. 11.7, 11.10) [1, 5, 9]:

- In normal operation and under accident conditions, keep releases of radioactivity into the environment within permissible limits;
- accommodate the heat stored in, and released from, the primary cooling system in a loss-of-coolant accident and remove it through active cooling systems together with part of the decay heat (afterheat);
- protect the primary system and steam generators against external impacts.

The design pressure of the containment is determined in terms of its ability to accommodate all the water released and evaporated from the primary system

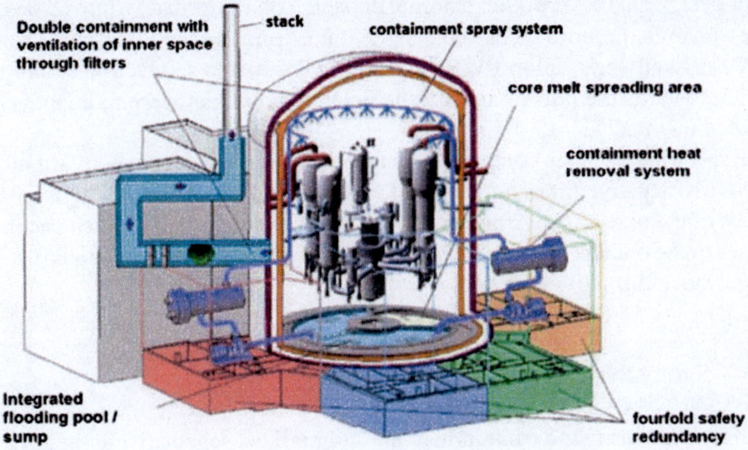

Fig. 11.10 EPR double containment with RPV, cooling systems and molten core spreading area [10]

(full-pressure containment). Moreover, the underlying assumption is that a steam generator fails in addition and its secondary-side water content is taken up by the containment. Finally, also the steam needs to be taken into account which is produced as a result of the emergency cooling water taking part of the energy stored in the secondary water of the steam generators [5, 9]. This leads to design values of approximately 0.6 MPa, depending on the volume of the containment. The design criteria of the outer containment with respect to primary and secondary stresses are similar or identical to those applying to the reactor pressure vessel. The required leak rate of 0.25% per day must be verified prior to commissioning and during reactor operation at prescribed intervals. Pressure tests must be conducted at regular intervals before startup and afterwards.

11.6.5.1 Different Designs for Reactor Containments

The containments of PWR and BWR plants have some characteristic differences in design:

- There are containments made of prestressed concrete with an inner steel liner, and containments made entirely out of steel (Sect. 5.1.1.4 and "SWR-1000 Containment and Passive Cooling Systems" in Sect. 5.1.2.2). For protection against external impacts, the reactor containment is additionally surrounded in Germany by a thick steel reinforced prestressed concrete shell which sustains
  - an impulse-type load associated with, e.g. a postulated airplane crash,
  - a pressure wave in connection with, e.g. a postulated chemical explosion.

This design also covers other external impacts, such as tornados, hurricanes, flooding or tsunamis. In more recent plants, the outer concrete shell has a wall thickness of 1.80 m (Kraftwerk Union PWR) or 2 m (EPR). It also serves as a shield against radiation towards the outside in case the inner containment were radioactively contaminated in an accident.

The space between the containment and the prestressed outer concrete shell is held at a slightly negative pressure relative to the internal pressure and the atmospheric pressure of the external environment by means of a blower. This makes uncontrolled leakages to the outside impossible in normal operation. The air taken in is discharged from the stack through filters.

11.6.5.2 Safety Systems in the Containment

The safety systems in the containment are summarized schematically in the illustration of Fig. 11.10 for the EPR containment.

- The containment spray system, which cools the atmosphere of the containment after a loss-of-coolant accident, condenses the steam released and thus accelerates pressure reduction. The water for spraying is taken from the in-containment refueling water storage tank (IRWST) (see Fig. 5.9).
- During the recirculation phase the low pressure emergency core cooling system, takes in the water from the containment sump for cooling (Kraftwerk Union PWR) or from the IWRST (EPR) (Fig. 5.9).
- A containment heat removal system decreases the temperature and pressure in the containment over the medium term (Fig. 5.9).
- Two redundant valves are used for containment isolation, one of which is installed inside, the other one outside the containment. Building isolation is initiated especially after a loss-of-coolant accident or when higher radioactivity levels are detected in the containment.

11.6.6 Analyses of Operating Transients (Safety Level 3, Design Basis Accidents)

The course of various operating transients must be studied for the following accidents with and without failure of the scram system [9]. Transients with failure of the scram system are also referred to as Anticipated Transients without Scram (ATWS). The operating transients listed below must be studied as a design basis:

- Failure of the main heat sink, e.g. as a result of closing of the main steam valve with the off-site (auxiliary) power supply functioning.
- Failure of the main heat sink with the off-site (auxiliary) power supply unavailable.
- Faulty opening of the main steam line valves.
- Complete failure of feedwater supply.

- Maximum reactivity increase as a consequence of withdrawal of control elements or groups of control elements at full power.
- Depressurization as a consequence of inadvertent opening of the pressure vessel safety valve.
- Maximum reduction of core inlet temperature due to disturbances on the steam generator secondary side.

In these accidents, the permissible stresses and temperatures of the reactor pressure vessel and the cooling system must not be exceeded. The boration system (secondary shutdown system) and the heat removal systems must be designed so that the reactor core can be shut down safely in these accidents and remains subcritical (Fig. 11.7).

11.6.6.1 Operating Transients of LWRs with the Reactor Shutdown System Functioning (Safety Level 3)

Disturbances of steady-state reactor operation arise from imbalances between heat generation and heat removal. This raises the coolant temperatures and coolant pressure. In all cases, the reactor protection system will shut down the reactor in the shortest possible time when limits of power or coolant temperatures have been exceeded or the pump speed limit of the reactor protection system has been underrun. For instance, the shut down rods of the reactor scram system, operating by the failsafe principle, drop into the reactor core from the top under gravity in about 2 s (PWR) or are pushed into the reactor core under pressure from below (BWR) in a similar time period. There must be another, diverse shutdown system (boric acid system) in case the first shutdown system were to fail (Sects. 5.1.1, 5.1.2).

One example is described below for a PWR. The accident sequence for a BWR is similar.

11.6.6.2 Loss of Off-Site (Auxiliary) Power Supply (Emergency Power Case) with Scram Functioning

Failure of off-site power supply causes the emergency power Diesel systems to start up and supply the most important components of the PWR with electricity [1, 2]. The instruments and some smaller electrical components are supplied from batteries. However, the power of the emergency Diesel systems is not sufficient to supply the large main coolant pumps of the primary system, the main feedwater pumps for the steam generators, and the main cooling water pumps of the turbine condenser. Figure 11.11 shows the key components of the steam circuit of a PWR.

In this accident, the turbine and the generator are separated first from each other. The main coolant pumps of the primary cooling circuits, the main feedwater pumps, and the main cooling water pumps loose speed and coast down. When 93% of the design speed of the main coolant pumps is underrun, the absorber-shutdown rods drop into the core (reactor scram) and the turbine emergency stop valve closes. Also the valve for the main steam bypass closes, as cooling the turbine condenser

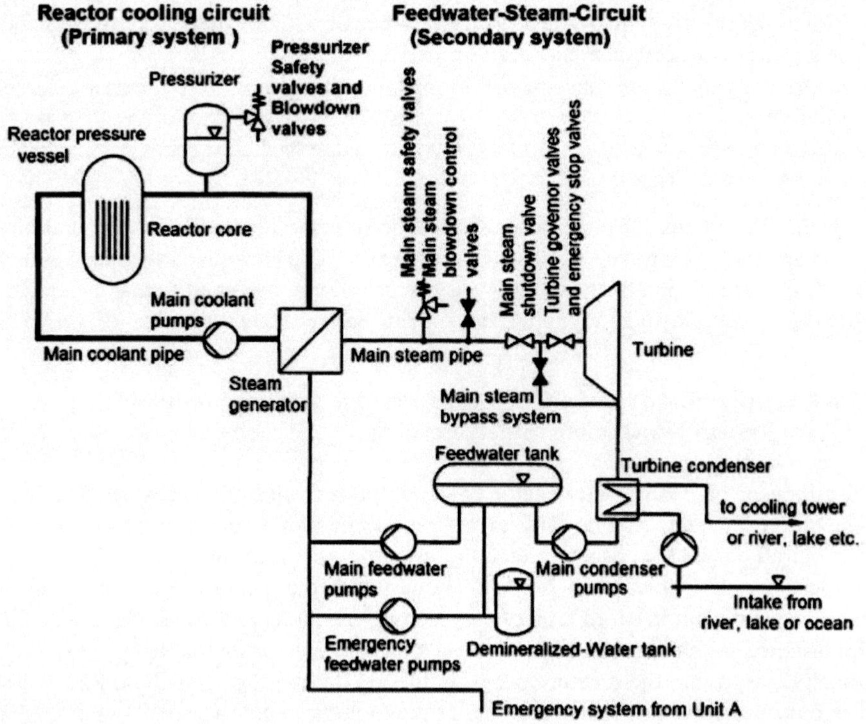

Fig. 11.11 PWR with the key components for steam production [1, 2]

fails when the cooling water pumps coast down. Closure of these valves (turbine, condenser) and steam production in the steam generators, which is continued for the time being, cause the main steam pressure to rise. However, this rise of steam pressure and coupled steam temperature on the secondary side of the steam generators can be limited by opening the main steam blowdown valves downstream of the steam generators when 7.0 MPa are exceeded. The escaping steam acts as a temporary heat sink. Initially, heat conduction in the steam generator pipes slightly raises the coolant temperatures and coolant pressure on the primary side. The pressurizer valve opens briefly, limiting the primary pressure. As a result of scram, the cooling water temperature in the reactor core decreases. Also the primary coolant pressure drops at the same time. On a medium term, the primary coolant temperature and the primary coolant pressure rise again slightly because the main coolant pumps coast down to lower speeds. Over the longer term, however, the primary coolant temperatures and the coolant pressure drop because the afterheat heat production decreases and afterheat cooling systems start working (Fig. 11.7).

This is shown in Fig. 11.12. At the same time, calculated results are compared with measured data for power, primary coolant pressure, and primary coolant temperatures as well as the water level in the pressurizer. The experiments were performed in a German 1.2 GWe PWR (Biblis) [1, 2].

Fig. 11.12 Power, temperatures and pressures in the reactor pressure vessel and cooling circuits in case of loss of off-site (auxiliary) power with scram functioning [1, 2]

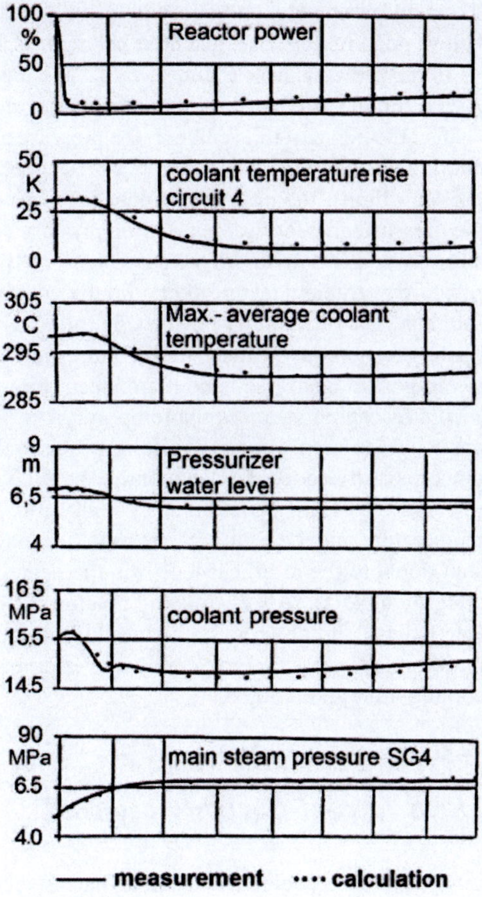

11.6.6.3 Computer Codes for Accident Calculations for PWR and BWR

Computer codes, e.g. ALMOD4 [11], RELAP [30] or RETRAN [1], TRAC [31–34] are available to compute the course of accidents in PWRs and BWRs. The theoretical models underlying these computer codes were verified repeatedly in out-of-pile test rigs.

11.6.7 Transients with Failure of Scram (Safety Level 3)

Safety level 3 of the safety concept also requires accidents including failure of the scram system to be considered [1, 2, 5]. This is why, again by way of example, the emergency power case with failure of the off-site (auxiliary) power supply for the main pumps etc. and subsequent failure of the scram system will be described below. Also in this case, accident behavior is similar for BWRs.

In the previous case (Sect. 11.6.6.2) the reactor scram was supposed to function. As the scram system now is supposed to fail and the nominal power will remain constant at the begin of the accident, the pressure and coolant temperature in the primary cooling system will rise. When the pressurizer relief valves would not open (lower pressure limit), the primary coolant pressure is limited to 17.6 MPa by opening of the pressurizer safety valves. As the primary coolant temperature (Fig. 11.13a) rises strongly together with the primary pressure (Fig. 11.13b), the negative coolant temperature coefficient takes effect, initially automatically reducing the reactor power to roughly 25% of nominal power (Fig. 11.13c). The high temperature in the primary cooling system also causes temperatures and pressure on the secondary side in the steam generator to rise. The main steam blowdown valves do limit the pressure to 7 MPa (escaping steam is the temporary heat sink), but the emergency feed pumps are unable to supply enough water as water for heat transfer of about 25% of nominal power is still needed. Consequently, the secondary steam temperature continues to rise and the steam generators gradually run dry. This causes the primary coolant temperature and the primary pressure to rise again. As a consequence, the reactor is shut down to the afterheat level via the negative coefficient of coolant temperature after about 450 s. Afterwards the reactor must be kept shut down by the boric acid (secondary) shutdown system. In addition the pressure must be further decreased until the low pressure emergency and afterheat cooling systems provide for further cooling after about 800 s.

11.6.8 Loss-of-Coolant Accidents

Loss-of-coolant accidents (LOCAs) can arise from breaks or cracks of pipes or from faulty sticking of valves in the open position. For large pipes, the leak size is assumed in accordance with fracture mechanics (Sect. 11.6.4) as 0.1 F (F = pipe cross section). However, for the design of the emergency core cooling system, the larger so-called 2F break (guillotine type rupture of pipe) is analyzed conservatively [9]. This highly conservative case will be described here by way of example in Fig. 11.14.

11.6.8.1 Loss-of-Coolant Accident Due to 2F Break of the Main Coolant Pipe

Large coolant leakages (2F break) do not necessarily require the scram system. Although the scram signal is initiated when the lower limit of primary pressure is underrun, void formation starting in the reactor core shuts down the reactor very quickly automatically via the negative reactivity produced.

When the leak has opened, a pressure relief wave passes through the reactor pressure vessel and piping system. At the leakage points, the critical outlet velocity of a

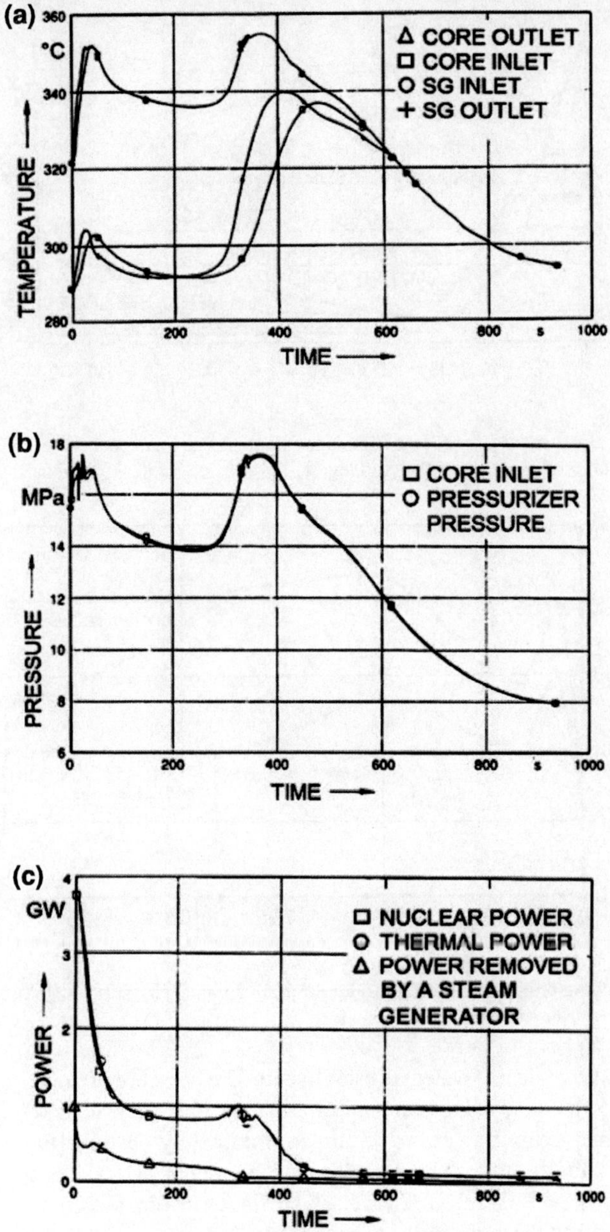

Fig. 11.13 Behavior of temperature (**a**), pressure (**b**), and power (**c**) as a function of time (s) in the core and steam generator (SG) after an emergency power case without scram [5]

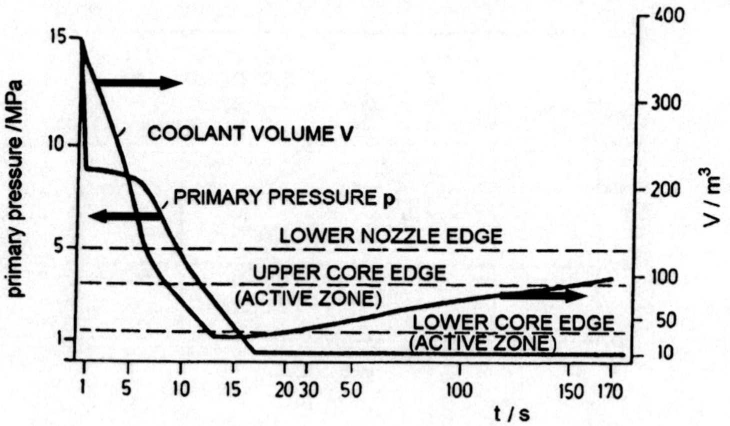

Fig. 11.14 Development as a function of time of coolant pressure and coolant volume in the reactor pressure vessel after a large leak in the cold leg [1, 2]

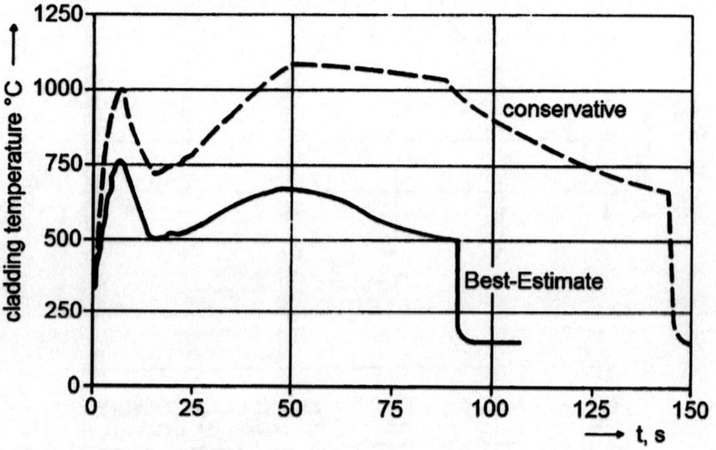

Fig. 11.15 Cladding temperature of fuel rod during 2F-Break accident (average fuel rod and fuel rod with highest conservative temperatures) [1, 2, 5]

two-phase mixture (steam–water) is established. The pressure of the primary cooling circuit drops to about 0.5 MPa within approximately 17 s. All primary coolant (water) leaves the primary cooling system within approximately 13–15 s (Fig. 11.14). Film boiling starts in the cooling channels of the fuel elements, and cladding temperatures rise very sharply to 750–1,000°C (Fig. 11.15). Afterwards, automatic shutdown of the reactor power to the level of decay heat power (after heat level), and simultaneous cooling by steam, cause a temporary decrease of temperature at the fuel cladding. However, this temperature again rises slightly until the borated cooling water of the pressure accumulators (Fig. 11.16) takes over core cooling after some 50–90 s. These pressure accumulators start feeding borated water when the primary pressure falls

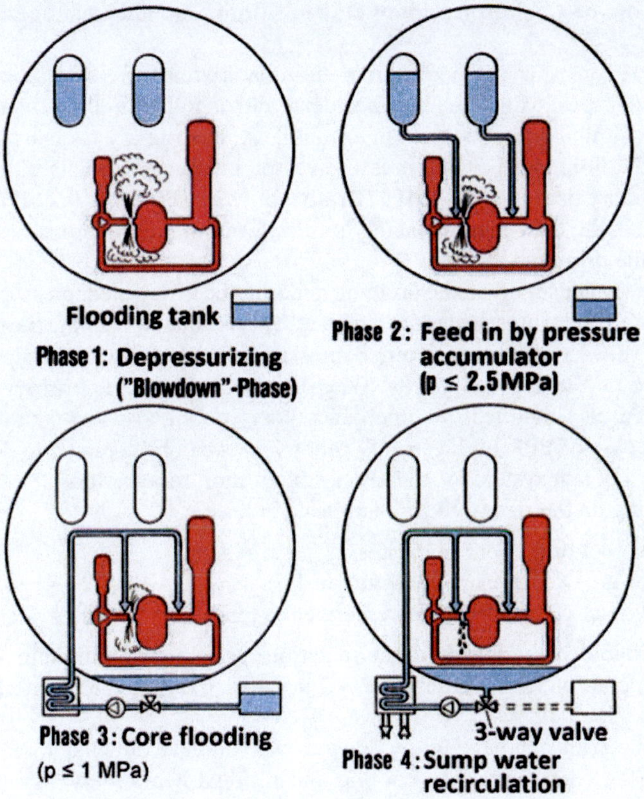

Fig. 11.16 Leak in hot primary pipe and injection of water into cold primary piping [35]

below 2.5 MPa (Phase 2, Fig. 11.16). The water level in the reactor pressure vessel rises up between 20 s and 170 s (Fig. 11.14) to the upper core edge again covering the core. It then fills the pressure vessel up to the lower inlet and outlet nozzle edges. (Height of the rising center level in the pressure vessel (Fig. 11.8) and the volume on the right scale of Fig. 11.14 are correlated.)

This borated water cools the reactor core and keeps the reactor subcritical. After further decrease of the primary pressure to <1.0 MPa the low pressure emergency water injection system takes over (Phase 3, Fig. 11.16). When the reservoir of borated water has been depleted, the water originating from the loss of coolant and collecting in the reactor building sump is taken in, cooled in the residual heat exchangers, and returned to the primary system (Phase 4, Fig. 11.16). In this way cooling by borated water from the pressure accumulators is supported by the low-pressure emergency core cooling systems (Kraftwerk Union PWR) provided their power supply is available. In case of EPR the low pressure emergency and residual heat system supply borated water from the IRWST.

11.6.8.2 Loss-of-Coolant Accident Due to Minor Coolant Leakages

The large 2F break is not necessarily the most extreme or most severe accident involving leakages. Minor leakage accidents rather follow a different sequence of events with similar consequences. In a small leak, the primary coolant pressure will drop, and the filling level in the pressure vessel will decrease. This initiates scram. When a primary pressure of 11 MPa (Kraftwerk Union PWR) or 9.2 MPa (EPR) has been reached, the high-pressure safety pumps feed borated water from the flooding tanks into the primary system.

When the water supply in the flooding tanks has been depleted, pressure in the primary cooling system must be reduced further. This is done by opening the main-steam blowdown valves (secondary system depressurization). This decreases pressure and temperature on the secondary side. When the pressure on the primary side drops below 1 MPa, the low-pressure emergency core cooling systems take over further cooling. In case of EPR high capacity relief valves can be actuated to depressurize the primary coolant system to <1 MPA within a short time period.

Depending on the size of the coolant leak, at least

- one or two out of the four high-pressure feed systems,
- one out of the two pressure accumulator feeds,
- one or two out of the four low-pressure emergency core cooling systems

must be available for feeding or for recirculation operation as a minimum requirement for accident control. In that case, the PWR in the long run can be transferred into the safely coolable mode.

However, serious damage to the reactor core can develop when three or all four systems of the emergency core cooling and residual heat removal systems or the emergency power supply fail. In that case, severe core damage will arise and the core will melt down.

11.7 Probabilistic Analyses and Risk Studies

Failure of all of these safety systems ultimately will cause a core meltdown accident. In probabilistic risk studies, such a failure of all systems is assumed conservatively.

Experience has shown that any component of specific safety systems can fail within a given period of time. In that case, event tree diagrams (fault tree analyses) and assignment of individual probabilities of component failure can be employed to compute the failure of major single systems and the overall probability of an accident sequence (probabilistic safety analysis).

11.7.1 General Procedure of a Probabilistic Risk Analysis

The first comprehensive study performed to determine the risk of LWRs by probabilistic methods, the US Reactor Safety Study WASH-1400 [36], was published in

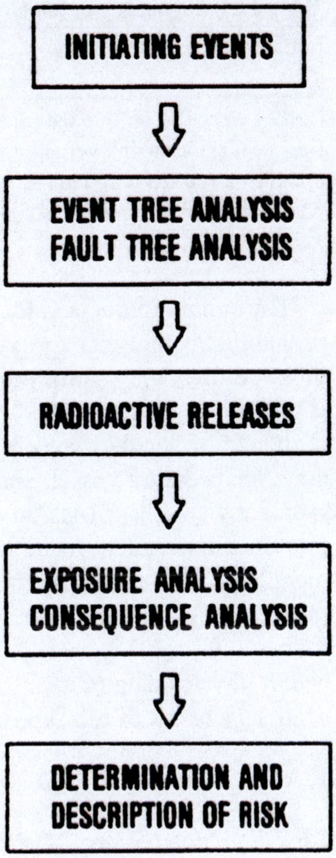

Fig. 11.17 Major steps in a reactor risk study [5]

1975 [5, 36–41]. Similar studies were performed later also in other countries, such as Germany [37].

The risk, R_m resulting from a type i accident initiated by a type m event (e.g. leak in a primary coolant pipe) in a reactor plant can be described in a simplified way by this relationship:

$$R_{m,i} = F_{m,i} \cdot D(C_{m,i})$$

where

- $F_{m,i}$ is the annual frequency of occurrence of a type i reactor accident initiated by a type m event,
- $C_{m,i}$ is the amount of radioactive material, expressed in Bq, released into the environment from the reactor outer containment during a type i accident initiated by a type m event,
- D is the damage resulting from the release $C_{m,i}$ of radioactivity. D depends on a number of other environmental parameters, such as atmospheric conditions, population distribution, etc. (Fig. 11.17).

Table 11.1 Frequency of initiating events PWR [5]

Initiating event	Frequency of initiating event per year
Loss of electrical off-site (auxiliary) power supply	0.13
Loss of main heat sink without loss of main feedwater supply	0.36
Small leak (80–200 cm^2) in main primary coolant pipe	9×10^{-5}
ATWS with loss of main heat sink and main feedwater supply	7.5×10^{-6}

The annual frequency, $F_{m,i}$ of occurrence of an accident is determined in detailed probabilistic analyses applying event tree and fault tree methods [1, 5, 11, 38]. In those studies, the failure probability of all relevant components of a safety system is taken into account (Fig. 11.18). In determining the radioactivity release $C_{m,i}$ the sequence of accident events must be assessed as a function of time in the reactor core, the pressure vessel, and the surrounding containment. This then results in the radioactivity, $C_{m,i}$, (fission products, activation products and actinides), released into the environment from the containment. Subsequently, meteorological data and models of atmospheric diffusion and aerosol deposition are used to determine the radioactivity concentration and the radiation dose to which individuals in the environment of the plant are exposed, countermeasures being taken into account. Finally, health physics data (Chap. 10) are used to determine the probability of disability or death as a result of the exposure dose.

11.7.2 Event Tree Method

An accident sequence is started by an initiating event, e.g. a leak in a pipe in the primary coolant system. The safety system of the reactor reacts to this initiating event, and the consequences of the sequence of accident events are controlled, provided that the safety system functions sufficiently well.

Table 11.1 shows some data for the frequency of initiating events used in risk studies.

Only if components of the safety system fail on a major scale, will there be a release of radioactivity. Figure 11.18 shows a simplified event tree for a loss-of-coolant accident in a PWR. In this case, the accident is initiated by the break of a pipe. This pipe rupture is assumed to occur with a frequency of f_m per reactor year.

The further development of this accident is then mainly determined by the availability of the electricity power supply. Failure of the electric power supply to operate the emergency core cooling system (ECCS) is assigned the probability of p_1. Since electricity is either available or not, the probability of power being available and the ECCS functioning properly is $(1 - p_1)$. If there is no electricity the ECCS will not work and the core, after having lost its coolant, will melt down partially or entirely for lack of cooling. In that case, there may well be major releases of radioactivity into the environment as a result of a failure of the containment.

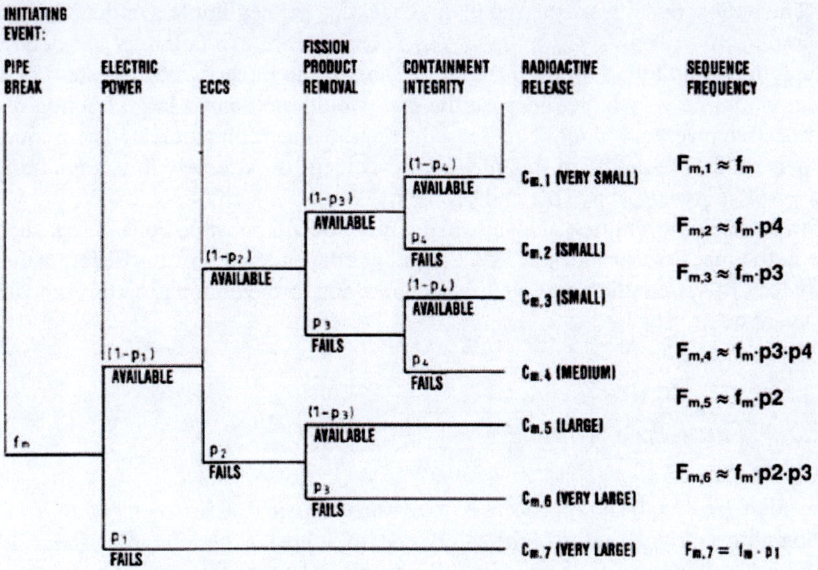

Fig. 11.18 Simplified event tree for a loss-of coolant accident in a water cooled reactor

If power is available, the next possible event will be a potential failure of the ECCS, which must be assigned the probability of p_2. The availability of the ECCS is again characterized by $(1 - p_2)$.

If fission products are released in the course of an accident, the fission product removal system mitigates the radioactivity release into the containment. The failure probability of this system is characterized by p_3, its availability $(1 - p_3)$.

The final barrier against the release of radioactivity is the leak tightness (integrity) of the outer containment. The probability of this containment function failing is called p_4 the availability of that function, $(1 - p_4)$. If the containment integrity is preserved, releases of radioactivity can only be slight, but if the containment leaks, radioactivity can escape into the environment, depending on the size of the leak.

From the results of this simplified event tree of Fig. 11.18 it can be seen that radioactivity releases can vary between very small and very large releases, depending on the level at which the safety systems fail.

Since the individual components of the safety system are characterized by their high availabilities and, consequently, very low failure probabilities ($p_1, p_2, p_3, p_4 \ll 1$), the probability of the availability of $(1-p_1)$ etc. of these safety components can each be assumed to be approximately equal to 1. Consequently, the sequence frequency at the upper end of the branching of the event tree of Fig. 11.18

$$F_{m,1} = f_m(1 - p_1)(1 - p_2)(1 - p_3)(1 - p_4) \approx f_m$$

The radioactive release caused in this case, $C_{m,1}$ is negligible. On the other hand, the radioactive releases, $C_{m,6}$ or $C_{m,7}$ as a consequence of a failure of the electricity supply followed by a failure of the ECCS and of the integrity of the outer containment would be very large, because the core would melt and a large fraction of the radioactive inventory would be released from the outer containment. The frequency of occurrence, however, of this maximum accident is extremely low, amounting to $F_{m,6} \approx f_m \cdot p_2 \cdot p_3 \approx p_2 \cdot p_3$, and $F_{m,7} = f_m \cdot p_1$.

In a detailed event tree analysis, many more details must be considered, such as the individual functions of the ECCS, etc. Interdependencies of the different events may lead to systematic consequential failures and to the elimination of branches in an event tree.

11.7.3 Fault Tree Analysis

The fault tree analysis approach is used for numerical assessment of the failure probabilities of larger units of the safety system. It breaks these larger systems down into single components, concluding about the failure probability of a larger unit from the failure probabilities of such individual components by taking into account the way in which the logical functions of the single components are interrelated. If common mode failures are possible they must be accounted for. Often, fault trees must be developed to such detail that available data on single equipment components or human error can be applied from experience. Uncertainties in reliability data are taken into account by entering not only single values, but distribution functions for the failure probabilities of single components. For other components, such as emergency power diesel systems, statistical data directly available from experience are applied. When determining the failure rate of the pressure vessel, methods of probabilistic fracture mechanics must be used in addition.

11.7.4 Releases of Fission Products from a Reactor Building Following a Core Meltdown Accident

11.7.4.1 Initiating Events

Initiating events controlled by the safety system will not contribute to risk. Accordingly, major contributions to the overall risk must be expected to arise only with large scale failure of fuel rod claddings and from those events in which the reactor core will melt down partially or completely because of an extensive failure of the safety systems. Event tree studies show that the occurrence of a major leak in a main coolant pipe followed by a failure of the respective safety systems (emergency cooling systems and afterheat removal systems) will cause the reactor core to melt down;

within a few hours the molten core can even penetrate the reactor pressure vessel. In an early superheated phase of the reactor core, hydrogen will be generated in a reaction between water and the zirconium in the fuel claddings.

After having penetrated through the reactor pressure vessel, the hot core material will contact the concrete foundation slab of the reactor building and gradually melt into the concrete. This will cause water bound in the concrete to be released and react with the melt, which will generate hydrogen. Depending on the type of concrete used, also CO, may be released. For the further sequence of accident events it was assumed in a pessimistic estimate in the US and German risk studies [5, 11, 36, 37] that the molten core contacts and evaporates the sump water, thus increasing the pressure in the containment.

11.7.4.2 Failure of the Containment

A number of penetrations through the containment building for locks, pipes and cables may develop leaks with a certain failure rate. In this case, radioactivity could escape to the outside. If, on the other hand, the containment is assumed to remain tight, i.e. preserve its integrity, the core meltdown accident described above would generate vapor, H_2, CO and CO_2, and raise the pressure in the containment so that the permissible design pressure of the outer containment could be exceeded. After failure of the outer containment integrity, radioactivity could be released into the environment.

In the US Reactor Safety Study, WASH-1400 [36], and the German Risk Study [5, 37], also the case of large scale hydrogen detonation and of a potential steam explosion resulting from a contact between molten hot core material and water was discussed. This was assumed to occur with a certain probability in the bottom part of the reactor pressure vessel.

11.7.4.3 Releases of Radioactivity

Radioactivity may be released from the reactor core in the following events:

- In cladding tube failures the gaseous and highly volatile fission products are released.
- As the fuel is heated to melting temperature and melts, fission products as well as chemical compounds with lower melting points will be released as aerosols.
- During interaction of the melt with the concrete, aerosols are generated [42].

In a pipe leak of the primary circuit, or if the reactor pressure vessel has been penetrated by molten core material, the gaseous and volatile fission products will enter the containment. They can be retained there by active removal (e.g. spray) systems and by diffusion, coagulation, condensation, sedimentation and thermophoretic processes of the aerosols. Radioactive decay makes the retention by the containment more effective, the longer its integrity can be ensured. Studies (German Risk Study,

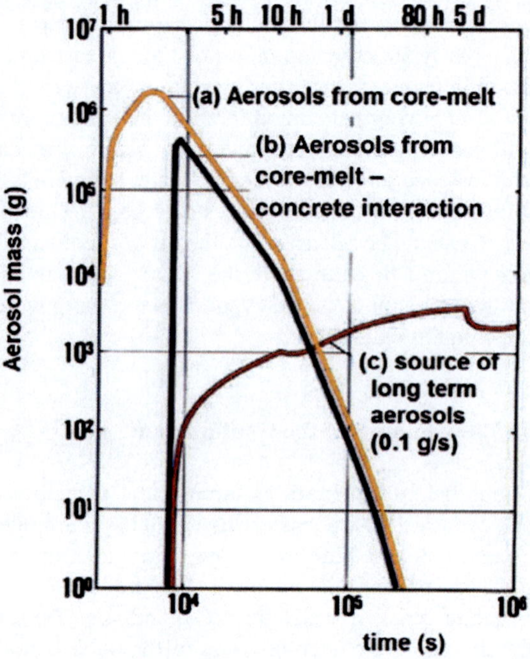

Fig. 11.19 Concentration of aerosols in the outer containment atmosphere as a function of time after release during a core melt accident [5, 35, 43]

Phase A [37]) showed that the time after which the maximum pressure would be exceeded and after which the containment fails would be like 5–12 days (depending on the concrete composition). Within such a time period of about 5 days or more, the concentration of airborne aerosols decreases already by orders of magnitude [44]. This is shown by Fig. 11.19.

11.7.4.4 Distribution of the Spread of Radioactivity After a Reactor Accident in the Environment

In reactor risk studies, releases of radioactivity are determined for various accident categories and all kinds of meteorological conditions at all the different reactor sites. This is used to determine a mean value for the consequences of radioactive exposure (early deaths, late consequences, soil contamination). However, an accident at a specific reactor site is determined only by the weather conditions prevailing at that time.

Computer codes, such as COSYMA [45] and RODOS [46], were written to describe this situation on the basis of the release of radioactive gases and aerosols during the accident and further spreading of this radioactivity in the atmosphere as a function of weather conditions and wind direction (Fig. 11.20). For each point at a certain distance from the reactor site it is possible in this way to determine

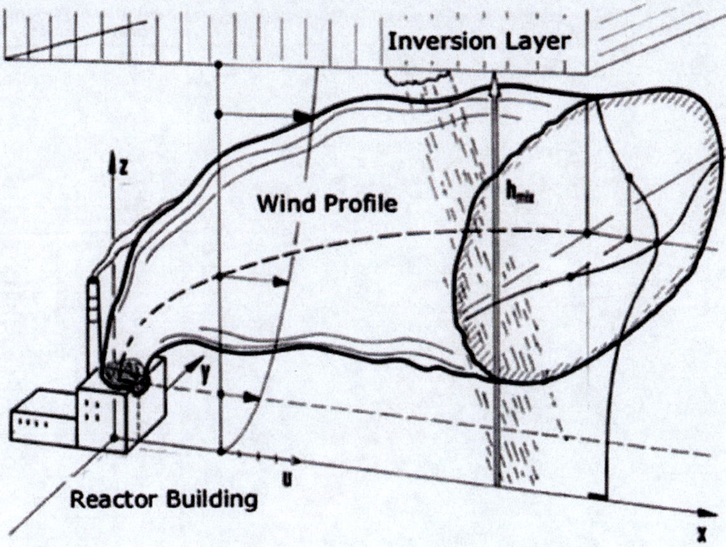

Fig. 11.20 Spread of radioactivity after a core melt accident with radioactivity release to the environment [47]

the radioactivity in the atmosphere and contamination of the soil. The radioactive exposure of the population and the environment is determined on this basis.

The radioactive cloud causes external and internal radiation exposures of persons. The external radiation exposure is the result of

- cloud-borne radiation (radioactive nuclides),
- ground-borne irradiation after surface contamination following precipitation of the radionuclides.

Internal exposure to radiation is the consequence of inhalation and ingestion of contaminated food items.

In Fig. 11.20, the cloud emitted, which is loaded with radioactive gases and aerosols, extends to the inversion layer h_{mix} and then spreads horizontally.

Figure 11.21 shows by way of example the spread of radioactivity released after a core melt accident into the environment. The release of radioactivity as a function of time (noble gases, iodine, aerosols), the wind speed and wind directions are assumed for this example. The different colors indicate areas where evacuation or sheltering of the population would be required.

11.7.4.5 Protection and Countermeasures

Radioactive exposure of the population can be affected by these protective measures and countermeasures:

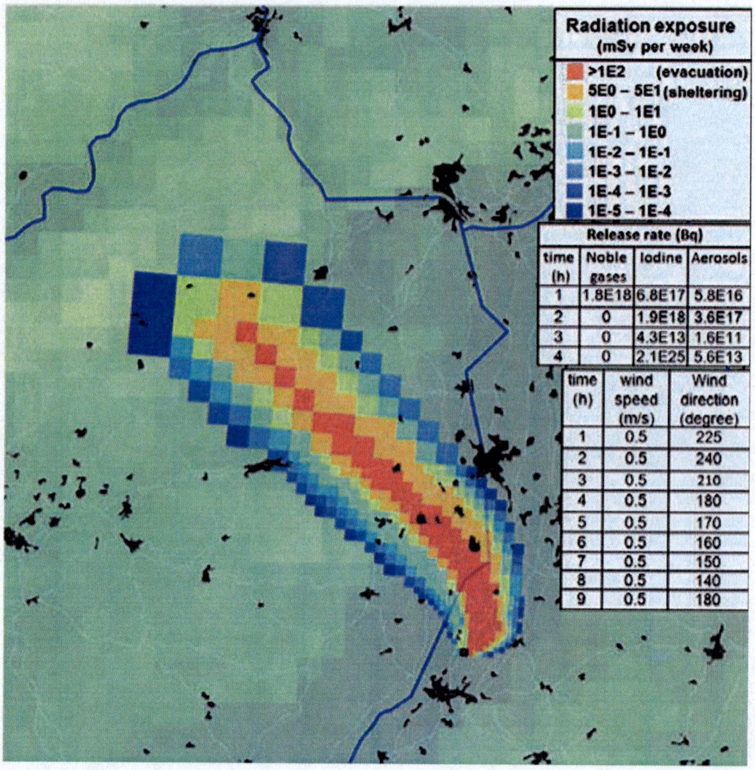

Fig. 11.21 Example for radioactivity release after a core melt accident with major release of radioactivity. The different *colors* indicate areas where the population would have to be evacuated or remain in shelters [48]

- sheltering in buildings and protective rooms,
- evacuation of the population,
- distribution of iodine tablets,
- ban on consuming contaminated food,
- relocation and blocking of areas,
- decontamination of urban areas and agricultural land.

In Europe, decisions by the authorities (Table 11.2) are based on these guiding values for various protective measures and countermeasures [47, 49, 50]. These are lower and upper limits—the population would receive during the first week—for which measures like sheltering, evacuation or relocation (1 month or 1 year) must be initiated. In addition there are limits for the different isotopes released during a reactor accident which do collect in the food like milk, vegetables, meat etc. These limits are given in Table 11.3 in case these limits are exceeded the food must be banned.

Table 11.2 Reference exposure dose values for initiation of protection and countermeasures [47, 49, 50]

Action	Reference dose	Reference value (mSv)	Integration time
Remain in house sheltering	Effective dose through inhalation and External radiation	10	7 days
Taking of Iodine tablets	Children <18 years	50	7 days
	Persons 18–45 years	250	7 days
Evacuation	Effective dose through inhalation and External radiation	100	7 days
Temporary Relocation	External radiation deposition of radioactivity	30	1 month
Longterm Relocation	External radiation deposition of radioactivity	100	1 year
Food ban	Effective equivalent dose by ingestion	5	1 year

Table 11.3 Upper limits for adults in Europe for concentration of radioactive materials in food [47, 49, 50]

Radioactive nuclide	Limits of radioactivity concentration (Bq/kg) or (Bq/l)	
	Milk products	Other food
Strontium isotopes especially Sr-90	125	750
Iodine isotopes especially I-131	500	2,000
α-emitters especially Pu-239, Am-241	20	80
Other nuclides with half-lives more than 10 days especially Cs-134, Cs-137	1,000	1,250

11.7.5 External Events

External events must be accounted for in the safety analysis of nuclear plants. This applies, for instance, to the frequencies of occurrence and the intensities of earthquakes and their repercussions upon important components determining the structural stability of the reactor building. For the site of a nuclear reactor plant, statistical data about past earthquake events are used for this purpose and also the geological structure of the site is taken into account. In addition, frequencies of floods, tornados, hurricanes, strokes of lightning, airplane crashes and explosion shock waves from nearby industrial plants or transport routes (rivers, pipelines, etc.) are determined and the potential release of radioactivity is assessed.

11.7.5.1 Seismic Design of Nuclear Power Plants

Nuclear power plants must be designed against anticipated earthquakes [9, 51–57]. Technical safety regulations make a distinction between

- operation basis earthquakes, which the nuclear power plant is to sustain undamaged (operation of the nuclear plant can be continued after an automatically induced reactor shutdown), and
- safe shutdown earthquakes, which may give rise to damage.

However, it must be possible to shutdown the nuclear power plant safely and keep the residual heat removal function intact. Containment of the radioactive fission products must be guaranteed (Protection goals must be fulfilled.).

On the basis of historical data from experience and geological and tectonic data, respectively, for the environment of a nuclear power plant site, criteria are derived for the horizontal accelerations to be expected as a consequence of an earthquake. The requirements to be met for safe shutdown earthquakes are more severe than those for operation basis earthquakes.

As the containment of a nuclear power plant is embedded in a soil rock structure, soil-building interactions must be taken into account. The seismic waves emanating from an earthquake are changed in both frequency and amplitude when passing through the soil structure in the site region. The horizontal oscillation frequency excited by an earthquake is in the range of 0.1 to a few Hertz. The US Regulatory Guide 1.60 [51–53] requires so-called horizontal and vertical response spectra to be taken as a basis for the analysis of the reactor building and all components (Fig. 11.22).

Aseismic design by installing damping elements on piping systems or underneath the reactor building are proposed for reactor plants to be built in areas endangered by strong earthquakes [58, 59].

The horizontal oscillations are transferred from the containment to the piping system of the cooling circuit. In that event, building structures must withstand the stresses generated. Moreover, the permissible stress limits arising in the piping system must not be exceeded. In addition to these detailed stress analyses of individual components of the cooling systems, supporting design measures may be employed, such as attaching to the piping systems components damping devices to avoid extreme oscillations. In regions with special seismic hazards, the entire baseplate of a nuclear power plant, like other building structures, may be placed on many thousands of damping elements (Fig. 11.23). Four nuclear power plants in France and two nuclear power plants in South Africa are placed on such damping elements [58].

The strongest earthquake affecting nuclear plants until 2010 (Nigata Chunetsu-Oki earthquake) near the Japanese nuclear power plant site of Kashiwazaka Kariwa showed measured maximum horizontal accelerations at the base plate of the nuclear power plant between 3.22 and 6.80 m/s^2. These were about a factor of 3.5 higher than assumed in the safe shutdown design of the plant. Nevertheless it was possible to shut down the plant safely. However, the plant was out of operation for repair for three years [58].

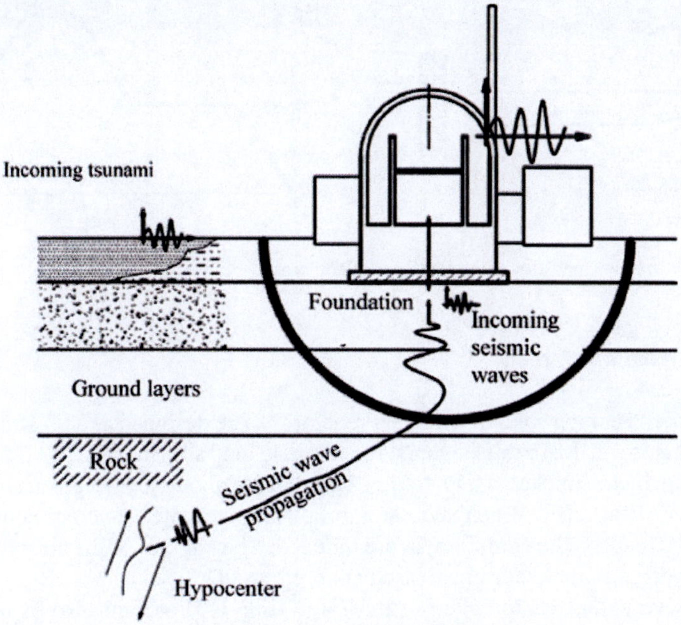

Fig. 11.22 Response spectra for nuclear power plant during earthquake

Fig. 11.23 Damping elements underneath the reactor building [58, 59]

In March 2011 an earthquake of intensity 9 on the Richter scale about 90 miles off the northeastern coast of the island of Honshu, Japan, shook the Fukushima nuclear power plant site. The reactors were shut down as foreseen, and the emergency Diesel power systems started up. However, they were hit by a tsunami wave of about 14 m height about 1 h later which put out of action the emergency Diesel power systems and the ultimate heat sink for afterheat removal (Sect. 11.8.3).

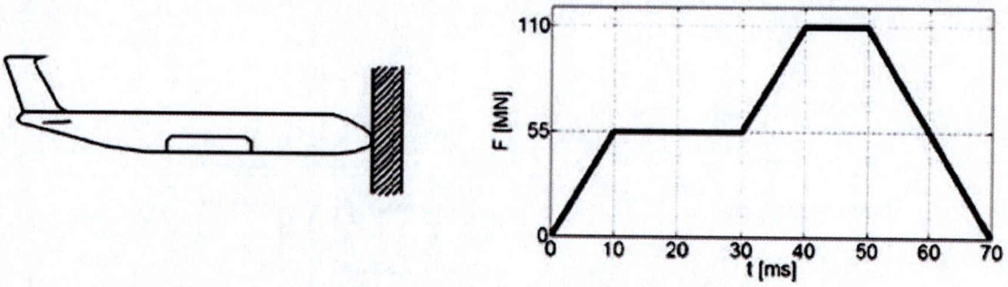

Fig. 11.24 Impulse load-time function for airplane crash (Germany) [9]

11.7.5.2 Airplane Crash

In Germany, the outer shell of a nuclear reactor is to be designed as a prestressed concrete containment 1.8 m thick against the impulse load shown in Fig. 11.24, which is caused, e.g. by the impact of a Phantom-2 airplane of about 20 t weight at an assumed speed of 774 km/h [9]. When such an aircraft hits the outer concrete containment, shock-type loads to the containment are induced. The engines of the aircraft may not penetrate through the outer prestressed concrete shell.

In this way, the reactor plant at the same time is protected also against other civilization-induced external impacts.

11.7.5.3 Chemical Explosions

Nuclear power plants must be designed in Germany against chemical explosion which could arise e.g. from chemical explosives transported by ships on nearby rivers [9].

11.7.5.4 Tornados, Hurricanes, Flooding and Tsunamis

Nuclear power plants must also be designed against tornados and external flooding [9]. If they are located at the ocean coast they must be designed against flooding by tsunamis caused by earthquakes deep under the ocean.

The regulations are different in the various countries operating nuclear reactors. In Germany flooding heights accounting for the highest floods of the past 10,000 years [60] must be obeyed.

11.7.6 Results of Reactor Safety Studies

11.7.6.1 Results of Event Tree and Fault Tree Analyses

The methods and analyses described above were applied to both PWRs and BWRs, e.g. in the US Reactor Safety Study [36] and in the German Risk Studies [37]. The

Table 11.4 Expected frequencies for core melt down (German Reactor Risk Study, Phase B) for 1.2 GW(e) PWR (Biblis B) [37]

	Initiating event	Frequency per year
Loss of coolant	Medium to large leak in primary coolant pipe >200 cm^2	$< 10^{-8}$
	Small leak in primary coolant pipe (25–200 cm^2)	8.1×10^{-7}
	Very small leak in primary coolant pipe (2–25 cm^2)	3.2×10^{-6}
	Leaks at pressurizer	3.0×10^{-6}
	Leak in annular zone (containment)	$<10^{-7}$
	Steam generator tube leaks	1.1×10^{-6}
Operational transients	Emergency electrical power supply loss of main heat sink with/without loss of main feedwater supply	1.5×10^{-5}
	Leak in main steam line	2.5×10^{-6}
	ATWS[a] (emergency electrical power supply, loss of main heat sink with/without loss of main feedwater supply)	2×10^{-7}
Plant internal	Fire	1.7×10^{-7}
	External flooding	$<10^{-7}$
Plant external	Earthquake with loss of main heat sink and main feedwater supply	3×10^{-6}
	Airplane crash in reactor containment	$< 10^{-7}$
Sum total		2.9×10^{-5}

[a] *ATWS* Anticipated Transients Without Scram

PWR and BWR designs built by Kraftwerk Union, as described in Sects. 5.1 and 5.2 were investigated in the German Risk Studies [5, 11, 37].

Table 11.4 lists the main findings of the event tree and fault tree analyses in the German Risk Study for a PWR operating in Germany at the River Rhine (Biblis B) [5, 37]. The Risk Study for a German BWR results (Table 11.5) in similar findings [11]. In analyzing all possible initiating events, mainly four groups of accident sequences were identified to result in core damage for the PWR:

- loss-of-coolant accidents initiated by a leak or break in the reactor coolant system,
- transients leading to an imbalance between the heat generated in the core and the heat removed from the core,
- reactor plant internal initiating events like fire or internal flooding,
- reactor plant external events e.g. earthquakes, airplane crash and flooding.

The data in Table 11.4 result from a detailed study of conceivable accident sequences. A loss of main coolant flow through small leaks in a coolant pipe with a frequency of occurrence of 3.2×10^{-6} per reactor year and leaks in the pressurizer valves (3.0×10^{-6} per reactor year) dominate the loss of coolant accident sequences. The largest contribution with 1.5×10^{-5} per year results from operational transient accident sequences where the loss of offsite power supply and the loss of emergency coolant supply for the steam generators play an important role. The contribution from a large leak in a main primary coolant pipe is relatively small (10^{-8} per year), because the safety systems have been optimized already to cope with this accident.

Other contributors are leaks in the main steam line and transients initiated by strong earthquakes exceeding the design characteristics of the nuclear plant.

The sum total of all contributions adds up to an overall core meltdown frequency of about 2.9×10^{-5} per reactor year. An earlier risk study called German Reactor Risk Study Phase A for the same PWR of 1.3 GW(e) power output had resulted in an overall core meltdown frequency of 9×10^{-5} per reactor year. The difference of a factor of 3 for the overall core meltdown frequency per reactor year is due to improvements of safety systems, the automatic reduction capabilities of steam generator pressure and temperatures (bleeding procedure) and due to more refined analysis of the different accident sequences. An even higher reduction of the overall core melt frequency per reactor year is possible by the introduction of so-called severe accident management safety measures (safety level 4, Fig. 11.2).

11.7.6.2 Severe Accident Management Measures (Safety Level 4)

The detailed analyses of accident sequences in the Phase B Risk Study [5] showed that it is possible, even after the failure of safety systems, to control accident sequences by introducing so-called severe accident management measures. These can then prevent core meltdown.

A number of accident sequences initiated by loss of coolant or loss of off-site (auxiliary) power supply are seen to have as their main cause the failure of cooling water supply to the steam generators. This failure of steam generator supply ultimately leads to core meltdown.

Accident management measures go beyond the **automatic** safety measures provided for in the reactor protection system [61, 62] allowing interventions by the operating personnel. Thus, for instance, the steam generators, in the late accident phase of running dry can be fed with water from the feed water tank or, by means of mobile pumps, from water reservoirs inside and outside the reactor. The decay heat (afterheat) can be removed by blowing off the steam through the main steam blowdown station (secondary feed and bleed procedure) (Fig. 11.11).

In addition to these secondary accident management measures, the pressurizer valves can be opened in the primary cooling system. In this way, pressure on the primary side can be lowered enough within about half an hour for

- the high-pressure safety feed pumps to be activated at 11 MPa,
- the pressure accumulators to start feeding at 2.5 MPa,
- and the residual heat removal systems to be activated at about 1.0 MPa.

In this way, for instance, core meltdown under high pressure can be avoided, or this accident sequence at least can be changed into core meltdown at low pressure. Other accident management measures include the actuation of relief valves or of pumps of auxiliary circuits by means of batteries. The internal battery capacity can be increased and reinforced accordingly.

German LWRs in addition were equipped with a so-called sheltered control room. It is possible to shut down the reactor to decay heat (afterheat) level from this com-

Table 11.5 Frequency of initiating events (BWR) [11]

Initiating event	Frequency of initiating event per year
Loss of electrical power supply from grid	0.04
Loss of main heat sink and loss of main feed-water supply	0.3
Inadvertent opening of turbine valve or by-pass valve	0.2
Failure to close a safety and relief valve	0.1
ATWS with loss of main heat sink	$<10^{-7}$

pletely independent control room in case the main control room can no longer be used, e.g. because of fire or radioactive gases. Additional emergency instrumentation, e.g. filling level probes in the reactor pressure vessel or measurements of radioactivity at various points of the cooling system, allow more precise knowledge to be obtained about the operating status of the reactor [5]. In addition, electric power can be supplied by underground cables from more distant power plants. Cooling water can be supplied by deep wells situated on plant site.

These **safety accident management measures** allowed the frequency of core melts to be reduced from 2.9×10^{-5} to **3.6×10^{-6} per reactor year.**

11.7.7 Results of Event Tree and Fault Tree Analyses for BWRs

Probabilistic risk analyses were carried out also for BWRs [11]. The main findings for the BWR-1300 (Sect. 5.1.2) will be summarized below. Table 11.5 lists some of the most important initiating events of accident sequences in the BWR-1300. They are in a similar range of frequency per year as in the PWR-1300.

Table 11.6 lists the frequencies per reactor year for the most important accident sequences which can result in core meltdown, such as

- loss-of-coolant accident,
- anticipated transients without scram (ATWS),
- operational transients.

While the frequencies of occurrence of loss-of-coolant and ATWS accident sequences are in the range of 10^{-6} per reactor year, the frequencies of occurrence initiated by operational transients—are in the range of 10^{-5} per reactor year. The frequencies of occurrence of core melts initiated by fire or earthquake are of a similar order of magnitude as in a PWR, i.e. in the range of 10^{-7} to 10^{-6} per reactor year. This leads to a total frequency per reactor year of 5×10^{-5}.

Accident management measures can reduce the frequency of occurrence per reactor year by roughly one order of magnitude to **4.4×10^{-6}**.

Such additional accident management measures [61, 62] are e.g.

- primary pressure relief by the pressure relief valves and steam blowdown to the water pools (condensation pool),

Table 11.6 Expected frequencies for core melt down (German Reactor Risk Study [11]) for 1.2 GW(e) BWR

	Initiating event	Frequency per year
Loss of coolant	Small leak (5–150 cm^2) in main steam line (outside containment)	2×10^{-7}
	Small leak (5–150 cm^2) in main steam line (inside containment)	4×10^{-7}
	Small leak (5–150 cm^2) in feed water line	3×10^{-7}
ATWS	Loss of main heat sink, no scram	10^{-6}
Operational transients	Loss of electrical power from grid	3.2×10^{-6}
	Loss of main feedwater supply	5.5×10^{-6}
	Loss of main heat sink	2.0×10^{-5}
	Loss of main heat sink and loss of main feedwater	1.5×10^{-5}
	Failure to close safety and relief valve	4.1×10^{-6}
Sum total		5×10^{-5}

- electrical power and water supply by an independent steam driven turbine-pump system,
- inertizing the inner containment with nitrogen to prevent hydrogen detonations.

11.7.8 Release of Radioactivity as a Consequence of Core Melt Down

The results in Tables 11.4 and 11.6 can be combined with results obtained from accident consequence analyses for assumed core meltdown and subsequent containment failure. This analysis can be simplified by grouping the results into release categories for accident sequences with the same containment failure mode. In addition to the frequencies of occurrence, also the schedule of radioactivity release from the outer containment after the onset of an accident and the fractions of fission products released can be determined. These fission product fractions refer to the total radioactive inventory of the PWR or BWR core.

In the Rasmussen Risk Study [36] and in the German Risk Study Phase A [37, 42] a very conservative approach was applied, e.g. it was assumed that a core meltdown accident is followed by a steam explosion with a certain probability of occurrence. For such a steam explosion it must be assumed that the molten core is mixing with water and fragmenting into very small particles which transfer heat very quickly to the water, thus rapidly producing a large amount of steam. This release category includes the highest radioactivity release which would occur at about 1 h after the initiating event with subsequent core meltdown. More recent studies described in Sect. 11.10.1 show that such large scale steam explosions with subsequent failure of the pressure vessel and of the outer containment can be considered to be impossible.

Other release categories of e.g. the German Study Phase A [36, 37, 42] did comprise core meltdown accidents with a subsequent large scale hydrogen-air detonation and failure of the outer containment (Sect. 11.10.2) or so-called containment bypass accidents (Sects. 11.10.3, 11.10.4). In these cases, openings in the outer containment of 25–300 mm equivalent diameter were assumed. Lower release categories of the Risk Studies [36, 37, 42] represented core meltdown with late containment overpressure failure. The lowest release categories cover loss-of-coolant accidents controlled by the emergency cooling systems. Since, in these cases, the reactor core is cooled sufficiently by the emergency core cooling system, the fuel element claddings will be damaged only partially and the reactor core will not melt.

11.7.9 Accident Consequences in Reactor Risk Studies

The US Reactor Safety Study, WASH-1400 [36], was performed for 100 reactor plants (PWRs and BWRs) in the United States. It was published in 1975, thus preceding the similar German Risk Study [37, 42] which was performed in 1979 for 25 German reactor plants. Compared with the results of the German Risk Study, and aside from slightly different safety designs of German LWRs, it was mainly the meteorological data, the population density, the purely linear dose-risk relationship as well as protective measures and countermeasures, which differed in the two studies.

It must also be emphasized that in these reactor risk studies the results are averaged over 68 different sites (WASH-1400 [36]) or 19 different sites (German Reactor Risk Study, Phase A [37]) with several hundred different weather conditions.

The basic statements included in the findings of these two studies are approximately identical.

The frequency of core meltdown accidents was determined to be 5×10^{-5} per reactor year in the US Reactor Safety Study [36]. The largest number of early fatalities was is approximately 3,300, with a probability of occurrence of 10^{-7} per reactor year. The cases of early illness are 45,000, with a frequency of occurrence of 10^{-9} per reactor year.

Figure 11.25 shows the complementary, cumulative frequency distribution of early fatalities, determined by the WASH-1400 Risk Study [36], which could be caused by radiation exposure after accidental radioactivity release. Results of the German Risk Study, Phase A [37] are similar.

Of course, a large number of early fatalities will occur if major releases are encountered on sites with high population densities, the wind blows in the direction of the sector with the highest population density, and rain falls in the immediate vicinity, thus creating high radioactivity concentration levels on the ground.

In the US and the German Risk Studies [36, 37, 42] late fatalities at all dose levels are reported since a linear dose rate (Sect. 10.5.1) without threshold value was assumed according to ICRP (1977). The occurrence of late fatalities is therefore not restricted to only the immediate vicinity of the reactor plant, as in case of early fatalities. Also a considerable fraction of the late fatalities determined in the Risk Studies [36, 37] was due to low radiation exposures of <50 mSv. (The average

Fig. 11.25 Results of the US-Risk Study WASH-1400 [36] for 100 nuclear reactor plants (*Source* WASH-1400)

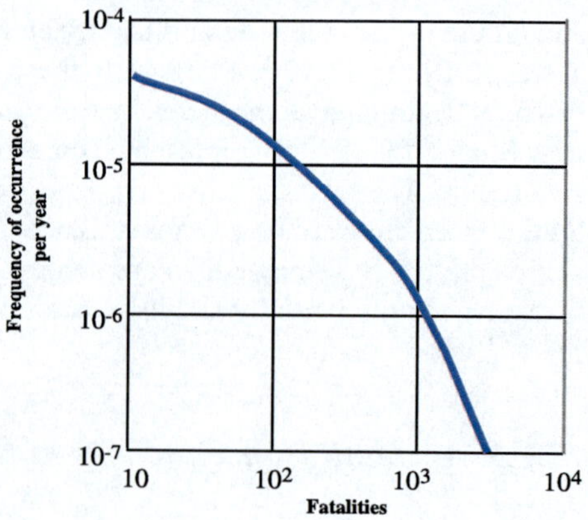

natural background radiation in Germany is about 2.1 mSv/year). Therefore, these results are still discussed controversially.

11.7.9.1 Comparison of the Risk of Nuclear Power Plants with the Risks of Other Technical Systems

The findings about the risk of LWRs as determined in the Rasmussen Risk Study (WASH-1400) [37] and the German Risk Study Phase A [37] were compared to the risks associated with other technical systems. Data from the experience of insurance companies for major accidents in coal mining, on oil platforms, tanker accidents in oil and gas transport, large fires in refineries, gas explosions, and risk studies for large chemical plants in England (Canvey Island) and France were used [63–67].

Figure 11.26 shows the number of deaths fatalities in major accidents as a function of the frequency of occurrence for dam failures, chlorine releases (chemical industry), explosions and fires as well as airplane crashes. The frequency of occurrence of major accidents in power generation by nuclear plants is seen to be below or in the vicinity of the figures for other systems of energy generation and for the chemical industry, respectively [36].

11.7.9.2 Major Accidents in the Power Industry

The worst accident in the coal mining industry occurred 1931 in Fushun (Manchuria, China) killing 3,000 persons. In the oil industry, the worst accident with 3,000 casualties occurred in 1987 when an oil tanker collided with a passenger ferry boat in the Philippines. In the natural gas industry, the severest accidents were recorded in Ufa, now Aserbeidjan, in 1989, with 600 persons killed, and in Ixhuatepec, Mexico, with 500 casualties [65].

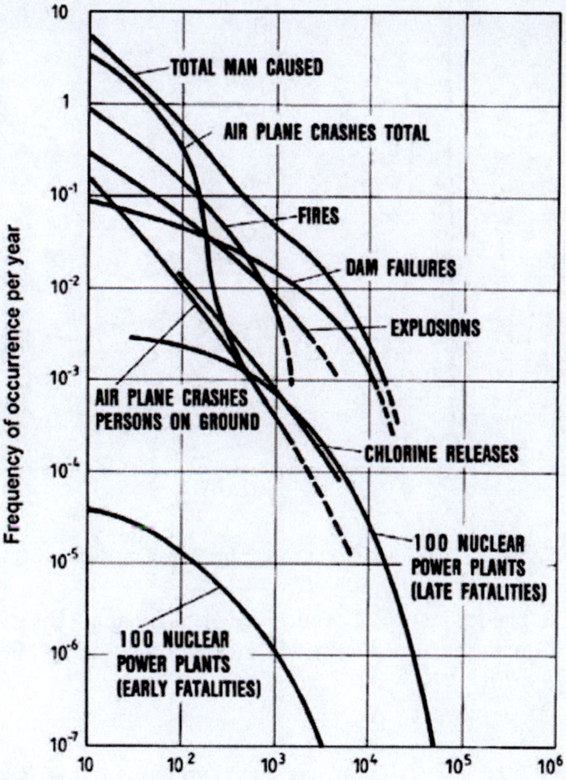

Fig. 11.26 Comparison of man caused fatalities of different technical systems with the results of WASH-1400 for 100 nuclear power plants in the USA [36]

When it comes to large dams for hydroelectric plants, the record shows 459 people killed in Frejus, France, in 1959. In Vajont, Italy, 1914 people were killed in 1967, and 2,500 people were killed at the Indian Machhu II dam in 1979.

The biggest accident in the chemical industry so far occurred in Bhopal, India, killing 3,400–8,000 persons [66].

11.7.9.3 Natural Disasters

It is interesting in this connection to have a look also at the consequences of natural disasters mankind must suffer as inescapable (Fig. 11.27) [68–70]. These are

natural disasters,
hurricanes,
floods, including tsunamis
earthquakes,
avalanches and landslides,
volcanic eruptions.

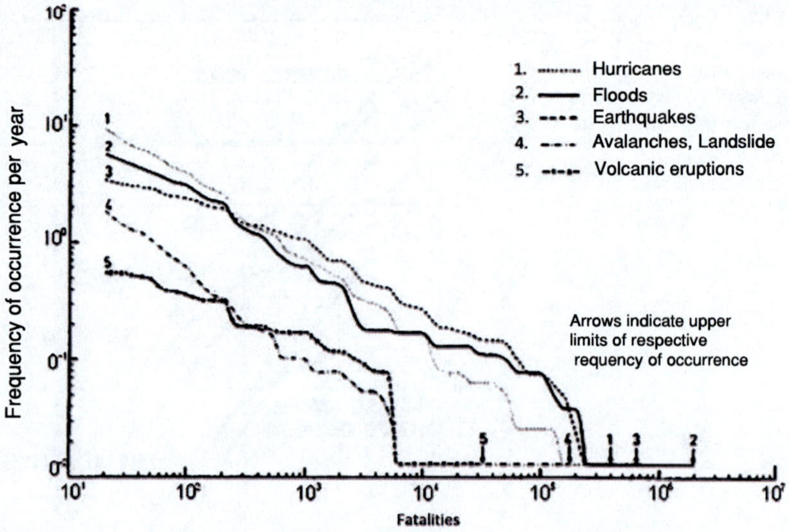

Fig. 11.27 Fatalities (deaths) occurring in natural disasters ([66, 69, 70] adapted)

The frequency of occurrence of such natural disasters is far higher than, e.g. that of major accidents in power generation systems. They can cause up to 10^5 deaths [68–70].

11.7.9.4 Safety Improvements Implemented in Reactor Plants After the Risk Studies

As mentioned in the previous sections, rather simplified and pessimistic assumptions were made in the German Risk Study and the US Reactor Safety Study [36, 37] about the sequence of accident events in which the molten core penetrates the bottom of the reactor pressure vessel leading to subsequent accident conditions which can lead to a relatively early loss of containment integrity. This leads to an overestimation of the accident consequences and risks. More recent theoretical studies and preliminary experimental results indicate that

- the assumed large steam explosions leading to early containment failure can be considered to be impossible (Sect. 11.10.1),
- the assumed early containment failure after core melt through under high primary coolant pressure can be counteracted by primary coolant depressurization or strengthening of the anchorage of the reactor pressure vessel (Sect. 11.10.5),
- the penetration of the molten core into the concrete does not necessarily lead to contacts with the sump water. However, if it contacts sump water, a loss of containment integrity would occur only after time periods of approximately 5–12 days. Exventing filters introduced after the publication of the reactor risk studies (WASH-1400 and German Risk Study Phase B) can avoid overpressurization

of the outer containment (Sect. 11.10.6.2). For reasons of aerosol physics [44], a considerable percentage of radioactive aerosols will have settled within the containment under fog or rainlike conditions and shortlived radioisotopes will have decayed away by the time containment failure was assumed to occur (Fig. 11.19). The release of radioactivity into the environment then decreases by several orders of magnitude.

These results were confirmed during the decades after the reactor risk studies appeared by large scale experiments as will be described in Sects. 11.10.1–11.10.8.

11.7.9.5 Risk Studies for Other Types of Reactors

Risk studies on other types of reactors have not yet been carried out to the same level of detail as on LWRs [71]. Only on LMFBR and HTR prototype reactors, the same methods were employed in similar studies [71].

The results of the LMFBR risk studies are not too different from those of the LWR risk studies, aside from the fact that the LMFBR risk studies had been conducted only for single prototype LMFBRs (CRBR, USA, and SNR 300, Germany). Compared with LWRs, LMFBRs have emergency core cooling systems that inherently dissipate the afterheat by natural convection. However, greater attention had to be devoted to potential failures of the shutdown systems.

Risk studies on an HTR prototype were performed in the Federal Republic of Germany on an HTR-1160 conceptual reactor design with block type fuel elements. For this type of reactor the main risk is dominated by accidents which are initiated by a failure of electrical power supply followed by temperature transients with failure of the integrity of the containment after about four days. The release of radioactivity in these cases is somewhat smaller than the results described for the LWR risk study.

11.7.9.6 Use of Results of Reactor Risk Studies

The value of Reactor Risk Studies should not be exaggerated or misinterpreted or even used for forecasts in which time periods core melt accidents could occur. As their methodology is based on probabilistic considerations their results can only be used for

- comparison with the results of risk studies for other energy production systems (coal, oil, gas etc.) which are based on the same probabilistic methodology,
- for the optimization of the design of the different safety systems to reach a well balanced overall safety concept for that nuclear plant for which the probabilistic safety analysis was performed.

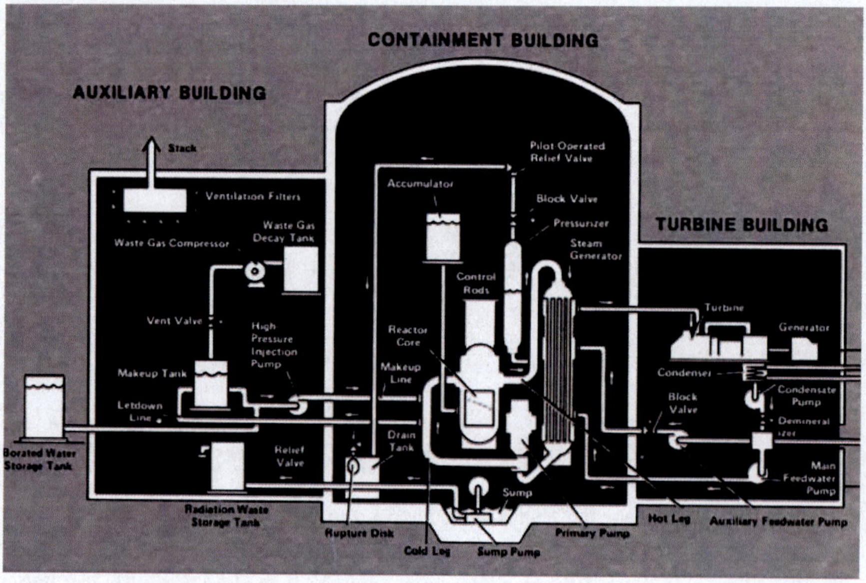

Fig. 11.28 Containment design of the Three Mile Island reactor [73]

11.8 Historical Occurrence of Severe Core Melt Accidents in Commercial Nuclear Power Plants

Three severe core destruction or melt down accidents occurred on March 28, 1979 at Three Mile Island, USA, on April 26, 1986 at Chernobyl, Ukraine and on March 11, 2011 at Fukushima, Japan. These three core melt or core destruction accidents are described here shortly with the objective to discuss their consequences and their repercussions on future safety concepts (Sect. 11.10).

11.8.1 Three Mile Island Accident

On March 28, 1979 the unit 2 PWR at Three Mile Island experienced a series of events which lead to partial core destruction [72, 73]. This reactor was a standard Babcock and Wilcox PWR with two coolant circuits A and B and a power output of 880 MW(e) (Fig. 11.28).

The event chain started with technical problems in the feedwater line to the steam generators. A feedwater pump trip was followed by automatic turbine trip and automatic start up of the auxiliary feedwater circuits. But valves in the auxiliary feedwater line were wrongly closed. This lead to pressure and temperature increase in the primary circuit as the heat sink potential in the steam generators decreased.

Due to the rising pressure in the primary circuits the pilot operated relief valve at the pressurizer opened and steam flowed to the drain tank in the containment building. Due to a high pressure signal the reactor was shut down automatically. The nuclear heat generation decreased to decay heat (afterheat) level and also the primary system pressure decreased.

Instead of re-closing automatically due to decreased primary system pressure the pilot operated pressurizer relief valve stuck open. However, unfortunately the operators received information to the contrary by the instrumentation. Primary coolant continued to escape through the pilot operated pressurizer relief valve. As a consequence of the loss of primary coolant, the primary pressure fell to the point where the high pressure injection system was activated automatically to compensate for the primary coolant loss.

The drain tank overflew and coolant escaped into the sump of the containment and was pumped to radioactive waste storage tanks in the auxiliary building.

After 5 min from the begin of the accident chain operators throttled the high pressure injection system due to high water level in the pressurizer. They drained water through the let down line. This lead to insufficient emergency coolant flow through the core due to the pressure decrease. After 6 min from the begin of the event chain the primary system pressure fell to the level at which the water in the core began to boil due to the decay heat (afterheat).

Despite of the fact that the operators succeeded to open the closed valves of the auxiliary feedwater line, the primary coolant loss and primary pressure decrease continued.

The operators did not understand anymore what is going on, due to the wrong information about stuck-open pressurizer relief valve. Between 15 and 30 min after begin the drain tank ruptured and the storage tanks did overflow. Some radioactive gases and aerosols escaped to auxiliary buildings and then to the environment through the ventilation filters and the stack.

At 1 h and 13 min after begin of the event chain the operators turned off the primary pumps in the primary loop B because the steam in the circuit caused the pumps to vibrate. Some 27 min later they also turned off the pumps of primary loop A for the same reasons. In the core fuel channels steam built up and the fuel rods heated up.

After 2 h and 20 min from the begin of the accident chain the operators succeeded to close a block valve upstream of the pilot operated pressurize relief valve. This halted the coolant loss from the pressurizer but stopped, unfortunately, also cooling of the core. Steam stopped streaming axially along the fuel rods. The fuel rods heated up to such high temperatures that steam reacted with the zircaloy claddings to produce ZrO_2 and hydrogen.

Some 34 min later the operators tried to start a primary pump but turned it off because it was not running properly. Now the operators opened the pressurizer block valve for 5 min and the escaping steam provided steam flow along the fuel rods and some cooling (bleeding procedure).

At 3 h and 20 min after begin the operators started the high pressure injection system for a few minutes (feed procedure) and covered the core with water. However, the coolant flow was blocked by steam and hydrogen.

Finally, the operators started a feed and bleed procedure by injection of water through the high pressure system and bleeding steam by the pilot operated pressurizer valve. This feed and bleed procedure, however, was hampered by the noncondensable hydrogen. The operators noted a pressure spike in the reactor containment 8 h and 20 min from the begin of the accident sequence, but did not interpret it as a hydrogen burn. Only 15 h and 50 min after begin of the accident chain the operators started to activate a primary pump and achieved forced convection again. The reactor cooling system reached relatively stable conditions.

11.8.1.1 Reactor Core Destruction

Analysis showed that about one third of all zircaloy of the core fuel elements had reacted with steam to produce hydrogen [72, 73]. The quenching effect of the water of the high pressure injection system lead to fragmentation of the hot cladding and some of the UO_2 fuel pellets. In addition, melting of considerable parts of the Ag-In-Cd control rods had occurred (Fig. 11.29). Almost all fission gas plena were destroyed. Most of the radioactive noble gases and iodine radionuclides were released into the reactor containment and auxiliary building. The iodine radionuclides were almost entirely retained in the filter system as aerosols since iodine formed chemical compounds, e.g. even with cesium. The radioactive noble gases were released into the environment.

The average individual radioactive exposure to the public living in the surroundings was determined to be 0.015 mSv. Concentrations of I-131 were found in milk by a factor of 300 lower than the Food and Drug Administration maximum permissible levels [72].

The Kemeny Commission [74] stated in its report to the President: "there will either be no case of cancer or the number of cases will be so small that it will never be possible to detect them".

11.8.1.2 Decontamination and Defueling of the Three Mile Island Plant

After the accident the plant was very highly contaminated. The plant was decontaminated over several years and then defueled. The actual state of the core was found as shown in Fig. 11.29. The cost of these efforts were in the range of about 1 billion $ [72].

11.8.2 Chernobyl Accident

The Chernobyl accident occurred on April 26, 1986 in unit 4 of four RMBK-1000 units (each 1 GW(e) power) located 130 km North of Kiev (Ukraine; officially: Ukrainian SSR belonging at the time to USSR) (Fig. 11.30). The RBMK-1000 reactors are graphite moderated and cooled by boiling water. They operate with

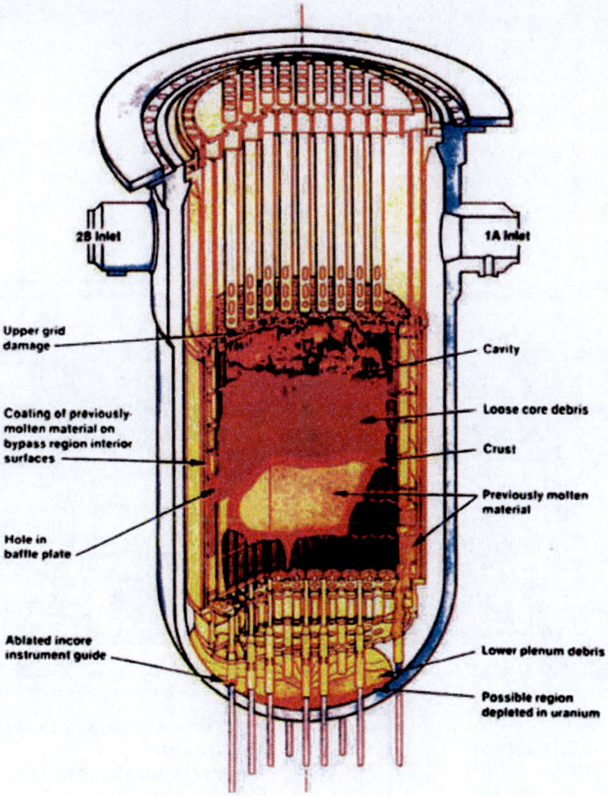

Fig. 11.29 TMI-2 core end-state configuration [72, 73]

low enriched (2% U-235) uranium fuel. This combination of moderator, coolant and fuel enrichment leads to a positive coolant temperature coefficient in the core.

An experiment had been prepared to use the rotational energy of the turbine during its coast down phase after an automatic trip of the reactor power as a source of emergency electrical power. The test was planned to begin at a power level of 700–1,000 MW(th).

11.8.2.1 Event Chain of the Chernobyl Accident

During the test procedure the power was lowered from 3,200 to 1,600 MW(th) by inserting the control rods axially into the core. However, the test had to be interrupted because the electrical grid load dispatcher requested to meet electricity requirements. At the same time the emergency core cooling systems were already shut off (preparation of the test). When the electrical grid load dispatcher allowed to resume the test, the power was lowered to the range of 700–1,000 MW(th) by inserting the control

1	**Reactor core**	**9**	**Fuel element charge / discharge machine**
2	**Pipeline to core entry**	**10**	**Reactor core**
3	**Lower radiation shield**	**11**	**Vertical channels for fuel elements**
4	**Collector – Distributor**	**12**	**Down-comer pipes**
5	**Side radiation shield**	**13**	**High pressure collector**
6	**Steam separator**	**14**	**Low pressure collector**
7	**Steam-water pipelines**	**15**	**Main coolant pumps**
8	**Upper radiation shield**		

Fig. 11.30 Design scheme of a Chernobyl reactor plant [75–77]

rods further, but this power level could not be maintained (computer problems) and fell to the range of about 30 MW(th). Negative reactivity due to Xenon poisoning built up during this power reduction phase. When the power was started up again it could only be risen to 200 MW(th) even with the majority of the control rods fully axially withdrawn.

The RBMK-1000 reactors have a negative fuel Doppler coefficient, but a positive coolant temperature coefficient. At a power of <20% of nominal power the positive coolant temperature coefficient becomes dominant over the negative fuel Doppler temperature coefficient. This causes the RBMK-1000 reactor to become unstable with a tendency to sudden power surges [76].

When the eight primary coolant pumps were started, the combination of low power and high coolant flow produced power instability problems which required manual adjustments. Therefore the operators turned off automatic trip signals of the reactor protection system (violating the regulations).

When the test began and the power was risen again the coolant temperature increased and caused a positive reactivity increase due to the positive coolant temperature coefficient. The operators noticed a too fast power increase and activated the insertion of all control rods for shut down of the reactor. However, the control/shut down rods had graphite followers preceding the neutron poison parts. This insertion of graphite followers replaced neutron absorbing water, thus generating additional positive reactivity which increased the power even further to about 100 times nominal power. This lead to bursting of the fuel rods, fragmentation of the fuel into very small particles and rapid thermal heat transfer to the coolant (**steam explosion**). The resulting pressure pushed up the massive top shield into an oblique position and the reactor core was opened to the atmosphere. The upper reactor building was destroyed. Fuel elements were thrown out of the reactor core. The graphite as moderator started burning, zirconium and steam produced combustible hydrogen. The hot gases and aerosols were blown up to 2,000 or even 10,000 m high into the atmosphere.

11.8.2.2 Consequences of the Chernobyl Accident

Some staff members (31 members) received lethal doses caused by very high radiation and thermal burns during fire fighting in the early phase after the accident. About 1,400 persons suffered various degrees of radiation sickness and health impairment [76]. In the literature very different numbers have been reported for additional people killed as a consequence of lethal radiation doses. According to IAEA [75, 76, 78, 79] another about 20 persons died up to 2011 of too high radiation exposure during the time period of the accident. This includes some children who died of thyroid cancer [80].

An additional about 4,000 casualties may occur during the life time of about 600,000 people under consideration [80]. This is doubted by some radiation experts [81].

The inhabitants of the nearby cities of Pripyat and Chernobyl (45,000 people) were evacuated 36 h after the accident. Their radiation exposure was estimated to 0.25 up to 0.5 Sv. In total 135,000 people were evacuated within one week. Changing weather conditions carried radioactivity to as far as Scandinavia and Western Europe [76, 77].

11.8.2.3 Chernobyl Plant Recovery

Hundreds of specialists and about 800,000 military personnel—so-called liquidators including helicopter pilots—were involved in removing fuel assemblies, stabilizing the reactor core by dropping sand, clay, lead and boron carbide and constructing radiation shields. Finally, the Chernobyl reactor plant unit 4 was entombed in some kind of a sarcophage and in an arch-shaped safe confinement [77].

Fig. 11.31 Cs-137 contamination from Chernobyl accident in different areas of the Ukraine, Belarus and Russia [75]

11.8.2.4 Contaminated Areas

Measurements revealed that changing weather conditions including rainfall during the first 10 days after the accident, lead to considerable contamination on the ground by Cs-137 (half-life 30 years) [75, 76]. (Cs-134 with a half-life of 2 years mostly decayed during the first 10 years.) I-131 (half-life of 8.2 days) was responsible for the radiation exposure of the population during the first months.

The three main spots of Cs-137 contamination are shown by Figs. 11.31 and 11.32. They are called C spot Bryansk-Belarus (B-spot) and Kaluga-Tula-Orel (K-spots). Depositions of Cs-137 of over 40 kBq/m^2 covered large areas of the Northern part of the Ukraine and the Southern part of Belarus. The most highly contaminated area in the Ukraine was the 30 km zone surrounding the RBMK-1000 reactor site, where the depositions exceeded 1,500 kBq/m^2.

In the Bryansk-Belarus (B-spot) area the highest contaminations were even found in some villages to be 5,000 kBq/m^2. In the Kaluga-Tula-Orel (K-spot) area the levels of depositions of Cs-137 were found to be less than 600 kBq/m^2.

Overall there were 3,100 km^2 contaminated by Cs-137 with levels exceeding 1,500 kBq/m^2, 7,200 km^2 with 600–1,500 kBq/m^2 and 103,000 km^2 with levels between 40 and 200 kBq/m^2.

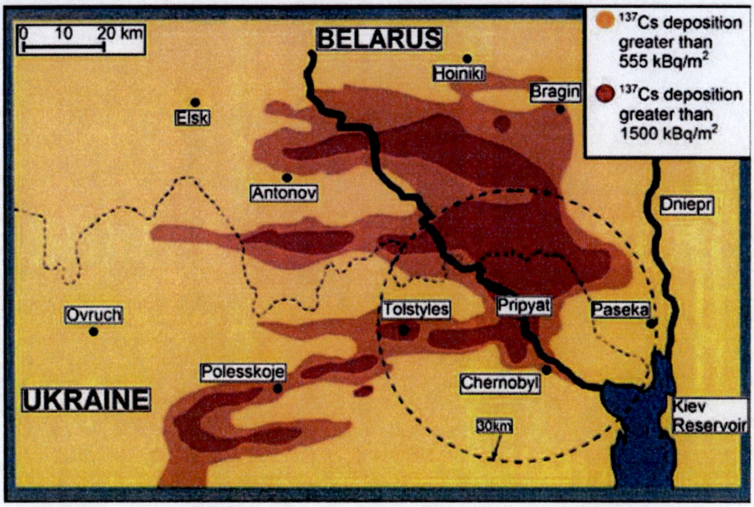

Fig. 11.32 Cs-137 contamination from Chernobyl accident in the Ukraine [75]

11.8.3 The Fukushima Reactor Accident in Japan

11.8.3.1 Event Chain of the Fukushima Reactor Accident

On March 11, 2011 an earthquake of unprecedented magnitude (magnitude 9.0 on the Richter scale) followed by a tsunami of unprecedented height of about 14 m led to a failure of emergency cooling in three BWRs at the Fukushima Daiichi reactor site north-east of Tokyo [82–88]. The earthquake occurred at 2:46 pm in the Pacific Ocean 80 miles east of the city of Sendai which is located north of Tokyo (Fig. 11.31). Four reactor sites were directly affected by the earthquake and tsunami. These are the reactor site of Onagawa, Fukushima Daiichi, Fukushima Daini and the nuclear research center Tokai-2 with a total of 14 BWRs (Fig. 11.33).

At the time of the earthquake the Fukushima Daiichi site the unit 4 was completely defueled and units 5 and 6 had been shut down for inspection and maintenance. All other 11 BWRs including the three BWRs of Fukushima-Daiichi were automatically shut down successfully as a consequence of the seismic ground acceleration and started automatic procedures to proceed to cold shut down. The diesel generators started to supply electricity to the emergency and decay heat (afterheat) cooling pumps.

However, this procedure was interrupted in units 1–3 of the Fukushima Daiichi reactor site, when 56 min after the earthquake a tsunami of about 14 m height hit these units and disabled their emergency power capabilities (off-site power and all on-site emergency diesel generators). The Diesel generators were located in the basement of the reactor building (Fig. 11.34) and were flooded by the tsunami as the doors of the reactor building were not water-tight. Some fuel tanks of the diesel generators were

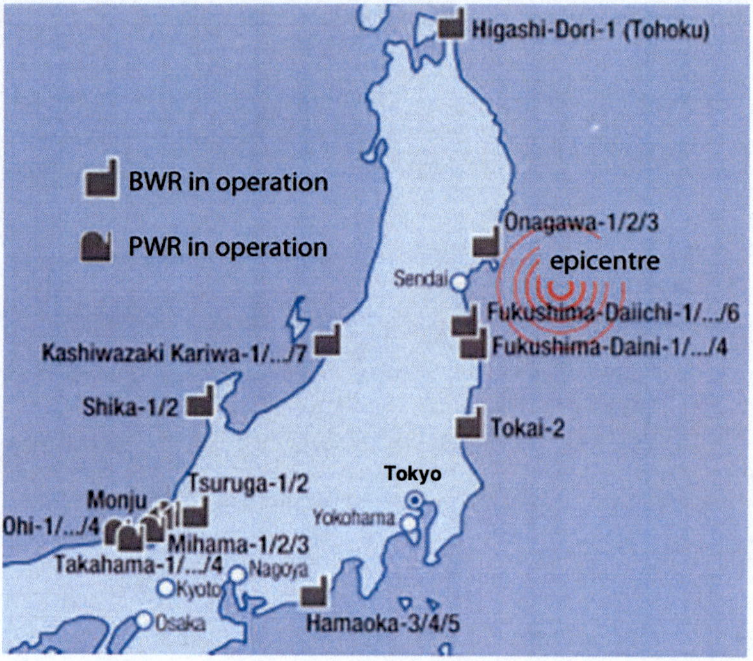

Fig. 11.33 Nuclear reactors operating in Japan (Island Honshu) when both the earthquake and the Tsunami hit the coast [88]

carried off by the receding tsunami. Water injection into the reactor pressure vessel by the emergency cooling system failed less than 1 h after arrival of the tsunami in units 1 (BWR, 431 MW(e)) and 2 (BWR, 760 MW(e)) and later also in unit 3 (BWR, 760 MW(e)). The tsunami also disabled the sea water pumps depriving these reactor units of their ultimate heat sink (Fig. 11.34). The station black-out at units 1–4 caused a complete loss of all instrumentation and control systems. The operators had to work in darkness with no communication systems available.

The Fukushima Daiichi reactor plants were originally designed for a 3 m tsunami and modified in 2000 to withstand a 5.7 m tsunami. The seismic acceleration was measured at unit 3 to be 507 cm/s^2. This exceeded the design acceleration of unit 3. The unit 3 plant had been designed for a lower design acceleration of 449 cm/s^2. For the reactor unit 1 and 2 the measured seismic acceleration was covered by the design (Fig. 11.35).

The plant staff started severe accident management measures as soon as possible. First the plant staff tried to maintain core cooling through battery power. In addition, a turbine driven pump powered by the steam of the reactor was used to inject water from the pressure suppression chamber and from the water storage tank into the reactor pressure vessel. However, as the batteries exhausted after 8 h this turbine pump system failed.

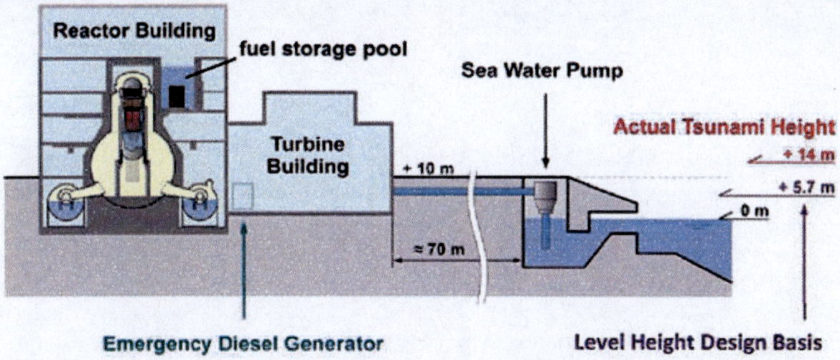

Fig. 11.34 At Fukushima Daiichi, countermeasures for tsunamis had been established with a design basis height of **5.7 m** above the lowest Osaka Bay water level [88]

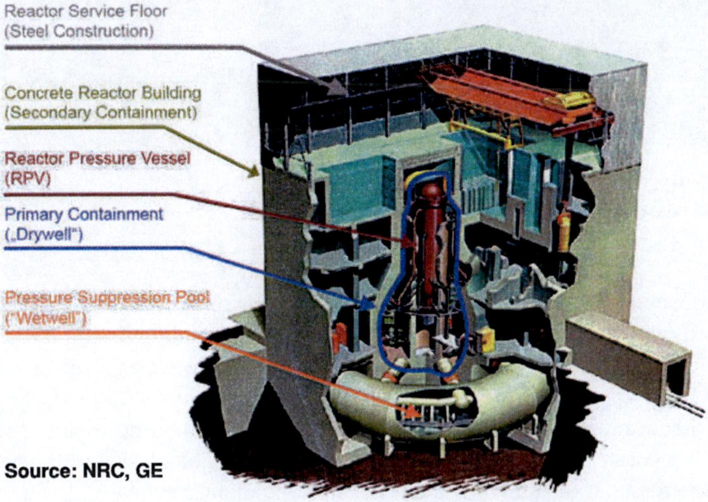

Fig. 11.35 Design scheme of the general electric BWR-3 reactor at the Fukushima-Daiichi nuclear plant [88]

As a consequence, the water in the core heated up and the pressure rise in the reactor pressure vessel caused the pressure relief valves to open 18 h after the earthquake and released steam into the pressure suppression chamber where the steam was condensed by the water (Fig. 11.35). As the water of the pressure suppression chamber could not be cooled either, this water heated up and the pressure increased. Within about 3 h the water level in the core had decreased such that the fuel rods heated up to the range where steam started to react with the zircaloy cladding to produce hydrogen

$$Zr + 2H_2O = ZrO_2 + 2H_2$$

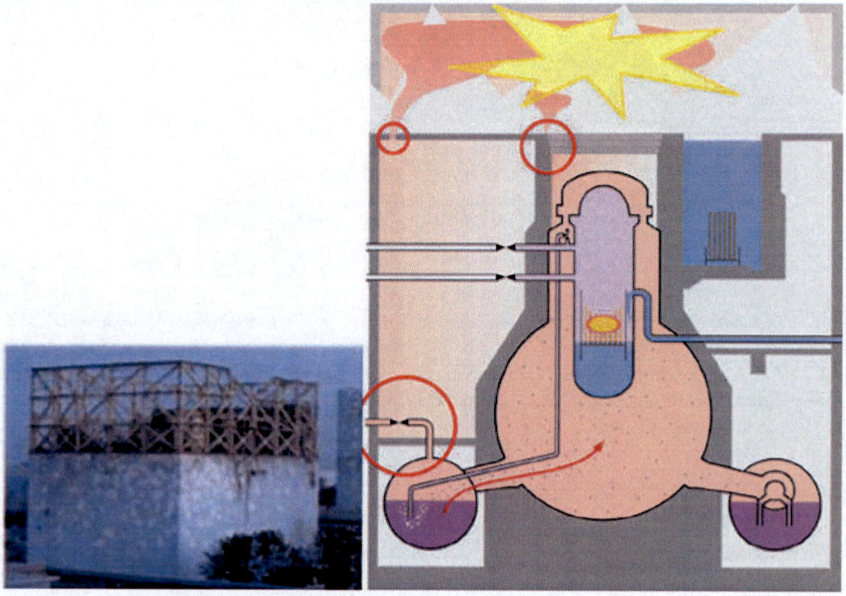

Fig. 11.36 Hydrogen explosion inside the reactor service floor leads to the destruction of the upper steel frame construction [85]

The fuel claddings failed, gaseous and volatile fission products of the fuel, mixed with steam and hydrogen, were released to the pressure suppression chamber and inner containment. After about four and a half hours the water level had dropped below the fuel upper head. The fuel temperature rose to above 2,800°C, the melting point of the uranium dioxide fuel, and melting began in the central part of the core. After 16 h most of the fuel, along with the control rods, had fallen to the bottom of the reactor vessel at the unit 1 of the Fukushima-Daiichi reactor.

The pressure in the pressure suppression chamber and in the primary containment rose above design limits and the plant staff had to initiate primary containment venting about 24 h after the earthquake. This involved steam, radioactive gases and hydrogen. The hydrogen self-ignited in the upper part of the reactor building and the hydrogen explosion blew out a large section of the roof of the service floor and the upper walls of reactor unit 1 (Fig. 11.36). Four technicians were injured and hospitalized. The hydrogen explosion did not damage the reactor pressure vessel or primary containment of the reactor plant.

As soon as it became possible the plant staff initiated severe accident management procedures and injected sea water mixed with boron into the core in order to cool the molten core.

Although the accident conditions at units 2 and 3 were slightly different in their time sequence, they followed similar accident paths with steam-hydrogen fission gas venting and hydrogen explosion.

Problems were then also experienced with the spent fuel storage pools which are located on the upper level of the reactor buildings (Fig. 11.34). Cooling in these spent fuel storage pools could not be maintained after the loss of off-site power and emergency power. Therefore the fuel and water temperatures did increase slowly and water started to evaporate. The spent fuel became uncovered. This appeared to be a particular problem in unit 4 where also a hydrogen explosion occurred. At the begin it was assumed that, a similar zirconium-steam chemical reaction as in the core of units 1–3 had occurred. Later it was assumed that hydrogen had been passed through a piping system from units 1–3 to unit 4. It turned out after later inspection that the spent fuel pool of unit 4 was always covered with water [85, 89, 90].

The problem with the spent fuel pools in units 1–3 could be counteracted by spraying water first from helicopters and then replenishing the water by special vehicles with concrete water pumps and long-extension-fire hoses.

11.8.3.2 Measurements of Radioactivity

About 200,000 people were evacuated from a 20 km zone around the Fukushima Daiichi reactor plant [86, 87]. The authorities distributed potassium-iodine tablets to protect the public from inner radiation damage by radioactive Iodine-131.

Measurements of the radioactivity in the atmosphere around the Fukushima-Daiichi nuclear plants were performed by especially equipped airplanes of the US National Nuclear Security Agency (NNSA) [87] (Fig. 11.37). These measurements showed the highest migration of radioactive material to the northwest of the Fukushima nuclear reactor plants.

In Fig. 11.37 these measurements are shown. They were used for a forecast of the radioactive exposure dose a person would receive staying for one year in a certain location (time measured from March 16, 2011 when the hydrogen explosion with radioactivity release occurred). Such forecasts are required for decisions to be made by authorities in which areas around the nuclear plants the population must be evacuated (Sect. 11.7.4.5).

11.8.3.3 Direct Health Effects

IAEA stated three months after the accident that no serious health effects had been reported in any person as a result of radiation exposure from the Fukushima nuclear accident [83]. Four members of the operational staff had been killed by the earthquake and by the following tsunami by drowning. Eight workers received radiation exposures of more than 170 mSv, two of them suffering beta radiation burns on their legs from contaminated water. None of the other workers of the staff reached radiation exposures of 250 mSv, the maximum radiation dose set by Japanese authorities [91].

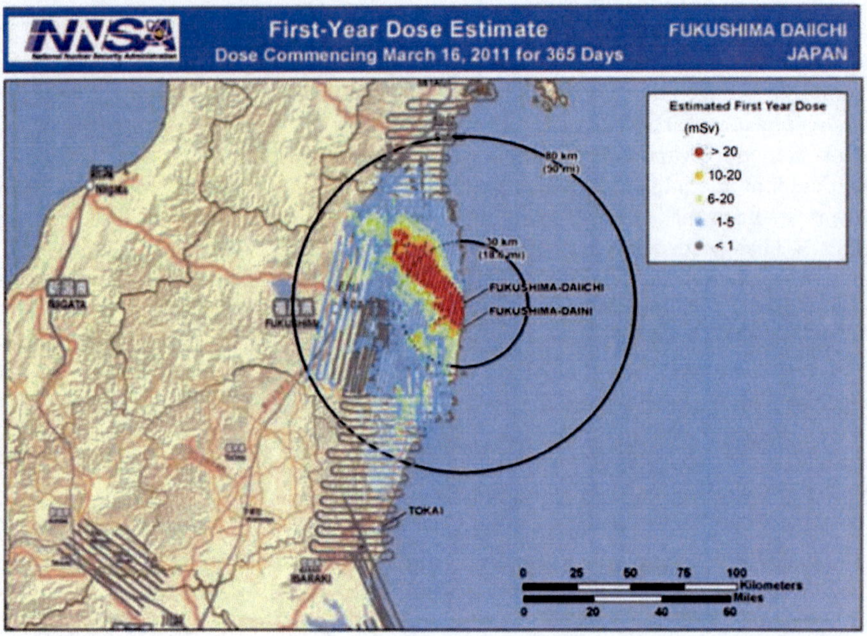

Fig. 11.37 Measured radioactivity release and forecast of radioactive exposure during the first year after radioactivity release on March 16, 2011 [86, 87]

11.8.3.4 Radioactive Exposures to the Public

Whereas the radioactive gases released e.g. I-131 (half-life 8.2 days) will decay relatively fast, the associated cesium isotopes will deposit on the ground [83, 86, 92, 89]. The isotopes Cs-134 (half-life 2 years) and Cs-137 (half-life 29 years) are then responsible for the further potential exposure of the public (direct external radiation, ingestion etc.) [89, 92]. Maps of calculated radioactive exposure doses which could be received by the Japanese population as a result of external radiation during the first year after the accident were published by IRSN (France), US DOE/NNSA and by the Japanese Ministry of Science and Technology (MEXT). Figure 11.38 shows a map of Cs-134 and Cs-137 deposits (Bq/m^2) for the areas around the Fukushima-Daiichi nuclear plants. This figure also gives the exposure dose levels e.g. 5 mSv/y or 10 mSv/y or 20 mSv/year a person would receive during the first year living in this area. These exposure dose levels are the basis for decisions by Japanese authorities on relocation (see also Table 11.3, [81, 92]).

In a report of IRSN [86] the affected population for this first year forecast is also given. About 3,100 people could receive more than 50 mSv and 2,200 people would have to stay out of the no-entry zone where they could receive 100–500 mSv during the first year.

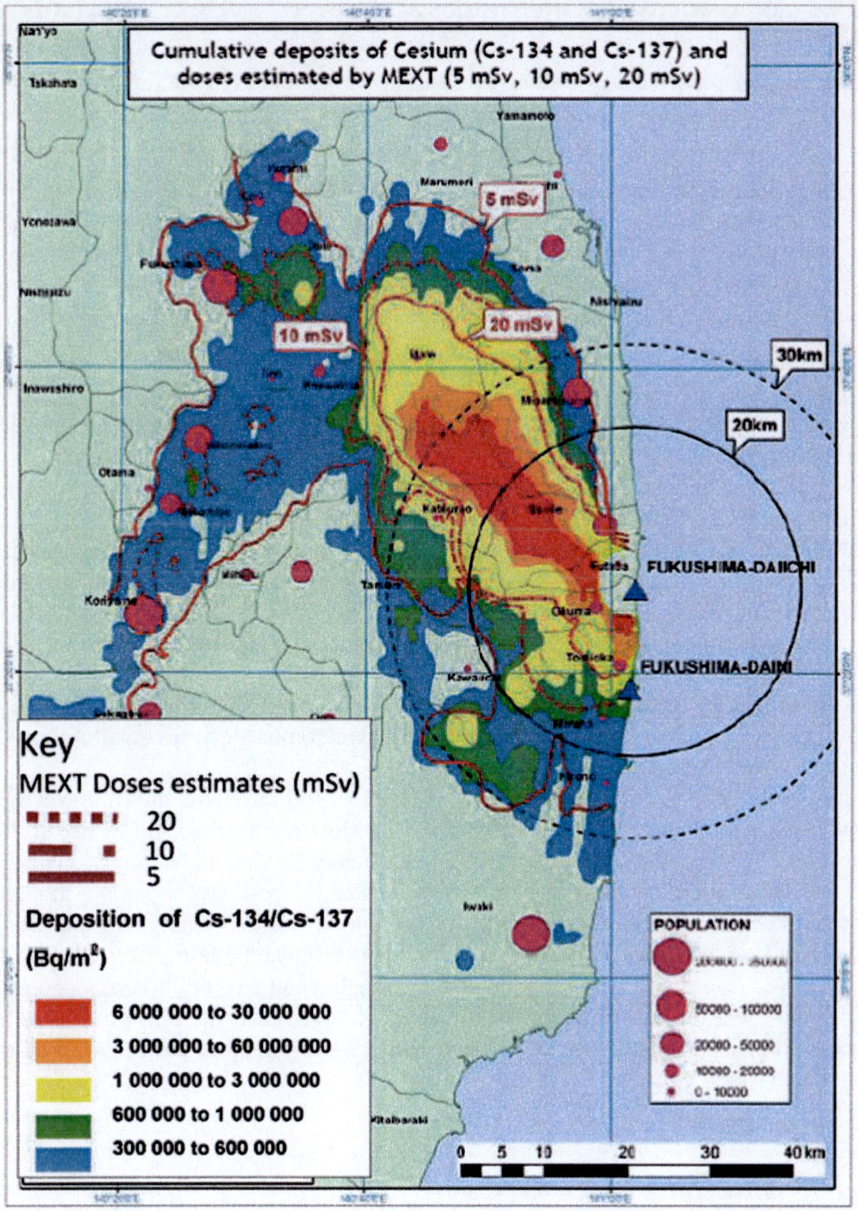

Fig. 11.38 Areas of Cs-134 and Cs-137 deposits in Bq/m^2 as well as estimates for radiation exposure of the Japanese public in certain locations during the first year after the Fukushima-Daiichi accident [86]

11.9 Assessment of Risk Studies and Severe Nuclear Accidents

The findings reported above show nuclear power generation in the range of damage of similar magnitude (deaths) as the other technical or power generation systems. Yet, the severe reactor accidents of Chernobyl and Fukushima have shown that

- large areas were contaminated with Cs-134 and Cs-137 such that food production either had to be restricted for long periods of time (the half-life of Cs-134 is 2 years and of Cs-137 is 29 years) or, where contamination was lower, must be monitored for radioactivity over long time;
- the population of large areas had to be evacuated and even be resettled.

While the economic losses arising, e.g. from major tanker accidents were estimated to run up to several billion dollars, the economic consequences of the Chernobyl accident are higher by orders of magnitude [63]. Besides the damage to health resulting from the radioactivity to which workers and the population were exposed, as well as associated psychosocial problems and psychosomatic disorders [81] it is the contamination of land by Cs-137 as well as the ban on food over prolonged periods of time (until Cs-137 has decayed to safe levels) which are a problem very specific to the use of nuclear power. Only accidents in the chemical industry involving highly toxic chemicals have similar characteristics.

In discussions about the risk of technical systems or by nuclear energy (Fig. 11.26) the following argument is often stressed:

The large damage is associated with extremely low probabilities of occurrence per year. Therefore the risk as the product of damage times probability of occurrence is small.

Around 1989, Kessler-Hennies-Eibl (KHE) [93–95] raised the question whether this risk argumentation and the associated findings of the risk studies had to be accepted as unavoidable for future LWRs or whether they could be improved upon.

European light water reactors are often located by large rivers passing through densely populated regions with cities of more than 100,000 inhabitants and large industrial plants. Cities that size are impossible to evacuate. Radioactive contamination of such cities and densely populated areas is beyond anybody's imagination.

From 1990 on, this thinking led to deeper research into the sequence and consequences of severe core meltdown accidents in LWRs and to the proposal of an extended safety concept.

11.9.1 Principles of the KHE Safety Concept for Future LWRs

The safety concept for future LWRs as proposed by KHE at the Research Center and Technical University of Karlsruhe, Germany is based on these considerations [67, 93–95]:

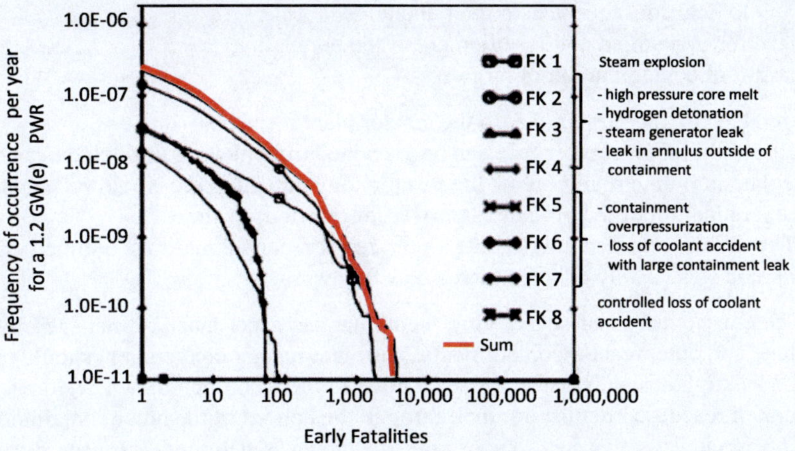

Fig. 11.39 Results of the German Reactor Safety Study, Phase A, shown with different contributions of those accident sequences resulting in the highest damage consequences [71]

- The major consequences of accidents as determined in WASH-1400 [36] and the German Reactor Risk Study Phase A [37] assume that
  - a steam explosion also called fuel coolant interaction (FCI) after core meltdown or
  - a major hydrogen detonation or
  - core melt-through under high primary pressure

  damage the outer reactor containment such that a large leak will release considerable amounts of airborne radioactivity (100% radioactive noble gases, 50–90% of the radioiodine, Cs, and Tc), roughly within 1 h after core melt down;
- a leak in the annulus between the outer containment and the concrete shell, or an uncontrolled steam generator accident with steam blowdown valves getting stuck in the open position, release somewhat smaller amounts of airborne radioactivity;
- the reactor core melting through the concrete baseplate followed by pressure buildup in the outer reactor containment result in delayed releases of radioactivity.

The different contributions of these different severe accident sequences to the overall damage consequences, i.e. early deaths (fatalities), are shown by Fig. 11.39 as the result of the German Safety Study, Phase A [37, 71]. These results are similar in terms of early deaths to the results of the US Reactor Safety Study WASH 1400 [36] as shown by Figs. 11.25 and 11.26.

In case it could be demonstrated that the above accident phenomena (and their consequences detailed in Fig. 11.39 [71]) occurring after a core meltdown result in neither early nor late failure of the outer containment i.e. the outer containment retains its integrity the

- radiological consequences to the population,
- need for evacuation and resettlement of the population,
- hazard of contamination of large areas

are reduced to a minimum, i.e. to the reactor plant itself.

If in addition the reactor core can be prevented from melting through the concrete baseplate into the ground below the reactor containment, there would be no danger of contamination of the groundwater over long periods of time.

The safety requirements of the outer reactor containment as outlined in the Karlsruhe KHE safety concept then are as follows:

- The consequences of severe core meltdown accidents must be managed by the inner and outer reactor containments. The outer reactor containment should retain the leaktightness of smaller 1% even after a severe core meltdown accident. The molten reactor core must not melt through the bottom of the outer containment.
- The frequency of occurrence of core meltdown is determined by the diversity and number of redundancies of the cooling circuits, safety systems, emergency core cooling systems, emergency power supply systems. The overall frequency of occurrence for a core meltdown should be around 10^{-6} per reactor year. It can hardly be reduced below this level for practical reasons.
- Probabilistic safety analyses only can serve for the optimization of safety systems and for demonstrating that the overall frequency of occurrence is in the range of 10^{-6} per reactor year. However, probabilistic safety analyses no longer can support the argument that risk—as a product of frequency of occurrence multiplied by damage—is so low that it can be accepted.

The results of the research program conducted at the Karlsruhe Research Center on the basis of this KHE safety concept are presented in the Sects. 11.10 and 11.11.

11.10 New Findings in Safety Research

This section will contain the new findings of safety research into the accident phenomena described in the section above which, in WASH-1400 [36] and the German Reactor Risk Study Phase A [37], still resulted in major accident consequences.

11.10.1 Steam Explosion (Molten Fuel–Water Interaction)

A steam explosion is the explosion-like evaporation of a liquid, such as water, in contact with very hot liquid fuel. This contact must allow a very fast heat input from the molten fuel to the coolant (water) in a very short time. Peak pressures may be in excess of 10 MPa. Steam explosions can also occur in combination with other liquids, e.g. in contacts between

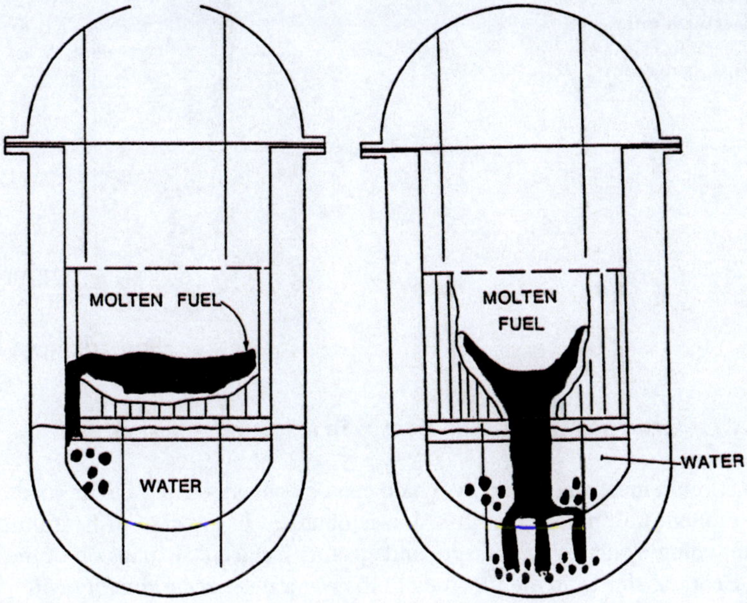

Fig. 11.40 Molten fuel pouring near the edge of a core (*left*) and molten pouring through the lower grid plate [98, 99]

- hot oil and water,
- molten aluminum and water,
- water and liquefied gas at cryogenic temperatures,
- hot lava (volcanic eruption) and water.

In an LWR, an accident sequence can include mainly two contact modes for a steam explosion:

- during a very fast superprompt critical power transient leading far beyond the nominal reactor power, the fuel rods may burst and the molten fuel will be injected as very finely dispersed fuel (with very large heat transfer area) into the cooling channel under high pressure and mixed with the cooling water. This happened, e.g. in the so-called SL-1 accident (water-cooled experimental reactor in USA) [96, 97] and in the Chernobyl accident (Ukraine);
- in the second mode of contact, molten fuel of the reactor core after a core meltdown accident can come into contact with remaining water (Fig. 11.40) either within the reactor pressure vessel, after melting through the grid plate, or outside the reactor pressure vessel, after melting through the bottom hemispherical head of the reactor pressure vessel.

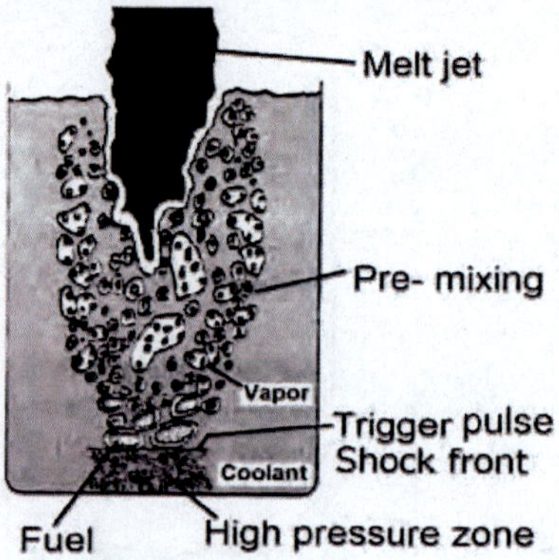

Fig. 11.41 Melt jet injected into water with different zones for premixing and fragmentation [98, 99]

11.10.1.1 Mechanically Released Energy in a Steam Explosion

The maximum mechanical energy which can be converted from the thermal energy of the molten fuel in a steam explosion is obtained in the case of heat transfer at constant volume and rising pressure and ensuing isentropic expansion of the steam [100]. Heat transfer from the fuel melt to the water must occur roughly within 1 ms. The ratio of volumes of the fuel melt and water in that case should be around 1. Theoretically, this would allow an efficiency of roughly 40% to be attained for the conversion of thermal into mechanical energy [101–103]. However, the efficiency measured in experiments with a simulated core melt (corium) on average nearly always is below 1%. The maximum efficiency of conversion in some experiments was 2–3% [100, 104, 96]. Only experiments with iron–aluminum–thermite and water resulted in efficiencies roughly a factor of 2 higher.

Estimates of the damage resulting in the SL-1 accident led to an efficiency of 10–15% [100]. However, the SL-1 accident is in the category of the first mode of contact between the molten fuel and water, which is initiated by a (superprompt critical) power transient with the fuel rods rupturing.

11.10.1.2 Description of a Steam Explosion Sequence

One of the most important preconditions of a steam explosion is fragmentation of the melt into many particles of 0.1 mm size to create the area necessary for fast heat transfer.

Estimates show that this kind of fragmentation from a large molten core mass into many particles of roughly 0.1 mm in size is not possible in one single step, as the energy input would be too large. The melt first can be split up only into larger droplets of cm in size (premixing). In this process, Rayleigh-Taylor or Kelvin-Helmholtz instabilities play a major role when the molten jet flows into the water (Fig. 11.41). These larger melt droplets roughly cm in size are surrounded by a vapor film.

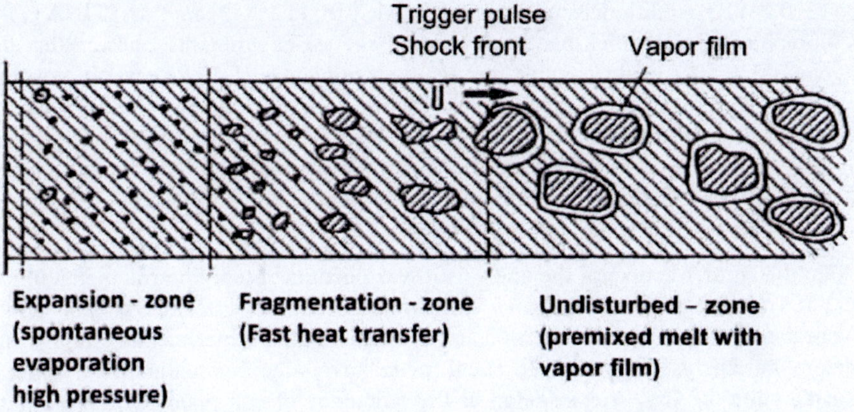

Fig. 11.42 Theoretical model for steam explosion with premixing, fragmentation (*trigger pulse*) and spontaneous evaporation (*expansion zone*) [98, 100]

A pressure pulse is required as the trigger initiating a steam explosion. The pressure pulse causes the vapor film to break down and very fine melt particles (Fig. 11.42) to be produced in a process of fine fragmentation. As a consequence, there is very fast evaporation and a steam explosion, respectively [98, 101, 102, 96, 105].

However, this ideal concept of the theoretical chain of events in a steam explosion will only occur in a random process in experiments including dissipation effects. Moreover, the core melt will never contact water as a bulk substance because melt-through processes will always proceed in an incoherent manner in terms of time and location.

11.10.1.3 Steam Explosion in the Reactor Pressure Vessel

Accounting for the above described discrepancies between experimental results and the ideal theoretical models, it is now postulated that the molten reactor core melting through the gridplate and falling as a molten jet into the water-filled region below the gridplate would give rise to a steam explosion producing a maximum amount of stress acting on the reactor head and its bolts. WASH-1400 [36] had assumed that a steam explosion would cause the head of the reactor vessel to be blown off and penetrate the outer reactor containment as a bullet (α-mode failure). This maximum accident was the subject of many research programs between 1980 and 2010.

These are the most important findings of the associated Karlsruhe Safety Research Program [100, 106]:

- Studies of the core meltdown accident by means of the SCADP/RELAP [107–110] or MELCOR [110, 111] computer codes led to a molten core mass of 110 t at a temperature of approximately 3,000 K and a pressure of 0.25 MPa above the baseplate (the 110 t corresponding to 85% of the core mass of a KWU-

1300 PWR). Additional studies using the MC3D [112, 113] and MATTINA [114] computer codes, which had been verified against experiments, showed that the core melt, after having molten through the gridplate (Fig. 11.41), would flow out in a molten jet of about 0.2 m^2 cross section. There would be premixing with a water volume fraction of 0.5–0.6. The **maximum** content of thermal energy in the larger melt droplet as premixing zone amounts to **roughly 3 GJ**. (On the average of all possible cases, the thermal energy content would be only 0.5–2.0 GJ with the corresponding water volume fractions of 0.2–0.5.)

In the further course of the analysis it was postulated that fine fragmentation to 0.2–0.3 mm, as measured in experiments, would occur and a steam explosion would be initiated. After careful inspection and assessment of all experimental findings and theoretical analyses against theoretical models available internationally, **a conservative value of 15% was selected** as the efficiency of conversion of thermal into mechanical energy. This results in a maximum mechanical energy release by the steam explosion of

$$3\,\text{GJ} \times 0.15 = \mathbf{0.45\,GJ}$$

(as the average of all possible cases, the result would only be 0.075–0.3 GJ).

11.10.1.4 Dynamic Mechanical Analysis of the Pressure Vessel

These results about the release of mechanical energy in the course of a steam explosion in the lower plenum (bottom hemispherical head) of the reactor pressure vessel were used to conduct dynamic mechanical stability analyses accompanied by 1:10 scale experiments (BERDA experiments) [115–118]. Theoretical models of similarity theory and accompanying strength analyses allowed the results of the 1:10 scale experiments to be transferred to the dimensions of the reactor pressure vessel.

As the steam explosion is initiated already while the melt jet is discharged, some 80 t of core melt would still be present on the baseplate. After rupture of the mechanical anchorage (break) of the gridplate, this volume must be accelerated upward together in the reactor pressure vessel. The upper part of the reactor pressure vessel of a PWR contains the internal structures with the guide tubes for the control and shutdown rods and for the in-core instrumentation (Fig. 11.43). These must be compressed by the core melt accelerated upward so as to be able to transfer the dynamic forces of the core melt to the head structures and head bolts. These internal structures and the head structures were simulated in great detail in the 1:10-scale experiments (BERDA).

This is the overall finding of all Karlsruhe BERDA experiments and theoretical analyses:

- Acceleration of the gridplate, with the core melt resting on it, up to the top internal structures of the head requires at least approximately 2 GJ.
- Another 0.8 GJ would be necessary to compress the internal structures and elongate the head bolts by a few mm. The compression of the upper internal structures results in a decrease of the impact forces and an extension of the time period for action

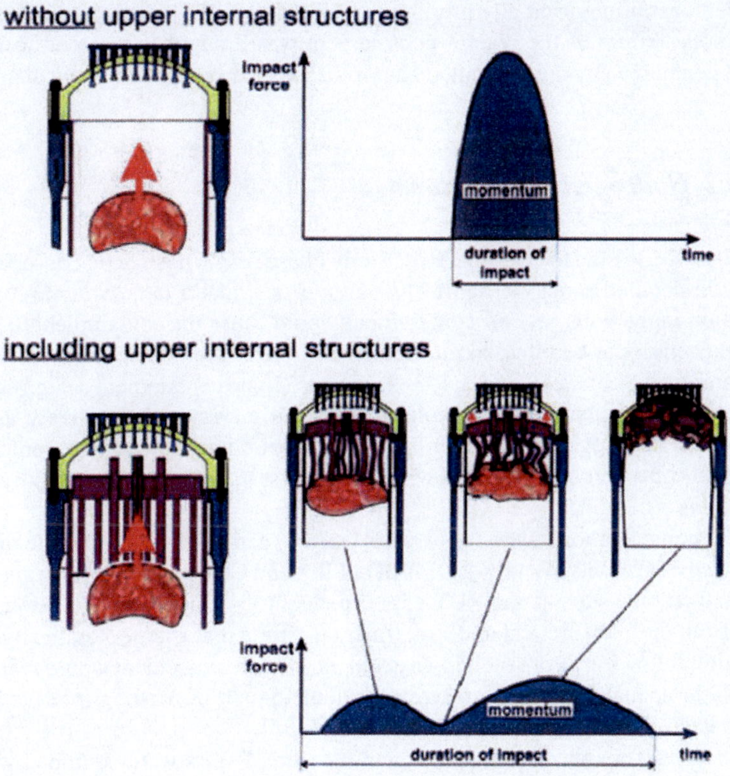

Fig. 11.43 BERDA experiments for melt slug impact on head of the reactor pressure vessel (without and with internal structures) [116, 119].

of these impact forces (Fig. 11.43). The reactor head would remain intact in the process.

- On the whole, at least **2.8 GJ**

of mechanical energy would have to be released in the steam explosion to accelerate the core melt plus the gridplate upward, compress the internal head structures of the reactor pressure vessel, and expand the head bolts by a few mm.

This 2.8 GJ must be compared to the 0.45 GJ occurring in the steam explosion at conservatively assumed efficiency of 15% for the conversion of thermal into mechanical energy. Even an efficiency of 40%, i.e. 1.2 GJ, for the conversion of thermal into mechanical energy in a steam explosion could not jeopardize the mechanical integrity of the reactor pressure vessel [119–123].

Consequently, the steam explosion with failure of the reactor pressure vessel and outer containment (α-mode failure) as assumed in WASH-1400 and in the German Risk Study Phase A can be considered impossible on grounds of physics.

A demonstration originally proposed in USA [124, 125] of the non-existence of the α-mode failure of the reactor containment was brought to a scientifically successful conclusion by these detailed Karlsruhe experiments and theoretical analyses.

11.10.2 Hydrogen Detonation

WASH-1400 [36] and the German Risk Study Phase A [37] postulated conservatively, with little detailed scientific and technical analysis, that a large-volume hydrogen detonation in the outer reactor containment would cause the containment to rupture and radioactivity to be released into the environment. This was doubted after some first theoretical estimates by KHE [93–95]. Appropriate containment design concepts were proposed which would be able to withstand a very conservatively assumed large-volume hydrogen detonation [95, 126]. This demonstrated that containment design concepts can be conceived which can resist to such large-volume hydrogen detonations.

These considerations were followed by many years of theoretical code development, such as GASFLOW [127], DET-3D [128], and COM3D [129] and experimental investigations, such as the RUT experiments [130–134]. The conclusion can be drawn from these theoretical and experimental efforts that a large-volume hydrogen detonation following a core meltdown accident (if such an accident would really happen) can be managed by the outer reactor containment of existing modern PWRs, like the KONVOI-PWRs of Kraftwerk Union [135].

As a first severe accident management measure, PWRs were equipped with so-called passive autocatalytic recombiners [136] able to reduce only slow release-rates of hydrogen release of approximately 0.5 kg H_2/s during core meltdown accidents. However, core meltdown accidents must be accounted for with higher rates of hydrogen release of up to 7 kg H_2/s. The related H_2-steam-air mixtures produced are capable of detonating [134, 135].

11.10.2.1 Load Carrying Capacity of a KWU-1300 PWR Containment in a Hydrogen Detonation

Analysis of the release of hydrogen in a core meltdown accident initiated by a small leak in the primary system will be described here as an example of the kind of analysis performed [135]. A small leak in the primary system with a delayed pressure drop in the secondary steam system results in water/steam and hydrogen release into the containment with a maximum release rate of 7 kg H_2/s over a certain period of time during the accident sequence [135]. The hydrogen is produced during the accident sequence by overheating of the Zircaloy claddings above 1,200°C and their chemical reaction with steam according to

$$Zr + 2H_2O \rightarrow ZrO_2 + 2H_2.$$

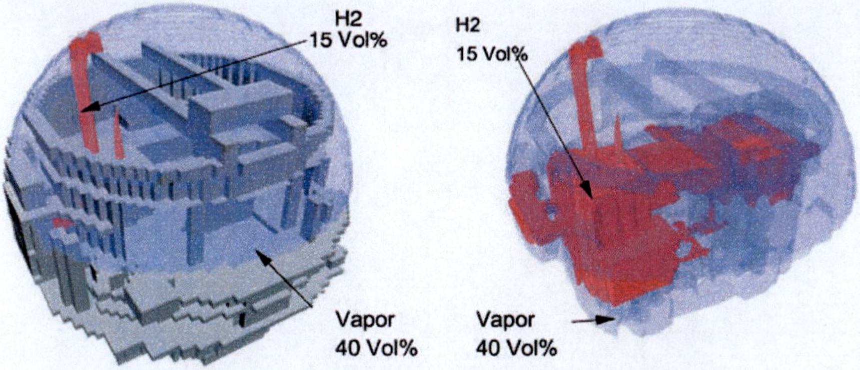

Fig. 11.44 Hydrogen-Steam-Air mixtures in the outer containment (two different *side views*) of a German PWR after an assumed core melt accident at the time of 7,950 s after begin of the accident sequence [135]

This hydrogen release is accompanied by steam release. The aggregate volume of hydrogen released eventually into the outer containment then is roughly 855 kg H_2 [135]. In the further course of the accident sequence, the hydrogen concentration in the air and steam mixture exceeds 15 Vol.% in parts of the reactor containment. The spatial hydrogen distribution and the hydrogen concentration within the mixture of air, steam and hydrogen within the outer containment is shown for the time of 7,950 s after begin of the accident sequence in Fig. 11.44. This mixture of air and steam with 550 kg of H_2, which is able to detonate, can be ignited, e.g. by an overloaded hydrogen recombiner. This was demonstrated in the RUT experiments [134, 135]. The calculations with the three-dimensional-time-dependent DET3D detonation code [135], which also takes into account shock wave reflections within the reactor containment, resulted in short time pressure peaks of 2 to 5 MPa and impulse-type loads of 10–30 kPa s. When the transient phase of the detonation is over, there remains a quasi-steady-state pressure of approximately 0.58 MPa and temperatures around 800°C over hours. This does not jeopardize containment integrity.

11.10.2.2 Structural Dynamics Response of the Spherical Steel Containment

The results of the detonation with respect to the impulse and pressures acting on the containment wall were used for analyses with the ABAQUS [137] and PLEXUS [138] codes. Figure 11.45 shows the deformations arising of the spherical outer steel containment of the reactor in the steel shell (magnified by a factor of 5) [135]. The largest deformations of approximately 4.6% occur in the vicinity of the materials transfer lock and close to the upper pole (2.4%). None of these deformations will cause the steel shell of the spherical reactor containment to fail [139, 140].

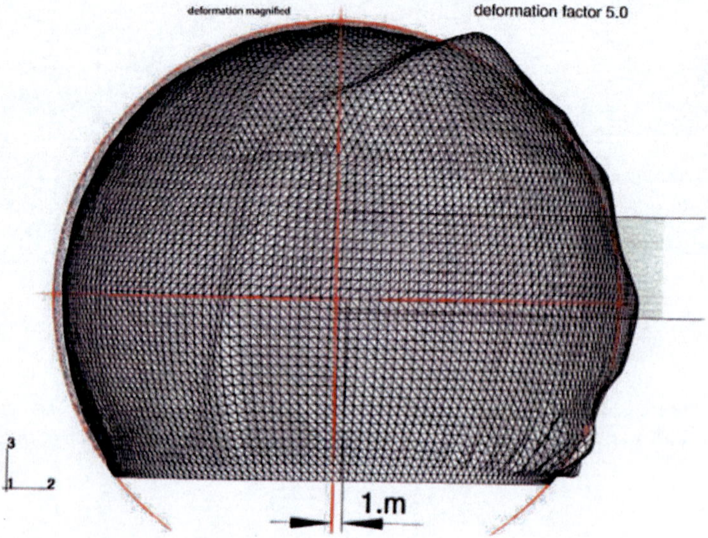

Fig. 11.45 Plastic strains (magnified) in the outer containment walls of a Kraftwerk Union PWR-1200 during a hydrogen detonation [135]

11.10.3 Break of a Pipe of the Residual Heat Removal System in the Annulus

The reference case for loss-of-coolant accidents with leaks in primary pipes in the annulus outside the outer containment as referred to in the German Risk Studies Phase A [37] and Phase B [5] was the assumed break of a pipe or the failure of valves of the residual heat removal system. This can ultimately cause core meltdown at low primary pressure. The radioactivity from the core melt in this case would bypass the leaktightness function of the outer containment and escape directly into the annulus and, through filters, on into the environment.

This weak spot in the design of existing early pressurized water reactors must be avoided in future PWRs by appropriate design measures, according to the KHE Safety Concept. To avoid such possible bypasses, the function of the multiple barrier system (outer containment) of retaining the radioactivity must be maintained for all pipes connected to the primary cooling system (double containment function) [1, 11, 67]. This is technically possible.

11.10.4 Core Meltdown after an Uncontrolled Large Scale Steam Generator Tube Break

In the highly unlikely case of a large scale steam generator tube break, primary coolant can flow to the secondary side. The loss of primary coolant causes the primary pressure to drop and the high-pressure safety feed system to be initiated. Should the necessary shutdown of the high pressure safety feed systems fail, this would ultimately cause overfeeding of the steam generators, and the main steam relief valves would open. When these main steam relief valves do not close again, primary coolant will flow straight into the environment. In the further course of the accident, there could be core meltdown, and the radioactivity released would escape directly into the environment.

Also this very rare core meltdown accident, which could become possible as a result of the present design of steam generators and main steam pipes, must be solved technically as a requirement of the KHE Safety Concept by appropriate routing of the steam pipes, use of the proper valves, which can close during such accidental situations, and by accident management measures such that it will no longer be of importance in modern PWRs, e.g. KWU PWR-1300 [67] and EPR (Sect. 5.11) as it was the case in the early risk studies e.g. the German Risk Studies Phase A and B [5, 37].

11.10.5 Core Meltdown Under High Primary Coolant Pressure

WASH-1400 [36] and the German Risk Study Phase A [37] had assumed core meltdown under high primary coolant pressure to lead to failure of the outer containment, followed by a major early radioactivity release, as a result of the reactor pressure vessel acting as a bullet. Core meltdown under high primary pressure could occur in an uncontrolled emergency power case (station blackout) or uncontrolled failure of the main feedwater supply [5, 37]. In both cases, the ultimate consequence is heating of the primary coolant plus primary pressure increase, thus causing the pressurizer relief valves to open. The reactor pressure vessel will be voided. The water level in the reactor core will drop, the fuel rods will heat up, and there will be a zirconium—steam reaction at the fuel rod claddings. This heats the reactor core still further, causing it to start melting. After roughly 1 h, the core will have molten more than 80%. After 3 h, the core will melt through the grid-plate in the reactor pressure vessel.

Molten fuel flows into the water contained in the bottom hemispherical head of the reactor pressure vessel. This water evaporates quickly. However, the coolant cannot be removed fast enough through the pressurizer relief valves, which causes the coolant pressure to rise. After approximately 3.5 h, the core will melt through the bottom of the reactor pressure vessel at a high primary pressure.

Very high buoyancy forces will arise which can accelerate the reactor pressure vessel upward.

In the German Risk Study Phase B [5] there had already been estimates of mechanical resistance offered by the anchorage of the reactor pressure vessel and the primary piping. The outcome had been that a primary internal pressure of >3 MPa during melt-through would cause the anchorage of the reactor pressure vessel to fail. However, the integrity of the outer reactor containment would be jeopardized by the

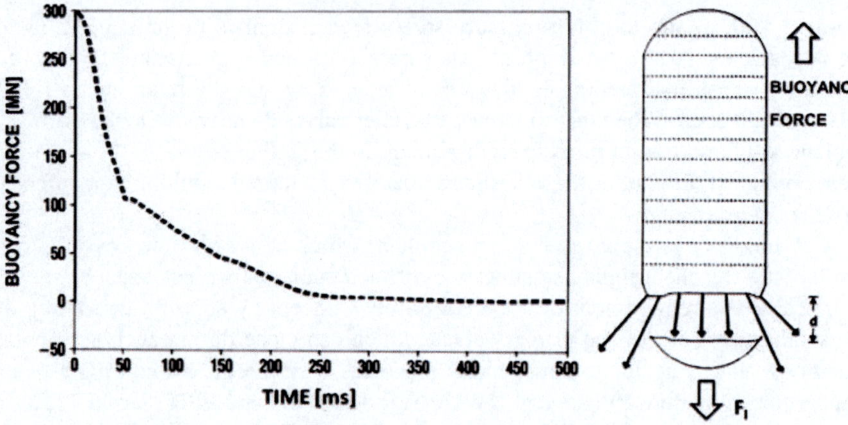

Fig. 11.46 Buoyancy force on pressure vessel in case lower head fails because of core melt through under operational pressure of 15.5 MPa [141]

reactor pressure vessel accelerated upward only above a primary internal pressure of 8–10 MPa.

As a conclusion drawn from all these findings, it was proposed for existing PWRs to reduce the primary pressure in the reactor pressure vessel by timely pressure relief by opening of a pressurizer relief valve (accident management measure), thus allowing the core to melt through at a low pressure as in a loss-of-coolant accident (LOCA). Failure of this accident management measure results in a frequency of occurrence roughly one order of magnitude lower per reactor year for core meltdown under high pressure [5].

Within the framework of the KHE Safety Concept, the RELAP-MOD-3 code [109] was used to determine the buoyancy forces arising from failure of the bottom hemispherical head at the full primary pressure of 15.5 MPa in the RPV [141].

This buoyancy force is shown by Fig. 11.46 as a function of time. It starts with 300 MN and decreases down within about 250 ms. A technical concept was proposed for reinforced anchorage of the reactor pressure vessel [93–95]. This stronger mechanical anchorage of the reactor pressure vessel—as an ultimate safety solution—would prevent the vessel from moving upwards even in the case of the reactor core melting through at a primary pressure of 17 MPa. The integrity of the outer containment is preserved even in this accident (core meltdown under high primary pressure). In this way it provides high flexibility in safety design to avoid the core melt down and high primary pressure.

In the EPR safety concept, the possibility was chosen to install two additional blowdown valves, with very high discharge capacities of 900 t/h which is to ensure depressurization to at least 2 MPa with a high reliability within a short time period. These blowdown valves can be actuated from the control room when the coolant outlet temperature exceeds 650°C. Depressurization to <2 MPa avoids significant melt dispersal (direct containment heating (Sect. 11.10.9)) in case of core melt break

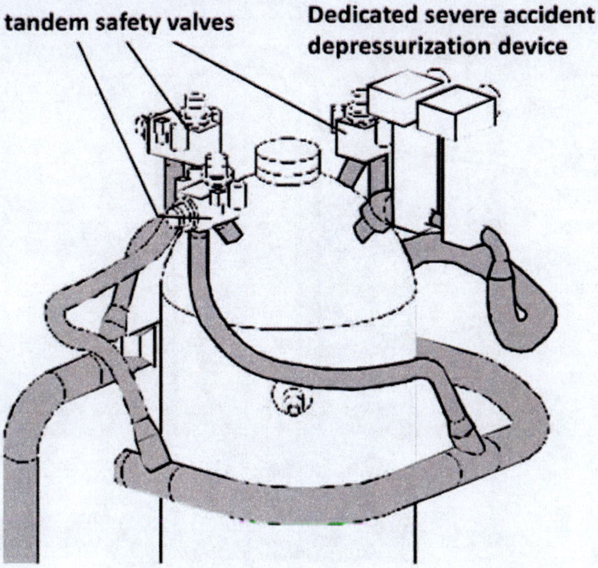

Fig. 11.47 Safety valves and dedicated depressurization devices of EPR [142]

through the bottom of the reactor pressure vessel [139]. The three tandem pressure relief and safety valves as well as the two blowdown valves for EPR are shown by Fig. 11.47.

This changes the core meltdown accident sequence under high pressure into a core meltdown under low pressure [142].

11.10.6 Medium Term Pressure Buildup in the Outer Containment

Even if no steam explosion occurs at low pressure in the bottom hemispherical head of the reactor pressure vessel (Sect. 11.10.1.3) during the influx of molten fuel, the residual water remaining there will evaporate very quickly. There are then two possibilities for the progress of the accident evolution and two safety design schools:

- Cooling the reactor pressure vessel by flooding with water from the outside as an accident management measure,
- No cooling of the reactor vessel by flooding with water from the outside because of the hazard of a steam explosion when the molten core would melt through. Instead installation of a molten core fuel retention and cooling device (core catcher), accommodating and retaining the core melt after core meltdown.

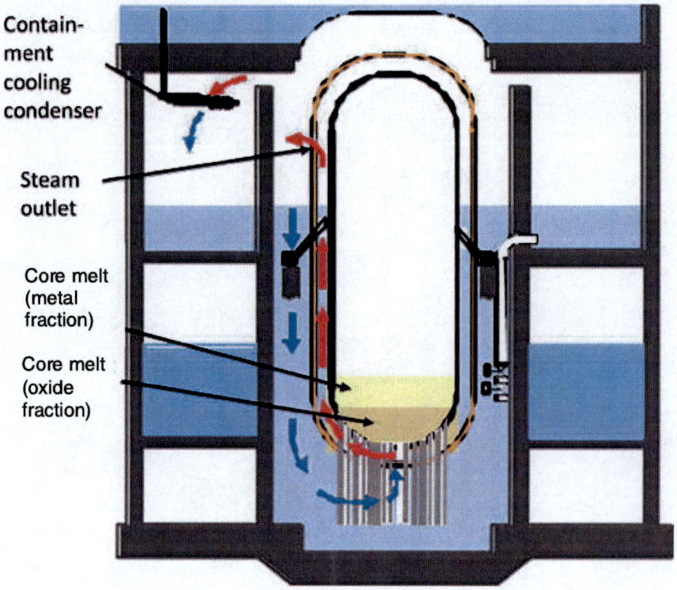

Fig. 11.48 Cooling of molten core by flooding the reactor pressure vessel with water on the outside [143]

11.10.6.1 Possibility of Cooling the Molten Reactor Core from the Outside of the Reactor Pressure Vessel

In this variant, the reactor pressure vessel is to be flooded with water from the outside as the reactor core is melting [143–149]. There are a number of theoretical and experimental investigations in the wake of the Three Mile Island accident in the USA which make this accident management measure appear successful. The heat fluxes from the melt to the wall of the bottom hemispherical head, heat conduction through the wall of the pressure vessel, the temperatures in the wall of the pressure vessel, and the stability of steel as a function of the wall temperature are taken into account. These research investigations in the USA demonstrated that the molten core will not melt through when the reactor pressure vessel is flooded with water from the outside. This is valid also for the case that the thermal insulation of the pressure vessel remains intact on the outside of the reactor pressure vessel.

Molten core cooling by flooding the reactor pressure vessel from the outside is also proposed for the SWR-1000 (KERENA) design (Fig. 11.48). The SWR-1000 bottom head of the pressure vessel has many penetrations (welded tubes for control rods etc.). Therefore, BWRs have a higher surface to volume ratio in this bottom part than PWRs. In addition, BWRs have a lower power density in the core melt than PWRs (see Tables 5.1, 5.2). The reactor pressure vessel of SWR-1000 can be flooded passively by water from the flooding pools. In addition to cooling of the

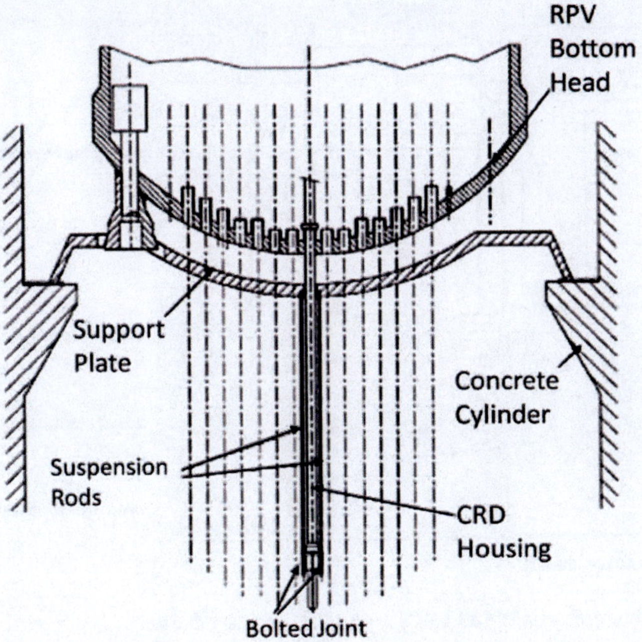

Fig. 11.49 Steel support plate acting as a back-up core catcher [143]

molten core in the lower head by outside flooding of the pressure vessel, there is backup possibility of retaining molten fuel (if it would melt through the bottom of the reactor pressure vessel) by a special steel support plate underneath. This support plate fixes the control rod drives in the bottom hemispherical head of the reactor pressure vessel. It could act as a back-up core catcher in case parts of the core would melt through the bottom head [149].

The decay heat (afterheat) of the core melt can be transferred by evaporating water. The resulting steam can be cooled and condensed by the containment cooling condenser (Figs. 5.19, 11.48) [10, 143].

The second safety design school proposes to keep the reactor cavity dry (no flooding with water) and to install a molten core fuel retention and cooling device (core catcher) outside of the reactor pressure vessel (Fig. 11.49) [10, 142].

11.10.6.2 Penetration of the Core Melt Through the Bottom Head of the Reactor Pressure Vessel

In case the reactor pressure vessel is not flooded with water from the outside, this would be the further course of accident events:

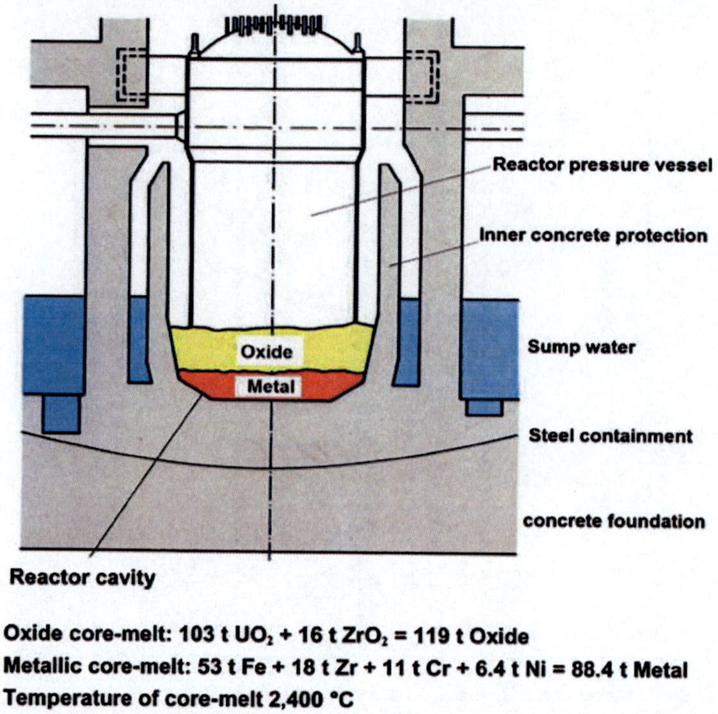

Fig. 11.50 Core melt in the reactor cavity after melting through the lower head of the reactor pressure vessel [35, 150, 151]

The molten mass of the reactor core will melt through the bottom of the reactor pressure vessel and drop into the dry reactor cavity. This is where the molten fuel will react with the concrete and its water of crystallization, giving rise to water vapor, hydrogen, carbon monoxide, and carbon dioxide. While these gases flow into the compartments above the reactor cavity, fuel melt moves downward into the concrete foundation (Fig. 11.50). The fuel melt also spreads in radial direction and is able to melt through the lateral biological shield. As the oxide melt mixes with concrete (mainly SiO_2) the oxide melt fraction becomes lighter than the metal melt fraction. In the process, the melt can contact the water of the building sump (Fig. 11.50). The sump water will then evaporate. This will cause the pressure in the outer reactor containment to rise gradually. After roughly five to six days, the fuel melt can eventually penetrate the concrete foundation of approximately 6 m [5, 152–154]. These predictions are backed by experiments and theoretical analyses (WECHSL code [5, 152]). After several days, the steam pressure can reach the design pressure of the outer reactor containment of 0.6 MPa [5, 36, 37].

In the German Risk Study, Phase A [37] it is concluded that the outer containment would develop a leak when exceeding this design pressure of 0.6 MPa. The

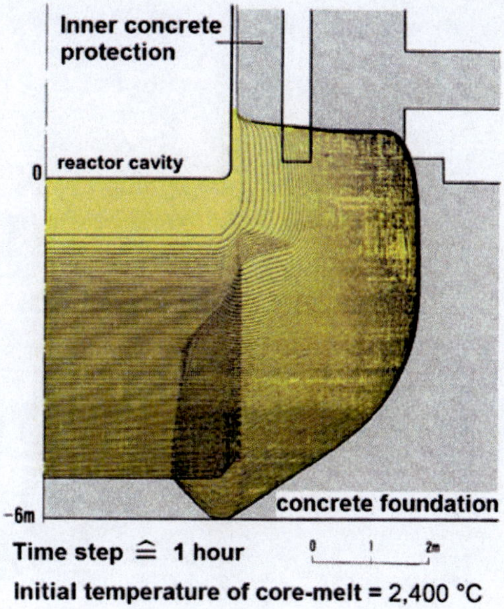

Fig. 11.51 Penetration of core melt into the concrete of the base-mat of the reactor building (calculational results of the WECHSL-code) [5, 153]

radioactive aerosols and radioactive gases would be released into the environment (Fig. 11.51).

More recent investigations [155, 156] showed, however, that the outer containment of KWU-1300 PWRs would develop a leak at the materials transfer lock not below some 1.2 MPa. In addition it was shown that dangerous overpressures in the outer reactor containment cannot be reached if a pressure reduction by a so-called ex-venting filter is achieved. Therefore, all German LWRs were equipped after the German Reactor Safety Study Phase B [5] with so-called aerosol ex-venting filters having a filter efficiency of 99.9% [61, 62]. The pressure will be relieved by opening a valve to this exventing filter and the radioactive gases and the radioactive aerosols are emitted through this filter (Fig. 11.52).

During pressure buildup over several days, however, most of the radioactive aerosols will be deposited in the outer reactor containment by sedimentation and thermophoresis etc. This decreases the amount of radioactive aerosols released to the environment by orders of magnitude (Fig. 11.19).

In the EPR safety concept a dome spray system can spray water from the dome of the containment and condense vapor in the containment. In this way depressurization of the containment is possible from 0.65 to about 0.2 MPa within one day. The water for the spray system is taken from the in-containment refuelling water storage tank (IRWST) and pumped through a heat exchanger to the spray system [10].

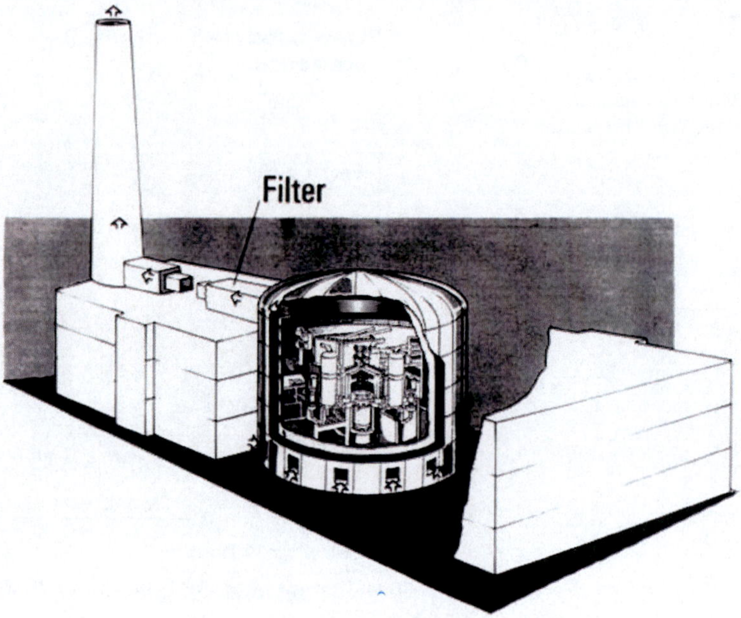

Fig. 11.52 Exventing filter of a German PWR for pressure relief in the outer containment [35]

11.10.6.3 The Core Melt Penetrating into the Subsoil Underneath the Reactor Building

Further penetration of the core melt into the ground below the outer reactor containment was neither studied in the WASH-1400 [36] nor in the German Risk Studies [5, 37]. In the Three Mile Island the core melt did not penetrate through the bottom head of the reactor vessel (Sect. 11.8.1.1), probably due to the fact that enough water was available timely enough for cooling. In the Chernobyl accident, dropping sand and lead from helicopters on the destroyed reactor core created a molten mass which ultimately did not melt through the bottom foundation of the reactor building despite fears that this might happen (Sect. 11.8.2). In the Fukushima accident the core melt caused some small holes in the bottom head of the reactor pressure vessel of unit 1 but the core melt did not penetrate further, probably because cooling could be provided early enough (Sect. 11.8.3).

All experimental and theoretical investigations culminate in the conclusion that, in a PWR for instance, the core melt—after having molten through the bottom head of the reactor vessel and through the concrete base plate—will move further into the subsoil below the foundation of the reactor building [157, 158]. In a matter of roughly 200 days, it would expand to a radius of approximately 12 m (Fig. 11.53). It would comprise a volume of 1,000 m^3 and consist of UO_2, ZrO_2, CrO_2, FeO_2, SiO_2, Al_2O_3 and CaO. The SiO_2 fraction would amount to roughly 75%. Ground-

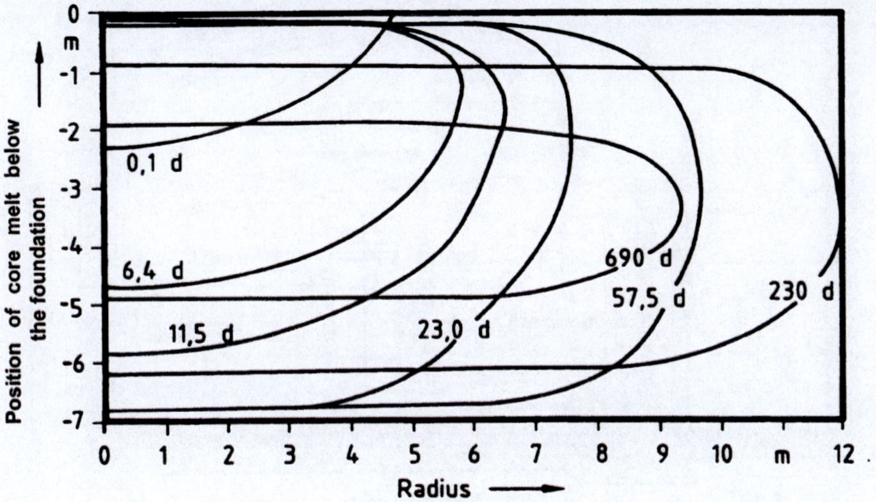

Fig. 11.53 Core melt penetrating into the subsoil below the foundation of the reactor building [157]

water would cool this enlarged molten mass and slow down and stop eventually its further penetration into the subsoil [157].

Slowly, the groundwater could dissolve fission products out of this originally molten and subsequently solidified mass. In a study [157] leaching rates roughly a factor of 100 higher than those applied for vitrified HLW in a deep geological repository waste were assumed. The leaching rate from the porous mass of the melt is determined by processes such as molecular diffusion, adsorption, desorption, ion exchange and colloid buildup. Further transport of the key radionuclides, Sr-90, Tc-99, and Cs-137, requires consideration of the hydrodynamic transport equations for advection and dispersion in the groundwater [157, 158]. The radionuclides could be carried through the groundwater to a well or into a river and then would move downriver.

Radiation exposure of the public mainly from Sr-90 and Cs-137 would then be possible through the intake of drinking water from the groundwater in the environment and downriver of the location of the core melt. Moreover, flooding by the river could cause the flooded regions to be contaminated as a consequence of sedimentation of radionuclides and drying up of the flooded regions.

11.10.6.4 Possible Countermeasures Against Core Melt in the Subsoil

Countermeasures against the spreading of radionuclides would be possible by

- installing sealing walls extending deep down to the contaminated groundwater,
- sinking wells to pump off radioactively contaminated groundwater.

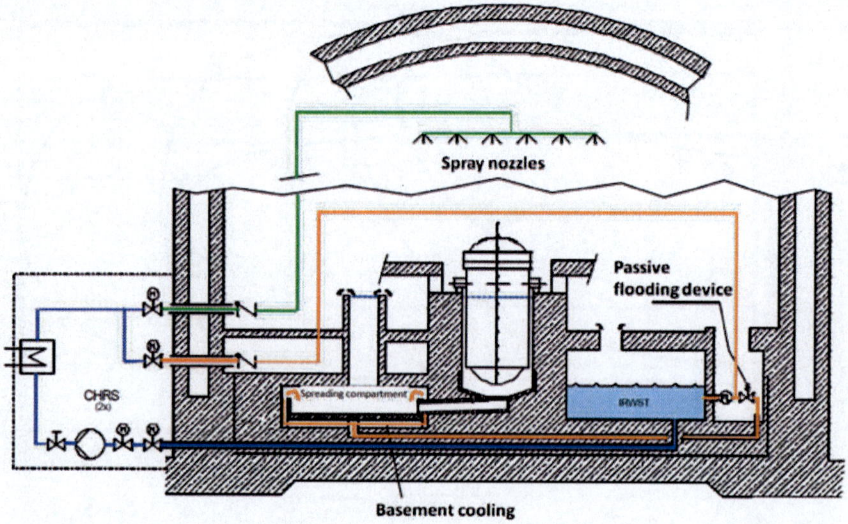

Fig. 11.54 Molten core fuel retention and cooling device (core catcher) together with spray nozzle system for containment atmosphere cooling [10]

If no such countermeasures would be taken, the groundwater of a relatively large area, and over long periods of time, would not be fit for use as drinking water or for irrigation of agricultural areas.

11.10.7 Molten Core Retention and Cooling Device (Core Catcher)

The countermeasures listed above for the period after melting of the reactor core through the concrete baseplate of the outer reactor containment, and the radioactive contamination of the groundwater and rivers in the vicinity, can be rendered superfluous by

- flooding the reactor pressure vessel on the outside vessel with water in case the core is going to melt down (accident management measures) (Sect. 11.10.6.1),
- a molten core cooling and retention device (core catcher) underneath the reactor pressure vessel. This cooling device for molten core masses is part of the EPR concept (Fig. 11.54), but not of any other PWR safety design concepts known up to now.

In the EPR design, the core melt is first kept in the reactor cavity for a short period of time so that core masses dropping slightly later can also be collected. After penetrating a melt plug (steel plate covered by a layer of concrete), the core melt flows through an inclined canal onto a dispersion area of approximately 170 m^2. The

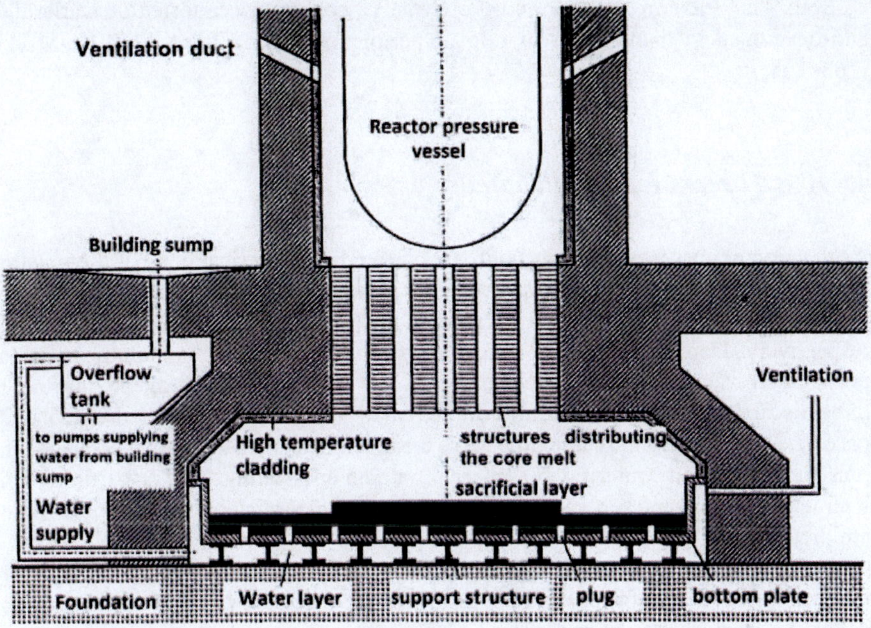

Fig. 11.55 The COMET core catcher and molten fuel heat removal concept [159]

core melt is allowed to spread there evenly to a thickness of roughly 30 cm. Then flooding of the melt with water is initiated passively by water flow from the IRWST. In this way, the melt is cooled from the top and solidifies in part. From the bottom, the melt is cooled by active bottom cooling and stabilized in this way (Fig. 11.54).

11.10.7.1 Other Core Catcher Designs

As a result of research programs, a number of other concepts were developed to cool core melts [160–162]. At this point, only the COMET concept developed at Karlsruhe will be explained briefly [162, 163]. In the COMET concept, the melt is to be flooded with water from below after erosion of a sacrificial layer. The rapidly evaporating water disrupts the melt, cooling its interior as a water-steam mix. Figure 11.55 shows the COMET concept. The melt is collected below the reactor pressure vessel and then first erodes a dry sacrificial layer of concrete 15 cm high. Afterwards the melt can spread completely and melt any cooling channels in the concrete layer from the top. This allows water to enter under the pressure of a overflow tank located high up. This is followed by an effective phase of melt cooling and fragmentation. The melt solidifies within a short period of time and can be flooded fully and cooled. The steam produced is cooled in a heat exchanger and condensed.

Both the EPR core catcher and the COMET concept were tested at Karlsruhe and developed in many years of pilot experiments (KAPOOL, KATS, COMET) [163–165].

11.10.8 Direct Heating Problem

As the reactor core is melting through the bottom hemispherical head of the reactor pressure vessel under high pressure, the core melt, in the form of a water vapor–melt spray, is driven into other compartments and into the outer reactor containment, respectively. Depending on the design of the inner and outer containments of a pressurized water reactor, this makes it possible for fine particles of the core melt to be distributed over large volumes of the reactor containment. As the droplets and particles of the melt at the same time carry the residual heat, this accident sequence in present PWRs is also referred to as direct containment heating. The Zr-particles can interact with steam and generate hydrogen. The hot fuel particles heat the containment atmosphere and increase the pressure.

These phenomena were studied in experimental programs which, in turn, allow the appropriate design proposals to be made for the reactor cavity and compartments below the reactor pressure vessel [166, 167].

In this way, problems of direct containment heating can be excluded for future reactors, such as the EPR, by a properly enforced flow of water vapor and melt spray into specific compartments.

11.10.9 Summary of Safety Research Findings About the KHE Safety Concept

The findings of recent safety research as outlined in Sect. 11.10 change decisively the results of the early WASH-1400 [36] and German Reactor Risk Study [37] in view of future reactor designs. The accident phenomena,

- steam explosion,
- hydrogen detonation,
- high-pressure core meltdown,
- containment bypass in the annulus and uncontrolled steam generator tube failure associated with radioactivity release through the main steam relief valves

which had the most severe accident consequences, are either controlled in present PWR designs (KWU PWR-1300) or can be managed in future LWRs by appropriate design measures. Examples of such future LWRs are the EPR (Sect. 5.1.1) and the BWR-1000 (KERENA) (Sect. 5.1.2). Both reactor lines were designed on the basis of the findings in Sect. 11.10 and along safety recommendations laid down by the

French and German reactor safety commissions [168]. They still need to undergo the licensing procedures required in those countries in which they are built and operated.

The results of the German Risk Study Phase A [37] for earlier German pressurized water reactors, which findings are outlined individually in Fig. 11.39, are not valid anymore for future LWR designs, e.g. EPR and SWR-1000 (KERENA). The results of Fig. 11.39 shrink to very small damage consequences [67]. **Core melt down accidents will no longer lead to large scale contamination of areas outside the reactor plant**.

Whereas these statements apply to the safety concept of future LWRs like EPR or SWR-1000 (KERENA) they do not apply to the majority of LWRs operating around 2012 in the world. Many of these LWRs were built from 1975 on and are constructed on safety design concepts and principles as they were analyzed in WASH-1400 [36] and the German Safety Studies Phase A [37] and Phase B [5]. A high percentage of these LWRs e.g. in the USA and France received an extension of their operating license after small design modifications up to 60 years operating life time. For these reactors only part of the research results reported in Sects. 11.9–11.10 are applicable.

Other reactor types, e.g. heavy water reactors or gas cooled reactors could not be discussed in this chapter.

11.11 The Safety of Facilities of the Nuclear Fuel Cycle

Unlike the fuel elements in the core, e.g. of a PWR-1300, where one fuel element produces approximately 19.6 MW(th) at full power, the power of that fuel element roughly one week after reactor shutdown has dropped to approximately 59 kW(th) [169]. After some three months the power will have dropped to roughly 20 kW(th); after some three years, to approximately 1.2 kW(th) (Sect. 11.1). The radioactivity of the spent fuel element will decrease similarly as a function of time.

11.11.1 In-Pile Fuel Element Storage Pool

After unloading from the reactor core, the fuel elements are first stored in deep intermediate fuel storage pools filled with water. These fuel element storage pools must be located within the outer reactor containment. They are clad by a stainless steel sheet. In addition, they are equipped with water heat exchangers for cooling and with water treatment systems in four fold redundancy. The heat exchangers can operate with forced or natural convection transferring the heat either to outside water or to the atmosphere. The water temperature in these storage pools thus will not exceed approximately 40°C.

The fuel elements stand in storage racks at defined spaces so that subcriticality is ensured at any time. Only failure of all cooling systems (failure of electricity supply and emergency power supply in the forced convection mode or failure of the ultimate heat sink) will allow the water temperature to reach roughly 60°C within 12 h and 100°C after about 48 h. This leaves time to restore emergency power or supply cooling water.

If cooling of the fuel elements could not be restored in that period of time, the intermediate store would dry up (Fukushima accident). Spent fuel elements still producing enough residual heat (just unloaded from the core) might fail, and the radioactive noble gases and fission product and fuel aerosols would enter the outer, leaktight reactor containment. The radioactivity would not be released to the environment. This confirms the necessity to arrange fuel element storage within the outer containment.

11.11.2 Wet and Dry Fuel Element Intermediate Stores

11.11.2.1 Wet Intermediate Stores

In some countries, wet intermediate stores are built outside the reactor for accommodating spent fuel elements. These wet intermediate stores must have the same safety design characteristics as the intermediate stores located within the reactor building (Sect. 11.11.1). However, they have a larger capacity of approximately 2,000–4,000 t of fuel. The fuel elements are unloaded from the in-reactor storage pool after some three to five years, transported and stored in the wet intermediate storage facilities until they are shipped either to the reprocessing plant or to a conditioning plant for direct waste disposal at a later point in time (Chap. 7).

11.11.2.2 Dry Intermediate Stores

In dry intermediate stores, the fuel elements are loaded into CASTOR casks (Sect. 7.1.2) roughly five years after having been unloaded from the reactor. These CASTOR casks are then set up in large dry intermediate storage halls (Sect. 7.1.2; Fig. 11.56).

The intermediate storage hall is cooled by ambient air circulated by natural convection. This passive type of cooling saves the need for redundancies of the kind required in active cooling, e.g. in wet intermediate stores.

The elements are put into a neutron-absorbing basket placed inside the CASTOR cask. Leaktightness of the cask is ensured by two lids arranged one above the other with metal gaskets and a sealing gas space filled with helium. Leaktightness is monitored by pressure measurements and checks for any helium leaking out. The heat production of approximately 1 kW(th) per PWR fuel element causes maximum temperatures in the fuel below 390°C. The heat is transferred by the helium gas to the wall of the cask and, at the outside wall of the cask, passed on to the air by natural convection from the fins arranged on the surface of the cask. At the fin surfaces, the

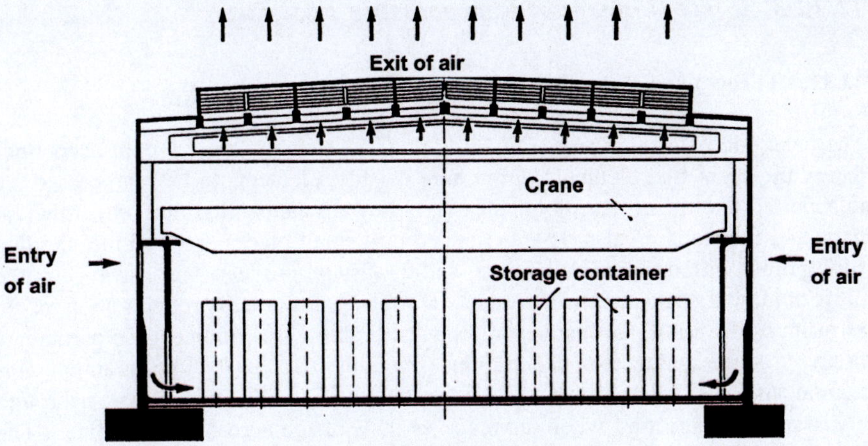

Fig. 11.56 Dry storage facility for CASTOR containers [170]

casks attain maximum temperatures of approximately 50–70°C at a thermal output of the fuel elements of roughly 1 kW(th).

As the fuel in the leaktight fuel rod claddings will not exceed a maximum temperature of 390°C caused by its thermal output, production of aerosols or dust of plutonium, americium, curium within the CASTOR casks is impossible. Fission products and, above all, plutonium and other transuranium elements cannot escape from the shipping casks because of the double lids of the cask and the monitoring for pressure and leaktightness, respectively, both under normal conditions and accident conditions. The intermediate store is continuously monitored for radioactivity releases. Should a leak occur at the lid of a CASTOR cask, another lid may be added and welded tight onto the barrel of the cask, or the gasket of the lid is replaced. In this way, there are always two barriers relative to the fuel. Each shipping cask is designed against external impacts (airplane crash, fire etc.)

Subcriticality of the fuel elements in the CASTOR cask remains ensured at all times. Even in an earthquake or a collapse of the intermediate storage hall, in case of a flood engulfing the storage casks or in an airplane crash on top of the hall, the storage cask will not be destroyed.

To obtain a license, the casks must undergo tests under extreme accident conditions (such as a drop test followed by a fire test and a test under water, respectively). The casks must still be leaktight after these tests. The tests guarantee that the casks will withstand all credible transport hazards, as has been demonstrated in many tests e.g. in the United States, the United Kingdom, and Germany.

11.11.3 Safety Concept of Reprocessing Facilities

11.11.3.1 Technical Processes

The head end of a reprocessing plant is the fuel element storage pool accommodating the spent fuel elements. From here the fuel elements to be reprocessed are transported into the disassembly cell where they are dismantled first into structural parts and fuel rods. Fuel rods are chopped into small pieces dropped into the dissolver filled with nitric acid. The fuel and the fission products are dissolved in the nitric acid, the gaseous fission products are released, the hulls remain undissolved as residues. Besides gaseous fission products (noble gases, Iodine) also particulate fission products and trace concentrations (fractions: 10^{-4} to 10^{-6}) of uranium, plutonium, and other transuranium elements enter the offgas. The offgases pass through a wet scrubbing section where as much as 99% of the aerosols are retained. The remaining 10^{-6} to 10^{-8} fractions are then passed through a dry scrubbing section with high-efficiency particulate air filters which cause another reduction by a factor of 10^3 in transuranium trace concentration. This step can be followed by more offgas cleaning for the other fission products (tritium, krypton, xenon, iodine, ruthenium). On the whole, it is seen that less than 10^{-9} parts of the inventory of fuel (uranium, plutonium, minor actinides) can get into the environment via the offgas (see Sect. 10.7.4).

11.11.3.2 Safety Design of Reprocessing Plants

In reprocessing plants, unlike nuclear power plants,

- a critical geometry is avoided by suitable technical design (no chain reaction),
- typical LWR fuel is processed at low power (only post-decay power approximately seven to ten years after reactor shutdown) and at atmospheric pressure,
- the activity of fission products and actinides has decreased already at least by a factor of 100.

Also in a reprocessing plant, use is made of the multiple-barrier (defense-in-depth) principle (e.g. vessel walls, building walls, pressure staggered in the direction of central building sections containing radioactivity) for containment of the radioactive materials (Fig. 11.57). The effectiveness of barriers is checked by continuous monitoring of the protective measures and of potential airborne or liquid-effluent-borne radioactive substances. Gaseous or liquid waste is retained, monitored for purity and discharged only when the licensed limits can be met.

During chemical reprocessing, the nuclear fuel is contained in concrete cells lined with stainless steel which must not be entered by the operating personnel and in which handling, maintenance and repair processes are carried out with tools fit for remote operation. The thick concrete cell walls shield against neutron and gamma radiation. These inner cells are kept at a pressure level below that of the surrounding building

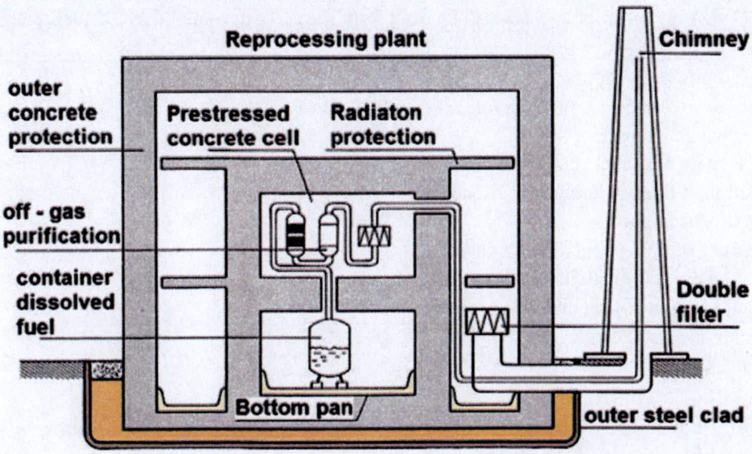

Fig. 11.57 Schematic diagram of multiple barriers in a reprocessing plant [170]

Table 11.7 Categories of accidents to be dealt with in safety design of a reprocessing plant (EPRI [111])

Reprocessing plant
• Drop of fuel assembly
• Loss of water in fuel storage pool
• Explosion and fire in a solvent treatment ion exchanger bed
• Criticality accident in processing cell
• Hydrogen explosion in a HLW tank
• Fire in LLW storage
• Explosion in HLW calciner
• Failure of krypton storage cylinder
• Earthquake and tornado in excess of design basis
• Aircraft crash into head end area

sections by staggered pressure levels, with the consequence that all leakages will be directed inward [170–172].

The full containers inside the concrete cells in the plant buildings are set up in floor troughs whose capacity is dimensioned so that they could accommodate the content of a container in case of a leakage. Pumps and other discharge systems would ensure safe transfer into other collection or reserve containers in case of a leakage. Radioactive aerosols potentially produced in a leakage or some other accident are retained down to trace concentrations in filter systems which remain intact.

Table 11.7 lists the main accident situations to be dealt with in a reprocessing facility.

Criticality safety of the radioactive fissile material solutions in the different sections of the plant is maintained by defining and monitoring (measuring) enrichment

Table 11.8 Categories of accidents to be dealt with in safety design of a MOX fuel refabrication plant (EPRI [111])

MOX fuel refabrication plant
• Hydrogen explosion in sintering furnace
• Ion exchange resin fire
• Dissolver explosion in scrap recovery
• Loaded final filter failure
• Criticality accident
• Plutonium shipping container damage
• Earthquake and tornado in excess of design basis
• Aircraft crash into head end area

in fissile material, by geometric dimensions of the components (diameter, volume), and by additional neutron absorbers installed (e.g. hafnium sheet).

Despite these precautions, reprocessing plants are designed against

- a criticality accident in a vessel holding the fissile material solution,
- the explosion (red-oil reaction) of an evaporator for HAW concentration.

The latter function is achieved by suitable offgas systems for cells or buildings containing the main activity, e.g. the main process building. Finally, the process plant buildings, as in nuclear power plants, are designed against earthquakes, airplane crash, chemical explosions, and third-party impacts.

11.11.4 Safety Concept of MOX Fuel Refabrication Plants

From the reprocessing plant, the so-called master mix, PuO_2/UO_2 powder, is transported to the MOX fabrication plant. This powder is compacted into pellets and sintered in a sintering furnace at 1,700°C. The fuel pellets are then ground to dimensions, filled into the cladding tubes which, in turn, are automatically welded tight with a noble gas after filling. This is done in such a way that the fuel rods are not contaminated with plutonium on their outer surfaces.

Table 11.8 lists the different accident situations to be dealt with in a MOX fuel refabrication plant.

Like nuclear power plants, MOX fuel fabrication plants are protected from external impacts by thick concrete walls. Subcriticality must be ensured by the geometric dimensions of the individual components for fuel processing. The plant must be designed against

- a criticality accident in the area of mixing PuO_2 and UO_2 powders,
- an explosion-like ammonium nitrate reaction in the fluidized-bed evaporator of the fuel processing stage.

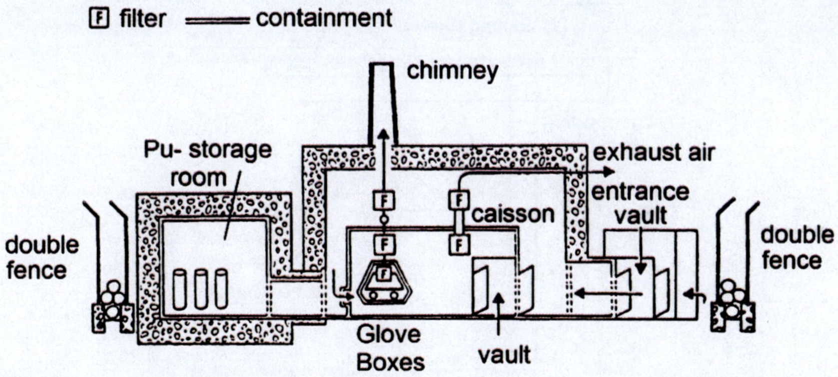

Fig. 11.58 Safety design concept of a MOX fuel refabrication plant [170–172]

The multiple-barrier principle is employed in the plant (Fig. 11.58). MOX powder, sintered pellets, and fuel rods in their unwelded state are handled in glove boxes working at negative pressures, which makes them leaktight to α-aerosols.

The staggered levels of negative pressure in the individual rooms are maintained outside in. The offgases from the gloveboxes and the tightly encapsulated sintering furnace are cleaned continuously by high-efficiency filters. As the fuel is handled far below its boiling point, dust and aerosol production is very low. Retention by filters therefore permits a factor in excess of 10^9 to be achieved (see Sect. 10.7.4.3) between the fuel inventory handled and the Pu aerosol volumes discharged through the stack.

Airborne aerosol concentrations in gloveboxes or in sintering furnaces are possible, according to the laws of physics, only in very low trace concentrations relative to the existing inventory ($\mu g/cm^3$). Any aerosols produced in MOX powder handling will very soon be precipitated again on the inner surfaces of the gloveboxes as a result of agglomeration and sedimentation [44]. The negative pressure maintained in the gloveboxes and sintering furnaces, and extraction through high-efficiency filters, achieve a degree of purity of the air in the overall plant and in the rooms accessible to personnel so high that the potential Pu aerosol content is below the limit of detection of 10^{-3} Bq/m^3.

Experience in the operation of MOX refabrication plants has shown the radioactive exposure of personnel to be below defined limits (Sect. 10.7.4.4).

11.11.5 Safety Design Concept of HAW Vitrification Plants

The safety requirements to be met by plants for vitrification of HAW are determined chiefly by radioactive liquids and by gases arising in the melter of the vitrification plant (Fig. 7.17; Sect. 7.5.2.1). Again, the principle of double confinement must be applied.

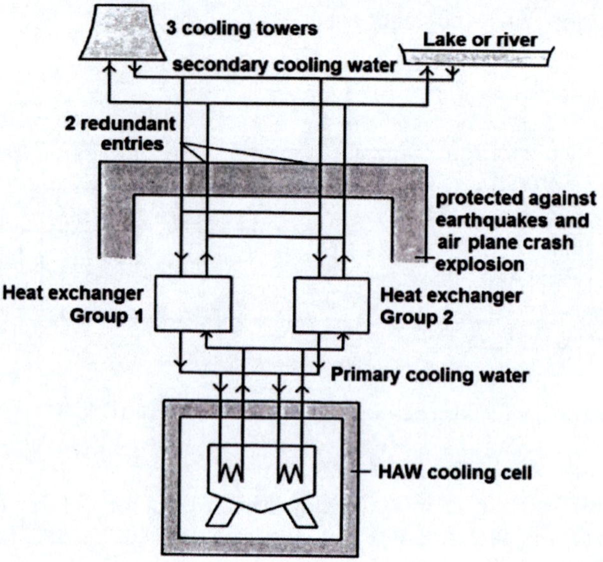

Fig. 11.59 Safety scheme of a plant for vitrification of high-level radioactive waste product solutions [170]

11.11.5.1 Technology of Vitrification of High-Level Radioactive Waste

The volume of high-level waste product solution from the reprocessing plant first is reduced by roughly a factor of 10 in an evaporator. This concentrated solution is stored for an intermediate period of time in double-walled and cooled stainless steel tanks. Their temperature is kept at <65°C by cooling (Fig. 11.59).

The concentrated high-level waste solution is conditioned by the process described in Sect. 7.5.2.1. In this process, the concentrated waste solution together with a glass frit is filled into a melter; the solution evaporates and the radioactive residue is fused into glass. For this purpose, e.g. a chemically highly stable borosilicate glass is used which also firmly immobilizes the waste constituents. The glass melt is filled into stainless steel (canisters) which are welded shut after controlled cooling. The offgases are cooled, cleaned by wet scrubbers and filter trains, and discharged in a controlled mode. According to experience in the operation of vitrification plants, their α-activity is more than a factor of 10^{12} lower than in the incoming waste streams, which is below the officially permitted emission levels (Sect. 10.7.4).

The liquid secondary waste of the chemical reprocessing step is added to the corresponding low and intermediate-level waste solutions and treated together with these.

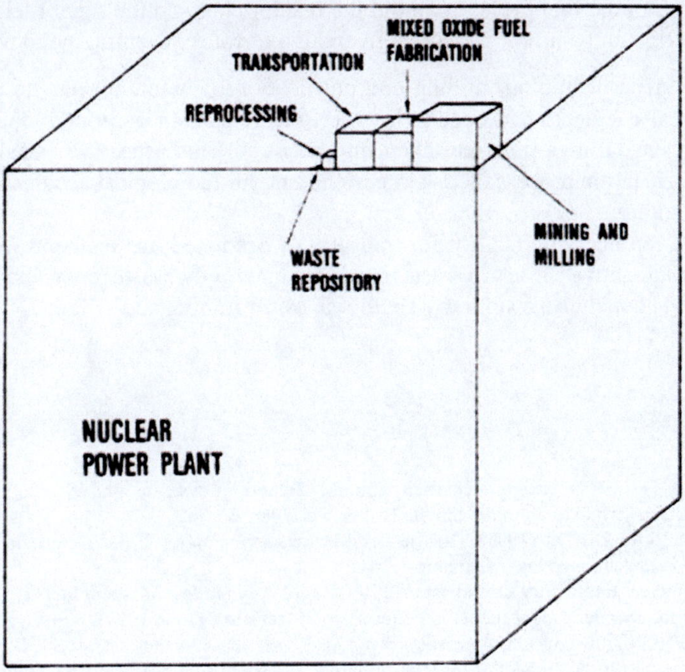

Fig. 11.60 Radiological risk of different parts of the fuel cycle [111]

11.11.6 Risk Studies of Fuel Cycle Plants

Besides risk studies of nuclear power plants, studies must also be performed on the whole fuel cycle. Until now, only the LWR U/Pu fuel cycle has been under development on an industrial scale. Accordingly, preliminary risk studies have so far been carried out on this LWR fuel cycle.

A study to cover the entire LWR fuel cycle by the same methodology as the US Reactor Safety Study (WASH-1400) was published by EPRI [111].

The result of that study is that the risk related to the fuel cycle is only approximately 1% of that associated with the reactor plant itself.

Figure 11.60 shows a comparison of risks associated with the reactor plant and other parts of the fuel cycle, the volumes of the cubes representing a measure of comparable risks.

The technical reasons underlying the lower nuclear risk of the fuel cycle compared with the LWR plant are mainly these:

- In certain parts of the fuel cycle (mining, milling, enrichment, fuel fabrication), there are no radioactive fission products.
- In other parts of the fuel cycle (reprocessing, waste disposal, transport), the radioactivity of fission products has already decayed by a sizable margin.

- Nowhere in the fuel cycle, except in the reactor core and the spent fuel pool, has the nuclear fuel a power density high enough to cause potential meltdown.

In the steps of mining, milling and purification of uranium ores, no particular radiological accidents can occur. However, the release of gaseous radon, dust of ores, and ore tailings must be taken into account. This aspect was dealt with in Sect. 10.7.1. In the process of U-235 enrichment, no radiological accidents must be expected either.

For the reprocessing facility, for transport of irradiated fuel elements, plutonium oxide or radioactive waste to a waste repository and for the waste repository the EPRI Study [111] concludes a similarly small risk as for mining and milling.

References

1. Smidt D (1979) Reaktorsicherheitstechnik, Sicherheitssysteme und Störfallanalyse für Leichtwasserreaktoren und Schnelle Brüter. Springer, Berlin
2. Emendörfer D et al (1993) Theorie der Kernreaktoren, Band 2: der instationäre Reaktor. BI Wissenschaftsverlag, Mannheim
3. US Nuclear Regulatory Commission (1999) Regulatory guide 3.54, spent fuel heat generation in an independent spent fuel storage installation (revision 1, Jan 1999). http://www.nrc.gov/NRC/RG/03/03-054rl.html
4. Xu Z et al (2005) Impact of high burnup on PWR spent fuel characteristics. Nucl Eng Des 151:261–273
5. Deutsche Risikostudie Kernkraftwerke Phase B (1990) Verlag TÜV Rheinland, Köln
6. Märkl M (1976) Core engineering and performance of pressurized water reactors. Kraftwerk Union AG, Erlangen
7. Tong LS et al (1979) Thermal analysis of pressurized water reactors. American Nuclear Society, LaGrange Park
8. Lahey RT et al (1977) The thermal hydraulics of a boiling water nuclear reactor. American Nuclear Society, LaGrange Park
9. RSK-Leitlinien für Druckwasserreaktoren (1996) Fassung 11.96, BAnz Nr. 214 vom 05.11.1996
10. Czech J et al (1999) European pressurized water reactor: safety objectives and principles. Nucl Eng Des 187:25–32
11. Kersting E et al (1993) Safety analysis for boiling water reactors, a summary, GRS-98. Gesellschaft für Anlagen- und Reaktorsicherheit, Garching
12. Aleite W et al (1987) Leistungsregeleinrichtungen und Begrenzungen von Druck- und Siedewasserreaktoren. Atomwirtschaft 32:129–134
13. Aleite W et al (1987) Leittechnik in Kernkraftwerken. Atomwirtschaft 32:122–128
14. Boland JF (1970) Nuclear reactor instrumentation (in-core). Gordon and Breach Science, New York
15. Bachmann G et al (1971) Leittechnik des Kernkraftwerks Stade. Atomwirtschaft 16:600–602
16. Aleite W et al (1971) Regeleinrichtungen des Kernkraftwerks Stade. Atomwirtschaft 16: 597–599
17. ASME Boiler and Pressure Vessel Code (2007) Rules for construction of nuclear power plant components, Sect III, Div 1. The American Society of Mechanical Engineers, New York
18. Laufs P (2006) Die Entwicklung der Sicherheitstechnik für Kernkraftwerke im politischen und technischen Umfeld der Bundesrepublik Deutschland seit 1955. Dissertation, Historisches Institut, Abteilung Geschichte der Naturwissenschaften, Universität Stuttgart, Germany

19. Merkle JG et al (1975) An evaluation of the HSST program intermediate pressure vessel tests in terms of light water reactor pressure vessel safety, ORNL-TM-5090
20. Derby RW et al (1974) Test of six-inch-thick pressure vessel, Series I: intermediate test vessels V-1 and V-2, ORNL-4895
21. Bryan RH et al (1975) Test of six-inch-thick pressure vessels, Series II: intermediate test vessels V-3, V-4, and V-6, ORNL-5059
22. Merkle JG et al (1976) Test of six-inch-thick-pressure vessels, Series III: intermediate test vessels V-7, ORNL-5059
23. Whitman GD (1976) Heavy section steel technology program quarterly progress report for January–March 1976, ORNL-TM-28
24. Griffith AA (1921) The phenomena of rupture and flow in solids. Philos Trans R Soc Lond A 221:163–198
25. Irwin GR (1958) Fracture. In: Flügge S (ed) Handbuch der Physik, Bd. VI, Elastizität und Plastizität. Springer, Berlin
26. Hahn HG (1976) Bruchmechanik. Teubner, Stuttgart
27. Kußmaul K (1978) Maßnahmen und Prüfkonzept zur weiteren Verbesserung der Qualität von Reaktordruckbehältern für Leichtwasser-Kernkraftwerke, 2. VGB-Kraftwerkstechnik 58(6):439–448
28. Hoffmann H et al (2007) Das Integritätskonzept für Rohrleitungen sowie Leck- und Bruchpostulate in deutschen Kernkraftwerken. VGB Power Technol 7:78–91
29. Kußmaul K (1984) German basis safety concept rules out possibility of catastrophic failure. Nucl Eng Int 12:41–46
30. Bruch CG (1976) RELAP4/MOD5, a computer program for transient thermal-hydraulic analysis of nuclear reactors and related systems, user's manual, vol 1: RELAP4/MOD5 description, SRD-113-76, Aerojet Nuclear Company, New York
31. Vigil JC et al (1981) Accident simulation with TRAC. Los Alamos Sci (Summer/Fall) 3:36–52
32. Liles DR et al (1978) TRAC-P1, an advanced best-estimate computer program for PWR LOCA analysis: methods, models, user information and programming details, LA-7279-MS, vol 1, NUREG/CR-0063
33. Liles DR et al (1979) TRAC-P1A, an advanced best-estimate computer program for PWR LOCA analysis. Report LA-7777-MS, NUREG/CR-0665, Los Alamos Scientific Laboratory
34. Liles DR et al (1981) TRAC-PD2, an advanced best-estimate computer program for pressurized water reactors loss-of-coolant accident analysis. Report LA-8709-MS, NUREG/CR-2054, Los Alamos Scientific Laboratory
35. Kuczera B (1990) Aktueller Stand der Reaktorsicherheitsforschung dargestellt anhand von Ergebnissen aus der deutschen Risikostudie Kernkraftwerke—Phase B, Radioaktivität–Risiko–Sicherheit, Herausgeber Kernforschungszentrum Karlsruhe (2. veränderte und aktualisierte Auflage 1991). VS Verlag für Sozialwissenschaften, Wiesbaden
36. Rasmussen NC (ed) (1975) Reactor safety study: an assessment of accidents risks in US commercial nuclear power plants. US Nuclear Regulatory Commission, Washington, WASH-1400 (NUREG-75/014)
37. Deutsche Risikostudie Kernkraftwerke Phase A (1980) Gesellschaft für Reaktorsicherheit (GRS), TÜV Rheinland, Köln
38. Lewis EE (1977) Nuclear power reactor safety. Wiley, New York
39. Farmer FR (1967) Reactor safety and siting: a proposed risk criterion. Nucl Saf 8:539–548
40. Dunster HJ (1980) The approach of a regulatory authority to the concept of risk. IAEA Bull 22(5/6):123–128
41. Starr C (1969) Social benefits versus technological risks. Science 165(169):1232–1238
42. Bayer A, Heuser FW (1981) Basic aspects and results of the German risk study. Nucl Saf 22:695–709
43. Hassmann K et al (1987) Spaltproduktfreisetzung bei Kernschmelzen. TÜV Rheinland GmbH, Köln
44. Bunz H et al (1981) The role of aerosol behavior in light water reactor core melt accidents. Nucl Technol 53:141–146

45. COSYMA (1991) A new program package for accident consequence assessment, EUR-13028. European Commission, Brussels
46. Ehrhardt J et al (2000) RODOS: decision support system for off-site nuclear emergency management in Europe. Report EUR 19144, European Commission, Brussels
47. Ehrhardt J (1991) Probabilistische Unfallfolgenabschätzungen. In: Radioaktivität/Risiko—Sicherheit, pp 51–61. Kernforschungszentrum Karlsruhe
48. Hasemann I (2011) Personal communication. Karlsruhe Institute of Technology (KIT), Karlsruhe
49. Rahmenempfehlungen für den Katastrophenschutz in der Umgebung kerntechnischer Anlagen (2008) GMBI Nr. 62/63 vom 19, Dec 2008. http://www.bmu.de/files/pdfs/allgemein/application/pdf/rahmenempfehlung_katastrophenschutz.pdf
50. Leitfaden für den Fachberater Strahlenschutz der Katastrophenschutzleitung bei kerntechnischen Notfällen (1989) Veröffentlichungen der Strahlenschutzkommission, Band 13. Gustav Fischer, Stuttgart; also: Radioaktivität und Strahlung Grenzwerte und Richtwerte. http://www.osiris22.pi-consult.de/userdata/I_20/P-105/Library/data/grenzwerte-und-richtwerte-04-03-internetversion.pdf
51. Regulatory Guide 1.60 (1973) Design response spectra for seismic design of nuclear power plants (Revision 1). http://www.orau.org/ptp/PTP%20Library/library/NRC/Reguide/01-060.pdf
52. Regulatory Guide 1.208 (2007) A performance-based approach to define the site-specific earthquake groundmotion, US Nuclear Regulatory Commission. http://www.nrc.gov/reading-rm/doc-collections/reg-guides/power-reactors/rg/01-208/01-208.pdf
53. McGuire RK et al (2001) New seismic design spectra for nuclear power plants. Elsevier Science B.V., Amsterdam
54. KTA 2201.4 (1990) Auslegung von Kernkraftwerken gegen seismische Einwirkungen—Teil 4: Anforderungen an Verfahren zum Nachweis der Erdbebensicherheit für maschinen- und elektrotechnische Anlagenteile, Fassung 6/90, Kerntechnischer Ausschuss (KTA). http://www.kta-gs.de/d/regeln/2200/2201_4.pdf
55. KTA 2201.1 (1990) Auslegung von Kernkraftwerken gegen seismische Einwirkungen, Teil 1. Grundsätze, Kerntechnischer Ausschuss (KTA). http://www.kta-gs.de/d/regeln/2200/2201_1.pdf
56. Safety Guide IAEA NS-G-33 (2002) Evaluation of seismic hazards for nuclear power plants. International Energy Agency, Vienna
57. Safety Guide IAEA NS-G-1.6 (2003) Seismic design and qualification for nuclear power plants. International Energy Agency, Vienna
58. Forni M et al (2009) Seismic isolation of the IRIS nuclear plant. In: Proceedings of the 2009 ASME pressure vessel and piping conference, RVP 2009, 26–30 July 2009, Prague Czech Republic
59. Nakamura N et al (2008) Analyses of reactor building by 3D nonlinear FEM models considering basemat uplift for simultaneous horizontal and vertical ground motions. Nucl Eng Design 238:3551–3560
60. KTA 2207 (2004) Schutz von Kernkraftwerken gegen Hochwasser, Sicherheitstechnische Regel des KTA (Kerntechnischer Ausschuss), Fassung 11/04. http://www.kta-gs.de/d/regeln/2200/2207n.pdf
61. Birkhofer A (1989) Anlageninterner Notfallschutz. In: Achtes deutsches Atomrechtssymposium, 1–3 März 1989, München. Carl Heymann Verlag KG, Köln
62. Schenk H (1990) Maßnahmen zum anlageninternen Notfallschutz. Atomwirtschaft 11:514–520
63. Fritsche AF (1988) Gesundheitsrisiken von Energieversorgungssystemen. TÜV Rheinland, Köln
64. Hubert P et al (1981) Les impacts sanitaires et écologiques de la production de l'électricité: le cas français, CEPN-R-34.3 (2ème version), Fontenay-aux-Roses, Paris
65. Rogner HH (1998) Comparing energy options, IAEA bulletin 40/1/1998. Sustainable Development, Nuclear Power, IAEA, Vienna

66. Frischknecht S et al (1998) Project GaBE, comprehensive assessment of energy systems, severe accidents in the energy sector, PSI-Bericht Nr. 98-16, ISS 1019-0643
67. Kessler G (2002) Requirements for nuclear energy in the 21st century, nuclear energy as a sustainable energy source. Prog Nucl Energy 60(3–4):309–325
68. Natural disaster. http://en.wikipedia.org/wiki/Natural_disaster
69. Le programme international sur la sécurité chimique, Aide-Mémoire number 87, révisé, Mars 1998. https://apps.who.int/inf-fs/fr/am87.html
70. Farmer R (1980) What is risk. Atom 282:108
71. Ehrhardt J et al (1982) Der Einsatz des Unfallfolgenmodells der Deutschen Risikostudie Kernkraftwerke bei Risikoabschätzungen zu verschiedenen Reaktortypen, KfK-Nachrichten. Jahrg 14-4(82):269–277
72. Knief RA (1992) Nuclear engineering, theory and technology of commercial nuclear power. Hemisphere Publishing Corporation, Washington, DC
73. Ireland JR et al (1981) Three Mile Island and multiple-failure accidents. Los Alamos Sci 3(Summer/Fall):75–85
74. Kemeny JG (1979) Report of The President's commission on the accident at Three Mile Island: the need for change: the legacy of TMI, ISBN 0935758003. The Commission, Washington, DC. http://www.threemileisland.org/downloads/188.pdf
75. Nuclear Energy Agency (1995) Chernobyl ten years on, radiological and health impact. Nuclear Energy Agency (OECD), Paris
76. IAEA (1991) The international Chernobyl project. Report by an International Advisory Committee, Vienna
77. Kovan D (2011) Chernobyl 25 years on: time for a giant leap forward. Nucl News (Am Nucl Soc) 54(5):57
78. Nuclear Energy Agency (2002) Update of Chernobyl: ten years on. Nuclear Energy Agency (OECD), Paris
79. Deaths due to the Chernobyl disasters (2011) http://www.en.wikipedia.org/wiki/Deaths_due_to_the_Chernobyl_disaster
80. International Atomic Energy Agency (2011) Chernobyl: answers to long standing questions. International Atomic Energy Agency (IAEA), Vienna. http://www.161.5.1.75/newscenter/focus/chernobyl/faqs.shtml
81. Jaworowski Z (1999) Radiation risk and ethics. Phys Today 52(9):24–29 (American Institute of Physics)
82. Nuclear News Special Report (2011) Fukushima Daiichi after the earthquake and tsunami. Nucl News 54(4):17
83. IAEA International Fact Finding Expert Mission of the Nuclear Accident Following the Great East Japan Earthquake and Tsunami (2011) Preliminary summary. IAEA, Vienna, 24 May 1 June 2011
84. Nuclear News (2011) TEPCO said that Fukushima Daiichi-I endured a melt down. American Nuclear Society, La Grange Park
85. Nuclear News (2011) Fukushima Daiichi accident offers lessons for all. American Nuclear Society, La Grange Park
86. IRSN (2011) Assessment on the 66th day of projected external doses for the populations living in the North-West fallout zone of the Fukushima nuclear accident. Report DRPH/2011-10, IRSN, Paris
87. DOE-NNSA (2011) DOE-NNSA Fukushima survey, PNG, 27–28 March 2011. http://www.en.wikipedia.org/wiki/File:DOE_NNSA_Fukushima_Survey_27-28.PNG
88. Mohrbach L et al (2011) Earthquake and tsunami in Japan on March 11, 2011 and consequences for Fukushima and other nuclear power plants. http://www.vgb.org/vgbmultimedia/News/Fukushima_VGB_rev16.pdf
89. Sekimura N (2011) Overview of the accident in Fukushima Daiichi nuclear power plants, IAEA, ICTP, Trieste, 8 Aug 2011. http://www.iaea.org/inisnkm/nkm/pages/2011/NEMschool2011/topics/topic0/fukushima%20Overview_Sekimura.pdf

90. World Nuclear News (2011) Theory for Fukushima Daiichi 4 explosion: regulation and safety, 17 May 2011. http://www.world-nuclear-news.org/newsarticle.aspx?id=30068
91. Iter Consult (2011) Independent technical evaluation and review, Fukushima Daiichi nuclear accident first considerations, preliminary report. http://www.iter-consult.it/ITER_Report_Fukushima_Accident.pdf
92. Michel R (2011) Fukushima: eine vorläufige Bilanz im Juni 2011. http://www.zsr.uni-hannover.de/dokument/fubiju11.pdf
93. Hennies HH, Kessler G, Eibl J (1989) Improved containment concept for future pressurized water reactors. In: 5th international conference on emerging nuclear energy systems (ICENES) 3–6 July 1989, Karlsruhe
94. Hennies HH, Kessler G, Eibl J (1992) Containments and core catchers in future reactors. Atomwirtschaft 37:238–247
95. Eibl J et al (1992) How to eliminate containment failure in tomorrows PWRs (pressurized water reactors). Nucl Eng Int 37(453):51–55
96. Berman M, Beck DF (1989), Steam explosion triggering and propagation: hypothesis and evidence. In: Proceedings of 3rd international seminar on containment of nuclear reactors, UCLA (SNL, report SAND89-1878C), Los Angeles, 10–11 Aug 1989
97. The SL-1 Reactor Accident. http://www.radiationworks.com/photos/sl1reactor1.htm; http://www.en.wikipedia.org/wiki/SL-1
98. Corradini ML et al (1988) Vapor explosion in light water reactors: a review of theory and modelling. Prog Nucl Energy 22:1–117
99. Magallon D (2005) FCI phenomena uncertainties impacting predictability of dynamic loading of reactor structures (from OECD SERENA programme). In: PSA-2 workshop, 7–9 Nov 2005, Aix-en-Provence
100. Jacobs H et al (1994) Untersuchungen zur Dampfexplosion, PSF Statusbericht, KfK 5326. Kernforschungszentrum Karlsruhe, pp 214–232, 23 März 1994
101. Cronenberg AW, Benz R (1980) Vapour explosion phenomena with respect to nuclear reactor safety assessment. Adv Nucl Sci Technol 12:247–335
102. Fletcher DF, Anderson RP (1990) A review of pressure-induced propagation models of the vapour explosion process. Prog Nucl Energy 23:137–179
103. Board SJ et al (1975) Detonation of coolant explosions. Nature 254(3):319–321
104. Diab A et al (2000) Long-term validation of the molten fuel-moderator interactions model. Nucl Technol 169:114–125
105. Berthoud G (2000) Vapor explosions. Annu Rev Fluid Mech 32:573–611
106. Struwe D et al (1999) Consequence evaluation of in-vessel fuel coolant interaction in the European pressurized water reactor, FZKA 6316, Forschungszentrum Karlsruhe
107. The RELAP5 Development Team (1995) RELAP5/mod3 code manual, vol 1–7, NUREG/CR-5535, INEL-95/1074. Idaho National Engineering Laboratory, Idaho Falls
108. Allison M et al (1993) SCDAP/RELAP5 mod3.1 code manual, vol I–IV, NUREG/CR-6150, EGG-2720. Idaho National Engineering Laboratory, Idaho Falls
109. Coryell E et al (1997) SCDAP/RELAP5 mod3.2 code manual, vol I–V, NUREG/CR-6150, INEL-96/0422. Idaho National Engineering Laboratory, Idaho Falls
110. SCDAP/RELAP5 mod 3.2. http://www.relap5.inel.gov/scdap/home.html
111. Power Research Institute (1979) InstituteStatus report on EPRI fuel cycle accident risk assessment, EPRI-NP-1128. Power Research Institute, Palo Alto
112. Valette M (1997) MC3D V3.0 directions for use. Commissariat ă l'énergie atomique Grenoble, STR/LTEM, STR-LTEM-96-52
113. Berthoud G, Valette M (1994) Development of a multidimensional model for the premixing phase of a fuel coolant interaction. Nucl Eng Des 149:409–418
114. Jacobs H et al (1995) Multifield simulations of premixing experiments. In: Proceedings of a multidisciplinary international seminar on intense multiphase interactions, pp 56–69, Santa Barbara, 9–13 June 1995
115. Krieg R (1995) Missiles caused by severe pressurized-water reactor accidents. Nucl Saf 36:299–309

116. Krieg R et al (1995) Slug impact loading on the vessel head during a postulated in-vessel steam explosion in pressurized water reactors—assessments and discussion of the investigation strategy. Nucl Technol 111:369–385
117. Hirt A (1998) Rechenmodell zum Aufprall von Kernschmelze auf die oberen Einbauten und den Deckel eines Reaktordruckbehälters, FZKA 6054, Forschungszentrum Karlsruhe
118. Malmberg T (1995) Aspects of similitude theory in solid mechanics, Part I: Deformation behavior, FZKA 5657, Forschungszentrum Karlsruhe
119. Krieg R et al (2003) Load carrying capacity of a reactor vessel under molten core slug impact. Final report including recent experimental findings. Nucl Eng Des 293:237–253
120. Stach T (1997) Zur Skalierung von Modellversuchen zum Aufprall flüssiger Massen auf deformierbare Strukturen, FZKA 5903, Forschungszentrum Karlsruhe
121. Krieg R et al (1980) Transient, three-dimensional potential flow problems and dynamic response of the surrounding structures, Part I. Comput Phys 8(2):139
122. Krieg R et al (2000) Load carrying capacity of a reactor vessel head under a corium slug impact from a postulated in-vessel steam explosion. Nucl Eng Des 202:179–196
123. Krieg R (1997) Mechanical efficiency of the energy release during a steam explosion. Nucl Technol 117:151–157
124. Theofanous B et al (1997) An assessment of steam explosion-induced containment failure, Part I: Probabilistic effects. Nucl Sci Eng 97:259–281
125. Amarasooriya WH et al (1987) An assessment of steam-explosion-induced containment failure. Part III: Expansion and energy partition. Nucl Sci Eng 97:296–315
126. Eibl J (1994) Zur bautechnischen Machbarkeit eines alternativen Containments für Druckwasserreaktoren—Stufe 3, KfK 5366, Kernforschungszentrum Karlsruhe
127. Travis JR et al (1998) GASFLOW-II: A three-dimensional-finite-volume fluid-dynamics code for calculating the transport, mixing, and combustion of flammable gases and aerosols in geometrically complex domains, theory and computational model, vol 1, FZKA-5994 and LA-13357-MS
128. Redlinger R (1999) DET3D: a code for calculating detonations in reactor containments. In: Proceedings of Jahrestagung Kerntechnik 99, Kerntechnische Gesellschaft e.V. Deutsches atomforum e.V. annual meeting on nuclear technology 99, Karlsruhe, 18–20 May 1999
129. Kotchourko AS et al (1999) Reactive flow simulations in complex 3D geometries using the COM3D code. In: Proceedings of Jahrestagung Kerntechnik 99, Kerntechnische Gesellschaft e.V. Deutsches atomforum e.V. annual meeting on nuclear technology 99, Karlsruhe, 18–20 May 1999
130. Veser A et al (1999) Experiments on turbulent combustion and COM3D verification. In: Proceedings of Jahrestagung Kerntechnik 99, Kerntechnische Gesellschaft e.V. Deutsches atomforum e.V. annual meeting on nuclear technology 99, Karlsruhe, 18–20 May 1999
131. Dorofeev SB et al (2001) Evaluation of limits for effective flame acceleration in hydrogen mixtures. J Loss Prev Process Ind 14:583–589
132. Dorofeev SB et al (1999) Effect of scale and mixture properties on behavior of turbulent flames in obstructed areas, FZKA 6268, Forschungszentrum Karlsruhe
133. Kuznetsov M et al (1999) Effect of obstacle geometry on behaviour of turbulent flames, FZKA 6328, Forschungszentrum Karlsruhe
134. Breitung W et al (2005) Innovative Methoden zur Analyse und Kontrolle des Wasserstoffverhaltens bei Kernschmelzunfällen, FZKA 7085, Forschungszentrum Karlsruhe
135. Krieg R et al (2003) Assessment of the load-carrying capacities of a spherical pressurized water reactor steel containment under a postulated hydrogen detonation. Nucl Technol 141:109–121
136. Rohde J et al (1997) Selection of representative accidents and evaluation of H_2-control measures in PWR containments. In: 141st session of RSK light water reactor safety committee, Jan 1997
137. ABAQUS (1989) A general purpuse linear and nonlinear finite element code, user manual standard 5.8. Hibbit, Karlson and Sorenson, Providence
138. Bung H et al (1993) A new method for the treatment of impact and mechanics in reactor technology (SMIRT12), Stuttgart

139. Krieg R (2005) Failure strains and proposed limit strains for a reactor pressure vessel under severe accident conditions. Nucl Eng Des 235:199–212
140. Jeschke J et al (2011) Critical strains and melting phenomena for different steel sheet specimens under uniaxial loading. Nucl Eng Des 241:2045–2052
141. Jacobs G (1995) Dynamic loads from reactor pressure vessel core melt through under high primary pressure. Nucl Technol 111:351–356
142. Plank H et al (2009) http://www.sacre.web.psi.ch/ISAMM2009/oecd-sami2001/Papers/p20-Plank/SAM-Paper-b.pdf
143. Stosic ZV et al (2008) Boiling water reactor with innovative safety concept: the generation III+ SWR-1000. Nucl Eng Des 238:1863–1901
144. Tong LS (1968) Core cooling in a hypothetical loss of cooling accident. Estimate of heat transfer in core meltdown. Nucl Eng Des 8:309–312
145. Henry RE et al (1993) External cooling of a reactor vessel under severe accident conditions. Nucl Eng Des 139:31–43
146. Rempe JL et al (1993) Light water reactor lower head failure, NUREG/CR-5642, EGG-2618. Idaho National Engineering Laboratory, Idaho
147. Thinnes GL et al (1989) Comparison of thermal and mechanical responses of the Three Mile Island unit 2 vessel. Nucl Technol 87:1036–1049
148. Hagen S et al (1987) LWR fuel rod behaviour during severe accidents. Nucl Eng Des 103:85–106
149. Kolev NI (2004) External cooling—the SWR-1000 severe accident management strategy. In: 12th international conference on nuclear engineering—ICONE-12, Arlington, 25–29 April 2004
150. Nazaré S et al (1975) Über theoretische und experimentelle Möglichkeiten zur Bestimmung der Stoffwerte von Corium, Abschlußbericht Teil II, KFK 2217
151. Schneider H et al (1975) Zur Bestimmung der Zusammensetzung verschiedener Corium-Schmelzen, KFK 2227
152. Reimann M et al (1981) The WECHSL-code: a computer program for the interaction of a core melt with concrete, KfK 2980, Kernforschungszentrum Karlsruhe
153. Reimann M (1987) Verification of the WECHSL-code on melt/concrete interaction and application to the core melt accident. Nucl Eng Des 103:127–137
154. Alsmeyer H et al (1987) BETA-experiments in verification of the WECHSL-code: experimental results on the melt-concrete interaction. Nucl Eng Des 103:115–125
155. Krieg R et al (1987) Failure pressure and failure mode of the latest type of German PWR containments. Nucl Eng Des 104:381–390
156. Göller B et al (1988) Failure pressure and failure mode of the bolted connection for the large component port in German PWR containments. Nucl Eng Des 106:35–45
157. Tromm W et al (1991) Radionuclide dispersion after core-concrete melt leaching by groundwater. Kerntechnik 56(6):7–12
158. Al-Omari I (1990) Abschätzung der Strahlenexposition infolge störfallbedingter Radionuklideinleitungen von kerntechnischen Anlagen in Fließgewässer unter Berücksichtigung der Zeitabhängigkeit relevanter Parameter, KfK 4793, Kernforschungszentrum Karlsruhe
159. Alsmeyer H (1989) Containment loadings from melt-concrete interaction. Nucl Eng Des 117:45–50
160. Turricchia A (1992) How to avoid molten core/concrete interaction (and steam explosions). In: Alsmeyer H (ed) Proceedings of 2nd OECD(NEA) CSNI specialist meeting on molten core debris-concrete interaction, KfK 5108, NEA/CSNI/R(92)10. KfK, Karlsruhe, p 503
161. Seiler JM et al (1992) Conceptual studies of core catchers for advanced LWRs. In: Proceedings of international conference on design and safety of advanced nuclear power plants, Tokyo, Oct 1992, vol III, p 23.3-1
162. Alsmeyer H et al (1998) Beherrschung und Kühlung von Kernschmelzen außerhalb des Druckbehälters. Nachrichten Forschungszentrum Karlsruhe, Jahrg 29(4):327–335
163. Tromm W et al (1993) Fragmentation of melts by water inlet from below. In: 6th international topical meeting on nuclear reactor thermal hydraulics (NURETH-6), Grenoble, 5–8 Oct 1993

164. Fieg G et al (1996) Simulation experiments on the spreading behavior of molten core melts. In: Proceedings of the 1996 national heat transfer conference, vol 9, Houston, 3–6 Aug 1996. American Nuclear Society, La Grange Park, pp 121–130
165. Lewis BJ (2008) Overview of experimental programs on core melt progression and fission product release behaviour. J Nucl Mater 380:126–143
166. Meyer L et al (2009) Direct containment heating integral effects tests in geometrics of European nuclear power plants. Nucl Eng Des 239:2070–2084
167. Meyer L et al (2003) Low-pressure corium dispersion experiments with simulant fluids in a scaled annular cavity. Nucl Technol 141:257–274
168. IPSN-GRS Proposals for the development of technical guidelines for future PWRs (1998) Structuring GPR-RSK recommendations as guidelines. Common report IPSN/GRS No. 42, vol 5, Institut de Protection et de Sûreté Nucléaire, Saclay, Gesellschaft für Reaktorsicherheit, Garching
169. Faude D et al (1989) Plutonium-Handhabung in Brennstoff-Kreislauf in Plutonium, KfK 4516, Kernforschungszentrum Karlsruhe
170. Schleisiek K (1980) Nukleare Sicherheit von Wiederaufarbeitungsanlagen. In: Grupe H (ed) Wie sicher ist die Entsorgung? Vortrag einer Informationsveranstaltung über Fragen der Kernenergie. Kernforschungszentrum Karlsruhe
171. Hennies HH et al (1976) Nukleare Sicherheit bei Wiederaufarbeitungsanlagen. Jahreskolloquium 1976 Projekt Nukleare Sicherheit, KfK 2399, Kernforschungszentrum Karlsruhe
172. Baumgaertel G et al (1988) Nukleare Sicherheit von Wiederaufarbeitungsanlagen. Grupe H (ed) Vortrag einer Informationsveranstaltung über Fragen der Kernenergie. Kernforschungszentrum Karlsruhe, letzte Aktualisierung
173. Summers RM et al (1995) MELCOR computer code manuals, vols 1–2 (Vers: 1.8.3), NUREG/CR-6119, SAND93-2185. Sandia National Laboratories, Albuquerque

Chapter 12
Safety Design Concept of Liquid Metal Cooled Fast Breeder Reactors (LMFBRs)

Abstract The safety design concept of LMFBRs follows the same basic principles (multiple barrier concept and four level safety concept) as they were developed for light water reactors. This holds despite of the fact that LMFBRs have different design characteristics (fast neutron spectrum, liquid metal as coolant, plutonium-uranium fuel). It has been shown that LMFBRs possess a strong negative power coefficient and good control stability. The main design characteristics of control and shut-off systems do not differ much from those of light water reactors. For sodium cooled fast reactors the sodium temperature and sodium void coefficient can become positive above a power output of the core above about 350 MW(th). Therefore, special design provisions are taken for future LMFBR designs, e.g. flat and heterogeneous cores. The excellent cooling and natural convection properties as well as the low system pressure of about 1 bar of liquid metal cooled fast breeder allow the safe decay heat removal by a number of ways. The consequences of sodium fires or sodium water reactions can be prevented or limited by special design provisions. On the other hand, lead and lead-bismuth-eutecticum (LBE) as coolant do not chemically react neither with oxygen of the atmosphere nor with water in failing tubes of a steam generator. Historically the characteristics of homogeneous sodium cooled cores with a positive sodium void coefficient of the early prototype fast breeder reactors have lead to the analysis of core disruptive accidents with the objective to find a basis for the main design requirements of the containment. The discovery of the negative control rod drive line expansion coefficient in the early 1980s, changed this situation and lead to a new safety design which avoids anticipated transients without scram (ATWS) for future LMFBRs. The high boiling points of lead with 1,740°C and LBE with 1,670°C offer an advantage with respect to safety concerns compared to sodium as coolant.

12.1 Introduction

The safety design concept of LMFBRs was developed in parallel to the LWR safety concept during construction and licensing of prototype LMFBRs, e.g. CRBR and FFTF in USA, PFR in the UK, Phénix and Superphénix in France, SNR 300 in

G. Kessler, *Sustainable and Safe Nuclear Fission Energy*, Power Systems,
DOI: 10.1007/978-3-642-11990-3_12,

Germany, MONJU in Japan and BN 600 in Russia [1, 2]. Most of these prototype LMFBRs were shut down after 35 years of successful operation. By the end of 2010 only BN 600 (Russia) was still in operation, whereas MONJU (Japan) was about to resume full power. The Russian BN 800 and the Indian PMFBR were under construction.

The safety design concept of LMFBRs follows the basic principles valid for LWRs, despite of the fact that LMFBRs have certain different design characteristics, e.g.

- a fast neutron spectrum
- liquid metal as coolant (sodium or lead-bismuth-eutectic (LBE))
- plutonium mixed oxide PuO_2/UO_2 (MOX) or U-Pu-Zr metal as fuel

12.2 Basic Principles of the Safety Design Concept of LMFBRs

As for LWRs also for LMFBRs the radioactive materials must be confined by a multiple barrier concept, i.e.

- the cristalline structure of the ceramic fuel, e.g. PuO_2/UO_2 or the structure of metallic fuel (IFR)
- steel cladding tubes (welded gas tight) containing the fuel
- the reactor double tank housing the core, the liquid metal coolant and intermediate heat exchangers (in case of pool type LMFBRs)
- the steel containment confining the reactor tank and the cooling systems
- an outer reinforced concrete containment protecting the LMFBR against external events (Chap. 6; Sect. 11.2).

In addition the requirements of the four safety levels (Sect. 11.3) must be fulfilled:

- accident prevention (level 1 and 2)
- design of the plant protection system to withstand the design basis accidents (level 3)
- accident management measures for accidents exceeding the design basis (level 4).

As will be shown, LMFBRs also have the safety potential to fulfill the requirements of the Karlsruhe KHE safety concept (Sect. 11.9.1).

12.3 Reactor Physics and Safety Related Characteristic Data of LMFR Cores

The reactor physics of FBRs has been thoroughly investigated between the years 1950 and 2000 in the USA, USSR, Europe and Japan. Zero power critical facilities were operated for the measurement of many reactor physics parameters (criticality, neutron flux, reactivity coefficients, reaction rate distributions etc.). The PuO_2/UO_2 fueled South West Fast Oxide reactor SEFOR reactor in the USA was particularly built and operated for the measurement of the Doppler coefficient, transient power behavior as well as control and stability characteristics. This is summarized, e.g. in [1, 2].

12.3.1 Safety Characteristics of LMFBRs

The safety characteristics of LMFBRs are characterized by a set of coefficients and constants that govern the related kinetic equations [3–6] (see also Chap. 3). These are the following:

- the effective fraction of delayed neutrons β_{eff}
- the effective decay constant of delayed neutrons λ_{eff}
- the effective lifetime of the prompt neutrons l_{eff}
- reactivity coefficients, especially the Doppler coefficient, the fuel expansion coefficient, the structural coefficient, the coolant temperature coefficient, and the coolant void coefficient (sodium cooled fast reactors)
- control rod worth and absorber material reactivity worths.

The effective fraction of delayed neutrons, β_{eff}, in a PuO_2/UO_2 fueled LMFBR core is about 0.0035, which is about half as much as the corresponding value of about 0.007 in thermal reactor cores with enriched UO_2 fuel and up to about 40% lower than than the corresponding value for the core of an LWR core (Pu-burner) with PuO_2/UO_2 (MOX) fuel.

This also means that the dollar unit of reactivity (Sect. 3.8.3.2), is about half as large in LMFBRs as in U235 fueled LWRs.

Due to the smaller cross sections in the 0.2 MeV neutron kinetic energy range (Sect. 3.2), the Δk_{eff} values are also smaller than in a thermal neutron spectrum, and the reactivity

$$\rho = \frac{\Delta k_{eff}}{k_{eff}}$$

for material movements or temperature changes is, therefore, similar to that in the case of a thermal reactor core [3].

The prompt neutron lifetime, l_{eff}, in fast spectrum cores is about two orders of magnitude smaller than in LWR cores (Table 12.1). This does not necessarily imply a disadvantage to the dynamic behavior of LMFBRs as will also be shown below (see Sect. 3.8.3.2).

As can be seen from Table 12.1 above the effective decay constants λ_{eff} for a one group approach of delayed neutrons for MOX fuel in LMFBRs is about $0.06\,s^{-1}$ and about $0.08\,s^{-1}$ for UOX fuel in PWRs. This similarity of the value for λ_{eff} for MOX fueled LMFBRs and UOX fueled PWRs is decisive for the subprompt critical regime despite the fact that the neutron life times l_{eff} and the β_{eff} are different (Table 12.1).

Table 12.1 Comparison of reactor kinetics parameters for LWR and LMFBR cores [3–8]

	Main fissile isotope	l_{eff} (s)	$\beta_{eff} \hat{=} 1$ (\$)	λ_{eff} (s^{-1})
PWR	U-235	2.5×10^{-5}	0.007	0.08
LMFBR	Pu-239	4.5×10^{-7}	0.0035	0.06

Table 12.2 Comparison of design characteristics for control and shut-off systems for PWR and LMFBR [15]

	PWR [1,300 MW(e)]	LMFBR SNR 300
Speed of control-rod (mm/s)	1	1.2
Movements (¢/s)	2.5	≤ 4.0
Speed of shut-off rod movements (cm/s)	156	150–190
Delay time prior to reaction of shut-off system (s)	0.2	0.2
Insertion depth into core (cm)	390	90–115
Time span for full insertion of shut-off rods in core (s)	2.5	0.6
Reactivity of shut-off rod systems (Δk (\$))	11	10
Reactivity of shim-rod of boron system (burn-up) (Δk (\$))	19	8

12.3.1.1 Subprompt Critical Regime

All control actions of nuclear power reactors are performed in the subpromt critical regime. For these subprompt critical range $0 < \rho < \beta_{eff}$ and ρ not close to β_{eff} the time behavior of the reactor power is given for all nuclear power reactors (including LMFBR) by Eq. 3.14 (Sect. 3.8.3.2).

$P(t)/P_o = \exp(t/T)$ with the steady asymption time period T given by

$$T \approx \frac{\beta_{eff}}{\lambda_{eff} \cdot \rho}$$

As β_{eff} is about a factor of 2 smaller, λ_{eff} about equal and ρ about equal in LMFBR compared to a PWR core, the control behavior of fast-reactor cores under subprompt-critical conditions is not much different from that of thermal reactors [4–6]. The design of the control and shut-off systems does not pose any difficulties, as can be seen from Table 12.2 where characteristic data of the control and shut-off systems for a PWR and a LMFBR are compared.

12.3.1.2 Superpromt Critical Regime

Under superprompt-critical conditions ρ becomes $> \beta_{eff}$. It was shown in [9] that the short neutron lifetime, $l_{eff} \approx 4.5 \times 10^{-7}$ s, is a fundamental safety problem only if the power/temperature coefficient would be positive. With a negative power/temperature coefficient the opposite is true. The negative Doppler coefficient and negative power coefficient of PuO_2/UO_2-fueled LMFBR cores lead to sharply limited narrow power bursts, until the delayed neutrons and eventually the shut down systems terminate the power burst (see Fig. 6.3, Sect. 6.3.8) [1, 9, 10]. With a significant negative Doppler effect the energy release within the power burst decreases with smaller neutron lifetime [9]. The half width of these power bursts is a function of the neutron lifetime of the reactor core. It is broader for LWRs with a neutron life time of about 2.5×10^{-5} s and even broader for graphite moderated thermal reactors with a neutron lifetime of 7×10^{-4} s [11] than for LMFBRs with a neutron life time of 4.5×10^{-7} s [10].

It has been shown by theory and by experiments [12] that large mixed-oxide-fueled LMFBR cores have a strong negative power coefficient and good control stability against positive reactivity input or coolant-flow oscillations [13]. In this case, the strong negative Doppler coefficient together with the negative structural and fuel expansion coefficients dominate a positive sodium void coefficient [3, 12, 14, 15]. A negative fuel-element-bowing coefficient is always assured now in large fast cores by the proper design of a steel-honeycomb-structure of the core with internal support plates and core restraint systems.

As the control behavior of fast reactors does not significantly differ from that of thermal reactors the electromechanical design of the scram and control devices, is very similar to that of thermal reactors (see Table 12.2).

12.3.2 *Reactivity Coefficients of the LMFBR Core*

The main reactivity coefficients affecting the transient power behavior of the LMFBR core are [1, 7]:

- the negative Doppler coefficient
- the coolant (sodium, lead or LBE) temperature coefficient,
- the sodium void coefficient in case of sodium boiling
- the negative axial fuel expansion coefficient
- the structural expansion coefficient.

12.3.2.1 Fuel Doppler and Fuel Expansion Coefficient

During fast power transients mainly the negative Doppler coefficient and the negative axial fuel expansion coefficient are prompt acting negative feedback coefficients [1, 7]. They are directly coupled to the fuel temperature which depends on the integral of the transient power development over time. In LMFBRs, with ceramic fuel the negative Doppler coefficient is dominating, whereas the axial fuel expansion coefficient is smaller (smaller axial expansion coefficient of ceramic fuel). In LMFBRs with metallic fuel the axial fuel expansion coefficient (higher axial expansion coefficient for metallic fuel) is dominating and the Doppler coefficient is smaller due to the harder neutron spectrum in a metallic fuel core compared to a ceramic fuel core.

The Doppler effect is predominantly caused by the resonance absorption of ^{238}U (contributions from ^{235}U, ^{239}Pu and other fuel isotopes and from structural materials usually contribute only a few per cent). In a typical fast reactor core the Doppler effect has about equal contributions from the range of resolved and statistical resonances, respectively, the main contributions coming from the range of neutron energies below about 75 keV. The temperature dependence of the Doppler effect can be sufficiently well approximated by the relation

$$\frac{1}{k}\frac{\partial k}{\partial T} = \frac{-A_D}{T^x}$$

where k means the criticality parameter, T the fuel temperature, A_D the Doppler constant, and x is usually close to unity for a typical fast reactor (there exists an upper bound for x: namely $x \leq 3/2$, which is representative for sufficiently hard neutron spectra) (see Sect. 3.8.2.1). For most applications $x = 1$ is an acceptable approximation.

From the above it is evident that changes of the neutron spectrum significantly influence the Doppler coefficient (e.g. by the addition of a solid moderator such as beryllium oxide or zirconium hydride to the core composition or by the removal as well as addition of absorber materials such as boron or fission products). In particular, it is important to note that a complete loss of coolant from the core may reduce the Doppler constant due to spectrum hardening by up to 50%.

12.3.2.2 The Sodium Temperature Coefficient

An increase in sodium temperature leads to expansion and to less sodium in a volume unit of the core. This effect is represented by three reactivity contributions [1–3, 7, 16]:

- less absorption (positive reactivity effect)
- increase in neutron leakage or higher transparency (negative reactivity effect)
- hardening of the neutron spectrum because of loss of moderation power of sodium (positive reactivity effect).

Figure 12.1 shows the core region of a large LMFBR with a positive contribution to the sodium temperature or sodium void coefficient. In the outer core region surrounding the zone of positive sodium temperature or void coefficient the sodium/temperature or sodium void coefficient is negative as neutron leakage effects dominate.

The hardening of the neutron spectrum due to loss of sodium can—as in case of the Doppler effect—also be affected by the addition of beryllium oxide or zirconium hydride.

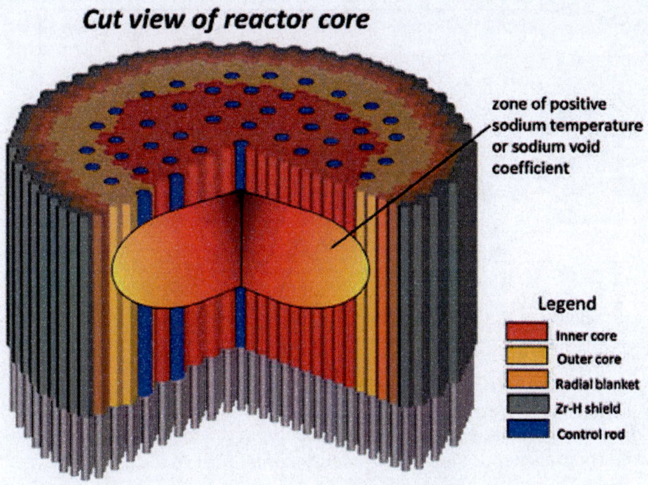

Fig. 12.1 Cut view of an LMFBR reactor core with central zone having a positive sodium temperature coefficient or positive sodium void coefficient [17–19] adapted

The effect of neutron leakage can be affected by the composition, the size and the configuration of the core (see Fig. 12.3).

12.3.2.3 Influence of Size and Geometry on the Sodium Temperature or Sodium Void Coefficient of the LMFBR Core

For small cores with a power output of about less than about 350 MW(th) or less, the sodium temperature coefficient is negative due to the high neutron leakage from the small core. This was shown by in a systematic analysis [16]. Figure 12.2 shows the sodium temperature coefficient for homogeneous and heterogeneous LMFBR cores (see Fig. 12.4) as a function of the power output in MW(th) or the size of the core (1 $¢ \triangleq 0.000035$).

For larger LMFBR cores with higher power output up to about 4,000 MW(th) the sodium temperature coefficient becomes positive in major parts of the core as shown by Figs. 12.1 and 12.2. For large cores with more than about 3,000 MW(th) or 1,200 MW(e) power output total voiding of the total core can lead to a total positive reactivity effect of about 5 $.

Different possible so-called unconventional core geometries for a minimization of the total positive void effect of large LMFBR cores were suggested in the USA by different companies in 1964 [14]. They are shown by Fig. 12.3. These are:

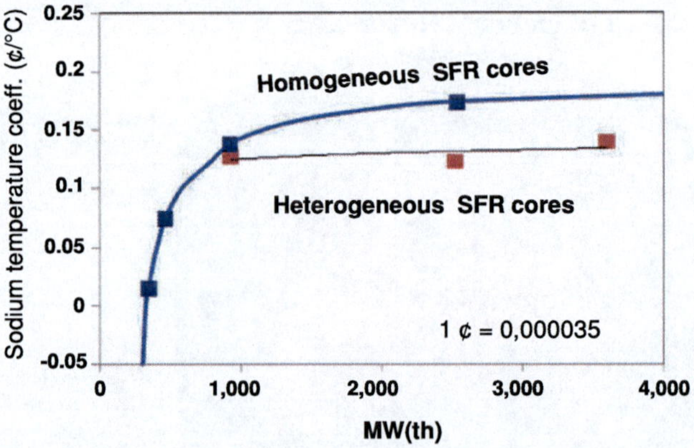

Fig. 12.2 Sodium temperature coefficient as a function of the size or power output in MW(th) of the LMFBR core (homogeneous and heterogeneous cores) [16]

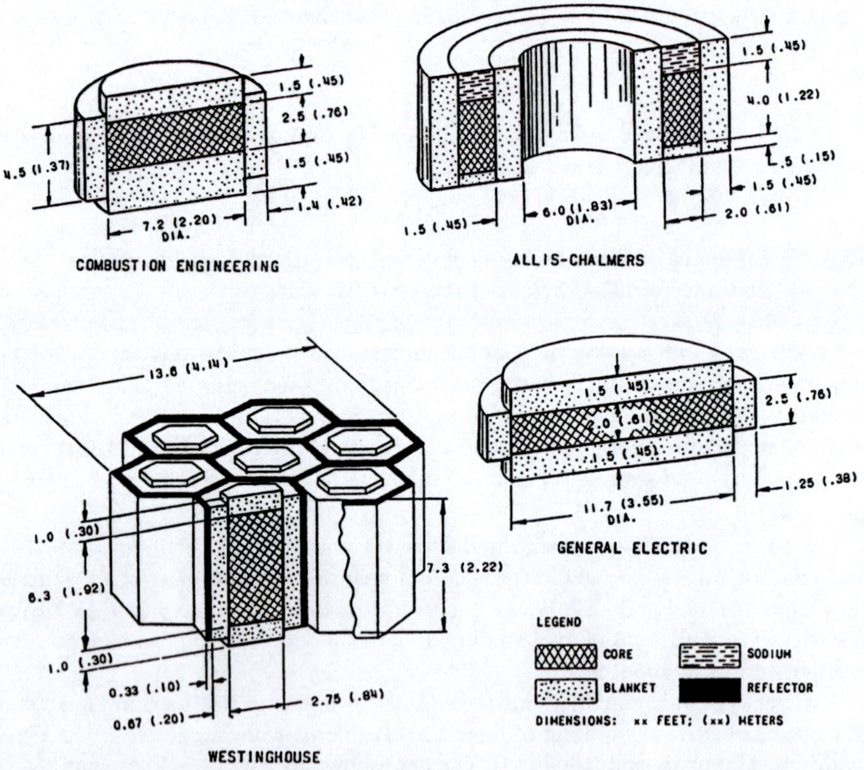

Fig. 12.3 Different possible LMFBR core geometries for minimization of the total sodium void effect proposed in the USA in 1964 [14]

- flat (pancake) cores with a high axial leakage (General Electric)
- annular cores (Allis Chalmers)
- modular cores with, e.g. 6 smaller reactor modules, arranged in one cluster of cores (Westinghouse).

(The proposed Combustion Engineering design had the highest total sodium void effect.) The most applied reactor core geometry is currently the so called radially heterogeneous core geometry, where some of the fertile fuel assemblies are arranged in a certain geometric pattern radially in the core surrounded by a relatively thin radial blanket (Fig. 12.4). With an additional decrease of the sodium coolant fraction in the subassembly this allows to decrease the total sodium void reactivity in a 1,000 MW(e) core to about +2.5 $.

The total sodium void reactivity of such a 1,000 MW(e) radially heterogeneous core can be further decreased by additional axial heterogeneous layers. If the axial upper and lower blankets are replaced by an upper and lower plenum and an absorber layer is placed above and below, then the total sodium void reactivity can become about zero [20].

12.3.2.4 The Structural Expansion Coefficient of the Core

The structural expansion coefficient is caused by several effects [1, 2]:

- increase of the coolant inlet temperature of the core (due to decreased cooling conditions in the intermediate heat exchangers) causes a radial expansion of the core grid plate. This leads to a decrease of fuel and cladding average density per unit core volume, i.e. to a negative reactivity effect of about -10^{-5} per °C.
- during power production in the core the axial and radial coolant and structural temperatures across the fuel subassemblies lead to bowing which is limited by the outer radial core restraint system (Fig. 12.5). The radial core restraint system together with the axial location of pads around the subassembly ducts must be designed such that the overall reactivity coefficient as a function of power variations and changes of the core inlet temperature is negative (around -10^{-5} per °C temperature difference between core outlet and inlet temperature (Sect. 6.3.7).
- Changes of the core outlet temperature as a function of power lead to temperature increases of the control rod drive line structures located above the upper axial blanket. This leads to axial expansion of these control rod structures. As the upper suspension of the control rod guide structures are fixed to the cover plate of the reactor vessel the control/absorber assemblies are forced to dive into the core. This causes a strong negative reactivity coefficient (Fig. 12.6).

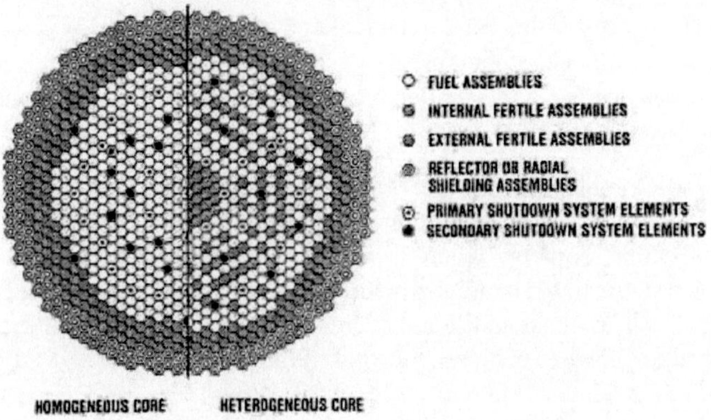

Fig. 12.4 Examples for radially homogeneous and heterogeneous large LMFBR cores [1, 7, 16]

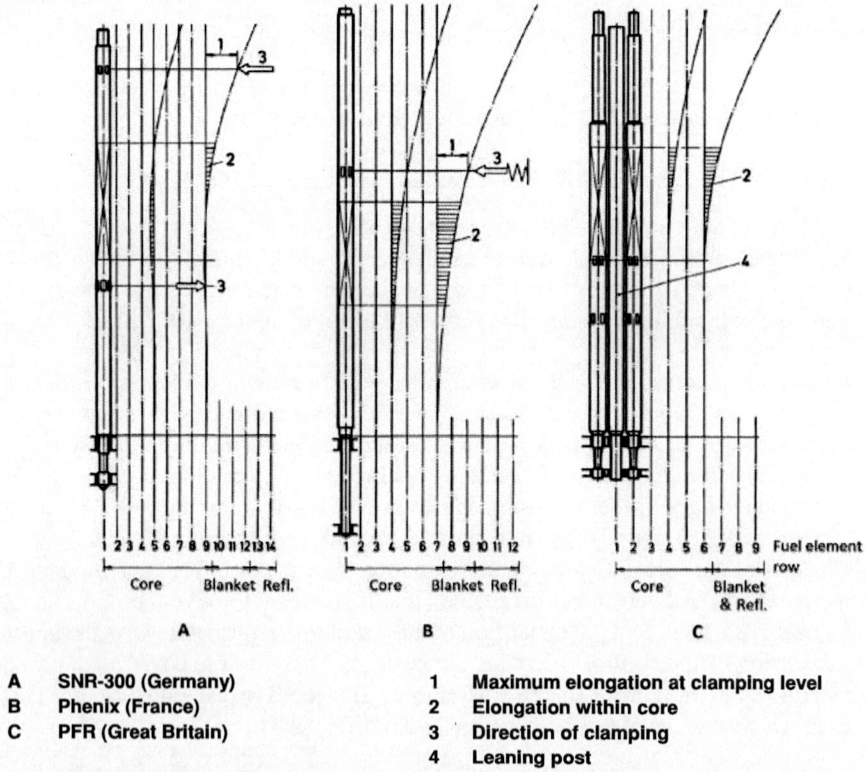

Fig. 12.5 Different clamping systems for fuel assemblies and bowing effects by radial temperature gradient [1, 21]

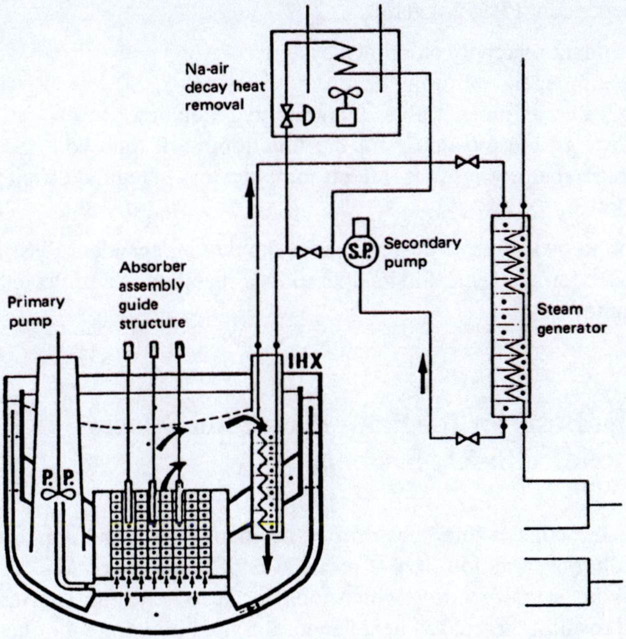

Fig. 12.6 Axial expansion of control rod drive mechanisms as a function of increasing core outlet temperature [1]

12.4 LMFBR Plant Protection System (PPS)

The LMFBR plant protection system initiates the rapid and safe shut-off of the plant in the event of abnormal transient conditions [1, 21, 22]. In addition it assures the safe afterheat removal from the core to the environment over a sufficiently long time period as well as the closure of the outer containment to avoid radioactivity releases into the environment in case of an accident.

Abnormal transient conditions can develop if the balance between heat generation in the core and heat removal from the core by the cooling systems is disturbed. This may occur either as a consequence of

- a positive reactivity input with a resultant power transient in the core at constant coolant flow conditions or
- as a consequence of coolant flow disturbances in the whole core at constant power.

The PPS monitors and processes the information of all signals relevant to the safety of the plant and triggers the respective safety actions. It comprises all installations of data acquisition and data processing, logic gatings and control interfaces including their supply of energy. For reasons of diversity and high reliability, LMFBRs shall be equipped with two spatially and functionally independent PPSs.

The two independent PPSs trigger:

- two independent reactivity shutdown systems,
- the different independent decay heat removal systems,
- the closure of containment valves (containment isolation),
- the assurance of energy supply for the functioning of the two PPSs and related safety systems (battery systems) after a complete loss of regular off-site and on-site power supply.

In addition to their diverse design, one of the two independent PPSs is fully protected against external events and its operation is installed in a protected emergency control building.

12.5 Design Basis for the Plant Protection System and Related Safety Systems

In analyzing the conceivable spectrum of accident conditions it is found that an imbalance in heat generation to heat removal is of primary concern. Consequently the focus must be on malfunctions which could lead to overheating (positive reactivity insertion and resulting increased heat generation) and impairment of heat removal.

Accident conditions beyond the boundaries of these design basis conditions leading to core melting and core destruction must be counteracted by containment design measures.

12.5.1 Design Basis Accidents Initiated by Positive Reactivity Input

Positive reactivity input into the core could arise when control rods are inadvertently withdrawn [1]. The analysis of the maximum speed designed for control rod movements and an assessment of the different interlocks for the prevention of inadvertent movement of several control assemblies lead to the overall result that in the worst case of combinations of errors the maximum conceivable positive reactivity insertion rate remains below 1 $/s.

Other positive reactivity insertion mechanisms could evolve from a fuel subassembly meltdown, from axial fuel subassembly movement as a consequence of loss of hydraulic hold-down from control rod ejection as a consequence of pressure buildup in the coolant channels of the control rods, from radial displacement of fuel subassemblies as a consequence of bridging effects within the core clamping system, from sodium voiding, and from the entering of cold sodium or oil (from pump leaks) or other hydrogeneous materials into the core.

Table 12.3 gives an impression of the reactivities involved in such an analysis for LMFBR cores. As a result of more detailed assessment, it is generally concluded

Table 12.3 Examples for potential reactivity worths to be considered in the analyses of design basis accidents [1]

	Superphenix
Fuel subassembly reactivity worth in a central core position	<1 $
Reactivity worth of one absorber assembly of the control, shim or shutdown systems	−1.8 $
Uncontrolled control rod withdrawal	<1 $/s
Reactivity difference from zero to full power	3.5 $
Burnup reactivity for one burnup cycle	9.5 $ (70,000 MWd/t)

that the PPS and related safety shutdown systems must be designed such that they can successfully counteract:

- positive reactivity ramp rates between 1 and 3 $/s, which could exceed a total reactivity level of several $.

12.5.2 Design Criteria for Shutdown Systems

The total negative reactivity worth to be covered by one of the two independent reactivity shutdown systems consists of the maximum conceivable positive reactivity insertion during accident conditions, and of the negative reactivity step necessary to bring the reactor core from full power to zero power and subcritical state. This necessitates overcoming the negative power coefficient and the coolant as well as the structural reactivity coefficients, when shutting down from full to zero power and down to coolant temperatures of about 200°C and associated fuel temperatures. In addition, the core should be kept subcritical at several $ negative reactivity.

Detailed studies show that the delay time between the monitoring of a positive reactivity incident by neutron monitors and the triggering of the shutdown system is in the range of 200 ms. The speed of the shutdown absorber rods must be in the range of about 2 m/s which leads to a full shutdown of the core in about 1 s. The speed of control rod movements can be in the range of one to several cm/s or ¢/s. This is achieved by similar electronic and mechanical equipment as used in LWRs. The absorber subassemblies can fall into the core by gravity or may additionally be accelerated by springs. Table 12.4 shows, by way of example, the technical data for the two independent SNR 300 reactivity shutdown systems.

The above design criteria and technical data for control and reactivity shut-down systems of LMFBRs are valid despite the fact that LMFBR cores, when compared with thermal reactor cores, have a lower fraction of delayed neutrons, β_{eff}, and a shorter prompt neutron lifetime, l_{eff} (by several orders of magnitude) (see Sect. 12.3).

Figure 12.7 shows the design principle of the two independent reactivity shutdown systems of an LMFBR. The primary shutdown system drops absorber rods into the

Table 12.4 Examples for technical data for the reactivity shut down systems of a 300 MW(e) LMFBR [1]

	First shutdown system	Second shutdown system
Number of absorber assemblies	9	3
Stroke (mm)	830	1,050
Scram time (full stroke) (ms)	700	700
Maximum scram velocity (mm/s)	1,800	2,100
Control velocity (mm/s)	0.7	1.53
Total reactivity ($)	23.9	10.5
Absorber material	B4C	B4C

1 $ corresponds to 3.5×10^{-3} ($\Delta k/k$)

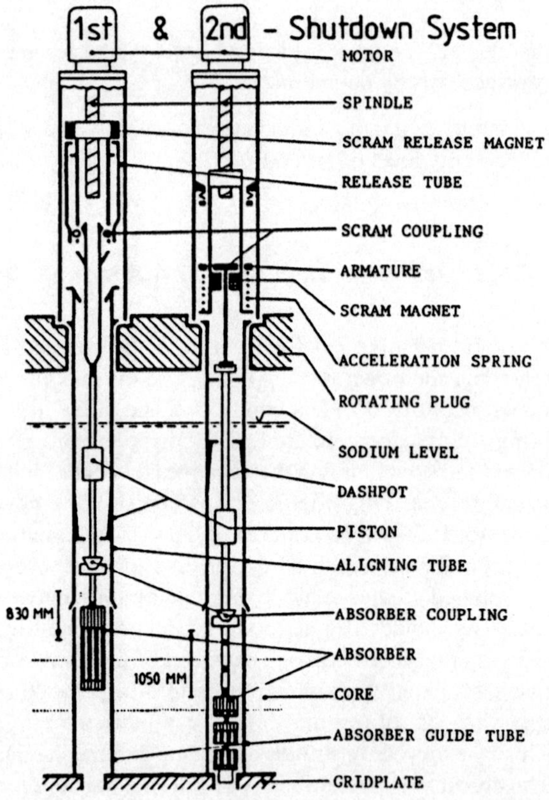

Fig. 12.7 Design scheme of two diverse and independent reactivity shutdown systems [1]

core. The secondary shutdown system pulls a flexible absorber chain into the core from below. Both systems are actuated by diverse electronic channels [1].

The primary shutdown system is released by interrupting the electrical supply to the electromagnetic coupling. This is followed by a drop of the release tube which allows the balls of the scram coupling to move outwards and release the scram rod, which falls down. The absorber connected to the scram rod drops into the core [1].

The absorber of the second shutdown system is a chain consisting of six links, the last three of which are absorber bundles. They are located in the region below the core and are pulled up by means of an acceleration spring. The spring is kept under load in the standby position by means of the armature and the scram magnet. When actuating the shutdown system, the electrical circuit has to be interrupted. To reach the standby position, the spring is put into tension by raising the magnet by means of a driving mechanism. After closing the electrical circuit of the scram magnet, the absorber is brought back into position below the core.

The probability of failure on demand of one of the shutdown systems was estimated to be between 10^{-4} and 10^{-5} per demand. With two redundant and diverse shutdown systems, as described above, and taking into account some probability for common cause failures, it was estimated that the unavailability per demand of present LMFBR reactivity shutdown systems is $<10^{-6}$ per reactor year [1, 22].

12.6 Mechanisms Leading to Impairment of Heat Removal

Loss of coolant flow in the primary loop leads to core coolant temperature increase at steady state power with insufficient fuel cooling or even to sodium boiling in the core. Such a loss of primary coolant flow may be initiated by loss of electrical power to the primary pumps, mechanical failure of the primary pumps or by a pipe break in the primary loop in case of a loop type LMFBR.

Similar incidents in the secondary sodium loop would lead to an increase of the coolant inlet temperature of the core as a consequence of impairment of heat removal in the intermediate heat exchangers. Again a reaction of the protection system, i.e. triggering the reactor shutdown systems, is needed.

Failures in the tertiary steam loops tend to result in less severe consequences to the reactor core, since the core inlet coolant temperature would increase only after a certain delay time. However, feedwater pump failures could rapidly lead to drying out of evaporators and steam generators, if the plant is not shut down in due time. Table 12.5 gives a list of design basis accidents leading to impairment of heat removal. These design basis accidents must be investigated in detail as part of the safety analysis of LMFBRs.

12.7 Instrumentation and Monitoring of the Protection System

The upper boundary of the reactor core and the reactor tank as well as the primary, secondary and tertiary coolant circuits contain specific instrumentation for the surveillance and the continuous monitoring of any deviation from steady state conditions [1, 22]. All readings are continuously processed and compared against preset levels. If the preset levels are exceeded, the protection system triggers the necessary actions.

Table 12.5 Design basis accident initiated by loss of coolant flow or loss of power

Loss of power
Loss of off-site electrical power
Loss of off-site and emergency diesel electric power
Primary coolant system
Loss of electric power to one primary pump
Continuous flow reduction by control valve malfunction
Mechanical failure of one primary pump
Major leak (pipe break) in one primary loop
Secondary coolant system
Loss of electrical power to one secondary pump
Mechanical failure of one secondary pump
Major leak (pipe break) in one secondary loop
Tertiary coolant system (water steam loops)
Feedwater pump failure
Feedwater pipe rupture
Steam line rupture
Inadvertent opening of steam generator outlet relief valve
Rupture of steam generator tubes with subsequent Na-H_2O reaction

The power level is detected through measurements of the neutron flux by means of reaction rates of suitable detectors. As soon as a certain preset power level, e.g. 112%, of nominal power is exceeded a trigger signal is set to the shutdown system. Two out of three circuits are used to cover the whole power range from zero to full power.

The temperature of the sodium in the core and the primary and secondary circuits is measured by thermocouples (Cr-Alumel or Ni-Cr) and Resistance Temperature Detectors.

For the sodium flow measurement in the primary and secondary circuit both the standard Venturi flowmeter and magnetic flowmeters are used. The pressure of liquid sodium is measured by instruments which transmit pressure via a diaphragm to a NaK column. Acoustic sensors and pressure transducers (Li-niobate crystals and capacitor microphones) have been developed as well.

The level of sodium in the reactor tank, in the IHX and other major tanks or components must be continuously monitored. Induction level probes have proved to be very reliable detectors [1, 23].

Sodium leaks can be detected by contact type sensors consisting of two electrodes extended to a location where leaking sodium may be expected to collect. The presence of sodium shorts out the electrode gap and produces a signal.

Sodium smoke detectors simply detect the presence of sodium aerosols, when sodium is leaking into an area where it comes into contact with air.

Minor leaks in steam generators are detected by hydrogen detectors using nickel membrane systems. Medium and large size leaks with ensuing sodium-water reactions are detected by pressure sensors. The destruction of rupture discs in case of large size leaks and the inrush of sodium is detected by controlsensors.

12.8 Analysis of Design Basis Accidents

12.8.1 Power Transients

Power transients can be initiated by reactivity or coolant flow imbalances and are counteracted by the PPS. Plant transients can be initiated by steam generator faults, pump failures etc.. They are counteracted by the PPS as well. For LMFBRs extensive analysis must be performed to ensure that none of these transients violate the system integrity limits. Several computer codes, e.g. SAS4A [24] were developed and are in use for such analyses. They describe the core and coolant circuit behavior by solving the differential equations for the core kinetics, the temperature field of the core, the feedback mechanisms of the core (core reactivity coefficients) and the heat transfer characteristics of intermediate heat exchangers, steam generators as well as the behavior of the turbine.

Power and plant transients are analyzed in three main areas:

- the reactor control systems must automatically keep the plant characteristics within the preset limits and avoid unnecessary and too frequent shutdown or too high steam temperature (safety level 1 and 2)
- the reactor shutdown systems must counteract positive reactivity insertions in due time, so that the core damage criteria are not exceeded, e.g. fuel melting in the centre of the fuel rod, cladding temperature >800°C, core coolant outlet temperature >620°C (safety levels 1 and 2)
- The PPS must counteract incidents causing heat transfer imbalances in the secondary and tertiary circuits to avoid too frequent and too high temperature gradients which may impair the plant components (plastic strains) and the lifetime of the plant (safety level 3).

12.8.2 Positive Reactivity and Power Transients

A positive reactivity ramp insertion of 10¢/s at full power of a 300 MW(e) LMFBR, will lead at 1.10 s after initiation of the insertion to shutdown, because the trip level of 112% reactor power is exceeded [1].

A steep positive reactivity ramp increase of 1.05 $/s at full power of a 300 MW(e) LMFBR would lead to a very sharp power increase up to 70 times of full power. However, the negative Doppler coefficient would turn the power around and bring it down again. At 0.4 s after initiation of the reactivity transient the shutdown system would be acting. Neither the fuel melting temperature in the highest rated fuel rods would be exceeded nor the sodium boiling temperature would be attained. The LMFBR would remain shut down and the afterheat removal system would be started.

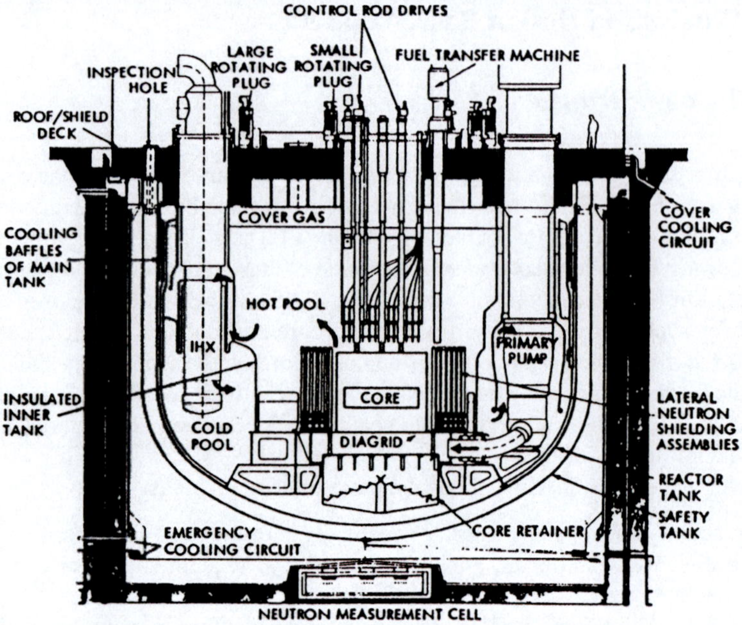

Fig. 12.8 Pool type LMFBR with primary and secondary coolant circuits [1]

12.8.3 Loss of Primary Flow

Loss of flow in the core is analyzed for the following cases [1]:

- one primary pump (out of three or four) is tripped or blocked,
- loss of off-site electric power; all pumps are tripped in this case.

The PPS reacts to these transients after the trip levels of the neutron flux/primary coolant flow, and the core outlet temperature are exceeded. As soon as the primary pump coast-down starts, the primary coolant flow decreases rapidly and the core coolant outlet temperature starts to rise. About 2 s after initiation of the incident, shutdown is initiated. As the power drops faster (within about 1 s to decay heat level) than the coolant flow (pump coast-down occurs in about 10–15 s), the core and reactor tank structures will be subject to temperature gradients in time and space (Fig. 12.8). LMFBRs pony motors, therefore, take over after pump run-down and ensure 5% of the full primary flow which is sufficient for further decay heat removal. If the power supply of the pony motors should also fail, natural convection would take over (see Sect. 12.9). In this case the sodium can heat up in the core and attain maximum cladding temperatures of up to about 800°C, until steady state conditions for heat removal through natural convection are established [1].

12.8.4 Transient Events in Sodium Steam Circuits

For the secondary sodium circuits and the tertiary water/steam circuits predominantly flow transients are investigated. As described above for the primary sodium circuits, also for the secondary coolant circuits the following cases are of interest:

- one secondary pump (out of three or four) is tripped or blocked,
- closure of one secondary valve is actuated,
- loss of electric power for all secondary pumps.

For the tertiary water steam system the following main transients are of interest:

- steam flow transients as a consequence of opening or closure of valves,
- steam turbine or generator failure,
- failure of the main condenser,
- transients in feedwater flow to an evaporator as a consequence of opening or closure of valves or failure of pumps,
- transients in feedwater temperature.

The PPS will react and shut the reactor down, when the different trip levels are exceeded. As a result of the analyses the temperatures and pressures (steam generators) for each component of the heat transfer circuit are obtained as a function of time. Stress and fatigue analyses are performed for the main important components to ensure that the plant can withstand a certain specified number of such incidents over its foreseen lifetime. This becomes necessary since most parts of the plant operate in the elastic-plastic range of austenitic steels at temperatures up to 550°C [1].

12.9 Decay Heat Removal in LMFBRs

The excellent cooling and natural convection properties of liquid metal coolants (Na, NaK, Pb or LBE) allow the removal of decay heat in various ways by natural convection only [1, 25, 26]. Figure 12.9 shows the removal of decay heat by the normal coolant circuits for Superphenix (primary circuit → secondary circuit → steam circuit), (feeding the steam not to the turbine, but to the main condenser). The condensed water is recycled. This solution assumes that at least one heat removal circuit out of three or four remains undamaged and that the heat can be transported by natural convection from the core to the steam condenser. This is feasible because of the good natural convection capabilities of sodium which has been demonstrated in several prototype LMFBRs (SEFOR, PHENIX, PFR etc.).

Figure 12.9 also shows another possibility of decay heat removal. A sodium/air heat exchanger circuit is coupled to the secondary coolant circuit. This sodium/air heat exchanger operates by natural convection flow.

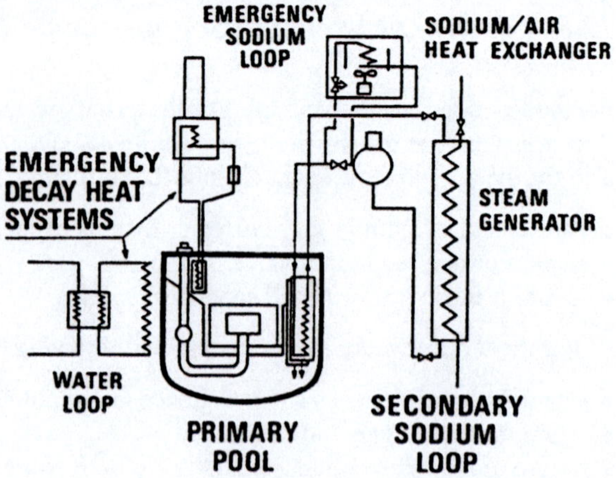

Fig. 12.9 Decay heat removal systems for a pool type reactor (Superphénix) [1]

A third possibility is shown in Fig. 12.9. The decay heat is removed by Na or NaK filled emergency coolers to air cooled heat exchangers. The emergency coolers are directly installed in the reactor tank and are independent of the main circuits. They are able to dissipate the decay heat to air coolers by natural convection. As an example, this design solution was applied in PFR, SNR 300, and in SUPERPHENIX [1].

Figure 12.10 shows a fourth possibility for decay heat removal. This possibility was first applied in PHENIX. The decay heat is transferred by radiation from the double wall reactor tank to a third, so-called safety tank which has a water filled tube coil system on its outer surface. This solution is also completely independent of the main circuits. Most LMFBRs have a combination of the concepts described above.

The decay heat removal system for SUPERPHENIX included sodium-to-air heat exchangers on the secondary loops plus four backup emergency sodium cooling circuits as decay heat removal systems. Each backup system included an immersion cooler in the reator tank, a closed sodium loop, and a sodium-to-air heat exchanger. In addition, the safety tank around the double-walled pool tank was cooled by water pipes as in PHENIX.

Demonstration of the safe decay heat removal is part of any commissioning programme of LMFBRs. During repair and maintenance, a plant must be operated with at least two circuits. This means that a four circuit plant with one coolant circuit under repair and one coolant circuit under maintenance can still operate at reduced power. Should the third coolant circuit fail, the fourth coolant circuit would be available for decay heat removal.

The emergency power and feedwater supplies must be protected from external impacts (tornadoes, floods, tsunamis, aircraft crashes, chemical explosions etc.).

Natural circulation experiments for the removal of the decay heat performed at PFR showed an excellent behavior of the plant. Similar experiments were performed at PHENIX, SEFOR and FFTF [21, 25]. An extensive review of experimental and

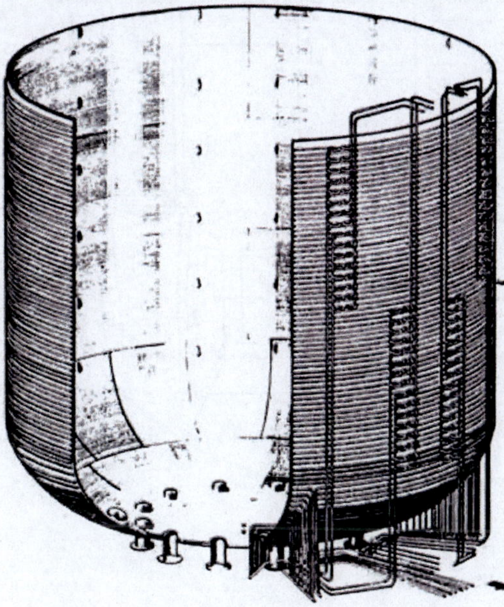

Fig. 12.10 PHENIX—ultimate emergency cooling system by water pipes on the outside of the reactor safety tank [1]

theoretical data on decay heat removal and natural convection in LMFBRs was published by [25]. Sophisticated computer codes, e.g. SASSYS [27], or COMMIX-1A [28] and out-of-pile experiments are used to predict the natural convection flow within the reactor tank [29, 30]. Figure 12.11 shows the complicated flow pattern within the reactor tank of a pool type reactor.

12.10 Anticipated Transients with Failure to Scram

Anticipated Transients Without Scram (ATWS)—as in LWR safety analysis (Sect. 11.6.7)—have to be considered for the safety analysis of LMFBRs, despite the fact that the two reactivity shut down systems of LMFBRs have a failure probability of about 10^{-6} per year. This ATWS analysis has to be shown for the two main accident initiating chains

- positive reactivity input
- loss of coolant flow.

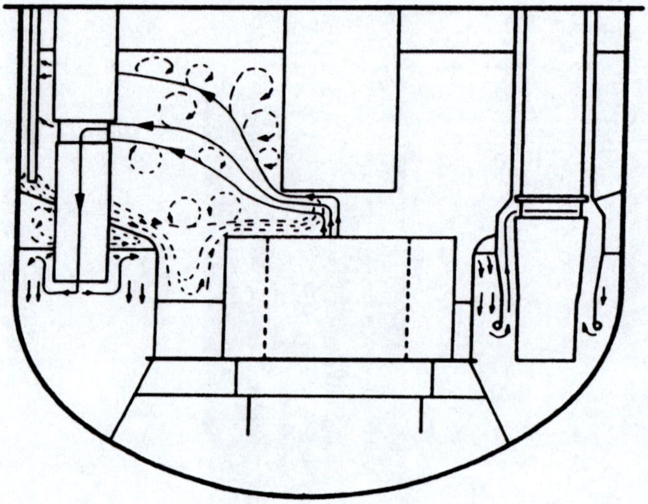

Fig. 12.11 Natural convection flow pattern in a pool type LMFBR [1]

Historically, there have been two phases, during which these ATWS accidents were treated differently. The reason is the control rod drive line expansion mechanism (Sect. 12.3.2.4, Fig. 12.6) as an inherent safety feature which was proven to exist not earlier than around 1985 by experiments at the British PFR and confirmed later also at Phenix etc.

12.10.1 Phase from 1970–1990

The safety discussions and the licensing approach for the LMFBR prototypes, e.g. CRBR (USA), SNR 300 (Germany) and MONJU (Japan) had led to the following approach regarding anticipated transients without scram (ATWS). The following accidents had to be considered:

- ULOF—Unprotected Loss of Flow initiated by primary flow coast down
- UTOP—Unprotected Transient OverPower initiated by small reactivity transients
- ULOHS—Unprotected Loss of Heat Sink initiated by failure of the heat removal system of the secondary or tertiary side.

Probability analysis for these ATWS accidents had shown that the failure probability of the two diverse and redundant shutdown systems (including common mode failures) are in the range of 10^{-5}–10^{-6}/years. Licensing authorities in the above mentioned countries required that these ATWS accidents be considered. All mechanical and thermal accidental consequences for the reactor tank and containment system had to be analyzed. Other limiting accident sequences like an immediate total flow blockage of one assembly, local blockage in an assembly and inert gas bubbles flowing through the core were also to be considered.

The analysis of the ATWS with the severe accident codes, such as SAS4A [24, 27] and SIMMER [31, 32], resulted in the following accident progression sequences for the ULOF accident:

- Boiling in the core would start in tens of seconds (about 15 s for SNR 300 and 25 s for Superphenix) after initiation of pump failure or flow coast down followed by failure to scram.
- Due to the positive sodium void coefficient the subsequent accident progression involved: sodium boiling → fuel rod failure and meltdown → molten fuel-sodium (thermal) interation → pressure build up → mechanical impact to the reactor tank [28, 33].
- For the description of the disintegrating fast reactor core, the description to the feedback mechanism and of the power transients the SIMMER code [31] was developed.
- For stress and strain analysis of the reactor tank, computer codes like REXCO [34, 35] were developed and applied to determine structural effects.

Experimental programs in research and test reactors like TREAT (USA), CABRI (France, etc.), Scarabee (France), provided experimental data for sodium boiling, fuel rod failure, meltdown. Research for molten core retention and cooling was performed [1].

For SNR 300, a mechanical energy load for the reactor tank of 300 MJ and also a core catacher were required by German licensing authorities. For Superphénix about 800 MJ were reported. Other international projects followed this approach more or less some relying more on the probabilistic approach, some assuming the deterministic approach as e.g. in Germany for the prototype LMFBR SNR 300.

12.10.2 The Phase from 1985–2010

Experiments and measurements at PFR, and little later at Phénix, showed that an additional negative reactivity coefficient (axial expansion of the control rod drive lines) existed, which had not been accounted for in the earlier ATWS analyses (ULOF, UTOP, ULOHS). This reactivity effect is in the range of -10^{-5} $\Delta k/K$ per degree increase of the core sodium outlet temperature (assuming constant vessel temperatures). In the late phase of PFR around 1989 this effect was demonstrated by shutting off the pumps without a scram. PFR stabilized itself at a higher sodium temperature and low power. A similar test was performed on Rapsodie during an ultimate test program.

This physical effect was soon accounted for in ATWS analyses by international groups. Analysis of the loss of flow accident without scram for Superphenix showed that including this effect of control rod drive expansion would just not lead to sodium boiling and the sodium outlet temperature would stabilize at about 750°C

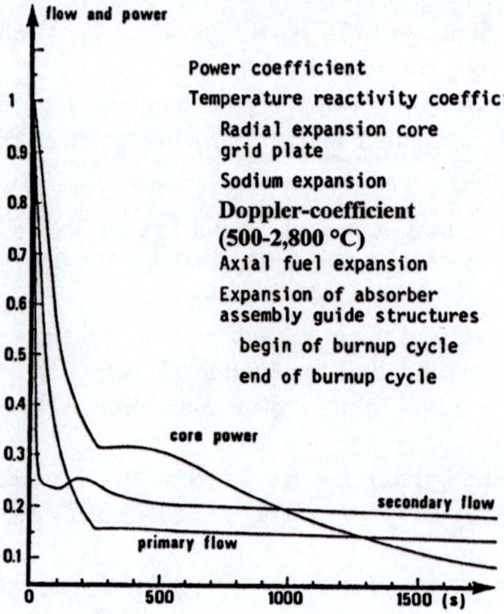

Fig. 12.12 Power and primary as well as secondary flow as functions of time for loss of flow (LOF) accident without scram for Superphenix (1,200 MW(e)) [36]

(see Figs. 12.12, 12.13, 12.14) [36]. Soon it was shown by several organizations that designing for an appropriate large negative Control Rod Drive Expansion effect (CRDE) the ATWS accidents i.e., ULOF, UTOP and ULOHS would result in equilibrium sodium temperatures of e.g. about 650–700°C. **They would not lead to sodium boiling with all following accidental events (fuel rods failure and melting).** This is valid for the existing positive total sodium void reactivities of large SFRs in the range of 3–5 $. The best suited designs are SFRs with metallic fuel (e.g. Integral Fast Reactor concept proposed by Argonne Laboratory). The high axial expansion coefficient of metallic fuel does not require special efforts for the design of the CRDE. ANL showed that this type of SFR shuts itself down in case of ULOF, UTOP, ULOHS and equlibrium sodium temperatures of about 650°C [16, 17, 37, 38].

For MOX fueled SFRs special engineered devices were developed (Athena-Karlsruhe [39]) which indirectly magnify the CRDE and in such a way achieve similar negative feedback-effects as in the case of Pu-U-metal fueled SFRs. In this way equilibrium sodium temperatures of about 700°C and self-shutdown of the reactor are also possible. With the SASS: ("Self-Actuated Shutdown System in the Japanese SFR core") [17, 32] the shut down function is initiated passively when the sodium outlet temperature exceeds the Curie point of the holding magnets [17, 32] (Fig. 12.15).

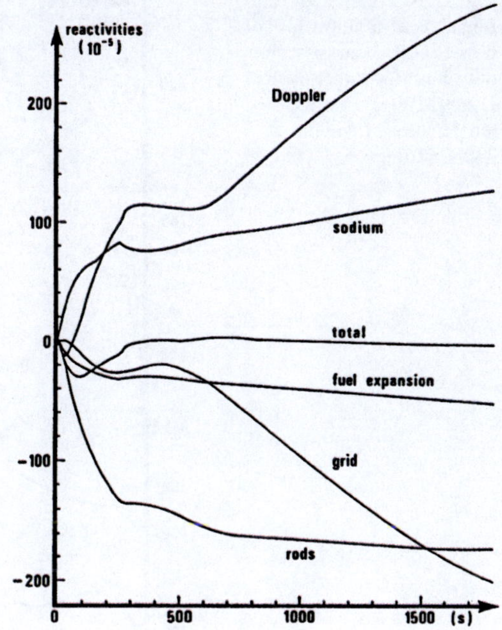

Fig. 12.13 Reactivity contributions as functions of time for loss of flow (LOF) accident without scram for Superphenix (1,200 MW(e)) [36]

12.11 Local Melting in Fuel Assemblies

The question whether there exists a realistic potential for coolant blockages in a fuel subassembly with subsequent melting of fuel rods and rapid propagation into significant damage of the LMFBR core has been subject of considerable research efforts (Fig. 12.16) after a subassembly blockage in the Enrico Fermi Fast Breeder Reactor (EFFBR) had occurred [1].

They were devoted to the following phenomena: local blockage formation, local boiling, clad dryout, melting including fission gas effects, mechanical and thermal damage propagation within the core [40, 41].

In blockage formation experiments for fuel assemblies with grid spacers, it was observed that the critical particle size would be 0.9–1 mm. Below this critical particle size no blockages were found at all. Particle blockages at grids in the core region can only be formed by fuel particles above this critical size, e.g. after fuel rod failure and fuel washout. Smaller size particles from the primary circuit may contribute to the blockage only after and in connection with significant fuel rod failures. Fission gas release from a failing fuel rod does not impair the cooling of neighboring fuel rods.

Experimental investigation of local porous blockages and propagation phenomena in 37 pin bundles (in-pile programme MOL 7C) showed no damage propagation. Inherent self-limitation of the fault was observed, in spite of severe damage of more than six fuel pins of the bundles including complete clad melting, partial fuel melting and the formation of a secondary blockage at the downstream spacer grid. In all in-pile

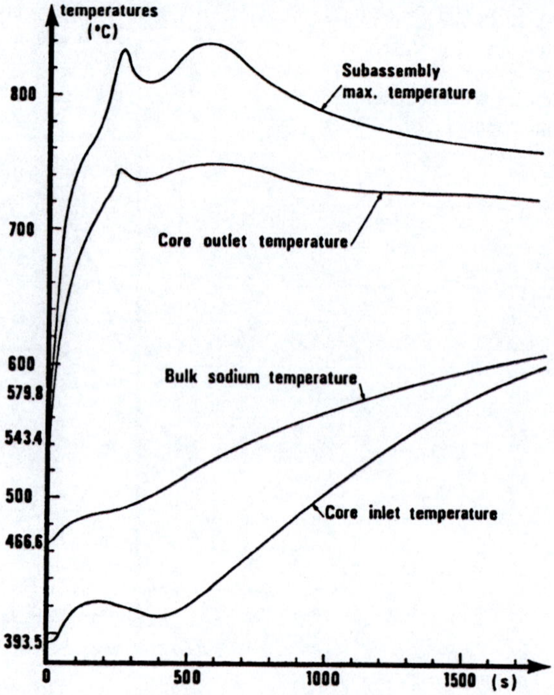

Fig. 12.14 Temperatures as functions of time for loss of flow (LOF) accident without scram for Superphenix (1,200 MW(e)). Boiling temperature of sodium at 840°C [36]

blockage experiments, significant signals were observed at delayed neutron monitors associated with the development of the local damage. An assessment of these results showed that an automatic trip of the safety systems is possible after a time delay of 28 s due to the transportation time of the sodium from the core to the delayed neutron monitors [40, 41].

Similar experimental results where obtained from the FEFPL (Fuel Element Failure Propagation Loop) experiments at Idaho [42].

No significant difference is to be expected between these experimental results and similar blockage conditions in LMFBR fuel assemblies with wire spacers.

Another possibility for fault propagation would be the melting-through of hot fuel from one subassembly to another (thermal propagation). Investigations show that a crust of solid fuel could form at the duct of the fuel assembly. The radial heat transport would be high enough that the wall would melt and sodium boiling and dryout would occur in the gap between two adjacent fuel assembly ducts. As a consequence, melting of the adjacent wall of the next fuel assembly would begin. Such radial melt-through would be a slow process which could be detected by the core instrumentation [40, 41].

Nevertheless, the Japanese EAGLE program is experimentally investigating these potential fault propagation sequences. In case they would turn out to be a serious safety problem, provisions by fuel assembly design (FAIDUS-Concept) could be

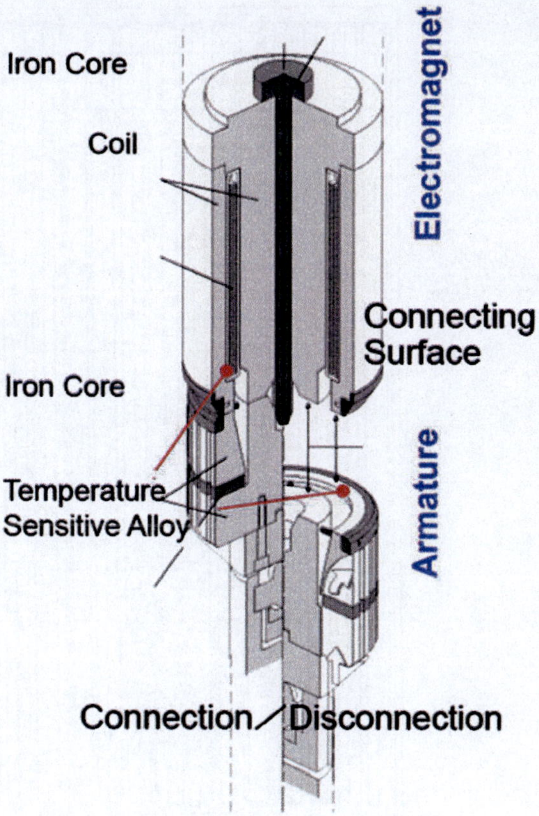

Fig. 12.15 Design scheme of the self-actuated shut down system (SASS) [17, 32]

made as shown by Fig. 12.17 [32]. Molten fuel from a fuel rod zone could move to the inner duct of the FAIDUS fuel element and be discharged out of the core zone.

12.12 Molten Core Cooling Device (Core Catcher) for LMFBRs

If large scale melting of the core of an LMFBR would occur the core fuel can be collected and distributed in a core catcher underneath the core grid plate at the bottom of the reactor tank.

Figure 12.18 shows such a core catcher design for a Japanese 1,500 MW(e) LMFBR design. The core catcher is arranged as a tray system where the coolant sodium can always enter in spaces underneath the trays and cool the core melt.

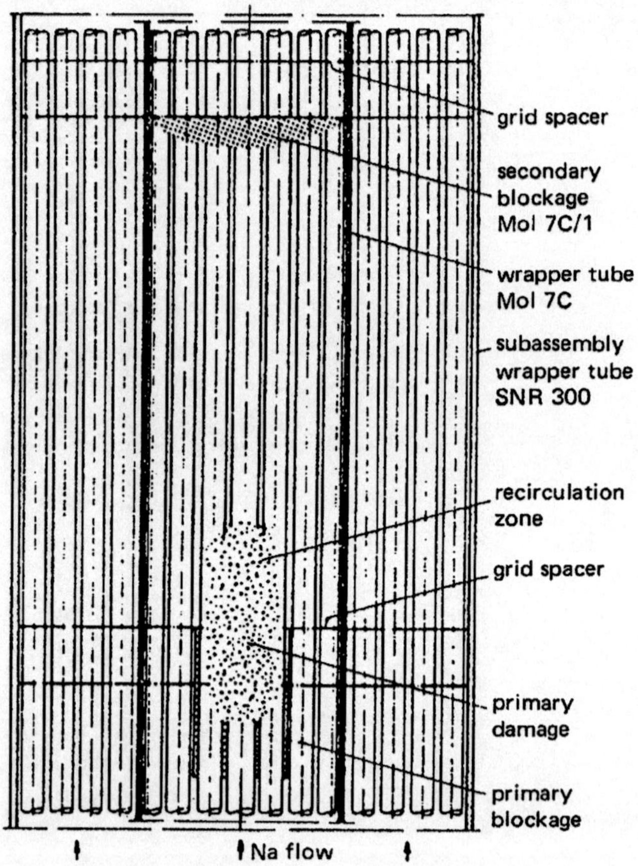

Fig. 12.16 Local blockage in an LMFBR fuel element [40]

12.13 Sodium Fires

Sodium leaks from pipes or components of the primary or secondary sodium circuits bring hot sodium in contact with either oxygen of the atmosphere of the inner containment or with oxygen from air in steam generator buildings. Such sodium leaks and sodium fires occurred in several LMFBR test reactors. One usually distinguishes between sodium pool and sodium spray fires. A pool fire would occur if liquid sodium is spilled onto the floor of the containment or into the steel lined cells of the inert inner containment enclosing the primary pumps or intermediate heat exchangers (loop-type reactors), thus mitigating the consequences of such an event. Spray fires can result from sodium being ejected from a pressurized pipe or component due to a leak or rupture. Other potential modes of occurrence for pool or spray fires can be envisaged within the containment as a result of an HCDA.

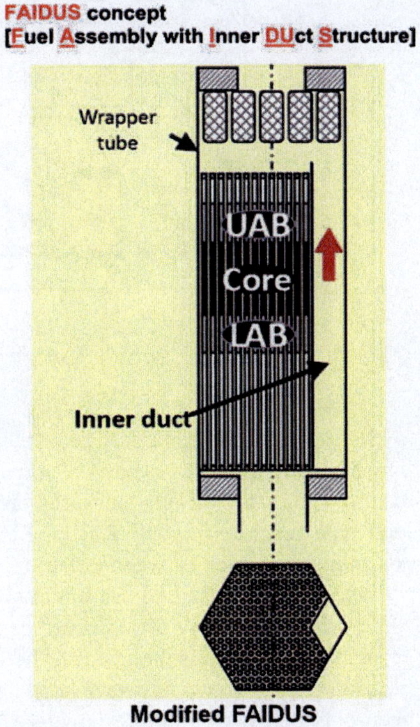

Fig. 12.17 FAIDUS fuel element concept for the prevention of fuel compaction by melting in the core zone [17–19, 32] (*UAB* upper axial blanket, *LAB* lower axial blanket)

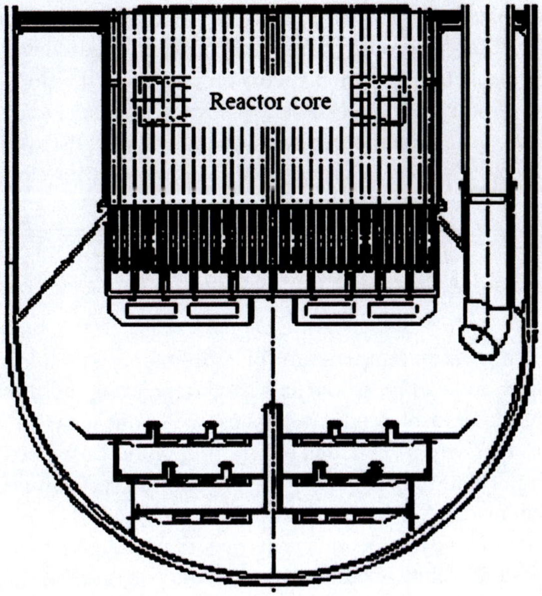

Fig. 12.18 Design scheme for a core catcher underneath the core grid plate of a Japanese 1,500 MW(e) LMFBR [17]

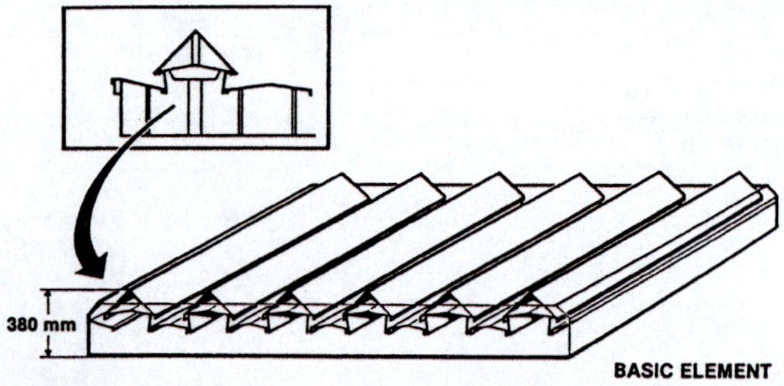

Fig. 12.19 SUPERPHENIX smothering catch pan system [1]

Sodium fires can be largely prevented by surrounding the sodium pipes and components with an inert gas (argon, nitrogen) in steel lined confinements (cells). An example for such design measures are the steel lined inner containment cells of loop-type reactors (e.g. SNR 300, FFTF etc.). The oxygen content in such inert confinements varies between 0.5 and 2%. Other design measures against pool-type sodium fires are so-called catch pan systems (Fig. 12.19). They collect leaking hot sodium in such a way that sodium can run through little holes into a leak recovery tank where it is isolated from the oxygen component of air. Different sodium fire extinguishing powders have been developed and especially good experience has been obtained with graphite powder for quick and effective extinction of sodium fires. Sodium leaks and sodium fires are automatically detected by contact sensors, smoke detectors, wire-bead-sensors and flame spectrometers.

A large amount of research effort has been devoted in all countries with LMFBR projects to experiments with sodium fires [43–46].

Apart from pool fire experiments extensive code modelling was also performed. Two typical computer programs are SOFIRE-II [46] and NABRAND [47]. A summary of codes developed in different countries can be found in [44].

12.13.1 Sodium Spray Fires

The ignition temperature for sodium spray fires is substantially lower than for pool fires and can be as low as 120°C depending on the sodium droplet size. The sodium oxide aerosol production rates are about a factor of 5 higher than for pool fires. Extensive experimental programmes have been performed on the same parameters of interest as for pool fires [48]. For sodium spray fires theoretical models and computer codes, e.g. SOMIX 2 and SOMIX 2C [43, 44, 49–51] were developed to describe and determine the main parameters needed for the containment design. A summary of codes developed in different countries can be found in [44, 47, 52].

12.13.2 Double Walled Piping

Interesting design solutions have been proposed within the Japanese JSFR project. The primary and secondary sodium coolant pipes are double walled as described in Sect. 6.5.3. The inner space between the two walls is filled with argon or nitrogen.

12.13.3 LBE as Coolant

LBE does not react with air. Therefore, no prevention for LBE fires will be necessary (Chap. 6).

12.14 Sodium-Water Interactions in Steam Generators

In the case of failure of a water filled pressurized steam generator tube, water is injected at high pressure into the sodium and a violent sodium-water interaction occurs [52–56]. This sodium-water interaction will give rise to peak pressures that must be accommodated by straining the steam generator shell and the pipelines in the secondary circuit. Moreover, if the pressure in the system gets too high, rupture discs will be destroyed, thereby relieving the pressure and ejecting sodium, sodium hydroxide, sodium oxide and hydrogen into a special collector vessel. This pressure suppression system is very important and protects the intermediate heat exchanger from damage. Figure 12.20 shows—as an example—the operating principle of the connections for the pressure suppression system at a straight tube steam generator. In the course of the sodium-water interaction, there is a violent short-term pressure rise which destroys a rupture disc and ejects sodium, hydrogen and NaOH into the pressure suppression line. The latter is filled with cover gas (nitrogen). The sodium-water interaction is immediately detected by instrumentation and the plant will be shut down if a large leak occurred [57].

The hydrogen will escape into the open atmosphere through a cyclone after having ruptured another rupture disc. Depending on the temperature, hydrogen may be self-ignited and burned there. Entrained sodium or NaOH will be separated in the cyclone. After pressure reduction on the water-vapor side, sodium and water of the steam will be drained. The steam generator can be removed and repaired.

Numerous experimental programmes concerning sodium-water interactions have been carried out on scaled-down steam generator models in all ongoing LMFBR projects [53, 54].

Theoretical methods have been developed to calculate time dependent pressures in the reaction zone, at any point in the steam generator, in the secondary sodium circuit, and in the intermediate heat exchanger. These models contain:

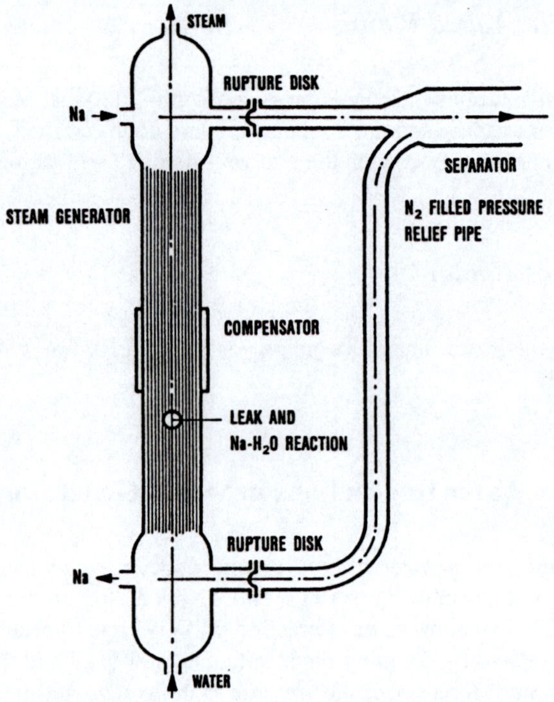

Fig. 12.20 Principle of pressure relief system of a straight tube steam generator [1]

- a chemical reaction model calculating the heat generation
- instantaneous hydrogen production for the pressure buildup and a bubble extension model
- the hydrodynamic equations for pressure propagation in the steam generator and in the pipeline system
- calculation of the strains in the steam generator and the piping system [58].

Computer codes describing pressure peaks in the circuit system are, e.g. TRANSWRAP [55] For precise analysis of the mechanical stresses and strains in the steam generator, the piping system and the intermediate heat exchanger mostly such additional computer codes as e.g. ICEPEL [56] are used.

12.14.1 Instrumentation of Steam Generators

Minor leaks in steam generators are detected by oxygen meters and by hydrogen detectors operating on the principle of nickel membrane systems. They are used for both hydrogen detection in the sodium and in the argon cover gas of the steam

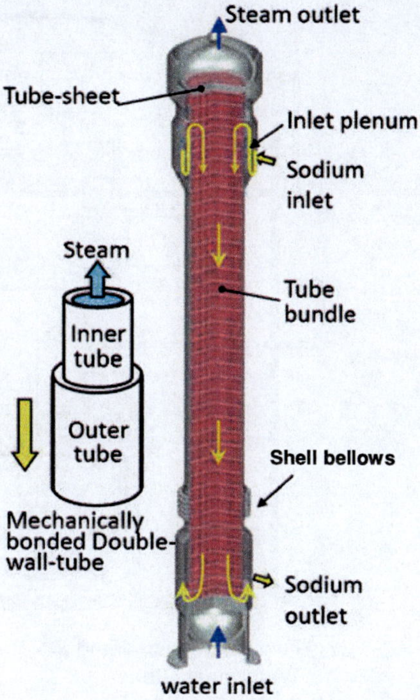

Fig. 12.21 Straight tube steam generator with double walled (mechanically bonded) steam generator tubes [17, 18, 58]

generator. The response time of these hydrogen detectors is a function of leak size and the transport time of hydrogen from the reaction zone to the detector. The response time will be of the order of minutes in the case of very small leaks. Acoustic detectors with wave guides eliminate the problem of long transport times. Medium and large size leaks with ensuing sodium-water interactions are detected by pressure sensors and the destruction of the rupture discs. In addition, the inrush of sodium into the pressure suppression lines is detected by contact sensors [57].

12.14.2 Double Walled Steam Generator Tubes

Interesting research efforts were reported for the Japanese JSFR design. Double walled steam generator tubes are being developed (Fig. 12.21). This would avoid sodium-water reactions in steam generators [17].

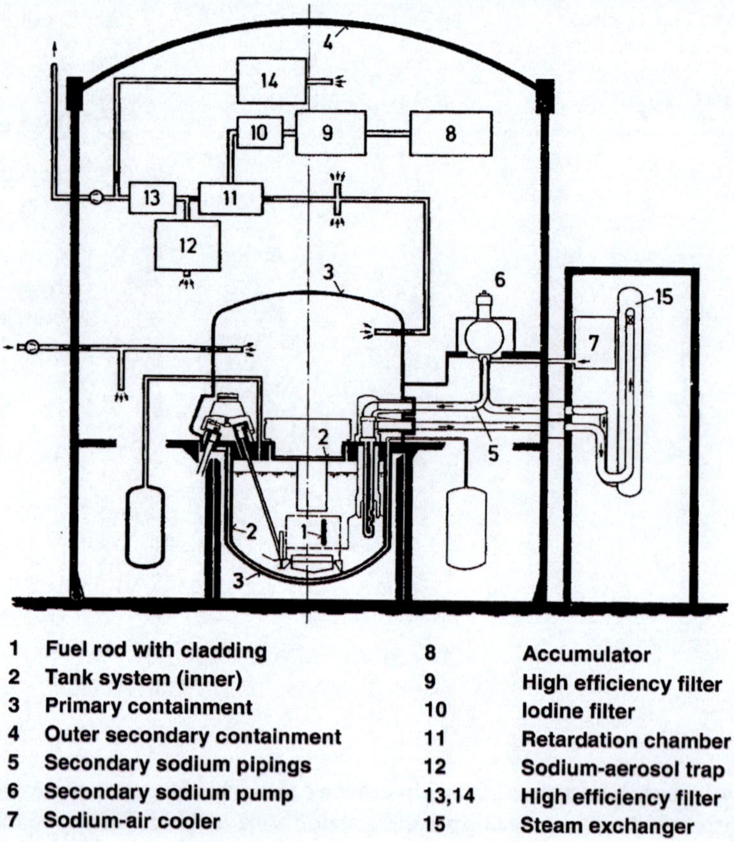

Fig. 12.22 Safety containment of Superphenix [1]

12.14.3 Steam Generators with LBE Coolant

As LBE as coolant does not chemically react with water no safety provisions are necessary for such LBE-steam generators.

12.15 Containment Safety Design Concept

Following the goals of protection defined in Sects. 10.2 and 10.3 for nuclear reactors the inner and outer containment structures represent the last barriers for preventing large radioactivity releases to the environment.

Figure 12.22 shows the double containment concept which was applied, e.g. to SUPERPHENIX. As shown by Fig. 12.22 the reactor tank system and the steel dome

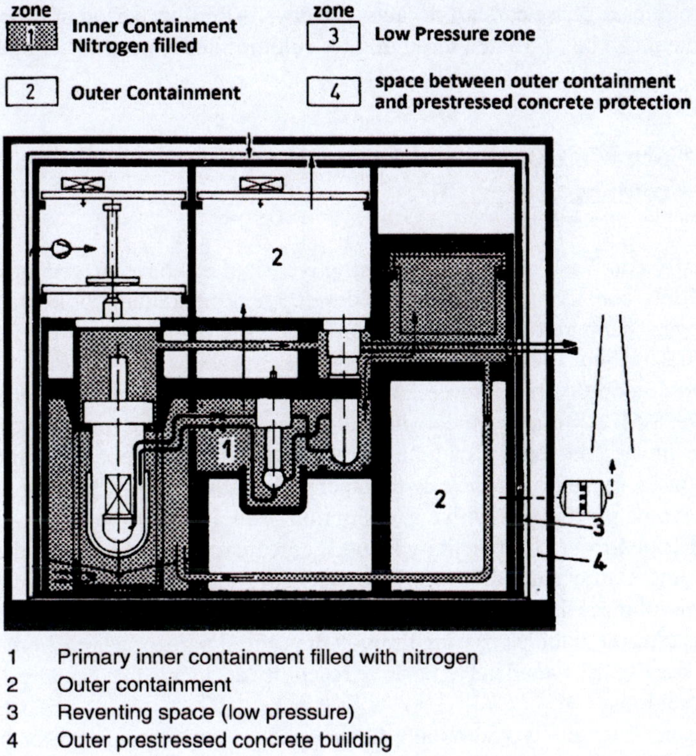

Fig. 12.23 Double containment with reventing capabilities for leaks from the outer containment (SNR 300) [1]

above the reactor cover represent the inner containment. The steel dome has a low leak rate (1 vol.% per day) into the outer containment. The air filled outer containment is kept at low pressure and can be vented through a filter system to a stack outside.

A double containment with pump-back or reventing capability is shown in Fig. 12.23. It was realized for SNR 300. It consists of the steel clad inner concrete containment with vaults for access built around the reactor tank with cover. This is filled with nitrogen (low oxygen content) and has a relatively low leak rate. The outer containment surrounds the inner containment and is filled with air. The outer containment consists of an inner concrete and an outer steel shell with a so-called reventing gap. Any air leaking from the inside of the outer containment and from the outside into the reventing gap is pumped back into the outer containment (revented) by reventing blowers. The lower pressure within the reventing gap guarantees that in the event of an accident no gaseous radioactivity would leak to the outside as long as the reventing blowers are in operation and as long as the pressure inside the outer containment does not exceed a certain maximum pressure (0.117 MPa).

This zero release phase can last for about 10 days. After this period of time the outer containment can be ex-vented through filter columns and a stack to the outside [59].

12.16 Core Melt Down and Core Disruptive Accidents in LMFBRs

Historically core melt down or core disruptive accidents have played an important role in safety considerations during the development of sodium cooled fast breeder technology. The main reason for these concerns is the high fissile enrichment of fast reactor fuel of about 15–25% of plutonium in the core zones of an LMFBR (Sect. 6.5). This leads theoretically to several critical masses in a fast breeder reactor core. Therefore, LMFBR cores are considered vulnerable to fuel compaction processes like core inward movement of fuel rods (bowing or movements, e.g. mechanical ruptures or earthquakes) or melting together of fuel. Such physical processes would lead to reactivity accidents with e.g. disruption of fuel rods in the core.

Another possibility for reactivity induced core disruptive accidents is the positive sodium temperature and sodium void coefficient of sodium cooled fast reactors with a total power of more than about 300 MW(th) (Sect. 12.3.2.2). Either sodium boiling or cover gas (Argon) bubbles passing through the central parts of a large homogeneous reactor core Fig. 12.1 can lead to positive reactivity accidents followed by disruption of core fuel rods.

However, for radially and axially heterogeneous large cores and for small fast reactors cooled by sodium or LBE, a zero or even a negative coolant temperature coefficient appears to be possible [20]. In addition the very high boiling temperature of the LBE (1,670°C) (Sect. 6.6.1) is favorable.

12.16.1 Core Melt Down and Core Disruptive Accidents for the Early Prototype Power LMFBRs

The characteristics of homogeneous sodium cooled fast reactor cores have led to the analysis of reactivity driven core disruptive accidents (CDAs) for prototype LMF-BRs (Sect. 12.10.1) and to the development of computer programs, e.g. SAS4A [24, 27, 28] and SIMMER [31, 32] and accompanying in-pile experimental programs. The result of such analyses was the thermal energy released during the disruptive accident which would be transferred to produce mechanical energy with a certain efficiency. This includes fast heat transfer during a so-called thermal sodium fuel interaction (SFI) or fuel coolant interaction (FCI) (similar to a steam explosion for LWRs). The resulting mechanical energy was then the basis for dynamic structural analysis of the reactor tank system (inner containment) with computer programs, e.g. REXCO [34, 35].

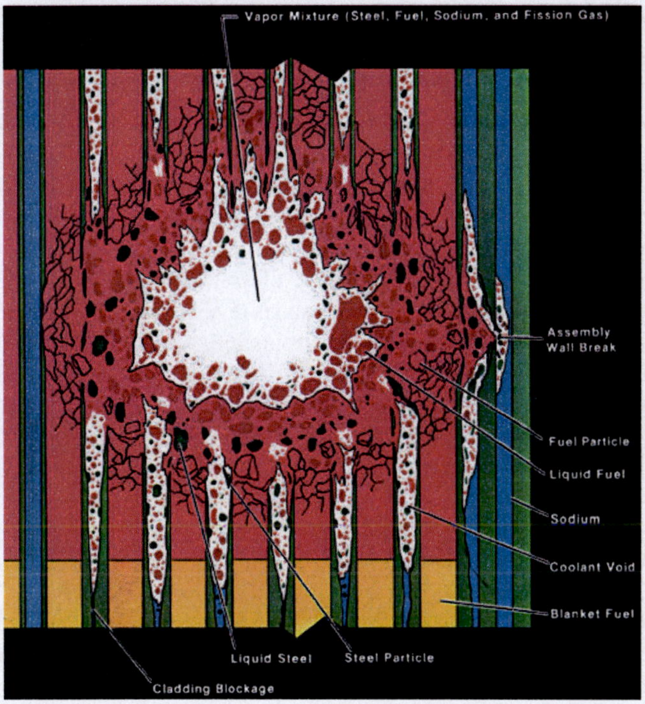

Fig. 12.24 Physics phenomena described by SIMMER code during core disruptive accident [31, 32]

This sequence of analyses was the basis for the requirement of licensing authorities for a certain mechanical work energy the reactor tank/cover system must withstand without exceeding certain strain limits, e.g. 2.5% plastic strain.

For SNR 300 with 300 MW(e) this mechanical work energy [60] was conservatively defined to

$$300\,\mathrm{MJ}$$

for Superphenix with 1,200 MW(e) it was set [61] to

$$800\,\mathrm{MJ}.$$

For the Japanese MONJU with 280 MW(e) [62–64] the mechanically work energy was similar as for SNR 300 (Fig. 12.24).

12.16.2 Core Melt Down and Core Disruptive Accidents for Future LMFBRs

The development of the SIMMER code to a 3D version [32] and further analysis revealed that these above mechanical work limits were very conservative. The SIMMER-code in 3D version is able to describe the physics phenomena in a disrupting core in sufficient details. It was shown that the main driving forces expanding the core during disruptive accident are caused primarily by fission gases and volatile fission products in the irradiated fuel and to a lesser extent by vapor pressure of the fuel itself [65–67]. New results in strain analysis revealed that the tank structures can be strained beyond 2.5 up to 25% plastic strains [68, 69].

The discovery of the negative control rod drive line expansion (CRDE) coefficient in the early 1980s and its demonstration at PFR, Rapsodie, FFTF and EBR-II lead to the design of special control rod designs with extended expansion and mechanical delatching mechanisms at temperatures of e.g. 650°C sodium outlet temperatures (Curie point effect of SASS) (see Sect. 12.10.2). These new designs solved the earlier safety concerns of anticipated transients without scram (ATWS) for the ULOF (unprotected loss of low) accident, the UTOP (unprotected transient overpower) accident and the ULOHS (unprotected loss of heat sink) accident (Sect. 12.10.1).

These ATWS or ULOF, UTOP and ULOHS accidents are no more of the same importance for the determination of mechanical work for the inner containment as they were for the early prototype power reactors (SNR, SUPERPHENIX and MONJU). The introduction of a corecatcher or a molten fuel retention and cooling device in the designs of SNR 300, BN 600 and in Japanese LMFBR designs assure the cooling of molten core masses (Sect. 12.12) and avoid the occurrence of possible recriticalities [60, 62, 63]. Decay heat removal is possible by natural convection under any accident condition (Sect. 12.9).

A reasonable definition of mechanical design loads for the inner containment of LMFBRs must be discussed and defined by licensing authorities in the future, when FBRs will be deployed on a large scale.

References

1. Status of liquid metal cooled fast breeder reactors (1985) Technical report series No. 246, IAEA, Vienna
2. Fast Reactor Data base (2006) update IAEA TECDOC Series 1531. IAEA, Vienna
3. Hummel HH et al (1970) Reactivity coefficients in large fast power reactors. American Nuclear Society, Hinsdale
4. Stacey W (2007) Nuclear reactor physics. Wiley, New York
5. Ash M (1965) Nuclear reactor kinetics. McGraw-Hill Book Company, New York
6. Graham J (1971) Fast reactor safety. Academic Press, New York
7. Waltar AE et al (1981) Fast breeder reactors. Pergamon Press, New York
8. Lewis EE (1977) Nuclear power reactor safety. Wiley, New York
9. Haefele W (1963) Prompt überkritische Leistungsexkursionen in schnellen Reaktoren. Nukleonik 5(5):201–208

10. Haefele W et al (1977) Fusion and fast breeder reactors, RR-77-8. International Institute for Applied Systems Analysis, Laxenburg
11. Emendörfer D et al (1993) Theorie der Kernreaktoren, Band 2: Der instationäre Reaktor. BI Wissenschaftsverlag, Mannheim
12. Greebler P et al (1972) SEFOR experimental results and application to LMFBRs. In: Proceedings international conference on engineering of fast reactors for safe and reliable operation, vol III. Kernforschungszentrum Karlsruhe, 9–13 October 1972, pp 1312–1329
13. Bogensberger HG et al (1975) Analysis of SEFOR experiments, KfK 2095. Gesellschaft für Kernforschung, Karlsruhe, FRG
14. USAEC (1964) An evaluation of four design studies of a 1000 MW ceramic fueled fast breeder reactor, COO-279. Reactor Engineering Division, Chicago Operation Office, US Atomic Energy Commission
15. Kessler G et al (1977) Safety of the liquid metal cooled fast breeder reactor and aspects of its fuel cycle. In: Proceedings of the international conference on nuclear power and its fuel cycle, vol 5. IAEA, Vienna, pp 661–674
16. Wade CD et al (1988) Trends vs. reactor size of passive reactivity shut down and control performance. In: International reactor physics conference, Jackson Hole, Wyoming, 18–21 Sept 1988
17. Kotake S et al (2010) Safety design features for Japan sodium cooled fast reactor, passive safety and CDA mitigation. Atomwirtschaft 56(6):386–392
18. Tobita Y (2010) JSFRs approach against core disruptive accident. In: 3rd International consulting meeting, Tokyo, 9–11 Sept 2010
19. Kotake S (2010) Safety design approaches of JSFR. In: 3rd International consulting meeting, Tokyo, 9–11 Sept 2010
20. Rineiski A et al (2011) Sodium void effect reduction and minor actinide incineration in ESFR. Trans Am Nucl Soc 104:339–343
21. Broomfield AM (1982) Safety related experience from the operation of PFR. In: Proceedings of the topical meeting on fast reactor safety, Lyon, vol 1. European Nuclear Society, American Nuclear Society, pp 231–245
22. Davies SM et al (1982), Design of the large developmental plant control and protection systems. In: Proceedings of the topical meeting on fast reactor safety, Lyon, vol 2. European Nuclear Society, American Nuclear Society, pp 471–483
23. Yevick JG et al (1966) Fast reactor technology: plant design. MIT Press, Cambridge
24. Wider HU et al (1982) Status and validation of the SAS4A accident analysis code system. In: Proceedings of the topical meeting on fast reactor safety, Lyon, vol 2. European Nuclear Society, American Nuclear Society , pp 13–23
25. Agrawal AK (1981) Decay heat removal and natural convection in FBRs. Hemisphere, Washington
26. Sha WT et al (1978) A three-dimensional transient single phase component computer program for thermal hydraulic analysis, ANL-77-96, NUREG-0415, Argonne National Laboratory
27. Dunn F et al (1983) The SASSYS LMFBR systems code, computers and engineering. In: ASME International conference Chicago 1983, CONF-830853-1
28. Heusener G et al (1973) Analysis of hypothetical accidents for SNR-300, KFK 1834. Kernforschungszentrum Karlsruhe, Germany
29. Costa J et al (1989) Buoyancy effects and natural circulation in liquid metal fast breeder reactor systems, NURETH-4. In: Proceedings of fourth international topical meeting of nuclear reactor thermal hydraulics, vol 1. pp 351–359
30. Hoffmann H et al (1994) Untersuchungen zur passiven Nachwärmeabfuhr für schnelle Reaktoren, KFK 5326. Kernforschungszentrum Karlsruhe, Germany
31. Smith LL et al (1976), SIMMER-I, An LMFBR disrupted core analysis code. In: Proceedings of the international meeting on fast reactor safety and related physics, CONF-871001:1195, Chicago, Ill
32. Kondo S et al (2002) Safety issues and approach for liquid metal reactors, OECD/NEA/CSNI-Workshop on Advanced Nuclear Reactor Safety Issues and Research Needs, OECD, Paris, 18–20 Feb 2002

33. Meyer R et al (1967) Fast reactor melt-down accident using Bethe-Tait analysis, GEAP 4809. AEC Research and development report, General Electric, Sunnyvale, CA
34. Chang YW (1975) A two-dimensional computer code for calculating the primary system response in fast reactors, ANL 75–19. Argonne National Laboratory, USA
35. Machertas A et al (1974) Nonlinear finite-element formulation for transient analysis of three-dimensional thin structures, ANL-8104. Argonne National Laboratory, USA
36. Gouriou A et al (1982), The dynamic behaviour of the SUPERPHENIX reactor under unprotected transient. In: Proceedings of the topical meeting on fast reactor safety, vol 2. European Nuclear Society, American Nuclear Society, Lyon ,pp 291–305
37. Wigeland RA et al (2011) Inherent prevention and mitigation of severe accident consequences in sodium-cooled fast reactors. J Nucl Sci Technol 48(4):516–523
38. Edelmann M et al (1995) Development of passive shut-down systems for the European Fast Reactor EFR, IAEA-TECDOC-884:68–78. International Energy Agency, Vienna, Austria
39. Kußmaul G (1994) Influence of control rod enhanced expansion devices on the course of unprotected transients in the EFR, KFK 5370. Kernforschungszentrum Karlsruhe, Germany
40. Schleisiek K (1982) Accommodation of subassembly faults in SNR 300. In: Proceedings of the topical meeting on fast reactor safety, vol 2. European Nuclear Society, American Nuclear Society, Lyon, pp 365–374
41. Gregory CV et al (1977) The study of local blockages in FBR subassemblies. J Br Nucl Energy 16(3):251
42. Ragland WA et al (1982) Sodium loop safety facility experiment P4, fast, thermal and fusion reactor experiments. In: Proceedings of the international conference, Salt Lake City
43. IAEA (1979) Sodium fires and prevention. In: IAEA/IWGFR specialists' meeting Cadarache, 1978, IWGFR/28. IAEA, Vienna
44. IAEA (1983) Sodium fires, design and testing. In: IAEA/IWGFR specialists' meeting Richland, WA, 1982, IWGFR/43. IAEA, Vienna
45. Alexas A et al (1979) Sodium fire aerosol experimental and analytical results—large scale tests in FAUNA. In: Fast reactor safety technology, Proceedings of the international meeting, vol 2. American Nuclear Society, Seattle, WA, pp 874–883
46. Beiriga P et al (1973) SOFIRE-II User's report. Atomics International, Canoga Park, CA, Rep. AI-AEC-13055
47. Alexas A (1976), Entwicklung eines Rechencodes zur Beschreibung von Natrium-Spritz- und Flächenbränden, I.: Die Codes SOFIRE II und NABRAND im Vergleich, KFK-2824, Kernforschungszentrum Karlsruhe
48. Cherdron W (1999) Abschätzung und experimentelle Untersuchungen der Auswirkungen von Natrium-Sprühbränden, Dissertation, Fakultät für Maschinenbau der Universität Karlsruhe, Germany
49. Morewitz H et al (1977) Experience on sodium fires and their aerosols. Nucl Eng Des 42: 123–135
50. Heisler MP et al (1981) SOMIX-2 Code, U.S. Department of Energy Fast Reactor Safety Program Progress Report, January–March 1981, ANL/TMC 81–2:176
51. Vaughan EU et al (1985) The SOMIX-2C sodium spray fire code. In: Proceedings of the international topical meeting on fast reactor safety, Knoxville, Tennessee, CONF-850410, vol 2. pp 937–942
52. Falgayrettes M (1973) Feux de sodium et réacteurs rapides, engineering of fast reactors for safe and reliable operation. In: Proceedings of the international conference Karlsruhe, vol 3. (1972), Kernforschungszentrum Karlsruhe, pp 1363–1385
53. Dumm K et al (1973) Experimentelle Untersuchungen zur Sicherheit der SNR-Geradrohrdampferzeuger bei Natrium-Wasser-Reaktionen, engineering of fast reactors for safe and reliable operation. In: Proceedings of the international conference Karlsruhe 1972, vol 1. Kernforschungszentrum Karlsruhe, pp 220–246
54. Hori M et al (1973) Sodium water reaction tests and analyses for MONJU-steam generator, engineering of fast reactors for safe and reliable operation. In: Proceedings of the international conference Karlsruhe 1972, vol 1. Kernforschungszentrum Karlsruhe, pp 203–219

55. Bell CR (1974) TRANSWRAP—A code for analyzing the system effects of large-leak sodium-water reactions in LMFBR steam generators, fast reactor safety. In: Proceedings of the international meeting Beverly Hills, CA 1974, vol 1. US Atomic Energy Commission, CONF-740401-P3, pp 124–133
56. Abdal-Moneim M (1975) ICEPEL—A two-dimensional computer program for transient analysis of a pipe elbow loop, ANL-7535. Argonne National Laboratory, USA
57. Gambillard E et al (1974) Générateurs de vapeur de Phénix: mesure de la concentration d'hydrogène du sodium pour la surveillance, Steam Generators for LMFBRs (IAEA/IWGFR Specialists Meeting, Bensberg, FRG 1974), IWGFR/1 159–167. IAEA, Vienna
58. Zeibig H et al (1985) Investigations on the integrity of fast breeder reactor piping and components. Nucl Eng Des 87:287–294
59. Oeynhausen, H et al. (1976), Design requirements for the SNR-300 containment system. In: Proceedings of the international meeting on fast reactor safety and related physics, Chicago 1976, vol 2. National Technical Information Service, Springfield, VA, CONF-761001, pp 452–459
60. Hennies HH et al (1990) The history of fast reactor safety—federal republic of Germany. In: Proceedings of the fast reactor safety meeting, Snowbird. Utah, USA, Aug 1990
61. Petit J et al (1979) Designers safety aspects for LMFBRs in France. In: Proceedings of the international meeting on fast reactor technology, Seattle, Washington, USA, August 1979
62. Nakai Y et al (1982) Safety aspects in the design of Monju. In: Proceedings of the topical meeting on fast reactor safety, Lyon, France
63. Sato I et al (2009) Elimination of severe recriticality events in the core disruptice accident of JSFR aiming at in-vessel retention of core materials. In: International conference on fast reactors and related fuel cycles: challenges and opportunities, Kyoto, Japan, 7–11 Dec 2009
64. Nakai R (2009) Design and assessment approach on advanced SFR safety with emphasis on core disruptive accident issue. In: International conference on fast reactors and related fuel cycles: challenges and opportunities—FR09, Kyoto, Japan, 7–11 Dec 2009
65. Breitung W et al (1989) Vapor pressure measurements on liquid uranium oxide and (U, Pu) mixed oxide. Nucl Sci and Eng 101:26–40
66. Fischer EA (1989) A new evaluation of the urania equation of state based on recent vapor pressure measurements. Nucl Sci Eng 101:97–116
67. Breitung W (1991) Analysis of measured total pressures from irradiated solid and liquid (U, Pu)-mixed oxide. Nucl Sci Eng 108:1–15
68. Krieg R (2005) Failure strains and proposed limit strains for a reactor pressure vessel under severe accident conditions. Nucl Eng Des 235:199–212
69. Jeschke J et al (2011) Critical strains and necking phenomena for different steel sheet specimens under uniaxial loading. Nucl Eng Des 241:2045–2052

Index

G. Kessler, *Sustainable and Safe Nuclear Fission Energy*, Power Systems,
DOI: 10.1007/978-3-642-11990-3,